# Numbers as Political Allies

*Numbers as Political Allies* analyses censuses of Jammu and Kashmir (J&K) as public goods, collective self-portraits and symbols of modernity and enriches the debates on the political economy of statistics. Using field interviews, archival resources and secondary data, the book tracks how censuses relate to their administrative, legal and political–economic contexts and captures their entire life cycle: from the political and administrative manoeuvring at the preparatory stage to the partisan use of data in policymaking and public debates.

The book argues that J&K's data deficit is shaped by, and shapes, ethno-regional, communal and scalar contests across different levels of governance, but the deteriorating quality of metadata limits our ability to evaluate the quality of census data. Further, comparing the experience of J&K with that of other states in India's ethno-geographic periphery, the book argues against resorting to legalistic and technocratic solutions to address the issue of data deficit and suggests possible measures to enhance public trust in the census.

**Vikas Kumar** is Associate Professor of Economics at Azim Premji University, Bengaluru. He is the co-author of *Numbers in India's Periphery: The Political Economy of Government Statistics* (2020) and the author of *Waiting for a Christmas Gift: Essays on Politics, Elections and Media in Nagaland* (2023). He curated the exhibition *Counting and Controlling Population: Postal Services, Census and Family Planning in Post-Colonial India, 1951–2011* at the Bangalore International Centre (January 2023).

# Numbers as Political Allies

The Census in Jammu and Kashmir

Vikas Kumar

Shaftesbury Road, Cambridge CB2 8EA, United Kingdom

One Liberty Plaza, 20th Floor, New York, NY 10006, USA

477 Williamstown Road, Port Melbourne, VIC 3207, Australia

314–321, 3rd Floor, Plot 3, Splendor Forum, Jasola District Centre, New Delhi – 110025, India

103 Penang Road, #05–06/07, Visioncrest Commercial, Singapore 238467

Cambridge University Press is part of Cambridge University Press & Assessment, a department of the University of Cambridge.

We share the University's mission to contribute to society through the pursuit of education, learning and research at the highest international levels of excellence.

www.cambridge.org
Information on this title: www.cambridge.org/9781009317214

First published 2023

Printed in India by Avantika Printers Pvt. Ltd.

*A catalogue record for this publication is available from the British Library*

ISBN 978-1-009-31721-4 Hardback

# Contents

# Figures

# Maps

# Tables

# Timelines

# Preface

In the first two decades after independence, the union government's role as the facilitator of interstate redistribution was closely linked to decennial population censuses. The allocation of seats in the parliament and federal funds tracked the most recent census data. In the mid-1970s, the growing concern over a rapidly increasing population amidst food and other scarcities forced a hasty uncoupling of the census and key federal policies to make room for more aggressive population control measures. This was, perhaps, necessary to protect the interests of states that had already achieved relatively lower levels of fertility. In the following decades, the census could not be conducted in Assam (1981) and Jammu and Kashmir (J&K) (1991) due to political disturbances. (In 1994, substantive changes were introduced in the Census Act, 1948, to expand the scope of punitive measures, among other things.) The year 2001 was therefore very important for the Census of India because the government was trying to enumerate the whole country once again after a gap of three decades, and there was an expectation that interstate redistribution of resources and power could be recoupled with the headcount. There was also a hype around the first census of the new millennium.

While the census managed to cover the entire country in 2001, it was marred by the politicisation of the headcount. The government had to postpone interstate delimitation to until after the first census taken after 2026. Six states – Arunachal Pradesh, Assam, J&K, Jharkhand, Manipur and Nagaland – could not even use the latest data for intrastate delimitation. In some of these states, the civil society and political parties alleged that the process of enumeration was subverted by vested interests and moved courts to challenge delimitation based on a flawed census. At least five others – Chhattisgarh, Meghalaya, Sikkim, Tripura and Uttarakhand – conducted intrastate delimitation under (political) constraints.

Soon after the 2001 census, Radhabinod Koijam, a former chief minister of Manipur, flagged the politically contentious nature of cartographic and demograhic statistics in his neighbourhood. In 2005, in an interview with Sanjoy Hazarika, chief minister Neiphiu Rio admitted that Nagaland's headcount was highly inflated due to political competition among tribes. As a student of engineering and later economics, I found it intriguing that a 'simple' measure such as headcount could be so deeply contested. The implication of Rio's statement became clear only after the state government managed to conduct a better census in 2011 that showed that the *reported* population of Nagaland contracted vis-à-vis 2001. That nudged me along with Ankush Agrawal to explore the politicisation of the census in Nagaland.

'It happens only in Nagaland …' was the common refrain of our interviewees in Nagaland. While a sense of their own exceptionalism that makes Nagas view most of their experiences as 'unique' may account for the refrain, it seemed implausible. Since the census is governed by the union laws and supervised by the union bureaucracy, lapses, if any, in enumeration are unlikely to be confined to just one state. So I turned to censuses of other states such as Assam, Manipur, Punjab and J&K and also other multiethnic, federal democracies to check if Nagaland's experience was, indeed, unique. Around this time Christophe Guilmoto and Irudaya Rajan wrote a paper on district-level fertility estimates based on the 2011 census in which they argued that the child population statistics of J&K were deeply flawed. They also drew attention to the work of Bashir Ahmad Bhat, who had flagged anomalies in J&K's headcount soon after the 2011 census. A report on the disruption of census-taking in Kashmir by Praveen Swami that appeared in the *Frontline* magazine in 2000 was another point of departure for the research that informs this book.

When I began my fieldwork in 2015, the plan was to replicate the research on Nagaland where I had been working since 2012. But I quickly realised that the case of J&K was quite different because of the non-synchronous nature of the census, several ad hoc and poorly documented changes in the reference dates, multiple categories of mobile population, the belated identification of the Scheduled Tribes, the intertwining of local and national politics of numbers and the non-availability of data for 1951 and 1991. So even though Nagaland and J&K faced a shared problem, its local determinants and manifestations were different in these two states. Moreover, the erstwhile J&K was at least five times larger than Nagaland. It was more than three ethnolinguistically and culturally different regions, each with its own local politics and a different relationship with New Delhi, packed into one administrative unit. The sudden reorganisation of the state in 2019 and the premature delimitation in 2022 added newer layers to the problem.

Since the data on the 0–6 age group were the most contentious part of the census in J&K, I visited health centres in villages across districts. All but one centre held records of only the recent births, deaths, pregnancies and immunisation, which did not reveal anomalies in the child sex ratio (CSR). Data from the only centre that maintained records since 2005–06 raised doubts about the 2011 census. Discussions with journalists, academics, retired and serving government officials, and civil society and political leaders not only added to the initial doubts, but also suggested that the impact of shifting reference dates on the process of enumeration and aggregation of headcounts was poorly understood. This was true even of senior (census) officials who, like everyone else, face a growing paucity and declining quality of metadata. So when I began writing my findings, I decided to clarify the process of enumeration in the state as it is quite different from that of the rest of the country and has changed erratically over time, and, also, discuss the declining quality of metadata.

In the beginning of this research, I had planned to cover only the political context of the census, but later it became clear that the legal and administrative contexts and the priors of the government about census-taking are equally important. Also, I had begun with the (statistical) assumption that each of the broad divisions of the state is dominated by one community, but multiple overlapping majorities became legible at different levels of aggregation during the fieldwork. This also meant an expansion of the geographical scope of the research to Ladakh, where the census data were not affected by any coverage error – that is, an error in the overall headcount. Together these extensions shaped the third part of the book that offers an extended discussion on the context of the production of census statistics in the erstwhile state of J&K and their consumption both within as well as outside that state. The examination of the context cleared the ground for a discussion on census reforms in the last chapter of the book.

In contrast to the existing literature that deals with the quality of census data in a piecemeal fashion, with contributions examining either coverage or, mostly, content errors, this book covers both types of errors and their context-dependence by exploring the entire life cycle of censuses: from the choice of enumerative categories to the use of data in public debates and policymaking. Even the contributions that examine the life cycle overlook crucial aspects such as the legal and administrative frameworks that govern enumeration and the self-image of census departments that is shaped by the self-imagination of developing countries such as India qua modern nation states and affects the quality of the data. This book tries to address such gaps. The impact of changes in data-processing technologies on the quality of data is one of the issues that could not be examined for want of information on the internal processes governing the transition from one technology to another.

The book also tries to fill in a gap in textbooks of statistics and econometrics that strip data of their context and deny students an opportunity to understand data as socially constructed objects with a life of their own. Instead of introducing a string of stand-alone examples from different places and periods, this book examines statistics about the same region (J&K) over an extended period (between 1951 and 2011, including the censuses that omitted J&K and the ones that were planned but had to be cancelled) and shows how data, development and democracy share a mutually constitutive relationship. While the book is focused on the erstwhile state of J&K, it will hopefully trigger a wider conversation on the context-dependence of the quality of census data as it offers extensive comparisons with other states, particularly Manipur, Nagaland and Punjab and, also, discusses the politicisation of numbers at the national level in India and other developing countries.

# Acknowledgements

When I started this research, Jammu and Kashmir (J&K) appeared to be an assortment of mutually inconsistent maps and statistics. B. S. Arun, R. N. Chhipa, Sanjoy Hazarika, Sevanti Ninan and Verghese Samuel introduced me to others who were based in Jammu and Srinagar or had worked there in the past, and then I gradually found my way through the state and added flesh to the barebones coordinates and numbers.

Census and other government officials in Jammu, Srinagar, Chandel, Guwahati, Imphal, Kohima, New Delhi, Phek and Senapati helped me understand the larger context of enumeration. In particular, Feroze Ahmed and Farooq Ahmad Factoo, the directors of census operations (DCOs) of J&K for the 2001 and 2011 censuses, respectively, and Metongmeren Ao and V. Hekali Zhimomi, the DCOs of Nagaland for the 2001 and 2011 censuses, respectively, were generous with their time and insights. Verghese Samuel patiently answered several rounds of questions. Senior officials in the Directorate of Health in Jammu and Srinagar, health department staff across the districts of Kashmir and Jammu, and officials at the Srinagar Municipal Corporation and the Tribal Affairs Department spared time to discuss population statistics. Various government officials helped with accommodation in remote locations.

Journalists and editors Harinder Baweja, David Devadas, Ahmad Ali Fayyaz, Altaf Hussain, Anuradha Bhasin Jamwal, Arun Joshi, Zulfiqar Majid and Praveen Swami offered crucial inputs on various developments they had covered in the state and helped me connect with others. Discussions with Shabeer Ahmad, Pramit Bhattacharya, Masood Butt, C. Chandramouli, Shah Mohammad Chaudhary, Rekha Chowdhary, Ajay Chrungoo, Rajiv Chuni, A. S. Dulat, Labha Ram Gandhi, Kunal Ghosh, M. S. Gill, N. Gopalaswami, Wajahat Habibullah, Mohammad Iqbal, P. T. Kunzang, Parveen Jaryal, R. K. Kalsotra, Feroze Khan, Zafar Ali Khatana, Vipul Kumar, Indha Mahoor, Taj Mohiudeen, Hari Om, N. A. Prasad, G. K. Pillai, Ali Mohammad Sagar, Satinder Sahani, Tsering Samphel, Ghulam Sayiden, Dipankar Sengupta, Javaid

Rahi, Nasir Shabani, Skandan Krishnan, P. Stobdan, Theja Therieh, S. Tsogphel, Nawang Tundup,  Himanshu Upadhyaya, Aditya Verghese and Lama Chosphel Zotpa, among others, enriched my understanding. Asgar Ali Karbalaie and Nawang Rigzin Jora spared time for discussions on a wide range of issues. Ankush Agrawal, Noor Ahmad Baba, Bashir Ahmad Bhat, Chakraverti Mahajan and Verghese Samuel offered feedback on various problems I faced vis-à-vis reconciling reported facts and field observations.

The manuscript developed over a long period. Several batches of students who took the course 'Political Economy of Government Statistics' at Azim Premji University, Bengaluru, served as sounding boards and offered valuable feedback on various ideas that animate this book. P. G. Babu, Sanjoy Hazarika, Chandan Gowda and Poonam Singh were always available for discussions and offered valuable advice at different stages.

After I completed a preliminary draft of the book, Manfred Holler created a much-needed opportunity to present the entire argument at the Adam Smith Seminar organised by the Centre of Conflict Resolution, Munich, where the audience offered helpful feedback. Gujarat National Law University, Gandhinagar, and Vellore Institute of Technology also offered opportunities to present parts of the fifth and seventh chapters, respectively. Charles Chasie of the Kohima Educational Society invited me to deliver a couple of lectures in Dimapur and Kohima on the politics of numbers, which allowed me to look at the manuscript as well as J&K from the other end of the country. A. Bimol Akoijam and P. G. Babu offered similar opportunities in Manipur and Mumbai, respectively. The imprint of the north-eastern detours is most clearly visible in the later chapters of the book.

B. S. Arun (*Deccan Herald*), Rama Lakshmi (*The Print*), Sevanti Ninan (*The Hoot*) and Sangeeta Barooah Pisharoty (*The Wire*) made room for contributions on statistics and public policy in their respective publications. East Asia Forum, too, served as a platform for sharing preliminary ideas in this regard. Parts of Chapters 5 and 7 draw on my contributions to the *Statistical Journal of the IAOS*, the journal of the International Association for Official Statistics.

Libraries at the Directorates of Census Operations in Srinagar, Jammu, Imphal and Kohima, and the Office of the Registrar General of India (ORGI), New Delhi, provided various census publications. The chief electoral officer's office in Jammu shared a compilation on electoral laws and delimitations. The Srinagar office of the Jammu and Kashmir National Conference provided copies of a recent edition of *Naya Kashmir*, while the Jammu office of the Jammu and Kashmir National Panthers Party provided copies of its manifestos and pamphlets. The archives at the Nehru Memorial Museum and Library, New Delhi, helped me access old pamphlets related to J&K, while the library of the Ministry of Defence provided access to publications on the 1965 and 1971 wars. The library of the University of Kashmir, Srinagar, and the archives of the *Kashmir Times* provided access to back issues of newspapers. Anuradha Bhasin Jamwal, Hari Om and Javaid Rahi, too, shared copies of older newspaper articles. Anuradha Bhasin Jamwal allowed us to use images from the *Kashmir Times* archive for designing the book cover. The library of Azim Premji University offered excellent support over the years.

Adil, Bilal, Sourabh Khajuria, Manzoor, Stanzin Phunchok, Rajesh Thapa and Umar offered much-needed travel support. The fieldwork that extended from 2015 to 2022 was punctuated by sporadic political unrest and, more recently, the Covid-19 pandemic. During such interruptions, several interviewees were kind enough to continue the conversations over the telephone and, in a few cases, email. I am grateful to Azim Premji University for institutional support throughout this period.

I first discussed the idea of writing about J&K's census with Anwesha Rana at Cambridge University Press several years ago. She offered invaluable support and encouragement over an extended period, putting up with several delays, without which I would not have managed to complete the book. She also created an opportunity to present a draft of a chapter at a conference. Two anonymous referees of the Press helped improve the manuscript by identifying sections that did not fit well and those that needed additional discussion. Priyanka Das carefully edited the manuscript and helped incorporate changes to capture the impact of the latest round of delimitation on demographic politics in J&K and insights from the reports of the 2011 census that were belatedly released. Anandadeep Roy worked with a large number of images I shared to help design a cover that captures the essence of the book.

In this short note, I have been able to acknowledge only a few of my hosts and interlocutors based in J&K, Ladakh and elsewhere in the country. Many more have been omitted to maintain anonymity. I remain indebted to all of them for their goodwill and warmth.

# Abbreviations

| | |
|---|---|
| AFSPA | Armed Forces (Jammu and Kashmir) Special Powers Act, 1990 |
| APDP | Association of Parents of Disappeared Persons |
| APSCC | All Parties Sikh Coordination Committee |
| BBC | British Broadcasting Corporation |
| BJP | Bharatiya Janata Party |
| BPL | below poverty line |
| BSP | Bahujan Samaj Party |
| CAG | Comptroller and Auditor General of India |
| CBR | crude birth rate |
| CDR | crude death rate |
| CMIE | Centre for Monitoring Indian Economy |
| CPS | Centre for Policy Studies, Chennai |
| CSO | Central Statistics Organisation/Central Statistics Office |
| CSR | child sex ratio |
| DCO | director of census operations |
| ECI | Election Commission of India |
| FPT | final population totals |
| FSU | first-stage unit |
| GDP | gross domestic product |
| GoI | Government of India |
| GoJK | Government of Jammu and Kashmir |
| GoUTJK | Government of the Union Territory of Jammu and Kashmir |
| GPT | general population totals |

| HMIS | Health Management Information System |
| IAS | Indian Administrative Service |
| IB | Intelligence Bureau |
| INC | Indian National Congress |
| J&K | Jammu and Kashmir |
| Jt DCO | joint director of census operations |
| JSPA | Janagoshthiya Samannay Parishad, Assam |
| LAHDC | Ladakh Autonomous Hill Development Council |
| LBA | Ladakh Buddhist Association |
| LoC | Line of Control |
| MHA | Ministry of Home Affairs |
| MOSPI | Ministry of Statistics and Programme Implementation |
| MPCE | monthly per capita consumption expenditure |
| NC | National Conference |
| NEFA | North-East Frontier Agency |
| NFHS | National Family Health Survey |
| NGO | non-governmental organisation |
| NGR | natural growth rate |
| NPC | National Population Commission, Nigeria |
| NPR | National Population Register |
| NRC | National Register of Citizens |
| NSC | National Statistical Commission |
| NSS | National Sample Survey |
| NSSO | National Sample Survey Organization/National Sample Survey Office |
| ORGI | Office of the Registrar General of India |
| ORGI&CC | Office of the Registrar General of India and Census Commissioner |
| PCA | primary census abstract |
| PDP | People's Democratic Party |
| RGI | registrar general of India |
| PEC | post-enumeration check |
| PEPSU | Patiala and East Punjab States Union |
| PES | post-enumeration survey |

| PLSI | Peoples' Linguistic Survey of India |
|------|-------------------------------------|
| PNDT | prenatal diagnostic techniques |
| PPT | provisional population totals |
| RSS | Rashtriya Swayamsevak Sangh |
| RTI | right to information |
| SASB | Shri Amarnathji Shrine Board |
| SC | Scheduled Caste |
| SCO | superintendent of census operations |
| SoI | Survey of India |
| SRB | sex ratio at birth |
| SRS | Sample Registration System |
| ST | Scheduled Tribe |
| TFR | total fertility rate |
| UA | urban agglomeration |
| UAE | United Arab Emirates |
| UN | United Nations |
| US | United States |
| UT | union territory |
| WPRAC | West Pakistan Refugees Action Committee |

# Part I
# Introduction

# 1

# Debating Numbers

Although the census is no alternative to self-determination, the local government employees must discharge their duties honestly to defeat the RSS–BJP [Rashtriya Swayamsevak Sangh–Bharatiya Janata Party] designs to change the demography of Jammu and Kashmir.

—Syed Ali Shah Geelani, Tehreek-e-Hurriyat (*New Indian Express* 2010)

Who in the UT [union territory] doesn't know that the population of Jammu province is more than Kashmir and that the figures of all the censuses held in and after 1961 were fudged to ensure Kashmiri domination over Jammu province?

—Hari Om, former dean, faculty of Social Sciences, University of Jammu (Hari Om 2021c)

[T]he rulers of Kashmir are anxious to pass off the District [Ladakh] as a Muslim majority one.... [We fear] the Buddhists would be officially relegated to the position of a minority in the Census of 1961.

—Bakula (1953: ii)

Yesterday [Ghulam Nabi Azad] was proudly saying that earlier there were more Buddhists in Ladakh and now Muslims are more. I have to say with regret that you tried to finish off Buddhists in Jammu and Kashmir by misusing Article 370.

—Jamyang Tsering Namgyal, member of parliament, Ladakh (Lok Sabha 2019: 160)

The Buddhists of Leh feel that they are dominated by Muslims in the J&K [Jammu and Kashmir] state. But Kargilis feel the same discrimination. The state government thinks we are Muslims, but Shias. The centre thinks we are Ladakhis, but Muslims.

—Asgar Ali Karbalaie, former member of legislative assembly, Kargil (Donthi 2019)

The Kashmiri Pandits with their population of 75,000 are represented by a Minister and a Deputy Minister in the Government of the State and in the Indian Parliament too, they are duly represented. But to the Buddhists the Kashmir Ministry like the Indian Parliament is forbidden ground....

—Bakula (1953: vi)

If 80,000 Ladakhi Buddhists can be given a hill council, why 7 lakh Kashmiri Hindus cannot be given a homeland?

—Panun Kashmir (n.d.2: 27)

Numbers games begin in the teacher's mind. Kashmiri teachers are not serious about enumerating our [Gujjar and Bakarwal] community.... We do not trust [the] census.

—Gujjar activist (interview, 4 December 2019)

Our present population is about one lakh but we claim three lakh. If Gujjars and Bakarwals can claim 25 lakh, why can't we?

—Tribal leader (interview, 5 December 2019)

[T]here is a longstanding and honourable tradition of cooking up figures in J&K [Jammu and Kashmir].... Unless proved otherwise, I would always assume data from J&K is cooked.

—Retired civil servant (interviews, 4 January 2016 and 23 February 2016)

Lack of reliable numbers about our population has been a handicap to our planners, policy makers and scholars.

—Farooq Abdullah, chief minister, Jammu and Kashmir (Government of India [GoI] 2001b: v)

Sadly, few independent demographers in the State or outside it, have seen it fit to intervene in this debate. It is possible that if they did, no one would be listening.

—Praveen Swami, journalist (Swami 2000a)

# Introduction

The sudden reorganisation of Jammu and Kashmir (J&K) into two union territories (UTs) after the repeal of Article 370 of the Constitution of India in 2019 triggered a vigorous debate on the political future of the erstwhile princely state. The potential threat of demographic change looms large in this debate.

Former chief minister Farooq Abdullah of the National Conference (NC) suggested that 'the new domicile law was intended to flood the [Kashmir] Valley with Hindus and create a Hindu majority' (Thapar 2020). Another former chief minister, Mehbooba Mufti of the People's Democratic Party (PDP), claimed, 'They [the BJP] just want to occupy our land and want to make this Muslim-majority state like any other state and reduce us to a minority and disempower us totally' (British Broadcasting Corporation [BBC] 2019). Haseeb Drabu, the finance minister in the last PDP government, argued that Kashmir faced the 'real possibility of the *demographic majority* being converted into a *political minority*' within a generation (Drabu 2020b, emphasis added). The Concerned Citizens' Group – comprising leading public figures, including former union minister Yashwant Sinha – which visited the state after the reorganisation, noted that Kashmiris 'believe that Indian government wants to marginalise them if not annihilate them. This fear is expressed most vividly as fear of demographic change by creating new settlements for outsiders', and that there 'is also fear of [the] National Register of Citizens [NRC] and how it could be used to legitimise settlers' (*Greater Kashmir* 2019). Such concerns predate the reorganisation of the state though. Two years before the reorganisation, *Greater Kashmir*, a Srinagar-based English daily, · summarised the Kashmiri Muslim concern about demographic change in an editorial:

> Be that the loss of fiscal autonomy, change of nomenclature from Sadré Riyasat to Governor … all of them are reversible…. However, the scrapping of article 35-A is a different ball game…. To change the demography of a conflict zone under your control, where United Nations is yet to hold a promised plebiscite, is a concern…. Its fall-out will dissolve the case of Self-determination of the people of Kashmir…. In case this article goes, Kashmiris will be reduced to a minority by the influx of outsiders and the question of a possible referendum will be buried forever…. The whole emotion of opposing any dilution of article 370 or 35-A in Kashmir is not about maintaining diversity in India. It's about preserving the resistance movement of Kashmir. It's about being able to vote in a referendum in Kashmir. Let's not hoodwink Indians by being hypocritical … the idea of complete freedom from India is deep-seated and cannot be expected to wane. (*Greater Kashmir* 2017)

*Greater Kashmir*, too, did not put forward a new argument, though. In May 2010, in the run-up to the 2011 census, Syed Ali Shah Geelani, the chairman of the Tehreek-e-Hurriyat, warned that a 'change in the religious composition of the State may help the Government of India in case India is compelled to hold plebiscite in J&K' (Bhat 2018: 181). As discussed later, this in turn was a reiteration of the tension between *mardamshumari* (census) and *raishumari* (plebiscite) during the 2001 census (Riaz-ud-din 2000). The obsession with the religious composition of the population is, in fact, as old as the Kashmir dispute. It is therefore not surprising that every administrative measure is vetted for its potential demographic impact. Everything, including even photographs,[1] is parsed for statistical proportionality and clues about any demographic fallout. A photograph that 'depicted [Lieutenant Governor Girish Chandra] Murmu holding a meeting with a battery of bureaucrats' that caught the attention of social media in Kashmir is a case in point. 'Of the 19 officials in the photo, there was just one *local* Muslim officer – Farooq Ahmad Lone, a former IAS [Indian Administrative Services] officer. This photograph, according to Kashmiri social media users, showed that the Union government was deliberately making demographic changes in the predominantly Muslim Kashmir Valley' (I. A. Malik 2020, emphasis added).

Another controversy surrounding a website managed by the Industries and Commerce Department, Government of Jammu and Kashmir (GoJK), offers further insights in this regard. An earlier version of the 'About J&K' section of the website InvestJK.in suggested that '[b]e it the Kashmir region or Jammu the population *is* predominantly Hindu. This *explains* the presence of [a] number of temples in the state' (GoJK 2020, emphasis added).[2] This section briefly described tourist destinations, including natural scenic attractions and temples. It evoked strong reactions in Kashmir amid 'fears of a demographic invasion'. The website was seen as reflecting 'the RSS [Rashtriya Swayamsevak Sangh] mindset that desires the erasure of Muslims' and as 'religiously motivated' because 'the Kashmir Valley is nearly entirely Muslim, [and] six of the ten districts of Jammu have Muslim-majority populations' (Mir 2021). Moreover, it was argued that

> Jammu province is not described in its entirety – it has a diverse geography and its largest area is inhabited by Muslims. Instead, the website focuses on the province's capital, Jammu. The city is described as being adorned by its founding Dogra rulers 'with numerous temples and shrines, [and is] now known as the city of Temples'. (Mir 2021)

The website mentioned Hindu places of worship across the state but not 'any of the hundreds of [Islamic] religious places in the Kashmir Valley' (ibid.). InvestJK. in seems to have been edited soon after questions were raised, and all religious and most of the demographic information were deleted. In Jammu, the hurried revision was seen as evidence of the presence of 'elements in the establishment who wilfully outrage the religious sensitivities of the Hindus' (Hari Om 2021a). While the initial post on the website was indeed blatantly flawed, there were other statistical errors as well that escaped the notice of *both* the sides. The website continued to mention incorrect estimates of the area and population of the UT of J&K. This episode highlights the narrow obsession with data on the religious composition of the population.[3]

The process of building inclusive and transparent democratic institutions that foster dialogue and reconciliation and ensure fair treatment of individuals and (minority) communities is slow, and its outcome is uncertain. The communal obsession with numbers pushes governments, political parties and people to resort to statistical proportionality as 'a favored legal and administrative tool' (Prewitt 2003: 16) and the ultimate arbiter of the fairness of institutions and policies. As a result, '[a]rguments about numerical quotas, availability pools and demographic imbalance become a substitute for democratic discussion of the principles of equity and justice' (Kenneth Prewitt quoted in Rose 1991: 680). In J&K, each side claims a majority within its territory of interest to justify its demand for a larger share of public resources and highlights divisions within rival camps even as it papers over its own internal differences. The debate in the parliament on the Jammu and Kashmir Reorganisation Bill, 2019, exemplifies this.

In a Rajya Sabha debate, Ram Das Athawale, a union minister, summed up the long-standing *national* perception of J&K's demography when he noted that 'in Ladakh there are more Buddhists, in Jammu there are more Hindus, in Kashmir there are more Muslims' (Rajya Sabha 2019b: 96).[4] The speeches of Ghulam Nabi Azad,[5] Hasnain Masoodi[6] and Jamyang Tsering Namgyal[7] in the parliament offered insights into the contesting *local* perspectives. Masoodi lamented that the union government 'had lost the trust of 1.25 crore people [that is, of J&K as a whole including Hindus, Buddhists and non-Kashmiri Muslims]' (Lok Sabha 2019: 142). Azad highlighted the divisions within Ladakh and Jammu but not Kashmir. He told the Rajya Sabha:

> When I talk about Jammu, I refer to [the] region ranging from Rajouri [and] Poonch to Jammu, from Kathua and Samba, from Udhampur to Ramban, Banihal and Kishtwar. I don't talk about Jammu city, Jammu is not only limited to the Jammu city but it has 10 districts and … Hindus, Muslims, Christians, Jains and Sikhs. (Rajya Sabha 2019b: 69)

Azad then commented on Ladakh's demography:

> Do you know that there are two districts in Ladakh – Leh and Kargil [?]. Do you know that the population of Shia Muslims is 52 per cent and the

population of Budd[h]ists is 48 per cent in Ladakh?[8]… *The history that you read twenty years ago is changed now.* It is no more called a Budd[h]ist State, population has grown tremendously after that. Now the situation has reversed. Ladakh is now a [M]uslim majority State. Does any of you know that for last thirty years our Budd[h]ist brothers from Leh have been wanting it to be declared a Union Territory and the people from Kargil wanted to leave Ladakh and become a part of Kashmir province? (ibid.: 72–73, emphasis added; see also Asaduddin Owaisi, Lok Sabha 2019: 208)

Namgyal sharply criticised Azad and Masoodi. He first questioned Azad:

> Yesterday [Azad] was proudly saying that earlier there were more Buddhists in Ladakh and now Muslims are more. I have to say with regret that you tried to finish off Buddhists in Jammu and Kashmir by misusing Article 370. Is this your demographic maintain [*sic*]? Is this your secularism?… In 1979, these families divided Ladakh and created Buddhist majority Leh and Muslim majority Kargil and made the two brothers fight till this day. (Lok Sabha 2019: 160)

Further, mimicking Azad's remarks on Jammu, Namgyal criticised Masoodi who had suggested that Kargil had shut down to protest the reorganisation (ibid.: 144). Namgyal said, 'These people have mistaken a road and a small market for Kargil. If you want to see Kargil, then go to Zanskar, go to Mulbek-Shargol there, go to Aryan Valley, see Derchiks-Garkon. Today, people of 70 per cent area of Kargil are welcoming this decision, this bill' (ibid.: 161). Asgar Ali Karbalaie[9] pointed out that the parliament and the national media did not notice that Namgyal mentioned only the Buddhist areas of the Muslim-majority Kargil district and omitted the Shia-dominated areas (interview, Kargil, 20 September 2019).[10]

What perhaps troubled Karbalaie more was that Namgyal claimed almost three-fourths of Kargil for Buddhists who account for about 15 per cent of its population.[11] More generally, Karbalaie has been concerned about the portrayal of Ladakh as a Buddhist land by the government as well as the media: 'When you say Ladakh, it is seen synonymous to monasteries, gompas … and Buddhists' (Donthi 2019).[12] Indeed, after the reorganisation of J&K, sections of the national media suggested that Ladakh was 'India's first Buddhist dominated union territory with a dominant Buddhist population' (Das 2019).[13] The Bahujan Samaj Party (BSP) president and former chief minister of Uttar Pradesh, Mayawati, noted that 'with Leh–Ladakh being declared as a separate Union Territory, the long pending demand of the Buddhist community in Jammu and Kashmir has been fulfilled. BSP welcomes this decision. From this decision, the whole country especially the followers of Dr Bhimrao Ambedkar who are Buddhist are very happy' (*Business Standard* 2019a). In fact, even Ranil Wickremesinghe, the then prime minister of Buddhist-majority Sri Lanka, tweeted, 'I understand Ladakh will finally become a Union Territory. With 70% of Ladakh's population being Buddhist, it will be the first Indian State with a Buddhist majority' (Srinivasan 2019).[14] Amidst the euphoria around the inauguration of a 'Buddhist' UT, a Leh-based Sunni leader suggested that reorganisation was not bad as it created two Muslim-majority UTs (interview, Leh, 21 September 2019).[15] Indeed, post-colonial censuses suggest that Buddhists never accounted for more than 54 per cent of Ladakh's population. Their population share has declined over the decades, and they no longer constitute a majority in Ladakh.

A similar divide was observed in the parliamentary debate on development.[16] The supporters of the reorganisation argued that the repeal of Article 370 was essential for development. Amit Shah, the union minister of home affairs, claimed that massive federal funding did not translate into development:

> In 2011–12, Government of India provided Rs 3,683 per head in India whereas in Jammu–Kashmir this amount was Rs 14,255 … in 2017–18 Government of India has sent on an average Rs 8,227 per head in India and in Jammu–Kashmir per head expenditure was Rs 27,358. But this money is not percolating to the ground level because Article 370 has created a monopoly there.[17] (Rajya Sabha 2019b: 136; see also Lok Sabha 2019: 250)

Masoodi responded, 'We do not need any development through revocation of Article 370. Our identity, our political aspirations, our autonomy is most dear to us' (Lok Sabha 2019: 151).[18] He added that the state was 'well-off' as reflected in, among other things, the 'highest per capita [income] in the country' and an absence of starvation deaths and farmer suicides (ibid.).[19] Contrary to Masoodi's claim, over the past half-a-century, J&K's per capita income has never exceeded 60 per cent of the highest per capita income of states in the country, with the peak being in the early 1980s before the onset of insurgency (Agrawal and Kumar 2020a: 265–73). Likewise, the state's very low poverty rate is partly an artefact of a large sample non-coverage in the National Sample Surveys (NSS) (ibid.). Otherwise, if we assume that the NSS data are correct, J&K supported a much higher level of per capita consumption relative to its per capita income than other states.

This discussion suggests that in J&K statistics serve as political weapons in ethno-regional, communal and scalar contests rather than as inputs to policymaking and administration. The statistical battles witnessed inside and outside the parliament in the first week of August 2019 are part and parcel of everyday public debates in and on J&K. However, hardly any stakeholder pays attention to the (quality of) statistics at the heart of the many conflicts of the state. Kashmiris are not worried about the quality of census as long as it does not challenge Kashmir's electoral grip over the state, while Jammu's concern with data quality is limited to undermining the dominance of Kashmir. The government is indifferent to the problem, while the academia has not examined it either.

The large literature on various aspects of the Kashmir conflict recognises the key role played by demographic concerns, but it does not examine the sources of population statistics used by the parties to the conflict. During the 2001 census, Swami (2000a) lamented that 'few independent demographers in the State or outside it, have seen it fit to intervene in this debate'. Post-2001 census contributions such as S. Bose (2003, 2007), Schofield (2003), Behera (2006), Swami (2007), Chowdhary (2010b, 2016, 2019), K. B. Ahmad (2017), Snedden (2017), Devadas (2018), Bhatia (2020), S. Hussain (2021) and Jamwal (2022) discuss how demographic concerns affect the political conflict but do not examine how the headcount is itself a product of the conflict. Researchers discuss the use of statistics in communal propaganda, note how statistics used in popular debates deviate from government statistics and even hint that the statistics may not be reliable, but do not ask how partisan politics shapes statistics.

Exceptions include Swami (2000a, 2014), who discussed the difficulties in conducting the 2001 census in J&K. Two scholarly assessments followed the 2011 census. Bhatt (2011, 2018) analysed the 2011 census data on child sex ratio (CSR) and discussed the immediate context of the headcount in the state. In their analysis of fertility trends across districts of India, Guilmoto and Rajan (2013) drew attention to the poor quality of the 2011 census data of J&K for the population aged 0–6 years. These analyses did not examine the quality of the larger body of census data of which the data on age and sex are a part. The mutually constitutive relationship between census data and their social and political-economic contexts was also left largely unexamined. This book analyses J&K's population statistics for the period 1951–2011. It unpacks the census data at different levels of aggregation and along different seams and tries to estimate coverage and content errors. In the process it uncovers a relationship between census statistics, on the one hand, and administrative, legal, social and political-economic processes, on the other, to arrive at a better understanding of how J&K's data deficit is shaped by, and shapes, the conflict and how the relation between state and statistics changes across tiers of government. J&K's experience is also compared with that of states such as Punjab, Nagaland and Manipur.

The remainder of this chapter delineates the scope of the book and elaborates the conceptual framework of the analysis. This is followed by an outline of the chapters and a note on terminology.

## Scope and Framework

In the first decade of the twenty-first century, J&K's decennial censuses reported unanticipated changes in the composition of the population, including a sharp decrease in CSR, an increase in child population share, a change in the relative population shares of the two main regions and religions of the state, a decline in the population share of Scheduled Castes (SCs), a contraction of the population of Shina speakers, a sharp increase in Kashmir's slum population and a sharp contraction of Jammu district's slum population (Figures 1.1a–1.1h). In addition, Ladakh reported very sharp changes in the population of the speakers of Tibetan, Ladakhi and Bhotia languages, whereas Gojri speakers were misclassified in parts of Kashmir. During the 1990s and 2000s, surveys conducted by the National Sample Survey Office (NSSO) revealed a sharp rise followed by a drop in urban monthly per capita consumer expenditure (MPCE) of J&K relative to the rest of the country even as its per capita income had been steadily declining relative to the rest of the country (Agrawal and Kumar 2020a: 267). In addition, there were discrepancies between the estimates of poverty based on the NSSO surveys and those based on the state government's below poverty line (BPL) survey (ibid.: 273). There were also discrepancies between the estimates of unemployment provided by the NSSO and those provided by employment bureaus

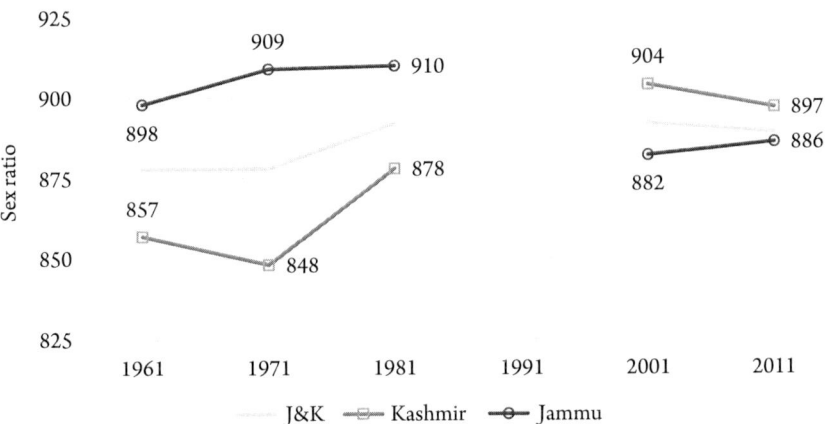

**Figure 1.1a**  Sex ratio of Jammu and Kashmir (J&K), 1961–2011

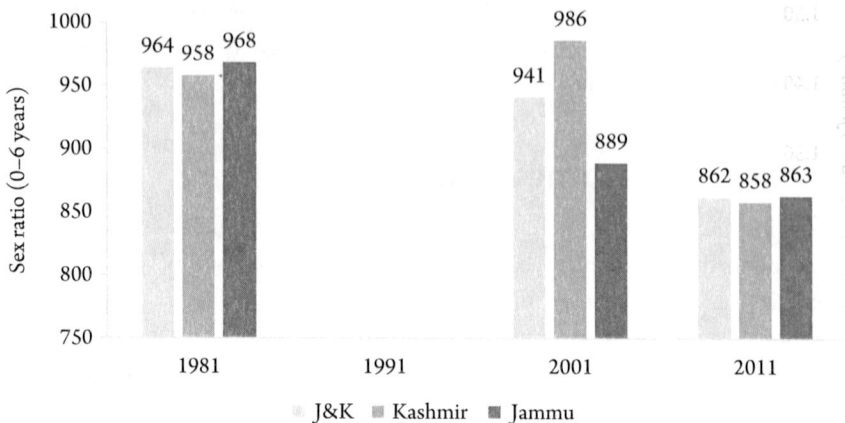

**Figure 1.1b**  Sex ratio of Jammu and Kashmir's (J&K) population aged 0–6 years, 1981–2011

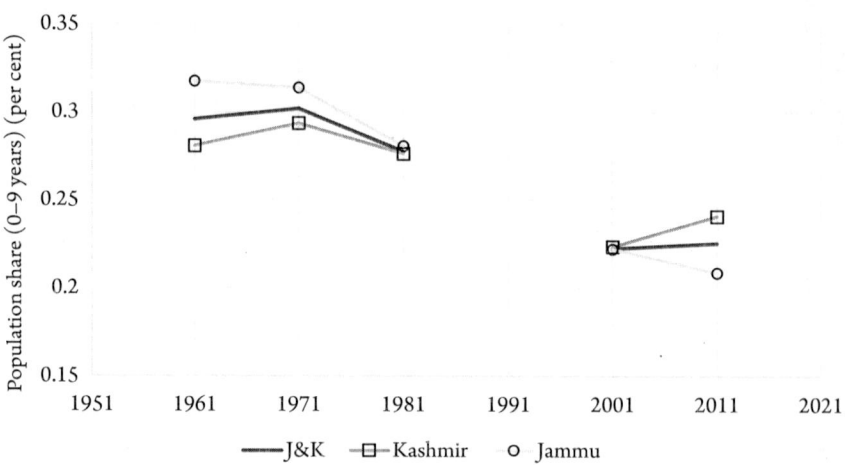

**Figure 1.1c**  Share of Jammu and Kashmir's (J&K) population aged 0–9 years, 1961–2011

(Government of India [GoI] 2011g: 12–13). The percentage of houses with sanitary toilets decreased in urban J&K between the 72nd (2014–15) and 75th (2017–18) rounds of the NSSO surveys. The level of registration of births and deaths in J&K was far below other states with comparable human development indicators (GoI 2014c).[20] Most recently, the new political map of the UT of Ladakh carved out of J&K omits some of its major internal borders (Kumar 2020b, 2020c).

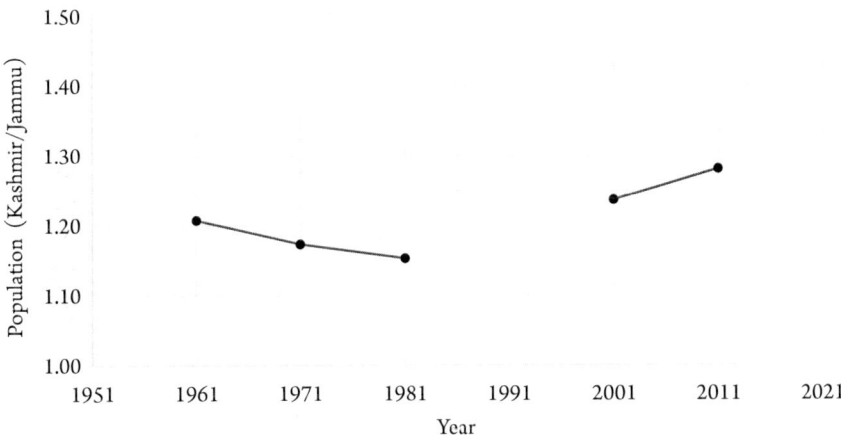

**Figure 1.1d**   Ratio of the population of Kashmir to Jammu, 1961–2011

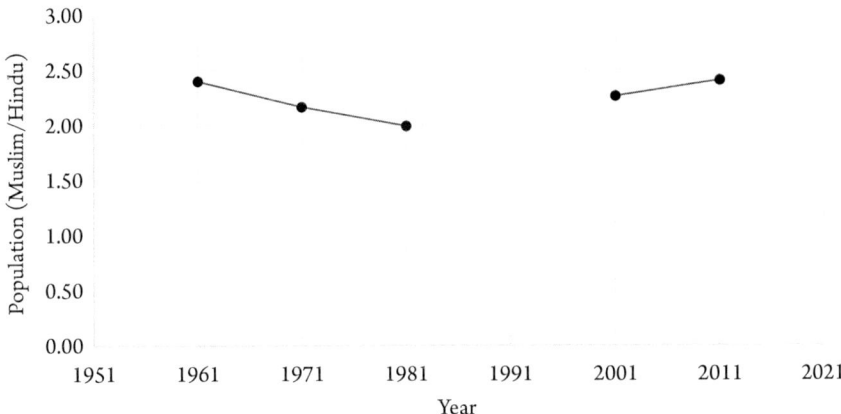

**Figure 1.1e**   Ratio of Muslim to Hindu population of Jammu and Kashmir (J&K), 1961–2011

In short, the quality of a whole range of key government sources of information is questionable in J&K.

We will examine decennial population censuses and refer to other sources of information such as maps insofar as they are essential for carrying out enumeration or analysing and representing census data. Survey statistics (from the NSS, the Sample Registration System [SRS] and the National Family Health Survey [NFHS]) and administrative statistics (on electorate and school enrolment) will be discussed to the extent they depend on the census or can help cross-check census

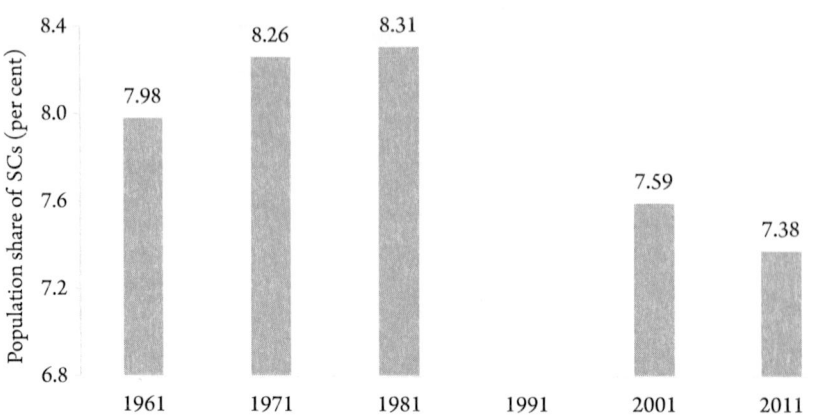

**Figure 1.1f**  Population share of Scheduled Castes (SCs) in Jammu and Kashmir (J&K), 1961–2011

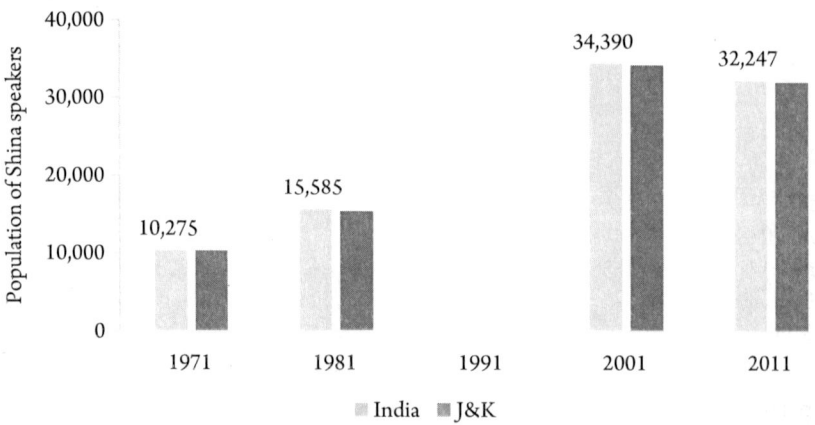

**Figure 1.1g**  Shina-speaking population in India and Jammu and Kashmir (J&K), 1971–2011

data and plug in gaps in the metadata on the census. In other words, the focus will be on population statistics and their context, which will be explored from three different perspectives by conceptualising the census as a symbol of modernity, collective self-portrait and public good.

## Insignia of Modernity

The use of statistics is unavoidable in modern states given the size and complexity of their operations, but it is also widely seen as desirable because it ostensibly

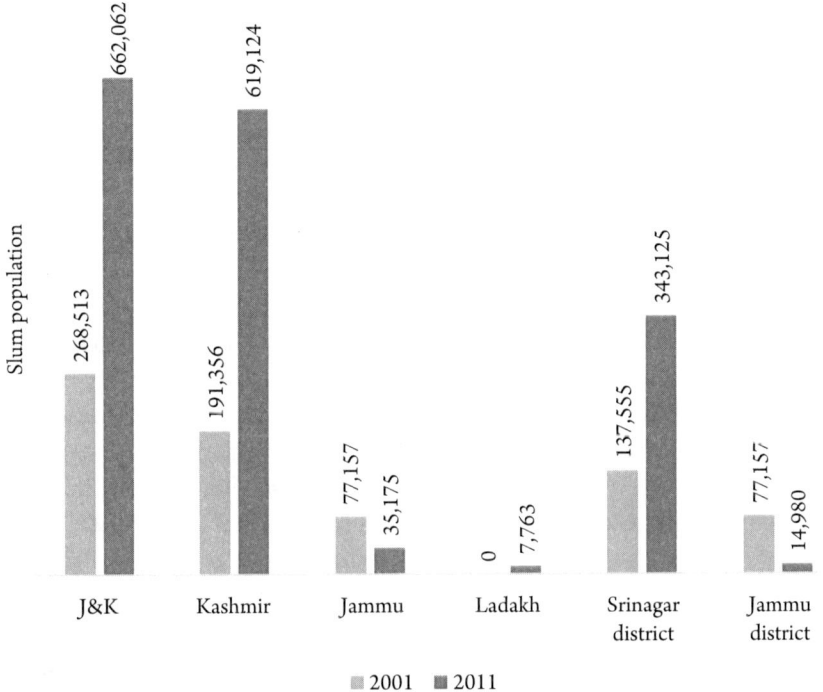

**Figure 1.1h**  Slum population in Jammu and Kashmir (J&K), 2001–11

*Source*: Respective primary census abstracts (PCA); Shina language: GoI (2007, 2018b).
*Note*: (*a*) Jammu and Kashmir (J&K) refers to the erstwhile state of J&K, whereas Kashmir and Jammu refer to the respective administrative divisions. (*b*) Census was not conducted in J&K in 1951 and 1991.

imparts objectivity and a scientific aura to policy interventions and their assessments. By the mid-nineteenth century, statistics had begun to transform, in popular imagination, from 'the eyes and ears of the government' (Rao 1999: 46) to an ally of public interest. French champions of 'progress through public information', for instance, proclaimed that 'wherever the struggle resurfaces between the champions of the general interest and that of private interest, you will find us at our post, armed and ready to march' (T. M. Porter 1995: 80). Around this time, there was also a 'shift towards willing participation [in state-sponsored surveys] on the part of the respondents' (Bookman 2013: 51),[21] while government statisticians began to view themselves 'as scientists rather than as administrators, as men who served their respective states less by following political guidelines than by employing science and its universal validity in the state's interest' (Göderle 2016: 69).[22] Statistics began to be seen as essential for 'the replacement of old relations of status, rank and dependence by those of objectivity and truth' (Rose 1991: 678; see also Woolf 1989: 602). For instance, for Karl Pearson, a pioneer of mathematical statistics, statistics were essential to rescue 'government and administration' from 'the hands of scientifically illiterate gentlemen and aristocrats' (T. M. Porter 1995: 20).

Statistics, though, did not merely demarcate the technical or scientific domain from the political. They were also essential for separating the *modern* from our 'savage past' (Wesley Mitchell quoted in Philipsen 2015: 61), 'barbaric darkness' (Keynes 1927: 565) and 'dark age' (Michael Boskin quoted in Philipsen 2015: 149) and for defeating 'superstition' (Rose 1991: 678). Statistics performed this function not only domestically over time within the West but also internationally across space by demarcating the West from the rest of the world. An entry on 'census' in the *Encyclopaedia Britannica* (2018) suggests that countries that had not been properly censused were engulfed in 'demographic darkness'. These theologically and anthropologically loaded descriptions of the pre-statistical era through the twentieth century reveal the importance of statistics for the self-image of modernity.

In such a milieu, where statistics were increasingly seen by common people, administrators and scholars alike as essential for recasting the society and polity to make them (scientifically) legible, data collection became part of routines that constituted a state in popular imagination. The ability to describe themselves and to justify their actions to both citizens and fellow states[23] in statistical terms was therefore constitutive of the self-imagination of nascent modern states. As a result, irrespective of the nature of their polity, the statistical practices of most European states converged over time (Woolf 1989: 590–91, 601–04; Tooze 2003: 2; Desrosières 2013: 10).

In a related development, government statistics began to be seen as both measures and attributes of the socio-economic and political advancement 'of the nation-states then under construction or seeking to assert themselves' (Desrosières 2013: 10), with 'nations with representative institutions' like Great Britain being capable of generating 'the most reliable and abundant statistics' (T. M. Porter 1995: 80). This understanding of statistics also informed nascent modern states in Asia and Africa, where the illegibility of colonised societies in statistical terms was identified as both a cause and a consequence of their backwardness and population censuses were an integral part of the colonial project of modernity.[24] Post-colonial leaders agreed that '[d]eveloping countries are not only economically backward, but also backward in basic relevant data' (Kansakar 1977). In fact, poor-quality statistics were seen not only as 'a sign of the underdevelopment' (Roger Riddel quoted in Jerven 2013: 1) but also as a source of 'collective shame' (Agrawal and Kumar 2020a: 286). So one of the first tasks of most newly independent states was to make themselves statistically legible to their own citizens as well as the rest of the world.[25] India enacted a law to govern headcounts – the Census Act, 1948 – two years before the adoption of the Constitution in 1950. Early Maoist China debated the appropriate means of erasing 'national *humiliation*' entailed by a 'lack of factual self-knowledge'[26] and making the country 'legible' and 'knowable' in modern terms while simultaneously distancing the new regime from the country's feudal past as well as imitators of 'shameful' Anglo-American bourgeois statistical methods (Ghosh 2020: 58–59, 112, emphasis added).[27] Elsewhere, after a few unsuccessful attempts at enumerating the country, a minister in early post-colonial Nigeria pointed out that '[t]he impressions of the manner in which a country conducts its affairs are one of the factors which earn for it the respect or disrespect of the rest of the world' and argued that '[i]t is our duty as a nation to see to it that we produce population census results which have been thoroughly conducted, verified and appraised, and therefore *acceptable, without any shadow of doubt, to all governments of the world and to all international bodies*' (Aluko 1965: 371, emphasis added).[28] A former census commissioner of Pakistan observed that 'frequent postponement of censuses by different governments became a domestic and international *embarrassment*' (Khan 1998: 481, emphasis added).

Just as the inability to generate internationally acceptable statistics is a matter of shame, the opposite seems to be a source of immense pride. According to the Registrar General and Census Commissioner of India:

> The Indian Census has a rich tradition and enjoys the reputation of being one of the *best* in the world…. It is remarkable that the great historical tradition of conducting a Census has been maintained in spite of several adversities like wars, epidemics, natural calamities, political unrest, etc. Very few countries in the world can boast of such a distinction…. The Indian Census is often billed as the *largest* peacetime mobilisation in the world. (GoI n.d.3: 1–2, emphasis added)

So as another census official puts it, 'We should be proud of our rich census heritage' (GoI 1981b: 8). This self-congratulatory and celebratory tone pervades government reports and is echoed in the public outreach of the census through, say, the postal network. Consider, for instance, three commemorative postage stamps on censuses, issued by the Indian Posts and Telegraphs Department (later Department of Posts): 'Census Centenary' (GoI 1971c), 'Census of India 2001'

(GoI 2001c) and 'Census of India 2011' (GoI 2011e).[29] The information brochures issued alongside these stamps throw light on how the census is integral to the Indian state's self-imagination qua a modern democracy.[30] The brochures offer two key insights into the relation between the census and the state. First, the unparalleled size and continuity of the operations are a source of national pride. Second, the brochures highlight the progress in the computerisation of the processing and analysis of census data. The repeated assertions about the unprecedented size of census operations have to be read along with the claim that the census is a 'modern'[31] and 'scientific' administrative exercise involving the use of advanced computational technologies (GoI 1957, 1971c, 1971d). The ability to successfully and efficiently execute a large-scale modern administrative exercise that involves 'extensive use of technology' (GoI 2011e) nourishes the Indian state's modern self.[32]

There is another aspect of India's positive self-perception that is closely tied to statistics and is stressed in census propaganda: 'India is known the world over as the world's largest and most vibrant Democracy. There is an intimate link between the democratic processes and the Census' (GoI n.d.3: 1; see also GoI 1971c, 2011e).[33] The census provides the most crucial input to the calculations necessary for the processes that constitute and animate a democracy.[34] In short, decennial censuses are integral to India's self-imagination as a modern and democratic country.

The political and symbolic significance of the census has an important implication. Territorial sovereignty entails the exclusive right as well as responsibility to map and enumerate the territory and its people.[35] The Constituent Assembly of India assigned 'The Survey of India, the Geological, Botanical, Zoological and Anthropological Surveys of India; Meteorological organisations' (item 68) and 'Census' (item 69) to the Union List (Constitution of India, Schedule 7; also Census Act, 1948, Section 3), which makes the union government exclusively responsible for collecting information in key fields. This is also true of other countries. In Nigeria, the chairman of the National Population Commission (NPC) warned that 'any state that publishes its own census … has committed a breach of the constitution' because '[t]he NPC is the only body mandated by law to conduct the census' (Yin 2007).

In 1997, when the non-sovereign Palestine tried to conduct the first systematic census of its territory, 'Hassan Abu Libdeh, the head of the Palestinian Central Bureau of Statistics, said the survey would give Palestinians a self-knowledge long denied [to] them under Israeli occupation' (Greenberg 1997). Palestinians were not merely seeking self-knowledge, though. They were seeking the right to enact the routines associated with modern states and viewed the census as 'a step toward statehood' (Greenberg 1997). The fledgling Palestinian authorities were trying to demonstrate their state-like capacity, which is why Israel tried to block the census as it did not want the exercise to proceed without any contestation.[36] Palestine is an exception though. In most parts of the world, non-state actors cannot realistically hope to replicate the state's data collection exercises. However, they can very easily disrupt the statistical routines of the state. Disrupting government surveys, including censuses, therefore emerged as an easy way to challenge post-colonial states and their claims to sovereignty. This explains the highly irregular and incomplete nature of censuses and other government surveys in most conflict zones across the world. Non-state actors have in the past tried to disrupt censuses and cartographic surveys in north-east India even as they failed to conduct parallel surveys.

Seen in this light, the insurgency in Kashmir disrupted the periodicity of both censuses and elections in the 1990s and challenged India's image as a modern democracy just when 'significant advancements in country's economy' had 'taken the country to the centre stage of world attention' (GoI 2011a: 3) and promised to reverse decades-old negative international perceptions dating back to the late 1950s and the mid-1960s. Since the 'acceptance and acquiescence of the territorial order … [supported by the state is] the prerequisite' (Göderle 2016: 86) for engaging the census,[37] the obstruction of or non-participation in enumeration by the supporters of Kashmir's independence was in line with their rejection of India's sovereignty over Kashmir. Resuming elections and censuses was therefore among the highest priorities of the union government in Kashmir in the late 1990s.

The hasty resumption of the census in 2000 was used to signal the return of normalcy in Muslim-majority J&K to both domestic and *international* audiences.[38] Commenting on the census, Mirwaiz Umar Farooq, the chairman of a faction of the All Parties Hurriyat Conference, cautioned that it must not be 'used as a tool to hoodwink the world community about the demographics of a state which is disputed' (Jameel 2010b).[39] The international dimension of domestic administrative choices in J&K was also alluded to in the parliamentary debate on the state's reorganisation by Shashi Tharoor,[40] among others. Tharoor said:

We know that we have spent decades – your Government, our Government, all Governments – to convey a message of normalcy to the world. We have had so many advisories by foreign Governments telling their citizens not to travel to Kashmir. All that, all those decades of efforts to *show* this is not a State in crisis have been undone by this action. (Lok Sabha 2019: 169–70, emphasis added)

Later, former union minister Kapil Sibal argued that the union government opened dialogue with mainstream political parties of Kashmir because it 'wants the world to *believe* that it is making serious efforts to restore democracy in J&K' (Sibal 2021, emphasis added). He added that the government wants 'to send a message to the world that the abrogation of Article 370 was not intended to throttle democracy but to ensure peace in a region sought to be destabilised by infiltrators and collaborators from within' (ibid.). The union government not only wants to convince the international community that conditions have improved in J&K but also wants Kashmir to believe that the rest of the world, including major Islamic countries, accepts New Delhi's position. The commentary around the launch of a direct flight between Srinagar and Sharjah after the repeal of Article 370 (Beigh 2021, Fareed 2022), the prime minister's remarks on the signing of a Comprehensive Economic Partnership Agreement (CEPA) with the United Arab Emirates (UAE) (S. Roy 2022), the prime minister's visit to J&K with a business delegation from the UAE (A. Sharma 2022), and one of the meetings of the G20 (Group of Twenty, a multilateral platform that brings together the largest economies of the world) in Kashmir (*Indian Express* 2022) is a case in point. Jammu-based journalist Zafar Choudhary suggested that such developments were 'endorsement of India's position on Kashmir, and evident rejection of Pakistan's stand' (Fareed 2022).

The government also needed a semblance of normalcy to reassure itself of its secular modern self. The perception of Kashmir as a touchstone of India's secular modernity dates back to the late 1940s. Mahatma Gandhi viewed it as a beacon of hope even as the rest of the country was being consumed by communal passions. As Sheikh Mohammed Abdullah, the second prime minister of J&K after 1947, put it:

> For Gandhiji, Kashmir was an issue about an ideal rather than territory.... He believed that even if diminished in size, its [India's] soul should remain unsullied.... He looked upon Kashmir from this perspective with its Muslim majority. He said that Kashmir was to be a touchstone for India's secularism. If India failed to *please* the Kashmiris, its *image* in the world would get tarnished, especially in the Muslim world. After the partition of India based on the two-nation theory, the world's attention was focussed [*sic*] on the experiment being made in Kashmir. Looked at from this perspective, Kashmir symbolized not only India's future but also its touchstone. (Abdullah 2016: 309, emphasis added; see also Lok Sabha 1986a: 327)[41]

Jawaharlal Nehru added that 'Kashmir is symbolic as it illustrates that we *are* a secular State, that Kashmir, with a large majority of Muslims, has nevertheless, of its own free will, wished to be associated with India' (Rai 2018: 213, emphasis added).[42] He feared that the developments in Kashmir would have an impact on Muslims in the rest of the country (Panun Kashmir n.d.2: 29). Agha Ashraf Ali argued that 'for Nehru, Kashmir was of prime importance because of the Muslims who did not leave for Pakistan; Kashmir was a *demonstration* that the Indian State could and would fairly treat its Muslim citizens' (Dulat 2010: 345, also 229, emphasis added). He added that '[i]f Sheikh Saheb [Sheikh Abdullah] had grasped this, he could have helped Nehru; if not fight some of the battles being waged in Delhi then at least not add to the skirmishes in Kashmir', but unfortunately Sheikh Abdullah 'was unable to fathom … what [Nehru] was trying to do in Kashmir to save Muslims in the rest of India' (ibid.). In the early 1970s, socialist leader Jayaprakash Narayan reportedly observed: 'It is the result of the decisive and enlightened leadership of Sheikh Mohammad Abdullah that today an Indian is able to cite Kashmir as a shining example of Indian secularism' (Abdullah 2016: 526). It has, in fact, been argued that 'India cannot detach itself from Kashmir because, tautologically, the continued inclusion of Kashmir as a majority Muslim state within India legitimates the state's ideational claim to pluralism' and 'validates New Delhi's claim to religious and cultural heterogeneity' (Hill and Motwani 2017: 123, 144).[43]

This view was widely shared by the mainstream leadership and intelligentsia within Kashmir and elsewhere. In a Lok Sabha debate on Kashmir, Somnath Chatterjee recalled Sheikh Abdullah's observation that the 'continued accession of Kashmir to India should however help in defeating this tendency' of India to tilt towards 'a religious state' (Lok Sabha 2000: 326). Saif-ud-Din Soz, a Kashmiri leader who later served as a union minister, suggested that being the 'lone Muslim majority State' Kashmir was 'one of the strongest elements of India's secular edifice' (Swami 2000a; see also Lok Sabha 1986a: 326–27; Soz 2018).[44] When asked by the late Pakistani leader Salman Taseer what would a plebiscite in Kashmir in favour of Pakistan mean to India, veteran journalist Shekhar Gupta said, 'It would be a mortal blow to the secular nationalism we are building as, thereon, all other Muslims will be seen as suspect, and may even be victimised' (S. Gupta 2011). More recently, an editorial in a leading national daily argued that 'the special status guaranteed to Jammu and Kashmir was not a partisan or personal decision of the founding fathers of the Indian republic … [but] a recognition of the role a Muslim majority state – its unique demography protected by the Constitution – would play in belying the claims on which Partition had taken place, and in strengthening the secular "idea of India"' (*Indian Express* 2019).

This discussion has a few key implications. First, restoring normalcy in Kashmir after a decade of unrest and, by implication, maintaining the territorial status quo was a paramount objective for the secular and modern state. The resumption of censuses and elections that are among the best-known routines of modern democracies was seen as a key measure of the return to normalcy. The quality of data was, understandably, of secondary importance under these circumstances. Interestingly, the government did not seem to be concerned that it was collecting data that it may ultimately not be able to use. Indeed, it could not use the census data for the delimitation of electoral constituencies as the exercise had to be postponed for 25 years due to serious differences within the state.[45] Second, as Rai (2018: 213) points out, the secular nationalist understanding of Kashmir's position within the union suffers from inherent contradictions: 'For the secularists in the Congress – Nehru among them – possession of Kashmir was vital to fulfilling India's national self-definition along lines no less tied to religious identity…. Paradoxically, then, emphasizing the Muslim-ness of Kashmir became instrumental to burnishing India's secularity.'[46] It is not clear if this translated into a pre-commitment to a non-decreasing share of Muslims in the state's population.

Third, because of the close association of statistics with 'modern' (and 'scientific') and their emergence as the preferred means through which the modern state communicates, attempts to engage the state or challenge its narrative have to be cast in statistical terms. In the early twentieth century when Basque nationalists were embracing 'the new age of statistics' in Spain (Urla 1993: 821), proto-Naga nationalists in north-east India had also begun to rely on statistics and statistical comparisons in their engagements with the state and the rest of the world (Agrawal and Kumar 2020a: 38–42). While 'urban business leaders, intellectuals, and professionals' drove the turn to maps and statistics among Basque nationalists (Urla 1993: 821), a handful of first-generation literates were at the forefront of similar developments in the Naga Hills (Agrawal and Kumar 2020a: 38). Around the same time, Kashmir's future leader Sheikh Abdullah, who was the first Kashmiri Muslim postgraduate in sciences, started collecting 'figures pertaining to the ratio of Muslim employees working in the administration', writing articles 'based on the collected statistics' on the plight of Muslims under the Dogra rulers and confronting the government 'armed with statistical figures' (Abdullah 2016: 36, 55, 58).[47] He shared a brief but an interesting description of this phase of his life:

> I left home early in the morning, mounted my bicycle and for the entire length
> of the day went from one department to another, collecting useful figures and

facts … often I had to go hungry…. At night, even a bird will make for its
nest. I too would return home. This went on for some months. (Ibid.: 62–63)

This turn to statistics in the early twentieth century was not confined to a
strategic or symbolic imitation of the state. Basque nationalists championed 'the
formation of an independent institute of social statistics' (Urla 1993: 821), while
Naga nationalists tried to conduct their own census and counter-mapped the
territory (Agrawal and Kumar 2020a: 36, 100). Sheikh Abdullah's *Naya Kashmir*
manifesto envisaged the formation of a statistical institute to support various state
institutions including the planning commission (National Conference 1950: 53).
There is, perhaps, a major difference between the public careers of statistics in
Basqueland, on the one hand, and Nagaland and Kashmir, on the other. In the
former, people encountered statistics in multiple domains, ranging from natural
sciences to politics and public policy. In contrast, Nagaland and Kashmir first
encountered statistics in the field of politics and identity formation. The path
dependence of public discourse resulted in the growing communalisation of
statistics with the passage of time.

With the turn to statistics, the majority–minority binary emerged as a key
point of departure for understanding plural societies within cartographically
bounded territories. This in turn foregrounded statistical proportionality as one
of the major influences on state intervention in the society and also as a yardstick
for challenging the government's claim to neutrality.[48] The dialogue between
J&K's constituent regions and between Kashmir and the rest of the country
is often phrased in statistical terms. The Hindus of Jammu draw attention
towards the persistent failure of the state government to satisfy statistical
proportionality at the level of administrative divisions. Kashmir responds by
highlighting the disparity between the Muslim- and Hindu-dominated districts
within the Hindu-majority Jammu division. The union government highlights
preferential federal transfers to J&K that far exceed its population share in
the country, cross-border infiltration of terrorists from the Pakistan-occupied
territory of J&K and casualties among civilians due to terror attacks. Kashmir
challenges the statistical hegemony of the union government by foregrounding
statistics about the impact of the long-drawn conflict on non-combatants.
The Kashmiri civil society invests in collecting and disseminating data globally
often in collaboration with international bodies and non-governmental
organisations (NGOs). This helps both in drawing attention to the suffering
of common people trapped in the conflict zone and in contesting the depiction
of casualties as unavoidable collateral damage resulting from counterterrorism
operations against cross-border infiltration. Hindu nationalists turn attention to

the exodus of the entire indigenous Kashmiri Hindu community from Kashmir and their continued inability to return to their ancestral land. In response, Kashmiri Muslims highlight their out-migration in the late *nineteenth* century during the (Hindu) Dogra period. Such statistical duels involving cherry-picked numbers are integral to the conflict in J&K.

## Collective Self-Portrait

The Registrar General of India (RGI) suggests that a population census 'provides an instantaneous photographic picture as it was of a community, which is valid at a particular moment of time ... [and] is called the "static aspect"' and, also, 'provides the trends in population characteristics, the "dynamic aspect" of the population' (GoI n.d.4). The director of census operations (DCO), Madhya Pradesh, added, 'Census is a reflection of truth and facts as they exist in a country about its people, their diversity ...' (GoI 2011a: 3).[49]

The understanding of census as a true photograph has to be read alongside the claim that it generates a collective portrait, which is stressed in the public outreach in the run-up to household enumeration. For instance, in 2000–01, pre-printed advertisements on postcards and inland letter cards presented the census as a 'mirror of the nation'[50] that produces a 'group photograph of the nation'[51] and as a device that converts individual information into a collective portrait of the nation (Figures 1.2a–1.2c). These advertisements featured on postal stationery in major Indian languages and were widely circulated through postal networks. Likewise, postage stamps and the cachets of their first-day covers released to commemorate censuses graphically emphasised the census as a portrait of the totality and diversity of India (Figures 1.3a–1.3c).[52]

The government presents enumerators who are key to generating the collective portrait as 'patriots' (GoI 1954b: vii–viii; see also GoI 2001b: xii). It also motivates people qua respondents to contribute to nation-building: 'Participation in the Census by the people of India is indeed a true reflection of the national spirit of unity in diversity ... participating in the Census is tantamount to participating in the Nation building process' (GoI n.d.3: 1).[53] Since the government invites people to participate in a census qua collective national photo session, its boycott amounts to rejecting the nation and accentuates the post-colonial anxiety of the society 'suspended forever in the space between the "former colony" and "not-yet-nation"' (Krishna 1994: 508).

In any case, the claim of photographic reproduction is misleading. Contrary to the vast evidence on the messy nature of modern censuses, the government wants the citizens to believe that the census is like a distant satellite taking

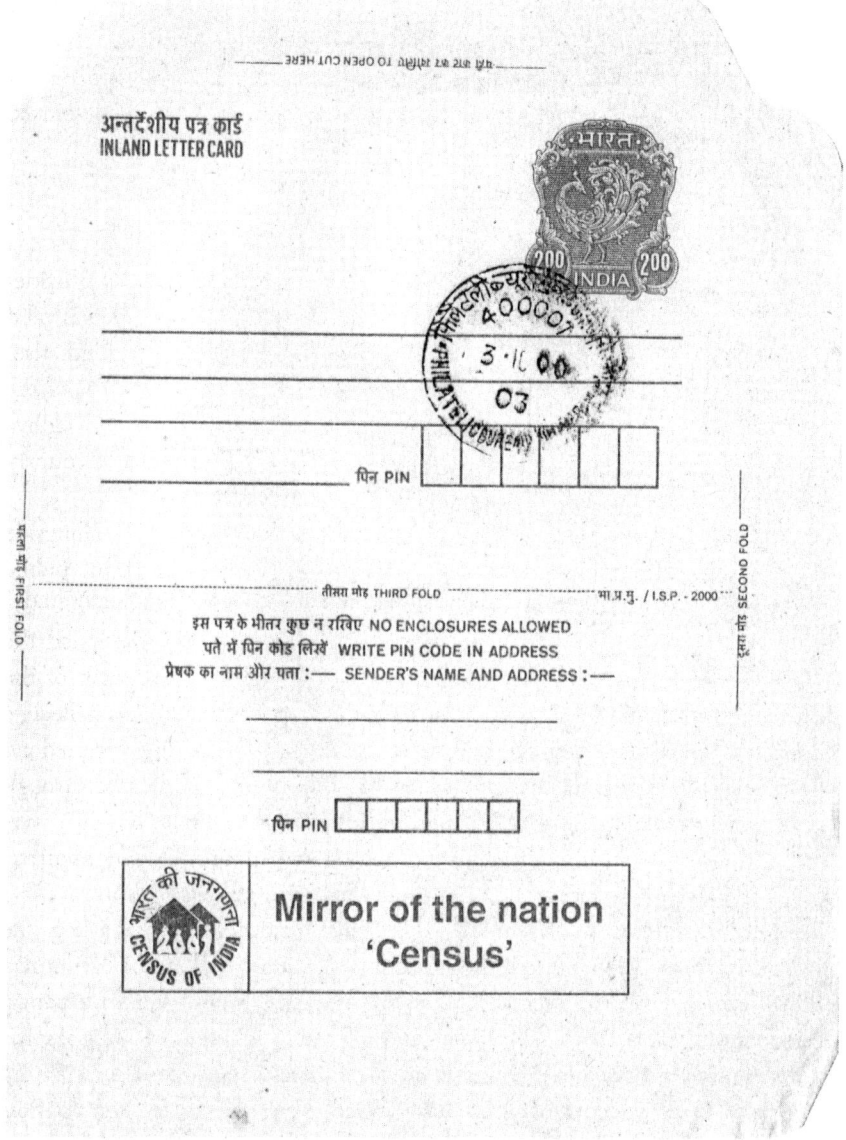

**Figure 1.2a** Inland letter card from 2000 advertising the upcoming 2001 census: 'mirror of the nation'

pictures of unsuspecting people. In reality, the political leadership, bureaucracy and respondents often have a priori expectations about how the collective self-portrait should look like. This is particularly true of conflict zones.

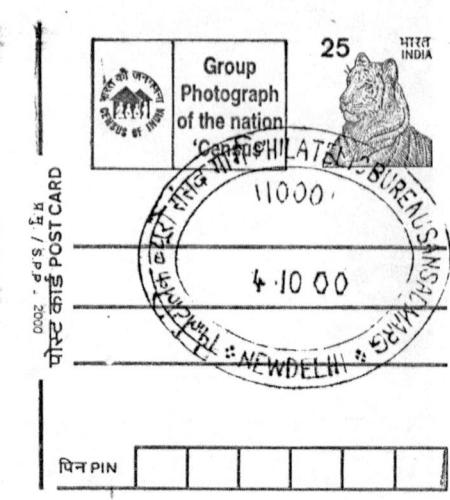

**Figure 1.2b**  Postcard from 2000 advertising the 2001 census: 'group photograph of the nation'

Since the photographer – the census bureaucrat – is one of us and, more precisely, belongs to one of the constituents of the collective,[54] and since different sections of the collective have different preferences over the contents of the photograph, the census becomes an arena of 'volatile confrontation between … enumerators equipped with tools that are presumed all-terrain and context-free, and myriads of agents actively promoting their local social and political agenda through this encounter' (Guilmoto and Rajan 2013: 69). This encounter generates a 'political document' (Brass 1974: 75, 77) or a 'social document' (Mohanty and Momin 1996) that involves 'negotiation' between different groups (Government of Nigeria n.d.; Okolo 1999: 323; Heine and Oltmanns 2016: 205; Aung 2018), building 'consensus' between 'multiple government departments' (Yu 2021) and 'political agreements between stakeholders from the spheres of politics, the economy and law' (Heine and Oltmanns 2016: 205). A rich diversity of possibilities arises depending on the extent of alignment of the preferences of the government, enumerators and respondents over who should be counted and how they should be represented in the collective portrait.

In the United States (US), the democrats and the republicans are deeply divided on whether 'unauthorized immigrants' should be included in 'population counts used for reapportionment' (Wines 2020). In Pakistan, 'the province of

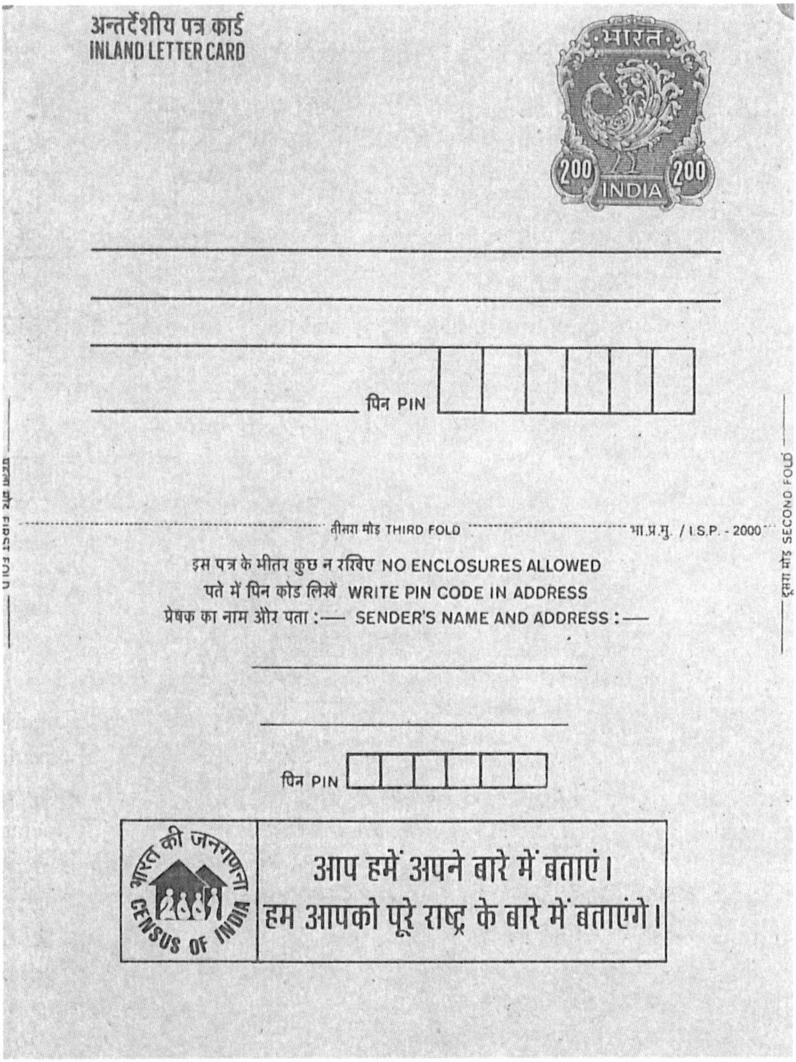

**Figure 1.2c**  Inland letter card from 2000 advertising the upcoming 2001 census: 'आप हमें अपने बारे में बताएं| हम आपको पूरे राष्ट्र के बारे में बताएंगे|' (You tell us about yourself. We will tell you about the whole nation.)

*Source*: Images by the author.

*Note*: As per Kotadia and Gandhi (2002a: 102, 2002b: 73), (*a*) the inland letter card in Figure 1.2a was printed at the India Security Press (ISP), Nashik and issued on 3 October 2000; (*b*) the postcard in Figure 1.2b was printed at the Security Printing Press (SPP), Hyderabad and issued on 4 October 2000; and (*c*) the inland letter card in Figure 1.2c was printed at the ISP and issued on 4 October 2000.

**Figure 1.3a** Commemorative postage stamp and the first-day cover for the 1971 census: 10 March 1971

**Figure 1.3b** Commemorative postage stamp and the first-day cover for the 2001 census: 10 February 2001

**Figure 1.3c** Commemorative postage stamp and the first-day cover for the 2011 census: 8 February 2011

*Source*: Images by the author.

Baluchistan and the city of Karachi share the concern that Afghan refugees might be counted as Pakistani citizens in the enumeration process' (Weiss 1999: 686), while indigenous Pathans complain that Balochs are over-counted (Horowitz 2000: 195). Such concerns are not restricted to international immigrants. In the hill states in north-east India, there is opposition to the enumeration of non-indigenous people from other parts of the country (Agrawal and Kumar 2020a: 199, 238). There has been opposition in Kashmir to the enumeration of West Pakistan refugees who reached Jammu in the late 1940s (Bhat 2018: 181).

To these examples of dispute over who should be included in the collective photograph, we can add instances of differences over the portrayal of those who are included. During the colonial period, Christians and Muslims opposed the enumeration of caste among them. On the other hand, 'by 1931, barring a few most downtrodden castes, viz. Bhangi, all other non-twice-born castes [among Hindus] had formed associations which claimed high ritual rank [including in the census], at par with twice-born castes' (Chaudhury 2005). In Rajasthan, Christians among tribes reported themselves as Hindus in government records to avoid the loss of benefits of affirmative action policies (Sahoo 2018: 34, 45). For similar reasons, non-tribal communities reported themselves as members of Scheduled Tribes (ST) in parts of Maharashtra (S. Kulkarni 1991; Guha 2003). Sumis of Nagaland switched en masse from the colonial exonym 'Sema' to the endonym 'Sumi' for their mother tongue even as the census persists with the older exonym (Agrawal and Kumar 2020a: 177–79). In Nagaland, Yimkhiungs objected to the enumeration of their Tikhir neighbours, an indigenous tribal community, as a separate ST (ibid.: 302). In Jammu, SCs opposed the use of older stigmatising exonyms in the census (GoI 1963a: 27). Gujjars allege that Kashmiri enumerators count them as Kashmiri-speaking non-tribal Muslims but not as Gojri-speaking tribal Muslims (see Chapter 3).

The bureaucracy that mediates between the lay-respondents or communities and the government may have its own preferences over how the collective photograph should look like. In most cases, it wants to ensure that post-enumeration surveys (PES) do not reveal a large error in the headcount and that the headcount does not vary too much vis-a-vis other administrative records. Further, officials may enumerate minorities or weaker groups in accordance with local stereotypes or the preferences of the dominant community. Legally speaking, the membership of STs, unlike that of SCs, is not conditional upon religious affiliation. However, in popular perception tribal identity is linked to religion. In states like Madhya Pradesh, in some areas tribal Christians may not have been recorded as STs (Venkatesan 2001). Likewise, non-Christian tribes may not have been counted as STs in parts of the north-east (Prabhakara 2012: 108). Commenting on the possible manipulation of headcount in Kashmir, Guilmoto and Rajan (2013: 64) note that given 'the scale of this exaggeration', 'active cooperation or initiative of local enumerators in the process of census manipulation cannot be ruled out'. However, even if bureaucrats are neutral, they may have to condone manipulation by other actors. As discussed in Chapter 5, in most cases they cannot defy politically motivated communities or manipulation of census categories by the government.

Governments may rework ethnic as well as social boundaries and headcounts to achieve an acceptable self-portrait of the country. State-sponsored demographic engineering can take various forms, including interference in population measurement, pronatalist policies, assimilationist policies, economic pressures, change in migration and domicile policies, boundary changes and population transfers (Bookman 2013: 32–34). Further, non-state actors can autonomously influence headcounts, and different tiers of the government might have conflicting preferences over outcomes (Brass 1974; Horowitz 2000; Morland 2018; Agrawal and Kumar 2020a).

Government interventions can change the data generation processes (through incentives that affect how people report themselves – for example, by restricting affirmative action policies to certain groups and leaving out closely related communities), the data collection processes (through alteration in the method of enumeration that can affect, say, how migrant, urban and homeless populations are counted), the collected data (through ex-post manipulation of data or their categorisation), the modes of interpretation (partisan treatment of outliers and missing values, partisan interpretation of trends and change in projection techniques) or even the facts on the ground (by changing the actual population through interventions that affect birth rate or migration or that facilitate assimilation) (Figure 1.4).[55]

As discussed in Chapter 6, the Indian government's pronatalist policies are limited to select micro-minorities. Likewise, it has also avoided other strategies mentioned earlier, except interfering with the definition and measurement of population, the 'least intrusive method of altering relative numbers', that '*de jure* result in a change in the ethnic population size, even if they may not *de facto* change that size' (Bookman 2013: 32–34). In 1951, the government decided that

> the only relevant question on caste or tribe incorporated in the Census Schedule was to enquire if the person enumerated was a member of any 'Scheduled Caste', or any 'Scheduled Tribe' or any other 'Backward class'…. While the data on scheduled castes and tribes were published, that on other backward castes was [collected but] not published…. From 1961 onwards, the information

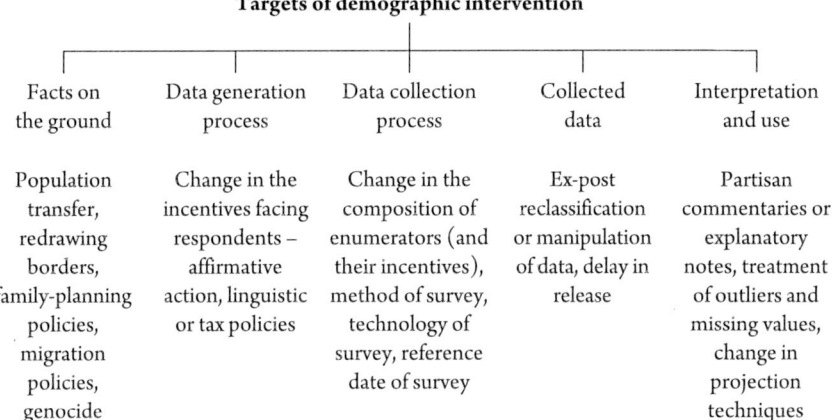

Figure 1.4   Demographic interventions by the government

*Source*: Prepared by the author.

was collected only for each Scheduled Caste and Scheduled Tribe. (GoI 2009a: 28; see also Maheshwari 1996: 142, 179; Roy Burman 1998)[56]

So the government changed the classification of data ex post. Similarly, after the 1971 census and the partition of Pakistan on (economic and) linguistic grounds, the government that normally celebrates ethnolinguistic diversity decided that only languages spoken by more than 10,000 persons would be reported separately (GoI 1976a: i), with the rest being grouped as 'Others' or as dialects of reported languages. Once again, in one stroke, the government changed the collective (linguistic) photograph of the country. Further, the cross-tabulation of data on religion was restricted right from the beginning, which 'reflects a conscious decision taken by the Indian government in 1949 not to cross-tabulate economic and ethnic data because it was felt that such tables had heightened communal and caste sensitivities during the colonial period' (Wilkinson 2004: 40; see also GoI 2004a: v, xiv–xv).[57] As a result of the government's reluctance to publish data on identity, 'the census organization has shown completely stoic indifference to the key cultural and ethnic variables' (Mohanty 1996: 165). In several other cases, ministries delayed or cancelled the release of inconvenient survey data and indefinitely suspended surveys (Agrawal and Kumar 2020c). In extreme cases, state governments did not identify certain communities in the data. The Uttar Pradesh government did not identify the ST population in the early post-colonial censuses even though certain communities were formally recognised as tribes (Verma 2013). In J&K, the state government did not even recognise STs until the late 1980s (Chapter 3), while Assam continues to deny ST status to long-standing tribal settlers from outside the region (Prabhakara 2012).

In Myanmar, the 'first nationwide census [in 2014] in more than 30 years allowed the respondents to choose from among 135 officially recognized ethnicities that were grouped under eight major "ethnic races"' (Aung 2018). Ahmadis are '[b]anned from calling themselves Muslim' in Pakistan and are not recorded in the census as Muslims (*Dawn* 2017b).[58] In the 1994 Census of Ethiopia, people 'were made to choose an ethnicity, even though many Ethiopians are of mixed heritage', because this helped dividing the country 'into regions that ostensibly followed ethnic boundaries' (*The Economist* 2019). The Sri Lankan government exhorted the Sinhala majority to not report internal divisions in the 1981 census, while the Tamil minority was further divided in the 2001 census (Morland 2018: 63–65). Israel's censuses have played with categories to enhance the Jewish majority (ibid.: 136–39). Blacks in apartheid-era South Africa (Lipton 1972), Tajiks in Uzbekistan (Abramson 2002), Tutsis in Rwanda (Uvin 2002) and Russians in Turkmenistan (Goble 2015) were undercounted in the census or sub-categorised due to state intervention. Similarly, the accounting of race in the West has been affected by state interventions (Bhagat 2003; Morland 2018: 151–58). Political transitions associated with wars and the rise and fall of communism offer further examples of sharp changes in the reported composition of population due to the re-categorisation of population through administrative fiat (Kertzer and Arel 2002a). In all these cases, censuses qua self-portraits conformed to a partisan visualisation of the nation or society.

In light of this discussion, we can argue that the union government was interested in somehow obtaining a 'photographic picture' of India *including* J&K. Supporters of greater autonomy or independence for Kashmir wanted a photograph that showed Kashmiris and Muslims as the most prominent group of the state. Jammu complained that New Delhi was in a rush to conduct the census and allowed Kashmir to manipulate figures. This, however, does not imply a congruence between the interests of New Delhi and those of Kashmir. The government and the supporters of autonomy or independence for Kashmir often differ on which figures better capture the conditions in the state. The government favours a portrait that reveals normalcy and explains departures as unavoidable but temporary and manageable consequences of cross-border terrorism. On the other hand, partisans of independence highlight data that question the official narrative – for example, low voting rates in elections, media blackouts and violation of human rights by the state including extrajudicial killings. In March 2011, when the government was celebrating the successful completion of census, local newspapers were debating the mismatch between two official statements in the state's legislative assembly on deaths during civil unrest (*Kashmir Times* 2011f). A week after the completion of the 2011 census, the Association of Parents of Disappeared Persons (APDP) 'begun count of the persons who went missing in the state during past 20 years ... under the United Nations (UN) Human Rights Project under its voluntary fund for victims of torture' (Dar 2011). This suggests that Kashmir's civil society resorts to 'scale jumping' (Herod 2011: 19) to overcome the constraints imposed by scalar politics within the Union of India. In short, both the desirability of the collective photograph and its contents are contested in J&K.

## Public Goods

Heine and Oltmanns (2016: 207) suggest that statistics can be treated as public goods because 'statistical data can be consumed on a non-competing basis' and the 'marginal costs of production are zero and therefore the price-mechanism does not work'. They further argue that 'the exclusion principle can be applied; in principle it is possible to limit the access of users to data. As a result, data are a good to which a property right can be assigned' (ibid.: 207). However, the exclusion principle does not apply in the case of government statistics in democracies. Access cannot be limited where an independent judiciary, a free press and provisions for mandatory disclosure – for example, the Right to Information (RTI) Act, 2005, in India – ensure that citizens can, in principle, access most information collected by the government.

Since government statistics are public goods, in divided settings such as J&K they are susceptible to problems generally faced in the provision of such goods insofar as

ethnic diversity – or more specifically, ethnic polarisation – leads to lower and more uneven levels of public goods provision. On the supply side, dominant ethnic groups are found to discriminate against other ethnic groups in the provision of public goods. On the demand side, ethnic polarisation fractures the possibility of inter-ethnic cooperation and undermines the possibility of popular mobilisation for universal goals such as public goods. (Gupta and Pushkar 2010: 65)

The supply of government statistics qua public goods is suboptimal also because incumbents do not see them merely as 'a policy-making tool' but also as 'a signal to supporters and detractors; a means for enhancing control over rents and revenues; and a source of pressure and competition to their position'

(Krätke and Byiers 2014: 26).[59] In other words, 'data producers and users have specific personal interests with regard to the scope and design of statistical infrastructure', and, as a result, 'individual rationality and collective rationality may fall apart and create welfare losses for society' (Heine and Oltmanns 2016: 201–02).

The supply and quality of politically salient government statistics have been affected in India due to, among other things, the growing politicisation of the administration and ethnic and religious polarisation. This is reflected in, say, the increasing delays in the release of government statistics on language, religion and migration despite significant improvements in the technology of data processing (Agrawal and Kumar 2020c). The politicisation has now reached a stage where it has begun to affect the timeliness of data collection rather than merely delay the publication of collected statistics. The indefinite and inexplicable delay in holding the 2021 census is a case in point. India's experience is shared by several other countries in Asia and Africa.

Most of India's neighbours have not been able to conduct censuses regularly due to ethnic conflict and political unrest, while very few African countries have managed to enumerate their populations on time. Pakistan has not been able to conduct regular censuses after 1961. It postponed the census five times between 1991 and 1998 and ultimately managed to produce figures that matched the expectations based on earlier censuses (Weiss 1999: 687, 691). Commenting on the 1998 census, Akhtar Hassan Khan, the census commissioner for the 1981 census, observed that 'no government [in Pakistan] was prepared to face census results which sharply changed inter-provincial ratios or rural-urban ratios, as these would have resulted in altering the seats allocated to different provinces in the national Assembly as well as the allocation of development funds' (A. H. Khan 1998: 481). Two decades later, commenting on the results of another much-delayed census, researchers lamented that 'in Pakistan [a census has] more to do with the domination of one province over others' and '[a]ll that the census is doing is maintaining the status quo' (S. G. Khan 2018). Sindh voted against the approval of the results of the 2017 census of Pakistan because it allegedly underestimated the province's population by more than 20 per cent (*Dawn* 2021a, 2021b). The inability of the government to conduct a post-enumeration survey after the latest census (Wazir and Goujon 2019) highlights the deepening politicisation of statistics in Pakistan. Bhutan's government intervened to ensure that the headcount and its ethnolinguistic composition conformed to its expectations (see notes 27 and 28 of Chapter 6).

In Ethiopia, the second most populous country of Africa, censuses have suffered repeated delays because of which 'it is often difficult to distinguish politics from demography' (*The Economist* 2019). In Nigeria, Africa's most populous country, conflicts over census were 'largely responsible for major landmarks in the ... political scene', starting with the constitutional crisis of 1962 followed by a series of coups that led to a situation where headcounts began to be seen as 'strong political weapons' wielded by rival regions and ethno-religious groups 'rather than statistical data to be used for planning for socio-economic development' (Adepoju 1981: 29, 35). More recently, the chairman of the NPC, Nigeria, lamented that 'even before the [2006] census was conducted, highly placed individuals and organizations in several states had already determined to the decimal point the population of a particular area or region' (Yin 2007). It has been argued that Nigeria's censuses reproduce historical population shares (Fawehinmi 2018). As discussed in Chapter 6, the Government of India, too, has intervened in the face of changing relative population shares of states. Its interventions have, however, been limited to freezing federal redistribution and delimitation formulae to protect the interests of states with lower fertility rates and restricting migration into sparsely populated tribal areas to prevent the marginalisation of indigenous communities.

Governments may not release politically inconvenient data and, in extreme cases, may even discontinue data collection. In 1951, the Government of India did not release caste data on backward classes that would have allegedly exacerbated social conflict and disaggregated language data for the Punjab, the Patiala and East Punjab States Union (PEPSU) and Himachal Pradesh that were affected by political mobilisation. Mauritania did not publish the results of the 1978 census, which allegedly showed that Kewri were a majority (Horowitz 2000: 195). In Sri Lanka, in the 1981 census, Sinhala respondents were exhorted 'to give up "Up Country" or "Low Country" self-categorisation, but crucially the results of this question were not published nor were they maintained in the public records' (Morland 2018: 63). Turkmenistan did not release the results of a census that did not conform to official propaganda about the country's ethnic composition (Goble 2015). In China, amidst concerns about the possible contraction of population, 'very sensitive' data from the 2020 census were not 'released until multiple government departments had reached a consensus on the data and its implications' (Yu 2021). Lebanon took the extreme step of not conducting a census after 1932 because it feared that 'taking one would reveal such changes in the religious composition of the population as to make the marvelously intricate political arrangements designed to balance sectarian interests unviable' (Geertz 1973: 275).

In conclusion, the manipulation of censuses is not implausible in ethnically and politically divided societies. Ethno-political contests undermine the provision of public goods, where

> the government monopolises access to basic goods and services valued by a majority of the population, and in which government officials have individualised discretion over how these basic goods and services are distributed … voters decide between politicians, not by assessing their policy positions, but by assessing whether a candidate will favour them in the distribution of patronage. (Chandra 2009)

Politicians and bureaucrats, too, face a 'trilemma' – 'the responsibility to be true to his office, to help members of [say] his clan, village, and tribe individually, and to bring forth development for his constituency as a whole' (Tinyi and Nienu, 2018: 174). In such settings, 'ethnic numbers game[s]' are fuelled by the 'close association between the desire for group hegemony and the democratic ethnos of universal suffrage and majority rule' that motivates 'groups in competition … to adjust their numerical ratios for sectional hegemonic interests' (Stephen Olugbemi quoted in Bookman 2013: 20–21).[60] Under these circumstances, 'an election can become an ethnic head count. Now it is clear that a census needs to be "won". So the election is a census, and the census is an election' (Horowitz 2000: 194–96).[61] In Kashmir, the census is seen as the precursor to delimitation of electoral constituencies and plebiscite that in turn generates pressure to fix the numbers in advance.

The conceptualisation of government statistics as public goods also helps understand data deficit in relation to development and democracy deficits. The unrepresentative character of institutions and curbs on their functioning in conflict zones such as J&K rob them of legitimacy and block inputs from the grassroots for planning. This undermines the legitimacy of development and public infrastructure projects, affects their design and hinders their implementation. Development deficit, in turn, lowers the opportunity cost of participation in disruptive and violent activities and is also associated with socio-economic inequalities that undermine democratic institutions. So democracy deficit is both a cause and a consequence of development deficit.

Data deficit hinders planning and aggravates development deficit. On the other hand, development deficit also contributes to data deficit by limiting resources available for the provision of public goods, including government statistics. Moreover, with the private sector crowded out of the economy by the public sector and stifled by long-standing political unrest and infrastructural bottlenecks, capturing the public pie by securing a favourable delimitation of electoral constituencies, demarcation of administrative units and distribution of affirmative action and social welfare benefits through the manipulation of data becomes an attractive 'economic' activity. The manipulation of government statistics, however, undermines the provision of public goods. In other words, data and development deficit are also intertwined.

Further, a lack of transparency in public institutions that is associated with democracy deficit erodes trust in government statistical institutions, while political instability and unrest physically disrupt data collection. Also, the government and the insurgents try to control the flow of information and mould the narrative, which further undermines the supply of unbiased information. The use of flawed data in public policy, in turn, erodes trust in institutions and triggers conflicts between the beneficiaries of such policies and others. So, as Agrawal and Kumar (2020a: 11, 313–19) suggest, data deficit is often enmeshed with democracy and development deficits.

We will argue that the growing communal and regional polarisation in J&K has affected post-1981 censuses. In addition to directly interfering with data collection, insurgency undermined the foundations of the general administration that is responsible for public goods including censuses. Further, both Kashmir and Jammu treat the census as a precursor to the delimitation of constituencies (and subsequent capture of the legislature) and plebiscite. While Kashmir views census as New Delhi's tool to alter the demography, Jammu believes that New Delhi uses the exercise to allay Kashmir's concerns by allowing it to manipulate the headcount. The lack of trust and the near-complete breakdown of the channels of communication between regions and communities[62] has meant that the government is unable to tackle various challenges facing the state, including repairing general administration and public institutions. In other words, while the machinery that delivers public goods fell into disrepair, the consensus on government statistics qua public goods has also been eroded.

There is another implication of accepting the embeddedness of data in the wider political and economic contexts – namely, measurements and the object of measurement are co-produced. So, as Tooze (2003: 3) puts it, statistics are 'an integral part of the economic and social world, which they seek to describe'. From this perspective, the poor quality of data is a product of the setting in which measurements are made, and, at least, in some cases, the problem may not be amenable to resolution through stand-alone legal, administrative or technological interventions. However, data quality, let alone its context dependence, is mostly ignored in the academic literature and administrative discussions – particularly in the case of developing countries. The lack of research in this regard can be seen from three perspectives.

First, countries and regions within countries with poor data quality are often politically marginal and account for a small share of population and/or income. The 'uneven statistical topography' arguably reflects the underlying distribution of political and economic power (Serra 2014). India's ethno-geographic periphery consists of 10 landlocked states and UTs, including J&K, Ladakh, Sikkim and 7 north-eastern states that accounted for about 15 per cent of its area but less than 5 per cent of the population and about 3.6 per cent of the national income in 2011. Together they control about 5.5 and 7.5 per cent of the seats in the lower and upper houses of the parliament, respectively. National policymaking is not adversely affected if surveys cannot be conducted in parts of the periphery or the quality of data from the periphery is not satisfactory.[63] Moreover, several union laws and policies were not applicable to J&K and are still not applicable to parts of the periphery covered by special constitutional provisions. This further reduces the urgency to understand the data deficit in the periphery. In most years between 1991 and 2012, the peripheral states (J&K) accounted for 70 (50) per cent of the sample non-coverage and almost the entire frame non-coverage in NSSO surveys, even though their population share was about 5 (1) per cent.[64] Likewise, the periphery accounts for all the major areas that could not be covered in post-colonial censuses and all but one state with less than 90 per cent enrolment in the Aadhaar (unique identity number) database. Indeed, evidence from across the world, including developed countries, suggests poorer coverage of the social and geographical margins.

Commenting on the geographical footprint of economic research, *The Economist* (2020) points out:

> The size of a country's economy is the most significant, accounting for nearly 80% of the variation in research attention.... The quality and availability of data matter too, though less than economic size, as does a country's use of English.... The 70 least-studied countries account for just 1% of all mentions in economics papers over the past three decades.

In a parallel development, there has been a narrowing down of focus on questions

> which can be answered with statistical analysis. An effort to pay more attention to the places least able to provide high-quality data, which often face the toughest roads to development, would force economists to grapple with qualitative matters. If critical contributions to development come from difficult-to-quantify variations in cultural factors, a geographically limited

discipline [restricted to areas with high-quality data] will find it hard to detect them. (Ibid.)

Not only are certain perspectives and questions left out when peripheral regions are excluded from studies, but the question of data quality is also overlooked.

In India, the smoothly varying national-level aggregates often conceal highly erratic sub-national figures – particularly in the periphery (Agrawal and Kumar 2020a: 304). Censuses and surveys in some of the peripheral states and UTs, which account for about a third of the subnational administrative units, are not only marked by irregularity and large coverage errors but also report sharp demographically inexplicable variations in, say, CSR, migration, urbanisation, literacy and slum population. Since some of the errors in data on peripheral states and UTs are non-random in nature and are shaped by their social and political-economic contexts, excluding them can bias statistical analyses of interstate differences. Greater engagement with the periphery will push governments to improve data quality and nudge academics to explore previously overlooked facets of the underlying society and political economy. Data deficit can also serve as a point of departure for understanding 'how a society adopts to the poor quality of the statistical data it produces; how can research, policymaking and administration thrive on figures of dubious plausibility, never fully accepted nor consistently rejected' (Begum and Miranda 1979: 80; see also Schwartzberg 1981: 57).[65]

Second, several scholars have attributed the gap in the literature on data quality to moral hazard in the academia. T. N. Srinivasan (1994: 4) suggests that 'researchers either are not aware of or, worse still, have chosen to ignore the fact that the published data … suffer from serious conceptual problems, measurement biases and errors, and lack of comparability over time within countries and across countries'. Shetty (2012: 41) laments that 'collection of accurate statistics has a low priority in policymaking today and the intellectual community which studies India's economic problems also shows no concern for the deteriorating quality of the Indian database, which they otherwise studiously use'. Commenting on the 'surprising gap between knowing innately that these numbers cannot be good and an unwillingness to study how bad they are', Jerven (2013: xiv), too, notes, 'The scholars who are best equipped to analyze the validity and reliability of economic statistics are often data users themselves and are thus reluctant to undermine the datasets that are the bread and butter of scholarly work' (ibid.: 8). Kelkar and Shah (2019) argue that 'few actors in the economy have an incentive to do a good job of measurement. As an example, academic economists are quite comfortable doing research with faulty data, because the academic economists who will review their work do not ask questions about data quality'.

The problem of moral hazard vis-à-vis research on data quality is not restricted to economics where the 'sub-field of the political economy of statistics is notable for its absence' (Wade 2012: 17) or is a 'largely unknown terrain' (Hartwig 2006: 535). Similar complaints are heard in, say, political science, where researchers 'make heavy use of census statistics but have given scant attention to the politics behind the production of those statistics' (Prewitt 2010: 237).

Third, gaps in university curriculum and training have also contributed to the crisis. Mahalanobis (1965: 43) stressed 'the need of cross-examining the data which is the first responsibility of a statistician' and lamented that the

> [c]ollection and scrutiny of primary information or processing and handling of data are usually considered 'dirty work', not fit to be done with one's own hands by statisticians and, therefore, also not fit subjects for training or for acquiring skill and experience.

As a result, there is 'lack of appreciation of the fact that a master's degree in statistics [of an abstract nature without any contact with data and without any clearly perceived purpose] … does not guarantee any knowledge or skill in the professional work of statisticians' (ibid.: 44). Pfeffermann (2015) draws attention to the widespread neglect of government statistics in university curriculum in various countries. We can add that leading textbooks of statistics and econometrics do not even pay lip service to the issue of data quality.

As a result of the above, '[i]n case of discrepancies it is often found easier to dismiss statistical data than engage them, leading researchers to either credulously accept statistics or indiscriminately ignore them' (Guilmoto and Rajan 2013: 69). Or, as Mahalanobis (1965: 43) puts it, 'Anything which is supplied or published by a Government office is accepted as reliable.' We will, however, engage with poor-quality data because discrepancies that symbolise the breakdown of statistical machinery or infrastructure make it less opaque. To put it differently, during moments of breakdown the statistical infrastructure and the larger context of data become 'analytically accessible for social science research' (R. Singh 2020: 56). The insights derived from the examination of J&K's dysfunctional census machinery will be relevant for other ethnolinguistically divided, politically restive, erstwhile special-category states in India's ethno-geographic periphery, where census-based delimitation was suspended in 2008. In fact, the insights will also be relevant for the 'mainland' because 'census' is a union subject (Article 246, List 1 [Union List], item 69) and is conducted across the country under the direct supervision of the Office of the Registrar General of India and Census Commissioner (ORGI&CC) that falls under the

Ministry of Home Affairs (MHA) of the union government (GoI 2019c: 91). In their capacity as census officers, state government officials report to the respective directors of census operations; the directors in turn report to the ORGI&CC, which is responsible for the entire exercise, from the design of schedules to the publication of results.

## Outline of the Book

The literature on census can be divided into two broad categories: assessments of the context of production and use of data (Alonso and Starr 1987; Horowitz 2000; Peabody 2001; Guha 2003; Anderson 2006; Bhagat 2001, 2003, 2006; Kertzer and Arel 2002a; Bookman 2013; Morland 2018) and assessments of the quality of data (Bose et al. 1977; Barrier 1981).[66] The literature on demography and politics (and conflict) suggests a bidirectional relationship between the two. On the one hand, conflict can affect fertility and migration (Smith 2012; Janus 2013; Morland 2018). On the other hand, demographic changes can reshape politics and precipitate conflict (Huntington 1996; Horowitz 2000). Commenting on the conflict in Kashmir, Marks (2006: 128) argued that 'there was a demographic tidal wave of unabsorbed youthful males appearing in the late 1980s, especially in Kashmir, just as political issues ... called into question the legitimacy of the existing order'. Swami (2010) added that while '[i]t is no one's case that the demographic tides are the sole force driving Kashmir's discontent.... There is little doubt though that Kashmir's youth bulge has provided the firmament for the crisis to flourish' as 'the economic gains of the first decades of independence had run up against a demographic wall' due to the high dependency rate in the state. After the provisional results of the 2011 census were published, Swami (2012) noted that

> Jammu and Kashmir is seeing the birth of the largest youth cohort in its history
> – another source of strain. Three in five Jammu and Kashmir residents are
> either under 19 or over 60 – and the young are growing fastest. This means
> there is great pressure on the productive age group, and an urgent need to
> create new jobs for those who will soon enter it.

This line of argument is problematic for two reasons. First, as discussed here, the quality of data is not exogenous to the conflict. Interestingly, Swami, too, did not cross-check the figures on age, gender and urbanisation even though he is among the very few in the media who have covered the problems in J&K's census. As a result, he assumed that Kashmir's dependency ratio that was high in the late

1980s dropped by 2001 'as the children reached the working age' (Swami 2010) and rose again by 2011 (Swami 2012). He did not ask if this change might be an artefact of the poor data quality.[67] Second, the boundaries of communities are not fixed. In fact, politics and conflict are often centred around identity formation and hardening of group boundaries in divided societies (Horowitz 2000). Comparisons based on surveys that do not account for the dynamism of the underlying categories can be misleading, particularly when the statistics suffer from flaws correlated with the political and economic developments under investigation.[68]

The aforementioned complexities notwithstanding, most analyses of the interaction between demography and politics are limited to a conceptual evaluation of the politicisation of data without an engagement with the quality of published data. The focus is on flawed categories, erroneous interpretations and the partisan use of data and how these intersect with political processes, identity formation and ethnic conflicts, but not on the actual errors in the underlying data. At the end of such analyses, one understands the context dependence of data, but not the extent to which the quality of existing data has been compromised and how that affects public policy. On the other hand, there are statistical analyses of data quality that overlook the social and political processes that might account for the anomalies. In the absence of conversations between these two strands of literature, we are faced with an explosive growth of statistical analyses that are not informed by the context dependence of data and their quality. This book tries to bridge these two approaches for a sub-national unit whose government statistics are deeply politicised but rarely studied. In fact, as discussed earlier, the literature on the Kashmir conflict recognises the centrality of demographic concerns but does not engage the sources of demographic information. This book estimates the magnitude of errors in census statistics, examines the mutually constitutive nature of statistics and the phenomena they capture and discusses the implications of the context dependence of data quality at different levels of aggregation in a federal polity.

The approach adopted here differs from that of the recent literature on issues related to the political economy of statistics and on demography and ethnic conflicts. Morland (2018) presents a theoretical framework to understand demographic engineering and illustrates it using cases from four countries, but the discussion is not based on a systematic analysis of data and their quality. Morland uses available data to study conflicts without directly analysing the context dependence of data, even though he refers to the politicisation of census. Agrawal and Kumar (2020a) examine data quality across much of the lifecycle of statistics, but they do not discuss the legal framework governing census, politics at the preparatory phase of census and the quality of metadata. Also, the choice of Nagaland as a field site meant that they dealt with a problem in which the union government and national politics were largely superfluous amidst pervasive grassroots manipulation. The choice of J&K in this book allows an examination of the roles of the government, political leadership and bureaucracy at the union level and helps understand the intertwined nature of the politicisation of statistics at the sub-national and national levels. The richer setting makes room for a fuller understanding of the context dependence of data and the importance of metadata. Further, the book examines J&K, which has been mostly studied from the perspectives of conflict and national security, through a novel lens – namely data deficit. This is particularly important because inferences about demographic change drawn from published census data play a crucial role in the Kashmir conflict, but the quality of data is rarely examined.

This book examines the whole range of contests over censuses through their lifecycle in J&K (1951–2011), from the grassroots to the level of union government and national politics, from political and administrative manoeuvring at the preparatory stages to the partisan use of published data in policymaking and public debate (Figure 1.5). It also examines the legal and administrative frameworks within which data are collected, how the use of census data by various actors – politicians, civil society leaders, journalists, academics and insurgents – amidst conflict affects and is also affected by the quality of data and how our ability to examine data quality is limited by the paucity of metadata. Finally, it discusses priors that shape the union government's choices vis-à-vis collection and evaluation of census statistics before discussing possible reforms to address some of the major organisational and conceptual problems facing the Census of India.

The first part of the book comprises the present chapter, 'Debating Numbers', which introduces recent demographic disputes in J&K. Given the uncertainty in the process of building inclusive and transparent democratic institutions, the obsession with numbers pushes political parties, the government and people to resort to statistical proportionality across salient cleavages – that is, religion and region – as the ultimate arbiter of the fairness of public institutions and policies. The obsession with the religious composition of the population has meant that every policy and administrative decision is evaluated for its potential demographic impact. This chapter puts together three vantage points to understand the politics of numbers – the census as an insignia of modernity, as a collective self-portrait and as a public good. Together they help explain the government's inability to

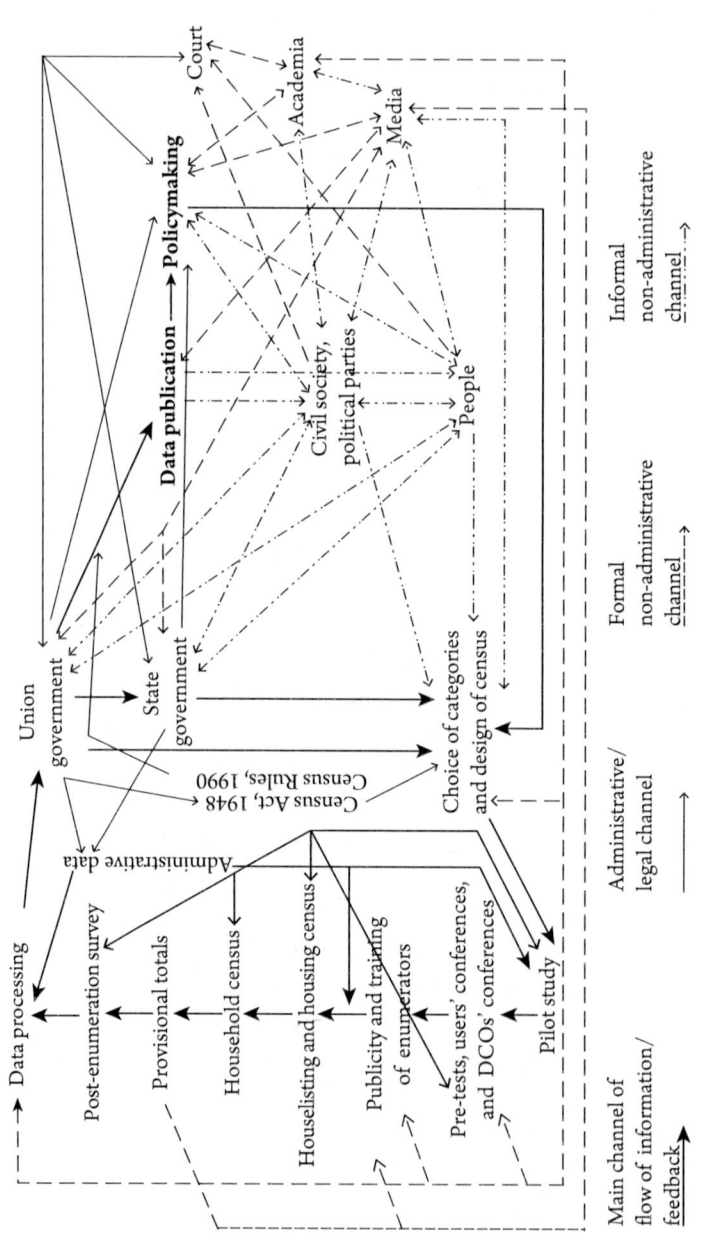

**Figure 1.5** Lifecycle of census statistics in India

*Source:* Prepared by the author.

*Note:* (*a*) The chart omits exchanges between international organisations such as the United Nations Population Fund and the Office of the Registrar General of India (ORGI). (*b*) Each stage of data collection shown in the left part of the chart includes both the preparation of maps, if any, and the collection of data. (*c*) The chart applies to the last two censuses insofar as data processing used to be a multi-stage activity in the earlier censuses. Computerisation has simplified that process, but mapping of entries onto pre-specified categories continues to be an elaborate process.

conduct a census in 1991, the undue haste in conducting the 2001 census and the flaws in the 2001 and 2011 censuses, and facilitate a multipronged assessment of census data across their lifecycle in the later chapters.

J&K has been a fixed point of national debates for three decades, but its geography and demography are poorly understood. The second part of the book contains two chapters that examine the cartographic and population statistics of the state. Chapter 2, 'Counting amidst Uncertainty', discusses maps and area statistics before examining various factors that affect enumeration in J&K. It argues that the inter-temporal comparability of censuses is rendered moot in J&K because of the changes in the borders of the state, the reference dates, the extent of snowbound territory, weather conditions, the method of accounting of mobile population and the quality of metadata.

Chapter 3, 'Inventing Boys and Miscounting Tribes and Languages', examines the quality of census data. It analyses abnormal changes in the CSR and child population share at different levels of aggregation. It rejects conventional demographic explanations of the abnormal changes in J&K's population statistics and shows that the overstatement of child population – particularly male children in Kashmir – and, to a lesser extent, the unusually large growth of Generic Tribes can account for a large part of coverage errors in the 2011 census. The coverage errors in the 2001 and 2011 data explain the declining population share of SCs in the state's population and the change in the relative population shares of the major regions and religions. The discussion on coverage errors is followed by an examination of content errors in the data on non-scheduled languages, dialects of scheduled languages, STs and miscellaneous categories of tribes and languages. The anomalous changes in Ladakh are related to the linguistic and religious polarisation between Bhotia-speaking Buddhists of Leh district and Zanskar *tehsil* (a sub-district administrative unit) of Kargil district on the one hand, and Purkhi (or Purigi)-speaking Shias of Kargil, on the other. The anomalies in Gurez may be related to the benefits of tribal status granted in 1989, but we do not have a clear understanding of the processes that undermined the data integrity in this case. In the case of Gujjars and Bakarwals, the anomalies are accounted for by the misreporting of their linguistic and tribal identity in Kashmir, the misclassification of data on their language and the abnormal growth in the population of Generic Tribes. The chapter then examines the changes in the population of speakers of Ponchi, wrongly reported under Gujarati, and Halam, a north-eastern language erroneously reported in J&K. The chapter also estimates the magnitudes of the aforesaid errors.

Given the persistent and growing errors in J&K's census, three chapters in the third part of the book examine the context within which population is accounted

– the political setting (Chapter 4), the legal and administrative frameworks (Chapter 5) and the government's prior expectations regarding headcount (Chapter 6) – and explain why the government fails to check the manipulation of the census.

Chapter 4, 'Anxious Majorities', examines the demographic politics involving three interlocked majorities – Kashmiri Muslims who constitute a majority in J&K, Dogra Hindus of Jammu who dominate Jammu and Hindus of 'mainland' India who constitute the single largest religious group in the country. Each of these majorities believes in its historical right to play a larger role in its chosen territory, which makes conflict unavoidable as the territories overlap. The resultant conflicts become visible in controversies over delimitation, which are discussed at length in this chapter. This chapter helps locate J&K's recent demographic controversies in the longer history of ethno-religious politics in north-western India with roots in the late nineteenth century. It also discusses the quality of census data on slums and urbanisation.

Chapter 5, 'The Limits of Law', discusses the legal and administrative framework of counting. It argues that it is neither desirable nor practical to provide security to enumerators. It explains why erring respondents and enumerators cannot be penalised for deliberate errors. It suggests that the punitive measures of the Census Act, 1948, serve as a hollow threat and compares India's experience with other common-law countries. It then discusses the absence of provisions in the Census Act, 1948, and the Census Rules, 1990, for the ex-post correction of published figures shown to be flawed.

After we uncover the communal politics that shapes headcounts and legal –administrative constraints that limit the government's ability to address the politicisation of the census, Chapter 6, 'Growth as Well-Being', locates the union government's indifference towards errors in J&K's census on a larger historical and global canvas. It argues that interpretations of population growth as a measure of well-being as well as progress and strong priors that rule out the possibility of over-enumeration in conflict zones inform the government's approach to census statistics in peripheral states dominated by religious and/ or ethnolinguistic minorities. It discusses coverage and content errors linked to state intervention and complements the discussion in Chapters 3 and 4. It argues that most of the anomalies in J&K are products of deliberate intervention and not mere aggregations of random errors at the grassroots and then discusses the nature of intervention responsible for the anomalies identified in the preceding chapters.

The last part comprises the concluding chapter, 'Reinventing the Census', which revisits the use of statistics in J&K, where numbers serve as political allies in an increasingly partisan public discourse. It locates J&K's data deficit in relation to its democracy and development deficits and discusses how data quality affects policymaking. This chapter also relates J&K's experience to similar problems in other states. It uses a simple game-theoretic framework to analyse the contrasting experience of J&K, Manipur and Nagaland. These three states, respectively, stand for the whole range of responses of state governments to the politicisation of data – namely, refusing to acknowledge the problem (J&K), blaming one community (Manipur) and resolving through dialogue (Nagaland). Given the limited efficacy of legal, technological and administrative interventions, this comparative discussion is used to emphasise the need for a participative approach to resolve the stalemate over census and suggest reforms.

## Terminology

Until August 2019, 'Jammu and Kashmir' referred to a state in the First Schedule of the Constitution (GoI 2015b: 279). It now refers to one of the successor UTs (GoI 2020a: 180–81). Since the bulk of our discussion is restricted to the period before reorganisation, unless otherwise specified J&K refers to the erstwhile state (Map 1.1). The successor UTs are known as the UT of J&K and the UT of Ladakh (Map 1.2).

'Region' is used to refer to the three geographically separable administrative divisions of J&K that existed between February and August 2019 – Ladakh, Kashmir and Jammu. There were two divisions before February 2019 as Ladakh was until then a part of Kashmir division. These two divisions used to be referred to as provinces in some of the earlier census reports (GoI n.d.15, 1965).[69] The three divisions are statistically separable from the perspective of the census because there has never been any transfer of territory among them due to formidable geographical barriers (Figures A.1–A.3). Jammu is divided into two 'natural divisions' or 'NSS regions'[70] by the NSSO: Outer Hills (Doda, Punch, Rajauri, Udhampur, Kishtwar, Ramban and Reasi districts) and Mountainous (Jammu, Kathua and Samba). We will treat all the districts of Jammu division as part of one region as that was and continues to be the actual unit of administration at the regional level. When required we will also refer to the two natural divisions of Jammu. There have been only minor transfers of territory or population between the Outer Hills and Mountainous regions after the 1970s, but even the earlier transfers involved very small populations. The Outer Hills can be separated into two subparts – namely, Punch and Rajouri on the one hand, and the old districts of Doda and Udhampur, which are now divided into five districts (Kishtwar, Doda, Ramban, Reasi and Udhampur), on the other.

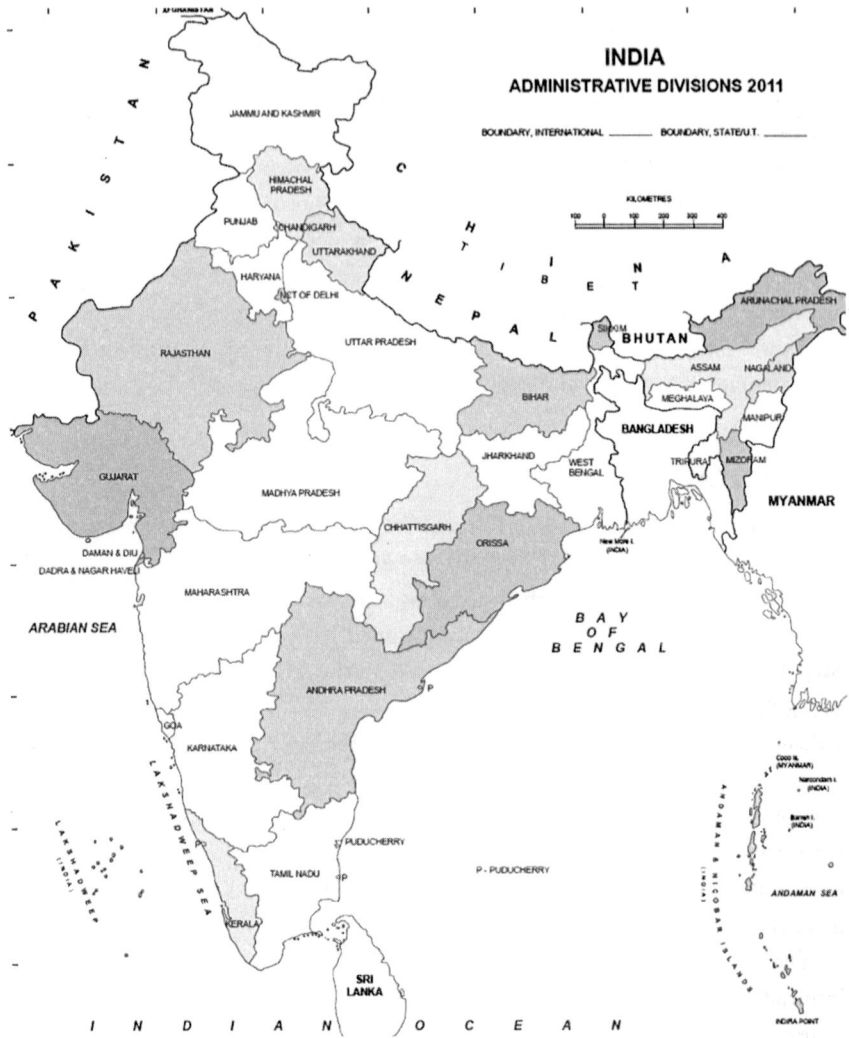

**Map 1.1**   States and union territories (UTs) of India, 2011

*Source*: 2011 census maps, https://censusindia.gov.in/2011census/maps/administrative_maps (accessed on 24 February 2021).

*Note*: The UT of Ladakh is included in Jammu and Kashmir (J&K) as it was a part of the state in 2011. Likewise, Telangana is included within Andhra Pradesh.

There has never been any transfer of territory between these two parts. Occasionally, we will also refer to the three parts of the Jhelum Valley: north (Bandipore, Baramula and Kupwara districts), central (Srinagar, Badgam and

**Map 1.2**    Union territories (UTs) of Jammu and Kashmir (J&K) and Ladakh, 2019

*Source*: 'Introduction', Administration of Union Territory of Ladakh, https://static.pib.gov.in/ WriteReadData/userfiles/Map%20of%20UTs.pdf (accessed on 13 February 2023).

Ganderbal) and south Kashmir (Pulwama, Anantnag, Shopian and Kulgam). Unless otherwise mentioned, 'Valley' refers to the Jhelum Valley of Kashmir. Also, following the Census of India, we use the spellings 'Baramula' and 'Punch' instead of 'Baramulla' and 'Poonch' used by the state government.

The landlocked ethnic or religious-minority states and UTs along the international border, including J&K, Ladakh, Sikkim and the seven north-eastern states, constitute India's ethno-geographic periphery.[71] These states are largely absent in federal discourses on statistics. The 'peripheral' states are governed by a different fiscal logic, and several union laws have limited applicability in some of these states. This diminishes the urgency to improve their statistical systems. For instance, Nagaland's forests and mineral resources are not governed by union laws because of Article 371A of the Constitution (GoI 2015b), and service tax was not levied in undivided J&K (GoI 2015c: 96).

According to the draft of the National Policy on Official Statistics, 'official statistics' refers to 'statistics derived by the Government agencies from statistical surveys, administrative and registration records and other forms and papers, the statistical analyses of which are published regularly, or planned to be published regularly, or could reasonably be published regularly' (GoI 2018c: 3).[72] We will use the phrase 'government statistics' to refer to what is widely known as 'official statistics' in academic and policy discussions – for example, *The Guide to Official Statistics* published by India's Central Statistics Organisation (CSO), the UN Economic and Social Council's *Fundamental Principles of National Official Statistics*, the International Association for Official Statistics and the *Journal of Official Statistics*. Non-governmental entities have gradually acquired the capacity to put together large datasets and are increasingly challenging the monopoly of the state over the production and dissemination of statistics. So the statistics produced by governments need to be explicitly identified. India has, indeed, come a long way from the 1960s, when Mahalanobis (1965: 44) lamented that in underdeveloped countries statistics 'remains a matter of formal or administrative sanctions; anything having the official stamp must be accepted as authoritative'. The Centre for Monitoring Indian Economy (CMIE), a business information company, has emerged as one of the most sought-after sources of data on unemployment after the demonetisation of high-denomination currency notes in November 2016, even as the NSSO continues to struggle with debilitating political interference and decaying organisational capacity.[73] *The Annual Status of Education Report* (*ASER*) released by Pratham, an NGO, is a widely used source of data on the school education system. The Peoples' Linguistic Survey of India (PLSI) has put together nearly 40 volumes on languages of India even as the corresponding government initiative languishes.[74] The Janagoshthiya Samannay Parishad, Assam (JSPA), tried to roll out an exercise that aspired to match the government's NRC in scale and details (JSPA 2021; Agarwala 2021). More generally, communities have begun to conduct surveys to generate self-knowledge to contest their representation in government data and also support their claims upon public resources. As Appadurai (2012: 639) puts it, 'self-enumeration, self-mapping and self-documentation' 'can become an irreversible force for stronger negotiations with those who see them as a burden, a blight or a vote bank' (see also Urla 1993: 818). Counter-mapping and ethno-cartography in J&K, Nagaland and Manipur and collection of statistics on human rights violations by security forces in J&K, the north-eastern states and Naxalism-affected areas are cases in point. These examples, though, relate to older-generation databases. In the age of big data, many private organisations hold more data in their respective fields than governments.[75] Moreover, the growing access to information technology over the last decade has enabled non-governmental bodies, data journalists and amateur fact-checkers to question the validity of government statistics (Kumar 2017a–c).

We use the phrase 'data collection' to refer to the actual act of enumeration or field survey as distinct from interventions that *construct what is going to be collected* or *shape what has been collected*.[76] Further, we mostly use the expressions 'Registrar General of India' (RGI) or the 'Office of the RGI' (ORGI) instead of the more appropriate 'Office of the Registrar General and Census Commissioner of India'.

Administrative statistics include 'statutory administrative returns and data derived as a by-product of general administration' (GoI 2001a: 448). Further, as defined by the National Data Sharing and Accessibility Policy, 2012, metadata is 'information that describes the data source and the time, place, and conditions under which the data were created. Metadata informs the user of who, when, what, where, why and how data were generated. Metadata allows the data to be traced to a known origin and know[n] quality' (GoI 2012: 11). We will include information about the legal and administrative frameworks that govern the collection and dissemination of data under metadata.[77]

Sex ratio is defined by the Census of India as the number of females per 1,000 males. The inverse of this, the number of males per 100 females, is also referred to as sex ratio (see, for instance, Organisation for Economic Cooperation and Development [OECD] 2007: 715). We will use the definition followed by the Indian census for the sake of consistency. Age refers to 'age on last birthday [before the reference date of the census, say, 00.00 hours on 1 March] in completed years' (GoI n.d.14: 42). Sex refers to male and female. In the 2011 census, Question 14 of the Houselisting and Housing Census Schedule allowed the head of a household to identify only as 'male' or 'female' (GoI n.d.18). However, Question 3 of the Household Census Schedule allowed individuals to identify as 'male', 'female' or 'other' (GoI n.d.19). The primary census abstract (PCA) for the 2011 census suggests that no one chose to identify or no one was recorded as 'other' because 'total population' is shown to be the sum of 'Total male' and 'Total female' (GoI n.d.17).[78]

Coverage error in census 'refers to either an under-count or over-count of units owing to omissions of persons/housing units or duplication/erroneous inclusion, respectively', while content error 'pertains to the error in the characteristics that are reported for the persons or housing units that are enumerated' (UN Secretariat 2010: 10). The former affects the accuracy of the overall headcount, while the latter affects the accuracy of the sub-categorisation of population.

# Notes

1. Everything including roads, signboards and the allotment of government accommodation is scrutinised in J&K from the perspective of the balance of power between communities and regions (see Chapter 4).

2. See M. Ashraf (2011) for a discussion of the implications of Kashmir's 'Hindu past and Muslim present' for Kashmiri Muslims, Hindus of India and Muslims of Pakistan.

3. This was not the only recent instance of sparring over statistics on government websites. A website of the Ministry of Home Affairs (MHA) indicated that 'the Kashmir Valley houses a total of 5,35,811 residents, while the Jammu region houses 69,07,623 residents.... However, according to Census 2011, the population of Kashmir is 69,07,623' (*Kashmir Times* 2020). This was most probably a typographical error that was corrected later. However, in a communally surcharged atmosphere such errors intensify fears of demographic change.

4. Interestingly, the undivided J&K was seen as the land of Muslims (of Kashmir), Hindus and Buddhists, but not that of Sikhs even though they were the third largest religious community of the state, which was once a part of the Sikh Empire. One could argue that Sikhs were not exclusively identified with any territory within the state. However, even the Shias of Kargil were not seen as constitutive of the state in governmental and popular imaginations.

5. A veteran parliamentarian, former union minister, former chief minister of J&K and member of the Indian National Congress (INC), Ghulam Nabi Azad was born in Bhaderwah in Doda district. He was the only chief minister of J&K born in the Jammu division.

6. Former judge of the J&K High Court (2009–16) and a member of the NC, Hasnain Masoodi represents the Anantnag constituency (2019–present) of Kashmir in the Lok Sabha at the time of writing this.

7. Former chief executive councillor of the Ladakh Autonomous Hill Development Council, Leh (2018–19), and a member of the Bharatiya Janata Party (BJP), Jamyang Tsering Namgyal represents the Ladakh constituency (2019–present) in the Lok Sabha at the time of writing this.

8. Azad confounded the population shares of the two districts of Ladakh with that of its two main religious communities. As per the 2011 census, Leh and Kargil districts accounted for 48.67 per cent and 51.33 per cent population of Ladakh, respectively. The corresponding shares of Muslims and Buddhists were 46.41 and 39.65 per cent, respectively.

9. Former chief executive councillor of the Ladakh Autonomous Hill Development Council, Kargil, and a member of the INC, Asgar Ali Karbalaie was the last member of legislative assembly from the Kargil constituency (2014–18).

10. The national media and New Delhi-based observers remain unclear about the region's demography. Wahab (2021), for instance, claimed, 'Contrary to popular perception, *even* the divisions of *Jammu* and Ladakh had Muslim majority, howsoever slim' (emphasis added). As per the 2011 census, the population of Jammu division was 53,78,538, out of which Hindus accounted for 33,64,618, which works out to 62.56 per cent, with the remainder being accounted for by Muslims, Sikhs, Buddhists and Jains.

11. A Buddhist leader claimed that Zanskar accounted for 75 per cent of Kargil's area (interview, New Delhi, 2 December 2019). It is not clear how Buddhists, let alone Zanskar, can account for 75 per cent of Kargil's area, unless we treat the whole of Zanskar constituency as a Buddhist area where Shias of Sanku *tehsil* outnumber Buddhists of Zanskar *tehsil*.

12. A report of the first post-colonial census in J&K identified Ladakh as 'the wonderland of Lamas' (GoI 1970: 1).

13. Commenting on the disillusionment in Kargil after reorganisation, Subramanian (2021) observed, 'The people of Kargil see themselves as a minority in Buddhist majority Ladakh'. Das (2019) claimed that the 'dominant Buddhist' community accounted for '1.33 lakh out of a total [population] of 2.74 lakh'. The 2011 census reported 108,761 (that is, 1.08 lakh rather than 1.33 lakh) Buddhists in Ladakh out of a total population of 274,289.

14. Several social media users pointed out the error in Ranil Wickremesinghe's tweet, but it was not corrected.

15. There are, in fact, three Muslim-majority union territories, including Lakshadweep, J&K and Ladakh.

16. After the reorganisation of J&K, its development indicators received both academic (S. Iqbal 2021) and journalistic (Radhakrishnan and Sen 2019) attention, but these debates overlooked the problem of poor quality of data.

17. Two years after the repeal of Article 370, senior BJP leader Ram Madhav changed gears and, contradicting the home minister, accepted that J&K enjoyed better development indicators. Instead of claiming that Article 370 blocked development, he argued that the state had developed because of extremely generous federal redistribution: 'For a population of just 13 million, its budget outlay for 2021–22 is around Rs 1,10,000 crore, whereas the same for UP, which has 20 times the population is Rs 5,50,000 crore' (Madhav 2021).

18. For the disconnect between New Delhi's development-centric understanding of Kashmir's problems and Kashmir's self-understanding, see Sikand (2010b, 2010d).

19. For a more detailed statement of Kashmiri assessment of the state of development, see Omar Abdullah's interview on the eve of the second anniversary of the reorganisation (Times Now 2021). The estimates of poverty (Agrawal and

Kumar 2020a: 273) and unemployment (GoI 2011g: 12–13) provided by the state government, including when Abdullah was the chief minister, are not in agreement with the figures he quoted in the interview. It is not the case that the state government's estimates are more reliable than the union government's statistics. However, the disagreement between the two should not be used as an opportunity to cherry-pick figures. Instead, it should lead to further investigation into both the quality of numbers and their meaning (see Chapter 7). It bears emphasising that the cherry-picking noted here seems irresistible to those who represent J&K. After the repeal of Article 370, the UT's bureaucratic leadership routinely uses the same old statistics to present a rosy picture of J&K.

20. The quality of vital statistics could be related to the levels of fertility and mortality – for example, jurisdictions 'with incomplete, defective or non-existent vital registration data are generally … [those] with higher mortality' (Chamie 2013: 138).

21. This trend seems to have undergone a reversal in the late twentieth century. Meyer, Mok and Sullivan (2015) report a steady rise in non-response rates in household surveys in the United States (US) over the past three decades. In Europe, another bastion of modern statistics, '[w]hile a very large majority of respondents consider that official statistics play a role in political decision-making, an absolute majority also do not trust these statistics' and there has been a 'downward trend in trust since 2007' (European Commission 2015: 19). The levels of trust were found to be low even in countries such as France and Germany.

22. The initial enthusiasm notwithstanding, the politicisation of bureaucracy, including statistical departments, eventually re-politicised policy knowledge (see, for instance, Boräng et al. 2017).

23. Other cues in the international arena buttressed domestic pressures to embrace statistics. For instance, Article 1 (also Articles 2 and 9) of the Montevideo Convention on Rights and Duties of States, 1933, defined a sovereign state as an entity with 'a permanent population and a defined territory', among other attributes (League of Nations 1936: 25). By the early twentieth century, 'population' and 'territory' were widely understood to be measurable entities. So statistical self-knowledge was integral to being a 'modern' state. Further, the emergence of various international bodies around the mid-twentieth century, which began to seek a wide range of data from states and offered adverse comments on the laggards, was another factor that pushed states toward statistics.

24. The relationship between statistics and colonised people was not, though, limited to a self-perception of shortcoming vis-à-vis an alien yardstick. There were, at least, four other strands in this relationship. First, as early as the first quarter of the nineteenth century, Indians had begun to use government

statistics to demand policy reforms (see, for instance, Ray 1832). Second, statistics began to seep into popular culture soon after the publication of the results of the first census in the early 1870s (Kumar 2015a). One of the best-known patriotic songs of India, Bankim Chandra's 'Vande Mataram', is a case in point: 'Terrible with the clamorous shout of seventy million throats, and the sharpness of swords raised in twice seventy million hands' (Bhattacharya 2003: 96–101). It contains one of the earliest Indian invocations of people as population. Not coincidentally, before he wrote the song, 'Bankim wrote an essay in 1872 on the population statistics of the census of 1871' (ibid.: 69). Third, there was in many quarters a willing embrace of this new language that was seen as integral to modernity. Commenting on the lack of agrarian reforms in colonial India, Nehru, who later served as the first prime minister (1947–64), observed: 'Even proper statistics had not been collected, in spite of insistent *popular* demand. This lack of statistics and surveys and necessary information had been a serious impediment in the way of progress' (Nehru 1994 [1946]: 372, emphasis added). India's first home minister, Vallabhbhai Patel (1947–50), recognised the 'vital role [census played] in the determination of many of our administrative policies' and its usefulness as a 'guide for the effectiveness or otherwise of our economic policies' (GoI 1957). Mahalanobis (1965: 45), the architect of India's Five-Year Plans, viewed statistics 'as a key technology for rapid economic growth in … developing countries'. Asok Mitra, a former census commissioner (1958–68), reminisced that Gobind Ballabh Pant, union minister of home affairs (1955–61), 'could quote from memory whole passages from the census reports of his state of Uttar Pradesh for the years 1911, 1921 and 1931' (Mitra 1994: 3209). Fourth, as discussed in Chapters 4 and 6, by the early twentieth century, political parties and social organisations had become willing participants in numbers games to promote communal interests. Discussions on population were not limited to communal calculus though. Population growth and economic policy was another field where Indians engaged with statistics in the late nineteenth century (see, for instance, Ranade 1906). During the late colonial period, there was also a vigorous debate from the perspective of women's health and their rights involving social reformers and political leaders like Raghunath Karve, B. R. Ambedkar and E. V. Ramasamy 'Periyar' (Basu 2009).

25. In several places, the process of decolonisation accentuated the politicisation of census. After independence, conflicts over delimitation of electoral constituencies and redistribution of federal resources often spilled over into the domain of census. Conflicts over census-taking were commonplace around the time of decolonisation in Asia and Africa so much so that in hindsight they appear to be 'rites of passage in the lives of modern states outside the West' (Kumar 2020a: 1143).

In fact, similar conflicts have also been reported around the formation of new sub-national administrative units. Soon after the formation of the state of Telangana, the state government launched a controversial Intensive Household Survey (*Times of India* 2014).

26. Half a century later, the head of the Palestinian Central Bureau of Statistics observed that census was necessary to obtain 'self-knowledge' (Greenberg 1997).

27. An entry on 'census' in the *Encyclopaedia Britannica* (2018) retrospectively throws light on one of the sources of statistical anxiety of the early Maoist China that was struggling to join the community of modern nations: 'When China reported a census in 1953, the last large part of the world was removed from demographic darkness.'

28. See Agrawal and Kumar (2020a: 285–87) on the association of poor-quality data with collective shame in twenty-first century Nagaland and how the public debate on the flawed 2001 census was quickly overtaken by a collective moral self-flagellation.

29. The 2011 commemorative postage stamp on the census was released by the president of India in the presence of the prime minister and several senior union cabinet ministers (see an advertisement in this regard in *Kashmir Times*, 8 February 2011).

30. Postal archives have so far not been studied from the perspective of the co-evolution of modern state and statistics in India. See Kumar (2023c, n.d.) for an attempt in this regard.

31. The emphasis on mastering a 'modern' function of the state in postal brochures is noteworthy because these brochures simultaneously push the boundaries of the history of census to the time of Kautilya, 'Third Century B.C.' (GoI 1971c). The brochures, in fact, reproduce the claim of introductory notes to most census reports that mechanically trace the origins to Kautilya (for example, GoI n.d.12; Srivastava 1972: 6). A rare exception, though, shrugs off the search for pre-colonial roots: 'For the origin of the institution of the Census as we know it today, one need not go as far back as Kautilya, Asoka, Vikramaditya or the Moghul emperors; it will suffice to refer to the Census of 1872 which was the first in the series of decennial Censuses held in India' (GoI 1954c: xxvii).

32. India was one of the first countries to use the image-based Automatic Form Processing technology (2001 census) and the Intelligent Character Recognition technology (2011 census) (Agrawal and Kumar 2020c: 219). However, the growing use of computers in Indian censuses has counter-intuitively been associated with increasing delays in the release of politically sensitive data on religion and language and even administratively important data on migration (ibid.).

33. The close connection between democracy and state-sponsored population censuses is not a uniquely Indian experience. In fact, the first major modern democratic state, the US, introduced decennial censuses in the late eighteenth century to aid decennial delimitation of electoral constituencies.

34. Rose (1991: 673) argues that 'democratic power is calculated power, calculating power and requiring citizens who calculate about power'.

35. See also note 23 of this chapter. Note that even Belgium qua trustee of Rwanda 'under the League of Nations and later the United Nations' was responsible for the collection of statistics (Uvin 2002: 149). In fact, even non-sovereign sub-national administrative units claim exclusive rights to map and enumerate their respective territories qua agents of the union government as evident from the conflicts over such activities in the disputed belt between Assam and Nagaland (Agrawal and Kumar 2020a: 67–68, 71–76).

36. India did not care to disrupt a plebiscite conducted by non-state actors in parts of the Naga Hills district in 1951. This had lasting political consequences.

37. For the close relation between maps and census, see Göderle (2016) for Austria and Agrawal and Kumar (2020a) for north-east India.

38. Feroze Ahmed, the DCO of J&K for 2001, was interviewed by the BBC (interview, Jammu, 4 December 2019). Farooq Ahmad Factoo, who served as the DCO for 2011, reminisced that MHA officials were 'keen to have census as it has national and international ramifications' (interview, Srinagar, 30 May 2022). Dulat (2010: 206), who was associated with various government initiatives in Kashmir during 1988–2004, clearly notes that resumption of elections preferably with the participation of separatist groups was among the highest priorities of the government. Devadas (2019a: 308, 331–32) suggests that '[e]lections were a priority' because 'repeatedly extending governor's rule was both cumbersome and *embarrassing*' (ibid.: 332, emphasis added). It is not clear if Prime Minister Atal Bihari Vajpayee's first state visit to the US, which was a major driver behind New Delhi's attempts to engage the Kashmir-based All Parties Hurriyat Conference (Baweja 2000b), also pushed the government to resume routine activities such as census and elections in J&K. Insurgents scaled up attacks around this time to demonstrate to the world community that the government had failed to restore normalcy. It is also not clear if the forthcoming delimitation, the first after the 1970s, too, exerted pressure to conduct a census in 2001. However, as late as 25 April 2003, the reference year for delimitation was 1991. In fact, the then census commissioner pointed out that his department had to put in considerable effort to reorganise the 2001 data after the reference year for delimitation was changed to 2001 (Banthia 2004: 3862). While the formal switch from 1991 to 2001 happened in 2003,

in several parts of the country local actors seem to have anticipated this before the 2001 census (Agrawal and Kumar 2020a).

39. Kashmiri opinion writers had voiced this concern during the 2001 census (see, for instance, Riaz-ud-din 2000). In fact, Kashmiri nationalists have long complained that New Delhi's concern about Kashmir has been limited to 'somehow to stage-manage a self-defence before the comity of nations' (Abdullah 2016: 447) and 'draw wool over the world's eyes' (ibid.: 535).

40. Shashi Tharoor, a three-term member of the parliament (2009 – present), was the under-secretary general of the UN for communications and public information (2002–07) and later India's minister of state for external affairs (2009–10).

41. Sheikh Abdullah maintained that Kashmir's choice to abandon Pakistan was a key moment in the history of secularism in the subcontinent. He was, however, wrong when he claimed, 'Even the secular character of the Indian constitution owed something to Kashmir's deep-seated faith in the oneness of all religions' (ibid.: 387). That choice was driven by an older consensus that emerged even before Kashmir appeared on the INC's radar.

42. Kashmiri Pandits, too, argue that their presence in Kashmir was essential for Kashmiri Muslims to 'maintain' their secular image (interview, Jammu, 21 July 2021; see also Panun Kashmir n.d.2: 9). Jamwal (2013: 27) suggests that 'the flight of the Pandits and other minorities ... eventually robbed the Valley of its secular character'.

43. Hill and Motwani (2017: 144) add that both 'India [as a secular nation] and Pakistan [as an Islamic nation] need to control Kashmir to validate themselves both internally and to the world'. See also M. Ashraf (2011).

44. A vocal section of Kashmiris, though, argues that maintaining the religious composition of the population 'in Kashmir is not about maintaining diversity in India. It's about preserving the resistance movement of Kashmir. It's about being able to vote in a referendum in Kashmir' (*Greater Kashmir* 2017; see also Bhat 2018: 181). The national media often endorses the demographic status quo in Kashmir as critical to the diversity and secularism of India without acknowledging the aforesaid position of a section of Kashmiris. Hindu nationalists accuse the liberal press and intelligentsia of being either ignorant or, worse, hypocritical vis-à-vis Kashmir's intent to preserve its Muslim majority character by all means, even as similar concerns of Hindus are condemned as repugnant to constitutional values. An exchange on this issue in the *Indian Express* is discussed in Chapter 4.

45. Amit Shah, the union minister of home affairs, alleged that delimitation could not be carried out in J&K due to Article 370, which restricted the applicability of laws enacted by the parliament in the state (Lok Sabha 2019: 247). However, given the intensity of regional and communal divides in the state, delimitation

would not have been carried out even in the absence of Article 370. See Agrawal and Kumar (2020a: 305–13) and Kumar (2022b) for the suspension of delimitation in several other states.

46. We can add that by trying to accentuate the Muslim-ness of Kashmir to bolster India's secular credentials, the government ended up nourishing both the Islamist constituency in Kashmir and the Hindu nationalists in Jammu and the rest of the country.

47. Before this, in 1924, Muslim notables had drafted a memorandum appealing to the viceroy that 'showed, with statistical details, the plight of Kashmiri Muslims' (Abdullah 2016: 43). Note that unlike the petitions of the 1920s, the later petitions began to specifically mention the population share of Muslims when demanding a share in public resources (cf. Kaur 1996: 127, 153).

48. Devadas (2019a: 401) suggests that 'Kashmiri-as-Muslim identity' struck roots in the 1920s when Deobandi influence began to spread in Kashmir. In the late 1920s and the early 1930s, Abdullah (2016: 40) visualised his fellow Muslims as 'the majority' in J&K. He reacted strongly to the characterisation of J&K as 'a Hindu state' by Prime Minister B. N. Rau and observed that '[t]he fact that the Maharaja is a Hindu will not alter the demographic composition of the state [with 85 per cent Muslim population]' (Abdullah 2016: 219). However, 'the majority' in his lexicon was not initially defined in reference to a clearly identified minority. His autobiography suggests that he began referring to 'majority' and 'minority' later. It would be interesting to ask if there was ever a possibility that Kashmiri Muslims could have viewed J&K as a state where Kashmiri speakers constituted a majority. Most states in the 'mainland' view themselves almost entirely in terms of the linguistic identity of the majority that also happens to be Hindu. In fact, at one point of time, sections of Kashmiri Pandits identified as members of the ethnolinguistic majority in the state, and they are still occasionally seen as such in Jammu (see note 51 of Chapter 4).

49. In 1971, the identity cards of census supervisors in J&K carried the following motto: 'Census is Truth' (GoI 1971b: app. xxiv).

50. The metaphor of 'mirror' is quite widely used in the case of government statistics. Recently, an advisor to a state government observed that 'Statistical Handbook under the publication of Economics & Statistics department is the mirror of the state' (*Morung Express* 2021b).

51. The Hindi variant reads, 'rāṣṭra kā sāmūhika citraṇa: "janagaṇā"' (collective portrayal of the nation: 'census'). The images on the cover pages of various reports on the census also highlight it as a collective portrait. Until the 1990s, the census used to publish a booklet for each state called *Portrait of Population*, containing an illustrated summary of the data accompanied by an accessible description (GoI 1974a, 1989a, 1997). A 1971 documentary titled *India*

*Counts Itself* produced by the Films Division presented the census as a self-portrait of the country (GoI 1971d).

52. The significance of the use of postal medium to advertise censuses lies in the fact that it was widely used for communication until a decade ago. The Indian Posts and Telegraphs Department printed three million copies of the 1971 stamp on the census, and several million postcards and letters were cancelled with census slogans in 1951, 1961 and 1971. The department printed 0.4 million copies each of the 2001 and 2011 stamps on the census. The scale of circulation of the 2001 census cancellation is, however, not clear (Kumar 2023c).

53. This is reminiscent of similar sentiments in Europe from the late eighteenth century to the mid-nineteenth century, where countries such as France, Austria and Italy had hoped that census would help consolidate the nascent nation states (T. M. Porter 1995: 35; Woolf 1989: 602; Göderle 2016: 64). More recently, the Palestinians viewed the 1997 census as a 'national project' that was important for 'laying the foundations for a Palestinian state' (Greenberg 1997).

54. It is not the case that the census was not shaped by the biases of bureaucrats in the colonial period when the political leadership and most of the senior bureaucracy did not belong to any indigenous faction (see Chapter 6).

55. Even when restrictive policies cannot be implemented, the state may condone stigmatising labels that reduce the political and economic bargaining power of communities. Bengali-speaking Muslims in Nagaland are routinely referred to as 'Bangladeshi', which is used in many parts of India as a shorthand for illegal international immigrants (Agrawal and Kumar 2020a: 86). Malays living in the south of Thailand for generations are still referred to as *khaek* – that is, 'visitor' (Horowitz 2000: 33).

56. The issue of enumeration of caste of the entire population has been debated in the run-up to censuses since the late 1990s (Roy Burman 1998; Sundar 2000; Deshpande and John 2010; Deshpande 2021; EPW 2021).

57. The government was, in fact, following in the footsteps of its colonial predecessor that frequently changed census categories in response to political exigencies. Caste data were not systematically collected for Christians after 1872 due to the strong opposition from Protestant missionaries (Oddie 1981: 137). Half a century later, the Muslim League 'was ideologically against the mention of "castes" among the Muslims', forcing the government to clarify that it had 'no intention ... to pursue a record of caste in case of Muslims' (Maheshwari 1996: 113–14). See Porter (1933: 422–23) for theological grounds on which sections of Muslims opposed enumeration of caste among them. The gradual turn to economic issues in line with global trends, too, played a role in marginalising identity in the colonial censuses (Sundar 2000: 116). Whatever the reasons, scholars of partition take note of this important shift with respect to the enumeration of Muslims (I. Ahmed 2020: 233).

58. In India, Ahmadis are shown in the census as a sect clubbed under Muslims. The 2011 census reported 119 Ahmadis in the country, which is obviously an undercount.

59. Taylor (2016: 13–14) suggests that self-interested politicians might want to invest in building autonomous statistical institutions because of long-term gains in monitoring and constraining successors and ensuring that they do not dismantle one's legacy at whim.

60. The principle of 'one person, one vote' that laid down the foundation of the enduring relationship between the modern state and statistics eventually politicised government statistics and undermined their reliability.

61. Censuses have been described variously as 'plebiscite' and 'political campaign' (Kertzer and Arel 2002b: 28), 'opinion polls' (Abramson 2002: 178) and 'a mandatory opinion poll' (Deshpande and John 2010: 41) because of the conflation between census and politics. It has also been suggested that censuses are treated as a 'weapon', an opportunity for 'show of strength' and an arena to wage 'war' in settings as different as the US (Siddiqui and McCarthy 2019), Northern Ireland (Anderson et al. n.d.), Palestine (Hassan Abu Libdeh of the Palestinian Central Bureau of Statistics quoted in Greenberg 1997), colonial India (Khan Bahadur Sheikh Fazl-i-Ilahi, superintendent of census operations [SCO], Delhi, quoted in I. Ahmed 1999: 124), independent India (Swami 2000a; Karmakar 2010; Bhat 2018: 181; Hekali Zhimomi, DCO, Nagaland, interview, Kohima, 25 June 2013) and Nigeria (Adepoju 1981: 35; Fawehinmi 2018).

62. A retired civil servant pointed out:

> In the old days, the Jammu elite had contacts with Kashmir and many of them would even spend their summers there. After 1990, hardly any person from Jammu excepting government servants goes to Kashmir at all. The younger generation in Jammu knows nothing about Kashmir. This is something I noticed when interviewing candidates for government jobs. I do not think there are any channels for communication between Kashmir and Jammu at the moment. Entire generations in both the divisions are growing up hating and fearing the 'other'.... Kashmiri Muslims and Jammu Hindus do not seem to talk to each other nowadays. Such contacts did exist in the past. The fact that they do not want to talk to each other makes the situation even more alarming. (Interview, 3 January 2016)

The author encountered people in both Kashmir and Jammu who are deeply ignorant of the other side and try to read the past with anachronistic lenses acquired after the outbreak of insurgency. (Some of the questions about

Hinduism one faced in rural Kashmir were reminiscent of questions in remote Naga villages along the Myanmar border – that is, areas with no known pre-colonial contact with Hindus.) In fact, even the Kashmiri leadership of Kashmir-based political parties does not have any credible channels to engage Jammu. The champions of Jammu's autonomy have not cared to maintain contact with Kashmir either, and a couple of well-placed interviewees even asked the author how it was possible to travel to non-tourist locations in rural Kashmir. Indeed, this rupture of communication within the state is the only, or at least the single biggest, 'achievement' of the insurgency. It must be admitted, though, that even earlier there was hardly any substantive dialogue between the two regions. B. Puri (1983a: 189), for instance, notes that 'there was never much trust between the Kashmiri leadership of the National Conference and its local leaders of Jammu. Even the group that campaigned [in Jammu] for and contributed to the abdication of the Maharaja was expelled from the National Conference without notice or explanation.' Just as New Delhi exploited divisions in Kashmir, the latter replicated the divisive tactics in Jammu and Ladakh.

63.  Elections could not be conducted in Assam in 1984 and Punjab in 1991, but the union government ensured that the exercise was completed within a year. Likewise, despite a fragile law-and-order situation in 1971, the government ensured the timely completion of the census in West Bengal.

64.  A sample survey suffers from frame non-coverage when a subgroup of population is not included in its sampling frame. On the other hand, sample non-coverage occurs when a subgroup of population that is included in the sampling frame is not surveyed. Non-coverage in sample surveys is different from coverage error in censuses. See the discussion on terminology in this chapter.

65.  Mudliar (2020) discusses how people on the margins deal with the poor quality of biometric databases that increasingly control access to welfare schemes. Agrawal and Kumar (2020: chs. 2, 7) discuss how public policy and public debates use and are also shaped by poor-quality data.

66.  Biemer (2010: 819) identifies the following dimensions of data quality: accuracy, credibility, comparability, usability (or interpretability), relevance, accessibility, timeliness, completeness and coherence. Discussions on data quality in India mostly focus on accuracy and comparability. Assessments of the accuracy of the Indian census mostly examine content errors in data on sub-categories of population: tribes (Kulkarni 1991; Guha 2003; Verma 2013; Agrawal and Kumar 2020a), castes (Bhagat 2006; Chaudhury 2012), religions (Bhagat 2001; Gill 2007) and languages (Brass 1974; Agrawal and Kumar 2020a). There are very few contributions on coverage errors in the census (Bhat

2011, 2018; Agrawal and Kumar 2012, 2020a, 2020b), which is the primary focus of this book. In addition, there are comparisons between the census and other databases such as the NSSO (Kasturi 2015); examinations of the quality of census data qua inputs to academic analyses (Guilmoto and Rajan 2013) and policymaking (Agrawal and Kumar 2020a); examinations of the interfaces between census and cartography (Agrawal and Kumar 2017a, 2020a), sample surveys (Agrawal and Kumar 2014, 2020a) and law (Kumar 2020a); and discussions of the public outreach of the census (Kumar 2020a, 2023c) and game-theoretic analysis of the politicisation of the census (Kumar 2023a). There are very few studies on other dimensions of the quality of India's government statistics such as timeliness and accessibility (Agrawal and Kumar 2020c) and usability (Kumar 2021a). There is a very rich literature on content errors in censuses of various countries (see the section on census as a collective self-portrait in this chapter and Chapter 6), but discussions of coverage errors are often limited to the undercount of minorities and migrants. There is a large literature on coverage errors in Nigerian censuses though (Aluko 1965; Adepoju 1981; Ahonsi 1988; Okolo 1999; Jerven 2013: 56-61). See also Zarkovich (1989) for coverage errors in erstwhile Yugoslavia.

67. We will briefly discuss unemployment figures in Chapter 7. Around 2010 the surge in violence in Kashmir involving young civilians nudged New Delhi to think of job creation as a conflict-mitigation strategy, but it did not have access to reliable statistics on unemployment.

68. The non-Kashmiri-speaking ethno-geographic periphery of Kashmir poses a particularly difficult problem with respect to the interpretation of census data on identity. See the discussion on Shina in Chapter 3.

69. The two main divisions continue to be referred to as provinces in government reports (GoI 2001b: 15) as well as popular discussions (*Kashmir Times* 2011e; Hari Om 2017a; Mir 2021).

70. The NSS regions (known as 'natural regions' earlier) are groups of districts within states that are as far as possible similar in terms of agro-ecological conditions. In practice, the NSS regions can include very different areas. Until the 65th round of the NSS (2008–09), Jhelum Valley used to nominally include Ladakh region (alongside Kashmir) even though the latter was excluded from the sampling frame.

71. India's periphery can be visualised from multiple perspectives: colonial political imagination (Curzon 1908 [1907]: 42), ethnolinguistic and religious composition (Agrawal and Kumar 2020a: 22–23) and distribution of electoral power (ibid.: 300). J&K belongs to the intersection of all these peripheries. Sanjoy Hazarika (2020) pointed out that the use of the word 'periphery' is problematic because the people of these areas may not see themselves as being

'peripheral'. Indeed, in discussions in Leh, Kargil and Srinagar, several interviewees stressed the centrality of the region in Asia's geopolitics. Sheikh Abdullah's autobiography abounds in references to the historical and contemporary importance and centrality of Kashmir (Abdullah 2016). A third point bears noting here. It can possibly be argued that only the Sixth Schedule areas of Assam should be counted in the periphery. If this distinction is followed, the share of the periphery in the national population will halve.

72. Pronab Sen, the first chief statistician of India, suggests that government statistics could also include statistics that were collected by NGOs but were endorsed by the government (Agrawal and Kumar 2020a: 37).

73. See Drèze and Somanchi (2021), Pais and Rawal (2021) and Pronab Sen quoted in Nahata (2021) for concerns about the quality of CMIE data.

74. See Agrawal and Kumar (2020a: 54, 191, 322) for remarks on the quality of the PLSI data.

75. Private sector counterparts of the government's administrative statistics have also begun to emerge as digital companies are generating large amounts of data as a by-product of their routine operations. Naukri.com's 'Naukri JobSpeak Index' is a case in point. Unlike the CMIE's survey-based estimates of unemployment, the aforementioned index measures 'hiring activities based on recruitment patterns on Naukri.com'.

76. Presser (2021) questions the metaphor of 'data collection' because data are constructed rather than being collected. As a result, the distinctions among the measure, measurement and the interpretation of measurement collapse.

77. See Kumar (2021a) for a discussion of the evolution of the notion of metadata and their quality in India.

78. Some of the earlier censuses instructed enumerators to record 'eunuchs and hermaphrodites' as males (GoI 1964a: 39). The approach of the census has changed over the past decade. The training manual of the 2011 Household Census foregrounds the male–female sub-categorisation, and 'other' appears as an afterthought to accommodate the residual category (GoI n.d.14: 38-39). A separate PCA was issued for 'others' at the national level, which reported a population of 487,803 (table PCA_OTH_0000_2011). Note that the population estimate reported in the aforementioned table includes 'any person who desired to record sex under the category of "Other"', but not transgenders who may have reported themselves as male or female (ibid.). However, as mentioned earlier, Table A-2, among others, reports only male and female populations that add up to the total population, which is difficult to understand in the absence of metadata. Question 12 of the 2021 Houselisting and Housing Census Schedule allows the heads of the household to identify as 'third gender' (GoI n.d.20). Question 3 of the 2021 Household Census Schedule (pre-test), too, makes room for 'third gender' (GoI n.d.21).

# Part II
# Counting People

# 2

# Counting amidst Uncertainty

[T]he second phase went off well till September 4, 2000 when a threat call against conducting of census by a militant organisation, Hizbul Mujahideen, appeared in all the newspapers including national dailies.
—Director of Census Operations, Jammu and Kashmir (GoI 2001b: 16)

## Introduction

Located in the north-west extreme of India's ethno-geographic periphery, Jammu and Kashmir (J&K) was initially a Category B state. Unlike other princely states belonging to Category B and Category C that were absorbed into larger states or reconstituted as union territories (UTs), J&K was directly granted the status of a separate state.[1] The state had two capitals: Srinagar (summer) and Jammu (winter). Before being reorganised into two UTs on 9 August 2019, the state comprised three administrative divisions: Jammu (26,293 square kilometres), Kashmir (15,948 square kilometres) and Ladakh (59,146 square kilometres). The last was until recently a part of the Kashmir division and was constituted as a separate administrative division in February 2019, less than six months before the reorganisation of the state.[2] The reorganisation marked the end of the seven-decade-old constitutional arrangement under Article 370[3] that guaranteed autonomy to the state to accommodate the restive Kashmir Valley amidst domestic and international challenges to the Muslim-majority princely state's accession to India (Timeline 2.1). International and domestic conflicts spawned by the unsettled political status have affected the whole range of government statistics in J&K, compounding the challenge posed by difficult weather and terrain.

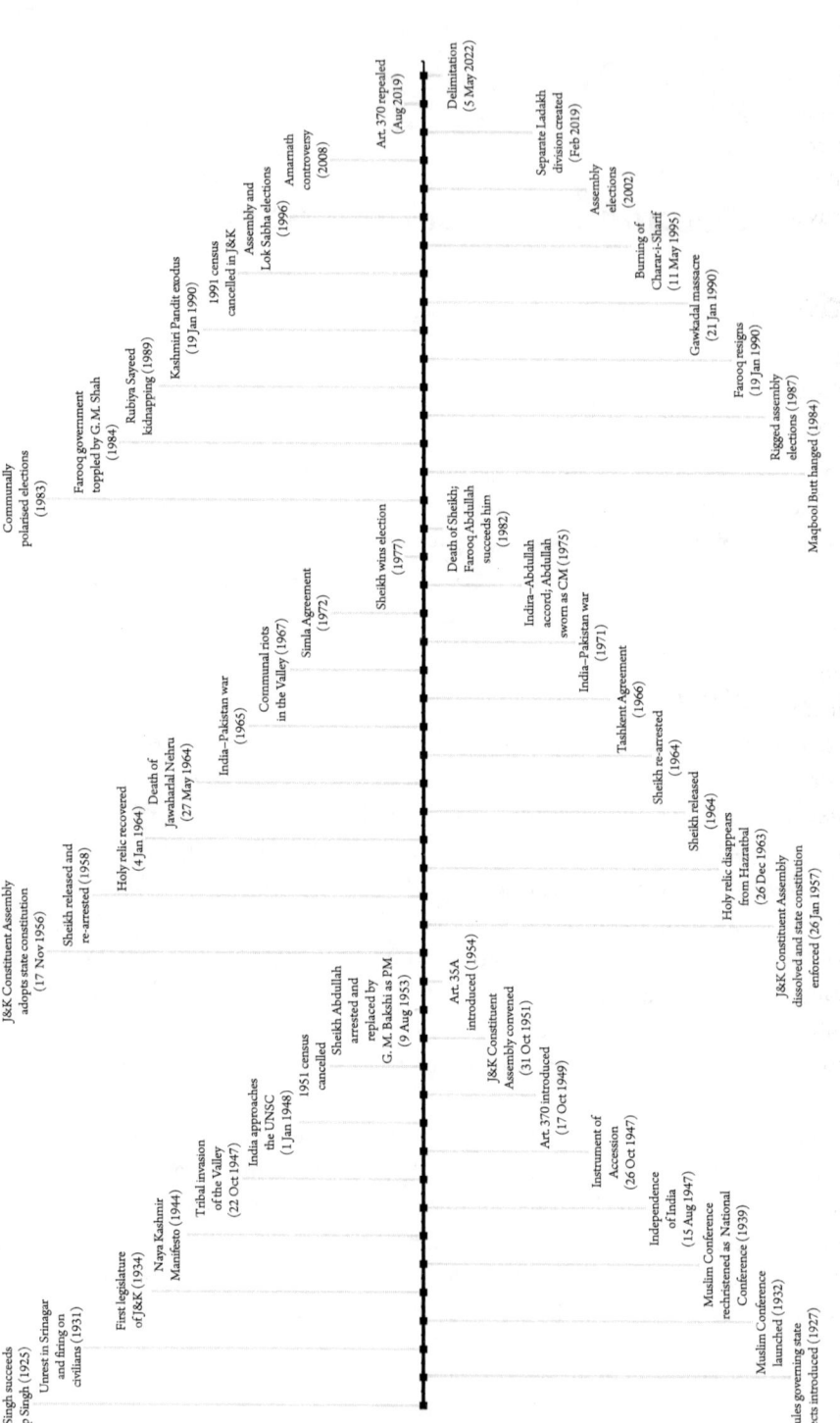

**Timeline 2.1** Major events in Jammu and Kashmir (J&K), 1925–2022

*Note:* See Timelines 2.2, 2.3 and 4.1 for further details relevant to censuses.

*Source:* Prepared by the author.

In this chapter, we will first discuss the less-than-satisfactory accounting of the area of J&K. This will be followed by a discussion of how political instability, uncertain weather, mobile population groups and shifting reference dates affect headcounts and how our ability to understand their impact on the quality of census data is constrained by the paucity of metadata.

# Indeterminate Borders

The territory of J&K under the Indian administration has remained largely unchanged after tribal militia backed by the Pakistani army seized parts of the princely state in 1947 and China occupied parts of Leh in the 1950s and early 1960s. The only officially recognised change on the western front happened under the Simla Agreement of 2 July 1972 after the 1971 Indo-Pakistan war that froze the 'line of control [LoC] resulting from the cease-fire of December 17, 1971' (GoI n.d.2). Census reports are deficient in information on international territorial changes, though.

The exchange of territory with Pakistan and its implications for inter-temporal comparisons are not properly recorded in census reports and, in some cases, are even at variance with other government reports.[4] According to the 1981 census, the territorial realignment in 1972 resulted in the gain of 11 villages and the loss of 41 full and 3 half villages to J&K that led to the addition of 818 square kilometres to the state's area (Table 2.1),[5] but this is not in agreement with the estimates provided by other government sources.[6]

There are no reliable estimates of the territory lost in the east to successive Chinese incursions in Leh. There was a decrease in Ladakh's area between the 1961 and 1971 censuses. It is not clear what explains this as the 1965 Indo-Pakistan war did not lead to any change in the status quo,[7] and the census reports are silent on the loss of territory, if any, to China. The 1971 census also reported minor but inexplicable increases in the area of Kashmir and Jammu. The change in the units of measurement from square miles to square kilometres cannot account for these inter-censal changes in area.[8] Also note that substantial, even if unpopulated, territory brought under the fuller control of India in the Siachen region through Operation Meghdoot (April 1984) is not reflected in the reported area under Indian administration in the 1991 census.[9]

At present, J&K's 'area figures include the area under unlawful occupation of Pakistan and China ... 78,114 Sq.km under illegal occupation of Pakistan, 5,180 Sq.km illegally handed over by Pakistan to China and 37,555 Sq.km

**Table 2.1**   Area of districts of Jammu and Kashmir (J&K), 1961–2011

| District | 1961* | 1961 | 1971 | 1981 | 2001 | 2011 |
|---|---|---|---|---|---|---|
| Doda | 4,380.2 | 11,345 | 11,691 | 11,691 | 11,691 | 8,912 |
| Kishtwar | | | | | | 1,644 |
| Ramban | | | | | | 1,329 |
| Reasi | | | | | | 1,719 |
| Udhampur | 1,731.6 | 4,485 | 4,549 | 4,550 | 4,550 | 2,637 |
| Punch | 1,689.1 | 4,375 | 1,658 | 1,674 | 1,674 | 1,674 |
| Rajouri | | | 2,681 | 2,630 | 2,630 | 2,630 |
| Jammu | 1,248.6 | 3,234 | 3,165 | 3,097 | 3,097 | 2,342 |
| Samba | | | | | | 904 |
| Kathua | 1,023.6 | 2,651 | 2,651 | 2,651 | 2,651 | 2,502 |
| **Jammu division** | **10,073** | **26,089** | **26,395** | **26,293** | **26,293** | **26,293** |
| Baramula | 2,536 | 6,568 | 7,458 | 4,588 | 4,588 | 4,243 |
| Bandipore | | | | | | 345 |
| Kupwara | | | | 2,379 | 2,379 | 2,379 |
| Srinagar | 1,205.1 | 3,121 | 3,013 | 2,228 | 2,228 | 1,979 |
| Ganderbal | | | | | | 259 |
| Badgam | | | | 1,371 | 1,371 | 1,361 |
| Anantnag | 2,096.9 | 5,431 | 5,382 | 3,984 | 3,984 | 3,574 |
| Kulgam | | | | | | 410 |
| Pulwama | | | | 1,398 | 1,398 | 1,086 |
| Shupiyan | | | | | | 312 |
| **Kashmir division** | **5,838** | **15,120** | **15,853** | **15,948** | **15,948** | **15,948** |
| Leh | | | | 82,665 | 45,110 | 45,110 |
| Kargil | | | | 14,036 | 14,036 | 14,036 |
| **Ladakh division** | **37,753.8** | **97,782** | **95,876** | **96,701** | **59,146** | **59,146** |
| J&K (India) | 53,665 | 138,992^ | 138,124 | 138,942 | 101,387 | 101,387 |
| J&K (Pakistan occupied)† | | | 78,932 | 78,114 | 78,114 | 78,114 |

*(Contd)*

**Table 2.1** (*Contd*)

| District | 1961* | 1961 | 1971 | 1981 | 2001 | 2011 |
|---|---|---|---|---|---|---|
| J&K (China occupied) | | | 37,555** | 37,555** | 37,555 | 37,555 |
| J&K (transferred by Pakistan to China)‡ | | | 5,180 | 5,180 | 5,180 | 5,180 |
| J&K (occupied) | 32,358.1 | 83,807*** | 121,667 | 120,849 | 120,849 | 120,849 |
| J&K (overall) | 86,023 | 222,799 | 222,236 | 222,236 | 222,236 | 222,236 |
| | Square miles | Square kilometres | | | | |

*Source*: General population totals (GPT), Census of India, 1961–2011.

*Note*: (a) *Area estimates were expressed in square miles in 1961 and in square kilometres in the later censuses. The table presents estimates in both units for 1961. (b) **Until 1981, census reports presented area estimates of Ladakh including 37,555 square kilometres 'under illegal occupation of China' but not 5,180 square kilometres 'illegally handed over by Pakistan to China'. J&K's area includes both these areas as well as 78,932 (78,114]) square kilometres 'under illegal occupation of Pakistan'. (c) ***The 1961 estimate of occupied territory seems to include the area of Shaksgam that was transferred by Pakistan to China but not the territory directly occupied by China. The latter was accounted for in Ladakh's area. (d) †Pakistan-occupied territory decreased due to the exchange of villages after the 1971 Indo-Pakistan war. (e) ‡Pakistan transferred 5,180 square kilometres to China under a border agreement in 1963. (f) ^The total area of the state is not equal to the sum of the area of the three divisions due to rounding off. (g) A few 1981 census publications report the 1971 area of Kashmir (for example, GoI 1989a: 1). The figures reported here aggregate district area estimates from the GPT. (h) Certain census publications do not include the occupied territory in India's estimated area (GoI n.d.5).

under illegal occupation of China' (table A-1, 2011 census).[10] The general population totals (GPT) of 2001 note that the density of 'the rural portion of Jammu & Kashmir has been worked out after excluding the entire area of the illegal occupied portion of Pakistan and China as its rural-urban breakup is not available' (GoI 2005a: 81). The 'General Notes to Provisional Population Totals [PPT]' of 2011 indicate that '[f]or working out density … the entire area and population of those portions of Jammu and Kashmir which are under illegal occupation of Pakistan and China have not been taken into account' (GoI 2011b: x).[11] Three points are noteworthy about these area estimates. First, the estimate of the area of Ladakh reported in the 2011 census (59,146 square kilometres) (Table 2.1) differs from the that provided by the Department of Jammu and Kashmir Affairs (33,554 square miles – that is, 86,904 square kilometres) (GoI 2011f). The figures released by the two departments operating under the Ministry of Home Affairs (MHA) cannot be reconciled even if we account for the area under Chinese occupation – that is, 37,555 square kilometres.[12] Second, until 1981, census estimates of Ladakh's area included '37,555 Sq.km under illegal occupation of China'. Similar corrections were not indicated for other partitioned districts – say, Punch in the Jammu division. Third, while representing data in maps, the census does not follow the caution observed in accounting area. The Pakistan-occupied J&K is shaded in grey, the colour that signifies 'data not available', but the territory occupied by China is not shaded in grey (GoI 2011c: 13) and, in fact, shares the colour of Leh district in choropleths (GoI 2014a: 10–11). Moreover, Ladakh's choropleths include the territory corresponding to '37,555 Sq.km under illegal occupation of China' but not 5,180 square kilometres 'illegally handed over by Pakistan to China'.

The discussion so far suggests that the government's estimates of the state's area are unreliable. We can add that while the state has been a fixed point of national security debates, even the broad contours of its territory are poorly understood. In 2020, a prolonged standoff with China in eastern Leh catapulted the remote Ladakh region into the limelight. After the initial clashes, the national news media took a couple of weeks to figure out the geography of the region, including counting and locating 'fingers' around Pangong Tso.[13] The government, too, continues to fumble through the region. In 2020, the weather bulletins of Doordarshan, India's public broadcaster, began to cover the Pakistan-occupied territories (Dutta 2020). The bulletins begin with Gilgit, which does not share a border with either Leh or Kargil but was part of the princely state of J&K, and then move on to Leh, Muzaffarabad (Pakistan-occupied), Srinagar, Mirpur (Pakistan-occupied) and Jammu (DD News 2020), skipping over Skardu, the capital of the large and crucial region of Baltistan that was part of Ladakh during the Dogra period.

More importantly, a series of new political maps of India issued by the Survey of India (SoI) – eighth (SoI 2019a), ninth (SoI 2019b) and tenth (SoI 2020) editions (Maps 2.1b–2.1d) – include Gilgit–Baltistan in the territory of Leh district of Ladakh (see also *Times of India* 2019). Earlier editions of this map (SoI 2017), too, included these regions in India (Map 2.1a), but they showed an awareness of the historical administrative boundaries between Gilgit and Ladakh that are overlooked in the most recent maps. The revised maps do not erase the internal boundaries of the Pakistan-occupied part of the UT of J&K though.

**Map 2.1a**   Survey of India map of Jammu and Kashmir (J&K), seventh edition, with borders separating Gilgit from Ladakh, 2017
*Source*: SoI (2017).

If for some reason the Pakistan-occupied territory adjoining the UT of Ladakh must be included in one of the existing districts on the Indian side, it should have been apportioned between Leh and Kargil districts on historical and cultural grounds. This 'error' in the maps is most likely a consequence of the long-standing practice of accounting the territory of Gilgit–Baltistan under Pakistan's occupation in Leh's area.[14] Earlier, the area of this territory was included in Ladakh. When Kargil district was carved out of Ladakh, the remaining area, including that under Pakistan's occupation, continued to be accounted under Leh (Table 2.1). While this bureaucratic slip is understandable,[15] the recent press releases and circulars of the government betray a genuine lack of clarity about the geography of the region. A press release on the new maps explains the territory of Leh and Kargil districts as follows:

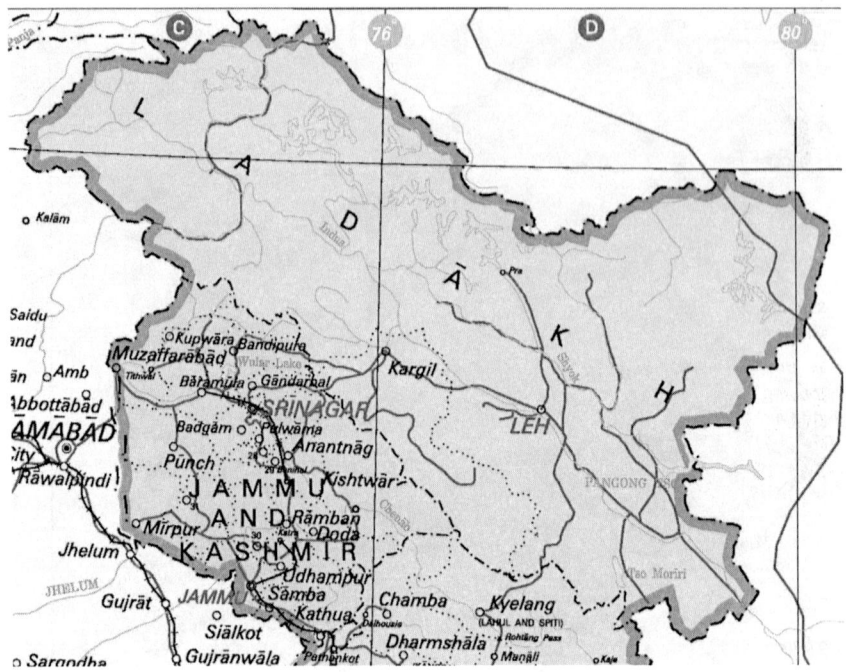

**Map 2.1b**   Survey of India map of Jammu and Kashmir (J&K), eighth edition, 2019
*Source*: SoI (2019a).

Kargil district was carved out from the area of Leh and Ladakh *district* [*sic*].
The Leh district of the new Union Territory of Ladakh has been defined in the
Jammu and Kashmir Reorganization (Removal of Difficulties) Second Order,
2019, issued by the President of India, to include the areas of the districts of
Gilgit, Gilgit Wazarat, Chilhas [*sic*] and Tribal Territory of 1947, in addition to
the remaining areas of Leh and Ladakh *districts* [*sic*] of 1947, after carving out
the Kargil District. (GoI 2019b, emphasis added)

It is not clear what 'Leh and Ladakh districts' refer to in recent documents (see,
for instance, GoI 2019a, 2019b). According to the *Administrative Atlas*, in 1961
and 1971, J&K included the district of Ladakh (GoI 2011c). The corresponding
entries for 1981 and 1991 refer to the districts of Ladakh and Kargil, which are
referred to as Leh (Ladakh) and Kargil in 2001 and 2011.[16]

   The cartographic confusion in Ladakh has serious political consequences.
As per the 1995 delimitation, there were 111 seats in the legislative assembly
of J&K, out of which 24 were left vacant 'until the area of the State under the

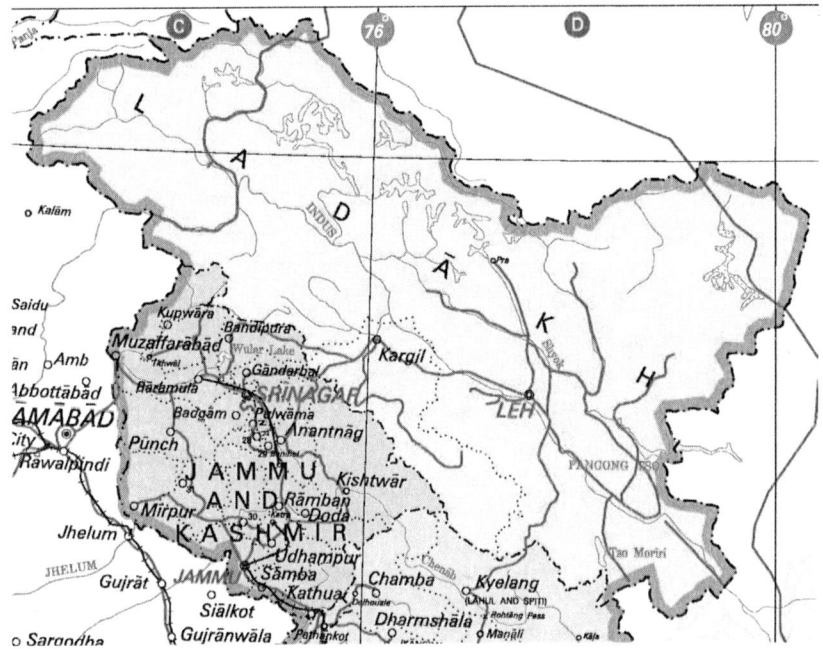

**Map 2.1c**   Survey of India map of Jammu and Kashmir (J&K), ninth edition, 2019
*Source*: SoI (2019b).

occupation of Pakistan ceases to be so occupied and the people residing in that area elect their representatives' (Constitution of Jammu and Kashmir [Twelfth Amendment] Act, 1975). There were 100 seats in the assembly, including 24 vacant seats between 1975 and 1995 and 25 vacant seats before that. As per the Jammu and Kashmir Reorganisation Act, 2019, all the vacant seats, including the seats of Gilgit–Baltistan, belong to the UT of J&K: '[T]wenty four seats in the Legislative Assembly of Union territory of Jammu and Kashmir shall remain vacant and shall not be taken into account for reckoning the total membership of the Assembly' (Section 14[4][a]).

This creates an anomalous situation because Gilgit–Baltistan falls in the UT of J&K for legislative and electoral purposes even though territorially it is part of the UT of Ladakh, which does not have a legislature. Discontinuing Gilgit–Baltistan's representation would have diluted the claim to the disputed territory. So the government seems to have devised the clumsy solution of mapping Gilgit–Baltistan to two different UTs. This solved one problem but created another, as now the people of Ladakh are denied a right that has been

**Map 2.1d**    Survey of India map of Jammu and Kashmir (J&K), tenth edition, 2020
*Source*: SoI (2020).

notionally extended to Gilgit–Baltistan.[17] A similar confusion prevails along
the internal borders of the territory under India's administration.[18] The work
of the delimitation commission constituted after the reorganisation of the state
was stalled due to 'haphazardly' created administrative units with imprecise
borders and overlapping jurisdictions (Pargal 2021). The cartographic confusion
examined in this section reflects the uncertainty generated by the underlying
conflict, which, as discussed next, also affects the census.

## Conflict

The history of the census of J&K is quite different from that of the rest of the
country. While 'a rough enumeration of the people inhabiting the various districts
or parganas … was carried out' in 1873, it 'was not regarded as a Census at any
time in the accepted sense of the word' (GoI 1968a: 36–37). 'In 1881 also, no
census was held in Jammu and Kashmir for reasons which are neither available
from the records of the State Government nor cited in any contemporary

chronicle' (ibid.).[19] Census was first conducted in J&K in 1891, and thereafter it was conducted regularly until 1941. Before 1947,

[c]ensus was not a central subject so far as Jammu and Kashmir State is concerned and the State Government then made their own arrangements for conducting the decennial Census, though the Operations were carried out under the guidance of the Census Commissioner of India and the tabulation was planned on the lines prescribed by the Centre. (GoI 1963a: 1)

The pre- and post-1947 censuses in J&K differ in terms of the tabulation of data. 'Before 1951, Religion was treated as the unit of tabulation and most of the demographic data such as marital status, literacy and other characteristics was tabulated by the communities. Further the classification of the population was being presented by religion *down to the village level*' (GoI 1965: 271, emphasis original). Moreover, the data on religion were subclassified by sects (GoI 1943b). Both these practices were discontinued after 1947.[20]

J&K was initially excluded from the purview of the Census Act, 1948. The Census Act was extended to the state 'vide Adaptation of Laws Order 1950' (GoI 1963a: 1).[21] Article 1(2) of the Census Act was amended on 7 May 1959 to drop the words 'except for the State of Jammu and Kashmir' and extend its applicability to the state, while 'respecting enactments not extending to Jammu and Kashmir' (GoI 2009a: 127), 'to facilitate the holding of a Census in that area in 1961 with the rest of India' (Das Gupta 1968: 266).[22] The population of Scheduled Tribes (STs) was reported for the first time in the 2001 census due to their belated recognition in the state.[23]

Unlike Assam, Bengal and Punjab that were also partitioned in the late 1940s,[24] J&K was not covered by the 1951 census 'due to abnormal conditions caused by tribal raids and disturbances in various parts of State' (GoI 1963a: 1, 14) and 'the subsequent splitting up of the State into two zones by what is known as the Cease-fire Line [later known as the LoC]' (GoI 1964b: i). The administrative report for the 1961 census notes that

the State was raided in 1947 and an area estimated at 32,358 square miles, comprising the tehsil [a sub-district administrative unit] of Muzaffarabad, part of tehsils of Uri and Karnah, tehsil of Skardu, a part of the tehsil of Kargil, districts of Astore, Gilgit and Gilgit Agency, the tehsils of Mirpur, Bagh, Palandari, Kotti [Kotli] and parts of tehsils of Bhimber, Haveli and Mendhar were occupied by the so-called Azad Kashmir Government. (GoI 1963a: 24)[25]

It adds that '[h]undreds of villages had been completely depopulated as a result of the tribal raids of 1947, but no list of areas which were thus rendered uninhabited was traceable' (ibid.: 21).

In October 1949 – that is, two years before the 1951 census, when preparations for the next census were already underway in the country – while introducing temporary provisions for J&K, N. Gopalaswami Ayyangar argued before the Constituent Assembly that the 'administration of the State should be geared to' tackle the 'unusual and abnormal' conditions arising out of the continued occupation of part of the territory by 'rebels and enemies' and the country's 'entanglement' with the United Nations (Noorani 2011: 87). Elaborating the reasons for excluding the state, a 1951 census report noted that 'it became clear that a complete count was impossible in the conditions prevailing in the state.... Government decided that the information likely to be secured by an incomplete count was not worth the effort, expense and very considerable strain on administrative resources which are necessarily involved in census-taking' (GoI 1957: 3). It is not clear if the weather too influenced the decision to not conduct the 1951 census.[26] Also note that while the census was not conducted in both 1951 and 1991 due to political instability, there are two key differences. First, there was no direct threat to the census in 1951.[27] Second, the census was planned in 1991, even though it had to be cancelled later.

These terse explanations of the cancellation of the 1951 census are insufficient though. Assam was also affected by 'widespread communal disturbances in Cachar, Goalpara, Kamrup, Nowgong and Darrang [that is, five out of eight districts], followed by a very heavy influx of refugees' in February and March 1950 (GoI 1954b: 5). These districts accounted for more than half of the area and population of Assam. In J&K, more than half of the state had not been affected by the raids or influx of refugees. In any case, the tribal raids had been repulsed months before preparations for the next census began, and the refugee crisis, too, had subsided by then. There are, at least, three reasons why census and other surveys could not be conducted in the early 1950s. First, the administration was in a disarray in Kashmir after the hasty retreat of the last Dogra ruler to Jammu along with senior officers, who were mostly from Jammu and Punjab, and Jammu was itself affected by communal violence. The National Conference (NC) tried to fill in the vacuum in Kashmir, but it did not have a presence in Jammu. Second, J&K continued to be in a state of flux for quite some time as the legal–constitutional framework guiding the government was not in place. The census was not yet part of the Union List in the case of J&K as the Instrument of Accession limited the union government's reach. Third, Sheikh Abdullah was trying to fulfil key commitments related to land redistribution, among other things, which would have strained the resources of the fledgling administration. Land reforms had, in fact, begun two years before the Constituent Assembly was formed (Devadas 2019a: 99).

The next three censuses – 1961, 1971 and 1981 – were held without any large-scale disruptions (Timeline 2.2). The household phase of the 1961 census overlapped with Ramzan and Holi, which necessitated some adjustments discussed in later sections. Note that the belated extension of the Census Act to the state might have affected the preparation for the 1961 census. Unlike other states, J&K had less than one year to prepare for the 1961 census from scratch as it was not covered in 1951. The household phase of the 1971 census was rescheduled across the country due to the Lok Sabha elections. The rescheduling along with cross-border migration triggered by the 1965 Indo-Pakistan war affected inter-temporal comparisons in J&K. The household phase of the 1981 census was delayed in the accessible areas 'due to the unfavourable weather conditions that prevailed in February 1981' (GoI 1985c: 3; see also GoI 1983a: 1). India's PPT for 1981 included 'the population figures of Jammu and Kashmir ... as determined by the Expert Committee on Population Projections' because of the delay in enumeration (GoI 1981d: 7).

There was a long-standing demand to recognise the ST communities of J&K. The government conducted a 'special census and community-wise specific studies in all the regions of the State' in 1986–87 to draw up the list of STs (Lok Sabha 1986b: 91–92). Ashish Bose (2000: 1433) referred to this exercise as 'an ad hoc census'. As per the union government, the 1986–87 exercise involved 'house to house enumeration ... on the lines of decennial census. Data was collected from each individual household concerning name, relationship to head of houschold, sex, age, religion, traditional and present occupation etc' (Lok Sabha 1991: 54). For the purpose of this 'special census', the state was divided into synchronous and non-synchronous areas (Lok Sabha 1986b: 91–92). The results were not formally released and, in any case, are not comparable with other censuses due to the unusual reference date.

The government appointed A. R. Parray as the director of census operations (DCO) in J&K for the 1991 census (GoI 1991a: 14; GoI 1995c: v). However, the exercise could not be conducted in the state. As a result, J&K became the first and only state to have missed the decennial census twice after 1947, once in 1951 and then in 1991.[28] In response to L. K. Advani's question, the union home minister informed the Lok Sabha that 'the field situation does not permit holding the 1991 Census' (Lok Sabha 1991: 54). Government reports are silent

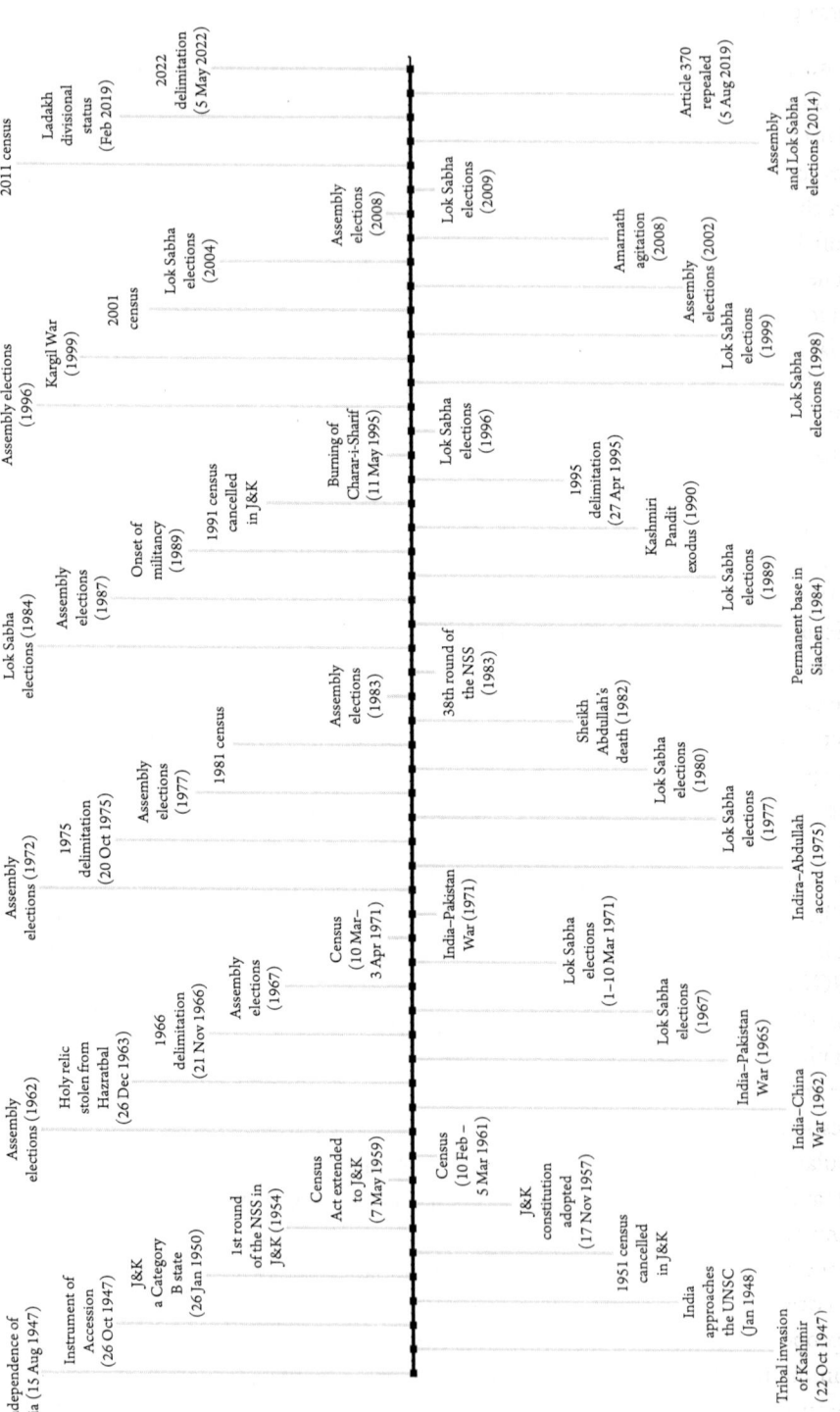

**Timeline 2.2** Censuses, surveys and major developments in Jammu and Kashmir (J&K), 1947–2022

*Source:* Prepared by the author.

on the exact details of the circumstances that led to the cancellation of the census in 1991. News reports suggest that 'threats by militant groups forced state authorities to cancel a census in 1991' (Rediff 2000a). Bhat (2018: 180), too, notes that '[c]ensus could not be conducted in J&K in 1991 due to opposition from the militant organizations and the consequent security concerns' (see also Schofield 2003: 231).[29]

The rationale behind cancelling the 1991 census in the state has been questioned. First, the exercise was conducted in neighbouring Punjab, where the fatalities per 1,000 population were much higher than J&K in both 1990 and 1991. In fact, 1990 and 1991 were the worst years in Punjab in terms of fatalities.[30] Census was also conducted in Nagaland and Manipur despite a marked deterioration of internal security after the split in the National Socialist Council of Nagaland (NSCN). It can, however, be argued that 1990 happened to be the second year of armed insurgency in J&K, whereas the administration was battle-hardened in the other states where insurgencies had begun long before the 1991 census.

Second, in 1990–91, insurgency was limited to Kashmir. It is not clear why Ladakh and (the southern districts of) Jammu that were largely free of violence were excluded from the census unless it is argued that the Srinagar-centric state government collapsed, leaving the whole administrative machinery in disarray, and Jammu, the other capital, was overwhelmed by an unprecedented influx of refugees. It is noteworthy that the 46th (July 1990–June 1991) and 47th (July 1991–December 1991) rounds of the National Sample Survey (NSS) excluded only Kashmir, while Ladakh was anyway not part of the sampling frame in the 1990s. In fact, even the half-yearly survey of the Sample Registration System (SRS), which is organised by the Office of the Registrar General of India (ORGI), was conducted in parts of the state not directly affected by insurgency (GoI 1992b: 80).

Third, the PPT for 1991 released on March 25 – that is, about three weeks after the completion of the revisional round of the household census in other states – noted that the estimate of '[p]opulation of India includes projected population of Jammu & Kashmir where the 1991 Census is yet to be held' (GoI 1991a: 1). The figures for J&K were based on 'projections prepared by the Standing Committee of Experts on Population Projections, October, 1989' (ibid.: 3). From the claim that the census 'is yet to be held', it is not clear if the exercise had been delayed due to insurgency. A few pages later it is clarified that enumeration 'is proposed to be conducted later this year' (ibid.: 11). The report adds, 'As in the earlier censuses, different dates have been prescribed for the census in Jammu & Kashmir State owing to weather conditions prevailing in that State during February–March, 1991. It is proposed to conduct the census in that State during August–October, 1991' (ibid.: 9). This means that at least until 25 March 1991, the government had not given up hope (see also A. Bose 1991: 31). However, on 1 August – that is, before the commencement of the census – the union government informed the parliament that the conditions were not conducive for enumeration (Lok Sabha 1991: 54). One of the first hints of the impending cancellation is available in the agenda notes of the Fifth Conference of the Directors of Census Operations (May 1992) held in New Delhi, which noted that there was '[n]o progress [in J&K] due to disturbed conditions' (GoI 1992b: 8). The final population totals (FPT) released later in that year throw more light on the cancellation: 'Due to unfavourable weather conditions prevailing in Jammu & Kashmir during February–March, 1991, it was proposed to conduct the 1991 census in that State in August–October, 1991, but this could not be taken due to disturbed conditions prevailing then' (GoI 1992a: 1). This report presented India's population excluding J&K as well as including projected figures for J&K (ibid.: 15).

Census publications do not adequately describe the political events and the administrative process that governed the cancellation of the exercise in J&K (see Timeline 2.3). The First Conference of the Directors of Census Operations was held in November 1989[31] – that is, more than a month after the targeted killings of prominent Pandits had begun, which in hindsight seem to mark the onset of the armed insurgency. The proceedings of the conference, which was attended by the deputy DCO and the director of the Bureau of Economics and Statistics from J&K, did not contain any hint of doubt about the feasibility of census and, in fact, noted that the ORGI gave a nod to the establishment of two new regional tabulation offices in J&K (GoI 1989c: 24). The proceedings also suggested that a different reference date would be followed in J&K even for the synchronous areas (ibid.: 5). Usually, when the ORGI fixes a non-synchronous reference date, it is chosen so that the entire country's figures can be released together. It is not clear if the decision to hold the census after the national reference date was dictated by the deteriorating security conditions. If inclement weather was the primary concern ex ante due to the experience of 1981, which was scheduled after the national reference date, the exercise should have been scheduled in 1990. The reference date of the 1961 census was pushed back in Kohima and Mokokchung districts of Nagaland due to the deterioration of the security conditions *during* enumeration (GoI 1966a: 1). The 1971 census was conducted *ahead* of schedule in West Bengal due to an *anticipation* of law-and-order problems (GoI 1975: 1; GoI 1977: 1).

The second conference was held in February 1990 – months after the kidnapping of a daughter of then union home minister, Mufti Mohammad Sayeed,

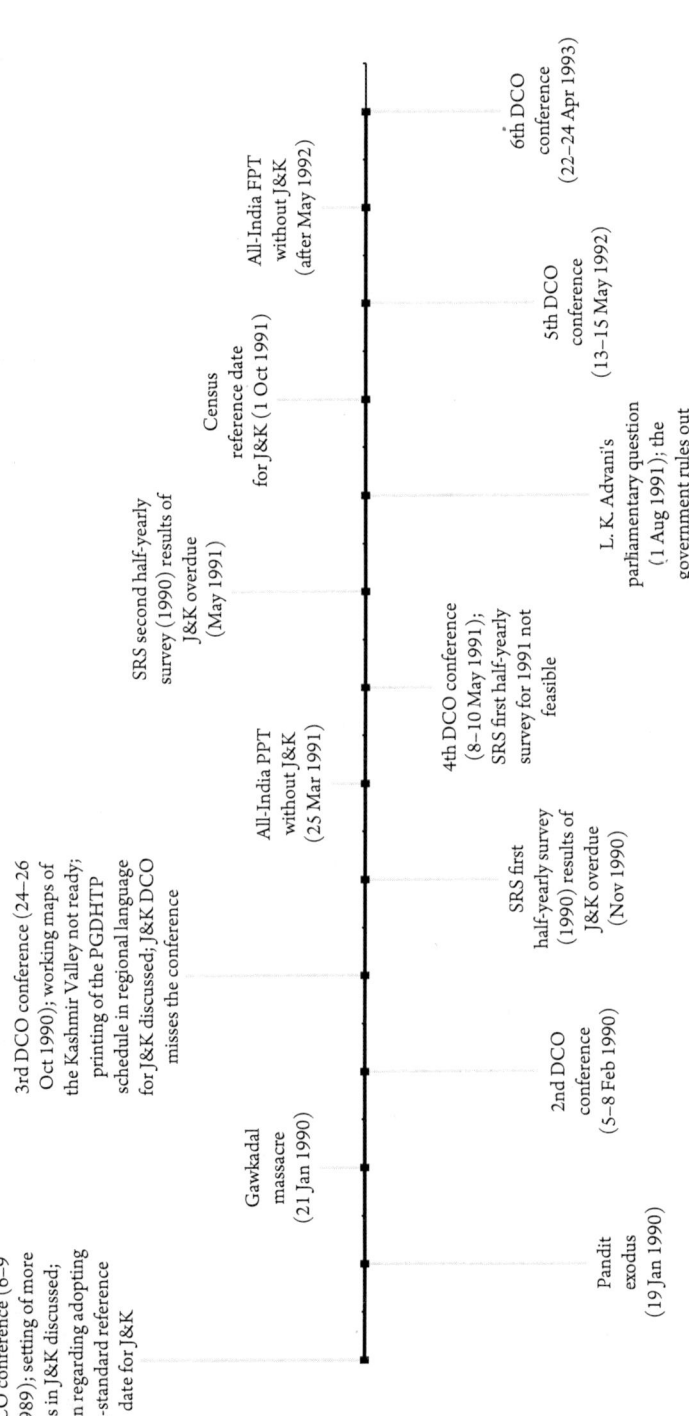

**Timeline 2.3**  The 1991 census of Jammu and Kashmir (J&K)

*Source:* Prepared by the author.

*Note:* (*a*) DCO: director of census operations; FPT: final population totals; PGDHTP: Post Graduate Degree Holders and Technical Personnel; PPT: provisional population totals; RTO: Regional Tabulation Office; SRS: Sample Registration System. (*b*) The 1991 census was eventually cancelled in the state.

that led to the release of terrorists, and a fortnight after the exodus of Pandits and a major incident of civilian deaths at Gawkadal in Srinagar. The proceedings noted that some non-synchronous areas, such as in J&K, would be enumerated in mid-1991 (GoI 1990b: 26), which agrees with the first conference that decided to 'notify the reference date of non-synchronous areas in ... Jammu & Kashmir', as the enumeration of the synchronous areas of the state was also different from that of the rest of the country (GoI 1989c: 5). There is no hint of the possibility of the cancellation of the census. Even the third conference held in October 1990, well after the commencement of the insurgency, noted that the houselisting would be carried out in 1991 (GoI 1990c: 1, 10) and discussed the printing of the 'Post Graduate Degree Holders and Technical Personnel' schedule in regional languages for different states, including J&K (ibid.: 39). The source of the optimism is not clear because by then there should have been sufficient clarity regarding the difficulty of holding a census in Kashmir. In fact, even the proceedings of the second and third conferences contain clear hints that all was not well. First, 'the enterprise list material in Urdu was destroyed in fire' in J&K (GoI 1990b: 14; see also note 29 of this chapter). The cause of fire is not specified, but arson cannot be ruled out as several government offices were attacked in that period. This is indirectly hinted by the Central Statistical Organisation that 'assured that fresh material would be soon sent' but added the suggestion that 'it could be stored in Jammu where it could be safer' (ibid.). There must have been reason to believe that the material would be relatively safer in Jammu, if it was destroyed by anti-state actors and not a fire accident. Second, 'working maps of [six] districts' of J&K were not ready as late as October 1990 (GoI 1990c: 38). The names of the districts are not mentioned, but in all likelihood these correspond to the six valley districts, which fell out of reach of other government surveys too. The base maps for the PPT report, too, were not ready (ibid.). In fact, the proceedings added that '[t]he position of Jammu & Kashmir could not be ascertained in the absence of representation from that State' (ibid.: 39). Moreover, by the time the all-India PPT was published, it was clear that it was going to be difficult to conduct the half-yearly survey of the SRS in Kashmir (GoI 1991b: 33), and even the reports of the two half-yearly surveys of 1990 had not yet reached the ORGI (GoI 1992b: 74). Yet the PPT suggested that the census was going to be held in the state later that year.

While the government failed to conduct a census, it is not the case that it did not have estimates of the population of J&K in the 1990s. The union home minister noted that 'in 1986–87 itself a house-to-house enumeration was done on the lines of decennial census.... It is proposed to use this data for preparing the projected population of J&K State in 1991' (Lok Sabha 1991: 54).

However, these data were most likely not used to arrive at projected population estimates for 1991. *The Digest of Statistics* published by the state government in the 1990s used figures 'as projected by the Standing Committee of Experts set up by GoI on Population Projections, October, 1989' (Government of Jammu and Kashmir [GoJK] 1993: 3). According to the 1991–92 digest, the estimated 1991 population of the state was 7,718,700 (ibid.: 3). The 1989 Standing Committee's estimate seems to have been based on the assumption that the state's population growth rate peaked in 1981. As per a parliamentary report, 'In 1991, census could not be held but according to approximate figures, there was population of 77 lakhs in 1991. In the Kashmir division there are 6 Districts having 40 lakh population. In Ladakh the population was 1 lakh 70 thousand and in Jammu it was 35 lakhs' (GoI 1995b: 5). This report does not identify the source of the projected figures. According to the Technical Group on Population Projections constituted by the Planning Commission, the state's population was 7.72 million in 1991, but it, too, does not indicate the source or method of projection (GoI 1996: 8–10). Table A-2 of the 2001 census estimated the 1991 population of the state to be 7,837,051 and noted that 'the population figures for 1991 of Jammu and Kashmir and its districts have been worked out by "Interpolation"'.

The 2001 census was conducted in J&K, but once again with a reference date that was different from that of the rest of the country. The agenda notes of the First Conference of the Directors of Census Operations noted, 'The reference date for the Census of India 2001 has been notified as 1st March, 2001 in all States and Union Territories except Jammu & Kashmir where the reference date will be … 1st October, 2000' (GoI 1999: 6). The DCO of J&K decided to conduct 'both Houselisting Operations and Population Enumerations in non-synchronous areas simultaneously in the month of September 2000' (GoI 2000c: 17–18; see also GoI 2000b: 9; GoI 2000d: 1). It is not clear whether the change in the reference date was necessitated by weather conditions or the potential threat of insurgency too played a role.[32] In any case, the field operations were eventually hindered by insurgency.

The 2001 PPT report says, 'Since there had been no census in 1991 in Jammu & Kashmir, there was a vacuum of last-time experience, expertise, documentary guidance (Administration Report etc.) coupled with an inept census hierarchy in the Directorate of Census Operations, J&K', and operations picked up 'momentum after taking over the charge of the Directorate by the Director of Census Operations on January 10, 2000 [that is, barely four months before the houselisting]' (GoI 2001b: 17). The PPT report notes the following on the threats issued by militant groups.

First phase of Census known as Houselisting had been carried out in Jammu & Kashmir alongwith [*sic*] the rest of India from May 16 to June 5, 2000 without much of hindrance or problem in spite of the threat call given by a militant out-fit against the conduct of census. Even the second phase went off well till September 4, 2000 when a threat call against conducting of census by a militant organisation, Hizbul Mujahideen, appeared in all the newspapers including national dailies[33].... Both census functionaries as well as the general public became scared of doing the census work and responding to the census questionnaire.... Thus while census was completed in Jammu Province barring a few pockets in Doda district [adjoining Kashmir] and Ladakh region, without much difficulty and fear, census process in Kashmir Valley remained disturbed. The enumerators faced with death threat from Hizbul Mujahideen asking them to stay away from the census operation, the date notified for completion of census had to be extended upto [*sic*] 15th December, 2000 in some of the places. (Ibid.: 16)

Confusion prevailed between 4 September (when 'dissociate with census or else' calls given by the Hizbul Mujahideen appeared in newspapers) and 21 September (when a high-level meeting was convened by the chief secretary). A meeting was held on 4 September to review the situation, and it was decided to go ahead with the exercise (*Times of India* 2000a). On 8 September, '[i]n view of open threats held out for the first time by the Hizbul Mujahideen and the Lashker-e-Toiba', Ashraf Qadiri, the president of State Employees Conference, told the press that he had 'directed the government employees not to take part in the census operation pending an executive meeting of the conference which will be convened within a day or two' and added that 'we will not allow government employees to become "qurbani ka bakra" (sacrificial lamb)' (*The Tribune* 2000b). Another employees' union, too, seems to have appealed against participation in the census (Rediff 2000a). However, as late as 10 September, the government appeared to be confident that it can roll out the exercise (*Hindustan Times* 2000). The next day the exercise failed to take off as very few employees reported for census duty, and the administration seemed to lack information on the areas that could not be covered on the first day (*Times of India* 2000b).

After being in a state of disarray 'for some days', the government stepped up to put the exercise back on track through a multipronged strategy. First, 'the Chief Secretary of the State, Mr. Ashok Jaitly ... declared in a State-level meeting that "come what may, the census in the State will be conducted"' (GoI 2001b: xii–xiii). His 'support and firmness ... was the main reason behind this success as it conveyed the message of Government's firm commitment to conduct the census' (ibid.). Second, Jaitly's message was reinforced by the registrar general. The 2001 PPT mentions the encouragement and motivation received by the employees by the visit of Registrar General, India, Mr. J.K. Banthia, to Srinagar Head Office, when he had a prolonged inter-action with the Census Employee's Association, wherein the employees assured the Registrar General that they were prepared to undertake any amount of risk to see the census-taking a success in the State, and they kept their word. (Ibid.)[34]

The ORGI also sent a team of four senior officials and one of its consultants who supported the DCO in Ladakh and a few districts of Jammu (ibid.).

Third, the 2001 PPT adds that

the team of officials comprising of S/Shri Abdul Aziz Andrabi and Nazir Ahmed Shah headed by Peer Bashir Ahmed, Investigator … visited each and every charge in the valley … to practically demonstrate to the enumerators that the threat perception should not come in the way of census-taking and if the enumerators have any kind of fear in their mind they will accompany them to such areas for enumeration…. Similarly, in Jammu Province also, the census staff visited the remote and hilly areas of Doda, Kathua, Rajauri and Punch districts to impart training and supervise the census operations. (Ibid.)

Fourth, the DCO told 'reluctant government officials … you will have to go to Assam on census duty' (interview, Jammu, 4 December 2019). The government also threatened to withhold salaries if employees failed to discharge their responsibilities (Rediff 2000a). Later news reports suggest that salaries were, indeed, withheld (*Kashmir Times* 2000a), but this also triggered protests by employees in, say, Pulwama against 'harassment by officials' (Ahmad 2000c). Last but not the least, the DCO 'told the people [that] your [Kashmiri] officers are doing census' (ibid.).

All these measures failed to calm the nerves of enumerators some of whom sought the approval of the Hurriyat Conference before resuming enumeration (*Kashmir Times* 2000i). In the end, despite 'places [like Baramula, Srinagar, Udhampur and Anantnag] where some difficult situations were either prevalent or caused', the exercise

succeeded to the extent that the employees who performed the census duty were with firm conviction and determination that they are going to participate in a national task, though at times, it was not without an element of risk to their lives, but they stood firm, and at the end their endeavour came to fruition. (GoI 2001b: xii)

The DCO later claimed that '[t]he 2001 Census was the best till then. Earlier enumerators filled in forms in village head's house. [In 2001] [e]very house was numbered, even in downtown Srinagar' (interview, Jammu, 4 December 2019 and 22 October 2021).

While the PPT report offers an unusually frank assessment of the problems faced during enumeration, it is silent on the impact of the disruptions on data quality. The exercise repeatedly came under attack, and enumeration had to be halted even after the extension of the reference date. For instance, on 25 October, the Hizbul Mujahideen raided several census offices across Kashmir and destroyed census material (Rediff 2000a). A week later, enumerators were still complaining that destruction of census records continued in many places (*Kashmir Times* 2000i). The state government managed to pat itself on the back for the completion of 'smooth' census operations only in mid-January 2001 – that is, three and a half months after the initially announced reference date (Joshi 2001). Census reports do not offer any estimate of the data lost due to the destruction of material or the method through which the gaps were filled in. It is not clear how the ad hoc changes in the reference date and different reference dates for different districts affected the accounting of the mobile population. Several news reports suggest that the mobile population was inadequately covered (*Kashmir Times* 2000j). In fact, there was considerable confusion even regarding the enumeration of the non-mobile population. In the last week of October, a principal census officer for Srinagar municipality claimed the completion of enumeration, while the DCO denied the same (*Kashmir Times* 2000d). Government officers who failed to report for census work received notices and a few teachers were even arrested in Tangmarg. The DCO was not aware of the notices, while a principal census officer suggested that notices were not needed as enumeration had been completed (ibid.). Indeed, as the DCO suggested, enumeration was not yet complete and continued until December.

Interestingly, almost all senior officials associated with the 2011 census in J&K suggested that the 2001 census could not be conducted properly (interviews, census official, Srinagar, 30 November 2015; retired census official, 28 July 2021 and 30 July 2021; former registrar general of India [RGI], 6 August 2021). Several other sources, too, question the reliability of the 2001 census. According to Bhat (2018: 180), militants 'opposed the conduct of census in 2001 and issued threats to the staff participating in census enumeration. This created a sense of insecurity both among the enumerating staff as well as the general public, and consequently, *lot of compromises were made in the conduct of Census 2001*' (emphasis added).[35] A Kashmiri journalist told the author that the

2001 [census] was not up to the mark. Militants had imposed a ban on census. Government employees did not go to houses. They filled in the form on their own at home. No one would risk his life. As there was no census in 1991 so they had to do in 2001. (Interview, Srinagar, 23 May 2016)

Schofield (2003: 231) suggests, 'Although the Indian authorities eventually announced that the census was completed, for a number of areas figures were reportedly based on assumptions, since the enumerators did not dare to move door to door because of the militant ban.' Restriction on free movement is a common problem identified by most observers. Indeed, contemporary news reports point out that the government had asked enumerators to approach households in a 'clandestine manner without giving the slightest impression that they were enumerators' (*Kashmir Times* 2000d, 2000i).

Interestingly, Jammu, too, did not want census in 2001 because of concerns about the quality of data collected amidst conflict. Interviewees from Jammu pointed out that members of the Bharatiya Janata Party (BJP) had cautioned the union government against conducting a census under the shadow of militancy, but the MHA, headed by Advani, was adamant (interview, Jammu, 2 June 2016). The 2001 PPT admits that the government at the highest level was determined to conduct census irrespective of the difficulties.[36]

It seems that in the second half of the 1990s, after an armed insurgency supported by cross-border terrorists had shaken the government, conducting elections and censuses became essential to reassert the authority of the state and more than that recreate a sense of normalcy in front of both the domestic and international audiences. After the completion of census, the then chief secretary of the state pointed out that this 'big achievement' followed successful *panchayat* polls (Joshi 2001). In other words, as pointed out in Chapter 1, the union government was impatient to resume censuses and elections that are among the routines associated with a modern state. In this context, the availability of census data was an end, with quality being of secondary importance. To appear to be modern, states must provide a statistical veneer to their self-description as well as policy choices. As a result, in most developing countries the optics of making data available override concerns about the usability of poor-quality data.[37]

A decade later, the 2011 census was affected by a natural calamity in Ladakh[38] and political unrest in Kashmir.[39] Unlike in 2001, the 2011 PPT is largely silent on the communal politics that affected the household phase in Kashmir, even though it notes that disturbances erupted in [the] Valley in the month of May/June resulting in turmoil in the Valley and few pockets of Jammu province. This continued till Oct/Nov 2010 and the period saw never ending Bandhs, Curfew restriction etc. making the public movement almost impossible. In the prevailing situation, the state government sought extension for completing the house listing operation up to 31.08.2010. As the situation did not show any sign of improvement another extension was sought up to 30.09.2010. The ORGI readily acceded to State Govt. request. The Directorate made all efforts to persuade field functionaries to complete the task and succeeded. (GoI 2011d: 1)

A crucial point is missing in this account. After the completion of the 2011 census, several commentators expressed satisfaction that pro-independence groups did not obstruct the exercise.[40] An opinion writer, for instance, observed that the 'positive point of this census exercise was that most extreme forces within our social and political system did not object to it and gave their nod for smooth conduct of this crucial exercise of head count and other parameters' (Iqbal 2011). However, these groups did not merely refrain from obstructing enumeration; rather, they mobilised respondents along communal lines. As per a news report:

The separatist leaders including Syed Ali Shah Geelani and Mirwaiz Umar Farooq had urged people to support the census staff and register themselves after they expressed their concern of changing the demography of Muslim majority state into a minority. The separatist[s] also alleged that the previous figures were distorted after most of the people in Kashmir remain abstain [*sic*] from the census 2001. (S. Yasir 2011)

Bhat (2018: 181) pointed out that 'militant organizations ... requested all Muslims of the State to get enumerated and the Census staff of the Valley was asked to fulfill all obligations towards its community, so that the Muslim-majority character of the state gets reflected in the Census 2011'. In response to the apprehension that the exercise could be used to alter the demography, Chief Minister Omar Abdullah told his audience at a rally in Rajouri that 'some people are creating noise on matters which have nothing to do with the issue of Kashmir or demographic character of our state' (Jameel 2010b). A retired civil servant suggested that in 2011, 'the insurgency and Muslim fundamentalism could quite possibly have affected the collection and compilation of data – not because the data collection would have been physically restricted but because of the biases affecting the collectors and compilers of data' (interview, 23 February 2016).

To conclude, most censuses in J&K have suffered one or the other problem that makes inter-temporal comparisons difficult.[41] Jammu, Kathua, Samba and

Udhampur are the only districts where the census was not hindered. Census data for the state are likely to be most reliable and inter-temporally comparable for these districts. This is also true of other government surveys. Jammu's 'Mountainous' region (Jammu, Kathua and Samba districts) is the only NSS region of the state for which we have an uninterrupted data series (Agrawal and Kumar 2017b). Except for a few rounds, Ladakh was not covered by the NSS until the 63rd round (2006–07), when it was included in the state sample. In the 69th round (2012), it was also included in the central sample. Ladakh was covered by the Urban Frame Survey for the first time in the 2007–12 phase (GoI n.d.9). Insurgency disrupted the NSS across Kashmir (Jhelum Valley) and the adjoining districts of Jammu (Outer Hills) that were irregularly or incompletely surveyed starting with the 45th round (1989–90), even though they were included in the sampling frame (Agrawal and Kumar 2017b: table 1).[42] The SRS did not provide results for J&K in the 1990s 'due to part-receipt of returns' as Kashmir could not be covered (see, for instance, *SRS Bulletin*, vol. 33, nos. 1–2; vol. 34, no. 1–2; and vol. 35, no. 1; see also GoI 1992b: 80). More recently, the SRS excluded the whole of 'the Jhelam Valley Natural Division' – that is, Kashmir in 2015 (vol. 50, no. 2) and 'the Ladakh Valley Natural Division' in 2016 (vol. 51, no. 1). The first round of the National Family Health Survey (NFHS) (1992–93) was restricted to the Jammu division. In the second round of the NFHS (1998–99), selected sampling units of Kargil district were replaced due to 'national security reasons' (S. Kulkarni 2004: 656) as the fieldwork overlapped with the Kargil War.[43]

Given the gaps in metadata on census, variations in the coverage of surveys such as the NSS can offer additional information on field conditions. During 1991–2012 (47th–68th NSS rounds), some of the quinquennial rounds of the NSS could not cover as much as 74 per cent of the state's population, including the entire population of Kashmir. The state that was home to barely 1 per cent of the country's population accounted for more than 51 (67) per cent of the un-surveyed rural (urban) first-stage units (FSUs) in the country during this period (Agrawal and Kumar 2020a: 258). Note that contrary to the experience of other parts of the country, Kashmir's urban areas suffered greater non-coverage in the NSS than its rural areas. Also, in 2001, the non-coverage was higher than during 1990–92 (Figure 2.1), which is explained by the spread of insurgency beyond Kashmir to the adjoining districts of Jammu.[44] The non-coverage spiked again in the run-up to the 2011 census, however, in Doda, Rajauri and Punch rather than in Kashmir. So the NSSO reports suggest that census enumerators, too, must have faced difficult field conditions, at least, in Kashmir and adjoining districts of Jammu in 2000–01 and in Doda, Rajauri and Punch in 2010–11. However, census reports are silent regarding the latter. The non-coverage in NSSO surveys

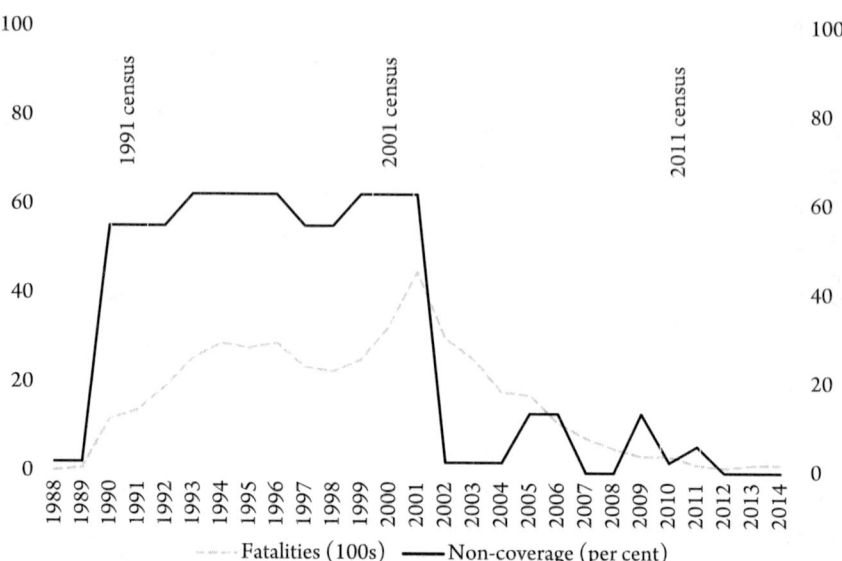

**Figure 2.1**  Armed conflict and non-coverage in the National Sample Surveys (NSS) in Jammu and Kashmir (J&K)

*Source*: Agrawal and Kumar (2017b: table 1) (non-coverage); South Asia Terrorism Portal, http://www.satp.org (accessed in December 2022) (fatalities).

*Note*: The annual fatalities (in 100s), indicated on the left axis, include civilian, security-force and terrorist or insurgent fatalities. The corresponding degree of non-coverage (in per cent) is indicated on the right axis.

also throws light on a few issues raised earlier in this chapter. The 1991 census was conducted in Punjab, where unlike J&K sample non-coverage was very low in NSSO surveys. It was not conducted even in the peaceful districts of Jammu, which were regularly covered by the NSSO (and also the SRS) surveys in the early 1990s, possibly because the general administration that is responsible for the census was tied down, among other things, to relief operations to deal with the unprecedented influx from Kashmir. In contrast, the NSSO surveys are conducted by a dedicated team that is not burdened by any administrative responsibility.

The data on assembly and parliamentary elections, too, show a marked drop in voting rates during the peak of insurgent violence (Figure 2.2a), which is in agreement with the spike in non-coverage in the NSS during this period. Further, the decrease in voting rates was driven by Kashmir (Figure 2.2b). In fact, election data throw further light on variations in the intensity of political unrest within Kashmir. In the elections around 2001, Srinagar, the summer capital of the state and divisional headquarter of Kashmir, reported some of the lowest voting rates,

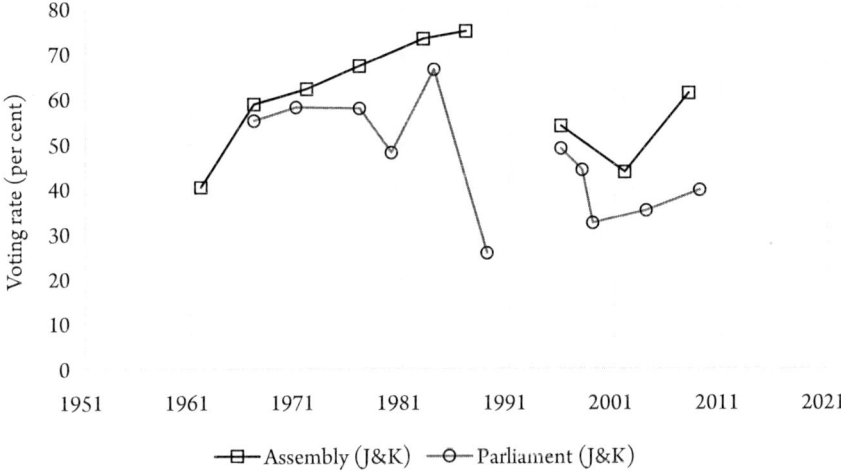

**Figure 2.2a** Voting rate in elections in Jammu and Kashmir (J&K), 1962–2009

*Source*: Prepared using statistical reports of the Election Commission of India (ECI).

*Note*: The report for the 1967 parliamentary elections mentions two different voting rates for J&K. The voting rate reported here is calculated using the figure for the electorate that is closer to the electorate of assembly elections held in the same year and for which the decomposition by sex is available.

which agrees with the higher non-coverage in the NSS in urban Kashmir. The information on the reach of NSS and elections corroborates the assessment of field conditions reported in the PPT of the 2001 census.

Yet another way in which conflict affects J&K's census relates to the accounting of armed forces. In the past, the *provisional* population estimates of the state did not adequately account for the armed forces. The 1961 administrative report highlights the communication gap between the armed forces and civilian authorities responsible for conducting census:

> In spite of all these precautions and advance arrangements, the provisional totals were reported by the District Officers on varying dates between 7th March, 1961 and 10th March, 1961. The delay was caused due to the non-availability of population figures of non-combatants of various army sectors, which did not, in spite of telegraphic and telephonic reminders issued both by this office and the District Census Officers, pay sufficient attention to this important work. This was brought to the notice of the higher army authorities and also reported to the Registrar General.... Anyhow, except for Headquarter Fox Trot Sector and 161 Infantry Brigade, Baramulla, which supplied the requisite data as late as April, 1961, the population figures of all areas, accessible and

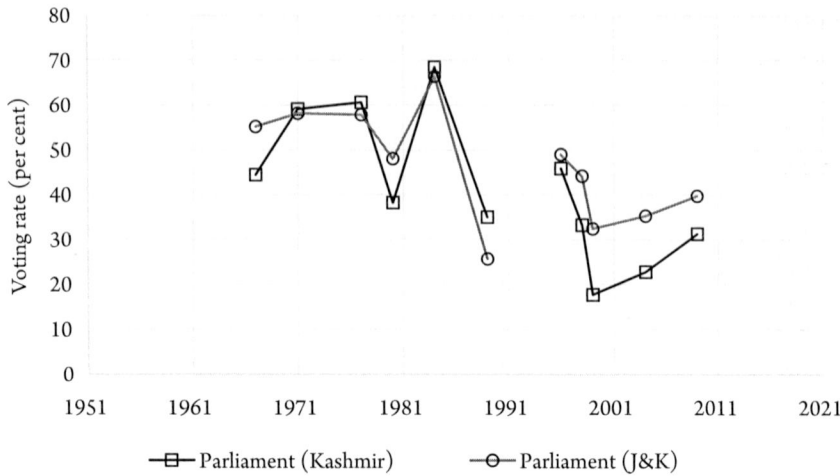

**Figure 2.2b**  Voting rate in parliamentary elections in the Kashmir division and Jammu and Kashmir (J&K), 1967–2009

*Source*: Prepared using statistical reports of the Election Commission of India (ECI).

*Note*: The report for the 1967 parliamentary elections mentions two different voting rates for J&K. The voting rate reported here is calculated using the figure for the electorate that is closer to the electorate of assembly elections held in the same year and for which the decomposition by sex is available.

> snowbound, whether inhabited by the civilians or non-combatants of the army were reported to the Registrar General by 10th March, 1961. The provisional totals, except those of the two army installations referred to above, which were fairly small, were released in a press conference held in my office on the evening of 13th March, 1961. (GoI 1963a: 40–41)

While later reports are silent in this regard, the communication gap has persisted, if not grown, over the decades due to the involvement of armed forces in counter-insurgency. In addition to late communication, there are also other ways in which armed forces affect census. First, their headcount is redistributed across the state so as to make it difficult to infer troop deployment patterns from the published data.[45] This does not help much in the very sparsely populated Ladakh, though, as the deployment reflects in the skewed sex ratio due to the inclusion of the predominantly male armed forces in the headcount.[46]

Second, it seems the questions on migration are not consistently canvassed in the case of military personnel. There is a lack of agreement between census reports in this regard. *The District Census Handbook* for Jammu for 1961, for instance, noted that 'the army personnel have been enumerated in the States where they were born irrespective of their being present there at the time of Census' (GoI 1966d: 4), which does not agree with GoI (1963a: 40-41) on the accounting of armed forces. More recent census publications are also ambivalent in this regard. The training manual for the 2011 census cautions enumerators that

> while listing the persons to be enumerated in the household, you will have to be extra cautious regarding the enumeration of persons who have already left their normal place of residence for studies or employment, etc.... Do not also include persons serving in the armed forces, etc., and living outside the place of enumeration ... even if their name continue to appear in the voters list of the area or in the ration card possessed by the household. (GoI n.d.14: 37–38)

However, the manual also instructs that questions related to '[t]ravel to place of work', '[p]lace of birth' and '[p]lace of last residence' were not supposed to be canvassed in the case of 'defence and similar paramilitary personnel' (ibid.: 82, 84, 90). Similar restrictions did not apply in the case of 'police lines', which are also treated as 'special charges' for the purpose of enumeration.

Third, there is a lack of clarity regarding the coverage of military camps in the houselisting phase of census. *The District Census Handbook* for 2011 notes that 'certain areas where the access of the civilian enumerators was not permissible due to security reasons termed as "Special Charges" such as the Defence and strictly Military/Para-Military areas, including operational areas were also covered. Such areas were not covered during the Houselisting & Housing Census' (GoI 2014e: 17). In short, there are differences between censuses, possibly even across districts of the state, with respect to the accounting of the armed forces.

## Mobile Populations

The administrative report of the 1961 census noted that 'the enumeration of mobile population [in J&K] is not as easy as elsewhere. We had to keep a regular track of the movements of migratory population to guard against double or non-enumeration' (GoI 1963a: 36, 92–96). There were four classes of mobile population in the state – nomads, boat population, forest labour and government officials – each posing a different kind of challenge. Changes in the usual reference date can alter the location of enumeration of mobile population and affect the population shares of the divisions.

## Nomadic Population

The nomadic communities constitute the largest class of the mobile population. They span not only more than one division of the state, but also both the snowbound and accessible areas. According to the 1961 administrative report:

> Nomad *tribes* like backerwals [*sic*] and shepherds usually move for summer from plains to the hills, where grazing facilities are available. They return to their *permanent* places of residence [that is, in Jammu] towards the end of September or beginning of October. (Ibid.: 36, emphasis added)

It is noteworthy that the Bakarwals are referred to as a 'nomad tribe' in the aforementioned report, even though they were officially recognised as an ST only in 1991,[47] and Jammu is identified as their 'permanent' place of residence.[48]

The census admits the possibility of coverage errors in the case of 'nomad tribes' due to 'some of them not being counted at all or being counted twice' (GoI 1971b: 183; see also GoI 1969: 38; GoI 1985b: 112). Therefore, elaborate arrangements were made to deal with the movement of people between snowbound areas and other parts of the state that are enumerated in different periods. The 1961 administrative report provides an extended discussion in this regard:

> The sector of the population consisting of bakerwals and shepherds[49] had to be divided into two categories; viz., those who were present at the time of enumeration in snowbound areas and others who had moved to places at lower heights in the accessible parts of the State. The former were enumerated in separate blocks outside the Revenue area and were given tokens certifying that they had been enumerated and should not, on their return to their *permanent* places of residence, be subjected to a fresh enumeration. The Charge Superintendents or accessible areas were particularly asked to verify before undertaking the enumeration whether any families had returned from hilly areas and to ascertain from them whether they had been already enumerated. As a precautionary measure to guard against the misstatements to be made by such people, the enumerators in snowbound areas were required to record on the reverse of the grazing permit of each nomad family that it had been enumerated. The Charge Superintendent of the accessible area was required to inspect this permit and to verify the statement whether or not any enumeration had been carried out previously. (GoI 1963a: 36, emphasis added)[50]

Likewise, precautions were taken to ensure that '[p]eople entering the snowbound areas after 1st October 1960 were enumerated at the check post by the officer incharge who was supplied with an adequate number of blank Enumeration Slip pads for this purpose' (GoI 1963a: 35). Similar arrangements were made in 1971 to enumerate the mobile population, with certificates of enumeration being issued to those moving out of snowbound areas after enumeration.

> Everyone present in these blocks was to be counted in the normal course along with the inhabitants of other non-accessible areas of the charge but anyone so counted and moving out from a forest block was to be issued a Certificate of Enumeration to safeguard against the possibility of he or she being counted elsewhere again. The entry under 'last place of residence' in the Individual Slip served as an additional check against double enumeration of any such individual. (GoI 1971b: 12)

The certificate was in addition to the entry 'made to this effect in their grazing permits which the enumerators in synchronous areas were instructed to examine before taking a count of any nomadic family in these areas' (GoI 1976b: 14; GoI 1971a: 1). A later census report also throws light on check posts where the entries were supposed to be made.

> Check posts should be set up from 1-10-1980 to 1-5-1981 on all recognised routes at the points of entry/exist from accessible to non-accessible areas. Where forest check-posts already exist, these should be utilised and wherever necessary additional check posts manned by revenue/forest officials should be set up. (GoI 1985b: 113)

This system designed to avoid double counting does not seem to have been sustained in later censuses.[51] It is also not clear how effective the arrangements were even earlier because the road network was not good and nomads often used shorter routes rather than 'recognised routes', where they would have met officials. Community leaders argue that their population was under-enumerated due to the non-availability of check posts on most routes. In fact, a 1961 census report noted that in 'Jammu and Kashmir there are already Forest checkposts which will be utilised as checkposts for Census also' and added that it was not 'possible for the Census organisation to establish any checkpost as *they have no jurisdiction in the State territories* for establishment of such checkposts' (GoI 1969: 38, emphasis added). The legal and logistical difficulties in setting up additional check posts were compounded by ad hoc changes to reference dates.

The problem of enumerating the nomadic population was addressed through two additional measures that stressed community involvement. First, '[i]n the selection of enumerators and supervisors, preference' was 'exercised in favour of officials belonging to [G]ujjar and [B]akarwal community and the teaching staff of the mobile schools moving with the nomads' (GoI 1985b: 113). But mobile schools were introduced in more recent decades, and their reach has varied across censuses. Second, 'consultations' were 'held with the local representatives of [G]ujjar and [B]akarwal community and their advice and cooperation sought in making the census count a complete success' (ibid.). A further innovation seems to have been introduced in 2001, when 'for the first time in the history of J&K Census 883 number of Bahaks [pastures] were demarcated into enumeration blocks in highland pastures' (GoI 2001b: xi).

Despite these arrangements, concerns have been voiced regarding the under-enumeration of the community, some of which have been recorded in official reports as well. The 1971 administrative report noted that

> some apprehensions about non-enumeration of nomadic population had been expressed by the now banned Plebiscite Front in a representation made to the Chief Secretary. However, after its Secretary had seen for himself the enumeration of [B]akarwals and [G]ujjars in Bhaderwah tehsil, he did not hesitate to acknowledge the efforts made by the Census Department in this behalf, publicly. (GoI 1971b: 13)

In the 2001 census, the first to be conducted after the recognition of Gujjars and Bakarwals as STs, the community approached the chief minister, and the DCO had to introduce new measures.

> Their main concern was that during summer months, they leave with their livestock in highland pastures where no enumerator approaches. Secondly, by the end of September, they will be mostly in move for their winter habitats. Therefore, they apprehend that their moving population will be left out.... A few suggestions made by them were agreed to ... with the result that for the first time ... Bahaks were demarcated into enumeration blocks in highland pastures. (GoI 2001b: xi)

Such neat ex-post accounts of successful resolution of problems often paper over a messy ground reality. The 1981 GPT inadvertently throws light on the confusion and resource constraint on the ground in J&K.

> If there is likely to be a very large number of houseless persons in your jurisdiction whom you may not be able enumerate single-handed in one night, you should report to your Supervisor in advance.... You should keep particular watch on the large settlements of nomadic population who are likely to camp on the outskirts of the village. These people will have to be covered on the

night of February 28, 1981. You should of course make sure that these persons have not been enumerated elsewhere. (GoI 1983a: 50)[52]

This adds another layer of confusion to the accounting of 'nomadic population' by suggesting that they are recorded as 'houseless' in Jammu, where they would be in the month of February. This fits oddly with other census reports. First, the houseless population was always a very small fraction of the population of the state, whereas the nomadic tribal community is the second-largest tribe in the state.[53] Second, earlier censuses report that their 'permanent places of residence' were in the Jammu division (GoI 1963a: 36), which was 'their normal abode' or 'their usual abode' (GoI 1981a: iii, 8).

The more recent census reports are largely silent about the enumeration of nomadic communities, even as Gujjar–Bakarwal activists 'expressed disappointment over the exclusion of nomadic community in National Population Register in first phase [of census] held between June and September 2010 in the state as all migratory tribes had migrated to upper reaches of Himalayas at that point of time' and demanded 'a special census of around six lakh nomadic, semi-nomadic and shelter-less Gujjars–Bakerwals and other migratory communities who had migrated to plains and warmer areas of state as part of annual migration' (*Kashmir Times* 2011a; see also GoI 2011d: 75).[54] They also demanded the introduction of a separate category – 'nomadic' or 'shelter-less' (*Kashmir Times* 2011a) – and the revision of the reference date of the census.[55] The possible double counting or omission of the part of the community that migrates out of the state in winters has not received attention from census officials and community activists.

## Boat Population

The boat population is the only category of the mobile population that is confined to just one division of the state and therefore has no impact on the distribution of population between Kashmir and Jammu. According to the 1961 administrative report, the boat population was spread between Srinagar and Bandipore.

> Boat population is chiefly found in Kashmir province except for a few cases in other parts of the State, where operators of ferries live in boats or in tents along-side the ferries…. In Kashmir province, the boat population is to be found throughout the navigable portion of the river Jhelum, in the Dal, Wullar, Anchar and Manasbal lakes and in the various creeks and canals connecting the lakes with the river. (GoI 1963a: 36–37)

Despite the small population, this category was the subject of elaborate discussions in earlier census reports, possibly because of its importance for the tourism sector.[56] The riverine territory of the boat population was first apportioned between administrative units as follows.

> Where a waterway forms the boundary between two tehsils or districts, those living in boats should be included with that tehsil with which the bank of the river to which the boat is tied is located. The boats found midstream should be tied to one bank of the river. The District Census Officers will take appropriate Steps to ensure that the boundaries of their jurisdictions over waterways are clearly defined and understood by the census staff. (Ibid.: 37)

The boat population was further mapped to land and boats according to differences in sleeping habits.

> Those found in boats but who actually reside on land should not be enumerated as boat population and their enumeration should be taken up by the Enumerator of the block in which they reside. In any case, special arrangements will have to be made in such cases to guard against the possibility of double enumeration or non-enumeration. This can be done by appointing special enumerators for recognized ghats and places, which should be attached to the General Village and Town Register. (Ibid.)

The administration provided 'boats for Census Field Agency' to ensure proper counting of 'population living on the fringes and islets of Dal Lake and other water-bound areas' (GoI 1971b: 22). Arrangements made to avoid the double counting of the boat population replicated the system of checks for the nomadic population: 'Passes should be issued by each Enumerator to the head of the family occupying a boat to the effect that the household has been enumerated so that if the boat moves from one ghat to another, the possibility of double enumeration may be avoided' (GoI 1963a: 37).

The elaborate conceptual demarcation of the boat population, however, did not guarantee a neat accounting. So the government had to take comfort in the fact that this population was likely to be small: 'As all these areas fall within the domain of accessible territory, the boat population at the time of enumeration, i.e., from 10th February, 1961 upto the sunrise of 1st March, 1961 will be at its minimum owing to climatic conditions obtaining during that part of the year' (ibid.: 36–37). The size of boat population must have been larger in 1971 and, more so, in 1981, when the household phase of census had to be delayed in accessible areas due to elections and bad weather, respectively.

This could also be true of the 2001 census that was scheduled in September 2000, but enumeration was delayed in Srinagar that accounts for a large proportion of the boat population.

### Forest Labour and Government Employees

The third category of mobile population included labour employed in forest blocks including in snowbound areas. The 1961 administrative report noted that this category overlapped with that of houseless population that

> mostly consisted of the labour employed by forest lessees in connection with the felling and other operations in remote forest areas.... almost the entire houseless population had to be enumerated in the various districts of Jammu province where, unlike Kashmir, climatic conditions were tolerably good at the time of census and work in the forests had not been suspended. (Ibid.: 36)

The last category of mobile population was related to the secretariat of the state government that moved between Srinagar and Jammu as part of a century-and-a-half-old Dogra tradition of 'Darbar Move'. Select offices moved from Srinagar to Jammu by November and returned to Srinagar by the end of April. The DCO for the 2001 census, Feroze Ahmed, suggested that 'over 15,000 employees and their families' relocated with the Darbar in the 1960s and the 1970s (Swami 2000a). But as discussed in Chapter 4, the flow of population as part of the Darbar Move was asymmetric after the late 1980s as more people travelled from Kashmir to Jammu than the other way round. Various sources suggest that about 10,000 officials moved with the Darbar in recent years (Chhibber 2020, Sidiq 2020), but this figure does not account for contractors, politicians and informal services providers and family members who moved along with the Darbar (see, for instance, GoI 1966d: 32). The tradition came to an end in July 2021 (Ashiq 2021), months after being deferred on grounds of the Covid-19 pandemic (*The Hindu* 2021). So the impact of the Darbar Move on the census varied between 1961–81 and 2001–11, and within each period the impact was contingent on the reference date, festival calendar and idiosyncrasies of weather.

# Forced Migration

International conflicts and armed insurgency that trigger forced migration and, also, limit the reach of the census present another challenge. In 1961, enumerators faced difficulty in directly accessing villages close to the border with Pakistan.

There are at least 10 to 12 villages situated between the Forward Defence Line and the Cease-fire Line where, due to security reasons, the Enumerators could not move about so freely as elsewhere in the State. In such cases, the District Census Officers and Charge Superintendents had to collect the data through the Lumberdars or the Head Village-man, but it was not possible to make any verification whether they had enumerated all persons in each such village. (GoI 1963a: 38)[57]

The problem has persisted over the years because of cross-border firing and infiltration, which have resulted in the securitisation of the border that restricts the reach of the administration[58] and the displacement of the settled population in some areas. Four wars with Pakistan have also caused extensive displacement.

Conflict-induced international migration, too, affects census.[59] The state received refugees from Pakistan-occupied territories (mostly in Jammu), West Punjab (in Jammu) and Xinjiang and Tibet (in Ladakh and Kashmir) before the 1961 census. The 1961 GPT for J&K notes that '[w]hile the enumeration work was in progress in the month of February, a large number of Central Asian Refugees [Khampas] entered Ladakh and encamped in various places in tehsil Leh. They were treated as houseless persons and enumerated along with the rest of the population of the district' (GoI 1964b: ix; see also GoI 1963a: 36).[60] The 'internal' migration from Pakistan-occupied territories, too, posed a challenge for enumerators.

Some difficulty was also experienced in respect of Question 4 (a), Birth Place. A number of persons declared their place of birth one of the areas now on the other side of the Cease-fire Line. *No instructions had been issued to the Enumerators in respect of such cases.* As soon as instances of this kind came to notice, telegraphic instructions were issued to all the Charge Superintendents, asking them to instruct the Enumerators to record against Question 4 (a) [Birthplace] the name of the district given by the person enumerated to be followed by the words and symbols (Jammu and Kashmir). (GoI 1963a: 38, emphasis added)

Further, certain villages were added to the state after the 1971 war – that is, after the 1971 census. These include villages added to Leh (Turtok, Chulungkha, Taksi and Thanga), Kargil (Bodagam, Hundermon, Hundermon Brok and Buzber), Punch (Nakarkote and Titri) and Kupwara (Gasla) (GoI 1983a: 85). But a substantial part of the population of these villages did not relocate to the Pakistan-occupied territory. On the other hand, 41 full and three half villages of

Akhnoor went to Pakistan, but their population seems to have relocated to other parts of the Jammu division.

Kashmiri Pandits had to abandon the Muslim-majority Valley in the early 1990s due to sustained communally motivated attacks and relocate to Jammu and other states of the country. The community complained during the 2001 and 2011 censuses that it was not properly enumerated (GoI 2001b: 29, 2011d: 75). A memorandum submitted by Panun Kashmir to the union government argued that '[the] Census 2001 cannot be expected to be correct in respect of exiled Kashmiri Pandits' and sought changes in the design of the census questionnaire to provide the community with 'an opportunity to link themselves to their roots in Kashmir' (GoI 2001b: 29).

Nearly a week before the completion of the 2011 Census, the All Parties Sikh Coordination Committee (APSCC) complained that 'the present Census in Kashmir' was 'inaccurate and infamous as actual figures and facts cannot be ascertained due to turmoil' (*Kashmir Times* 2011b). The APSCC added that

> this is the only state in India where maximum number of youth are out of their state either for jobs, education, business etc. Unfortunately it's the only state in world where maximum numbers of people are untraced.... 80 percent of minority communities, the Sikhs in Kashmir and border areas, have made temporary migration from village to city for security, employment and for education purposes. This temporary migration of Sikhs has not been included in the past Census – like in the case of Kashmiri Pandits (Migrants) which is the open violation of Human and Constitutional Rights given to every citizen. (Ibid.)

The APSCC demanded 'a special cell ... to pen down the correct population of Sikhs living in village and also those who have moved due to turmoil of 1989 from villages to cities' (ibid.).

## Shifting Reference Dates

Pre-1947 censuses divided J&K into three parts: synchronous areas that followed the normal census calendar; non-synchronous areas that were further divided into 'difficult and dangerous localities such, as snowy mountains, impenetrable forests and turbulent mountain streams and Nallahs'; and 'regions, called the remote and distant areas whence for want of communications and speedy means of transport the totals cannot be brought to the tahsil head-quarters or the nearest telegraph office except after a journey of several days by the fastest means' (GoJK 1933c: 17). Later censuses divided J&K into snowbound areas,[61] which are inaccessible in February and March when household enumeration is carried out in the rest of the country, and areas that are accessible throughout the year (Table 2.2). Kargil and Leh in Ladakh were the only entirely snowbound districts. Karnah in Kupwara and Gurez in Bandipora were only entirely snowbound tehsils in Kashmir.

The Household Census Schedule is canvassed in two rounds in the state. Except in 1981 and 2001, census was conducted in the accessible areas as per the schedule followed in the rest of the country. In the snowbound areas, the household phase of census was conducted immediately after the houselisting phase in the year before the reference date for the household phase in the rest of the country (Table 2.3). While the recent census reports do not clearly identify the actual schedule of houselisting in snowbound areas, field operations begin in these areas later than in the rest of the state (GoI 2011d: 73; *The Tribune* 2000a). This means that there is lesser time for training in the snowbound areas. Also, the duration within which the two stages must be completed is lesser than suggested by the nominal schedule and the two stages may have to be carried out simultaneously in some of the snowbound areas (GoI 2000c: 17–18).

Snowbound areas pose several challenges for maintaining the quality of the census.[62] First, the enumeration has to be conducted in a much shorter time frame that strains the usual checks and balances. The 1961 administrative report notes that 'certain operations had to be conducted … in quick succession and even simultaneously' (GoI 1963a: 14; see also GoI 2000c: 17–18). Second, the census needs to account for the population that moves between snowbound areas and accessible areas after houselisting is completed but before household enumeration begins (GoI 1969: 38). Third, the snowbound areas were 'excluded from the purview of the scheme [PEC], as they were not accessible during March–April, 1961' (GoI 1963a: 41). It is not clear if post-enumeration surveys (PES) covered the snowbound areas in 2001 and 2011. Fourth, during the earlier censuses the government did not have a permanent presence in all snowbound areas.[63] So enumerators found it difficult to access the remote villages. A. Bose (2003: 2933), for instance, recalled that in the 1971 census, 'census enumerators (mostly poor primary school teachers) had complained to me that they could reach remote villages in the mountains only on mule-back but were not given any "mule allowance"'. Last but not least, the process of identification of non-synchronous areas is not clear, and the changes in the distribution of snowbound villages across censuses are not explained. The particularly sharp increase in the number of snowbound villages between 1961 and 1971, even as the total number

**Table 2.2** Snowbound villages of Jammu and Kashmir (J&K), 1961–81

| Census | Number of inhabited (uninhabited) villages/towns | | | | Snowbound villages | | | |
|---|---|---|---|---|---|---|---|---|
| | J&K | Kashmir | Jammu | Ladakh | J&K | Kashmir | Jammu | Ladakh |
| 1961 | 6,659 (167)/43 | 2,922 (81)/18 | 3,400 (85)/24 | 237 (1)/1 | 648 | 85 | 324 | 239 (including one town) |
| 1971 | 6,503* (239)/45 | 2,874 (87)/17 | 3,394 (149)/26 | 235 (3)/2 | 782 | 121 (including one NA) | 421 | 240 (including two NA) |
| 1981 | 6,477 (281)/58 | 2,816 (83)/24 | 3,422 (195)/32 | 239 (3)/2 | 789 | 119 (including one town) | 426 | 244 (including two towns)** |

*Source:* GoI (1968a: 39, 1976b: 10, 1985b: 54–68) and general population totals (GPT) reports of the respective years.

*Note:* (*a*) Ladakh is entirely snowbound. (*b*) In 1971, the administrative report and the general report provided two different counts of snowbound villages. The figures reported in the table are based on the general report. The corresponding figures from the administrative report are 117 (Kashmir), 404 (Jammu) and 238 (Ladakh) (GoI 1971b: 30–31). (*c*) *Excludes 7 inhabited villages (Srinagar district 4 and Jammu district 3) which form the part of Urban Agglomeration of Srinagar and Jammu respectively' (GoI 1972a: 28). (*d*) **Leh town included nine villages in 1981. (*e*) Snowbound villages included eight and ten forest blocks in 1971 and 1981, respectively. (*f*) Information on snowbound areas is not available in the reports of the subsequent censuses. (*g*) NA stands for Notified Area.

**Table 2.3**    Census calendar for Jammu and Kashmir (J&K), 1961–2011

| Census | Enumeration + revisional round* | | Elections | Remarks |
|---|---|---|---|---|
| | Snowbound areas | Accessible areas | | |
| 1961 | 10–30 September 1960 + 1–3 October 1960 | 10–28 February 1961 + 1–5 March 1961 | | Revisional round extended in accessible areas due to Holi |
| 1971 | 12–30 September 1970 + 1–3 October 1970 | 10–31 March 1971 + 1–3 April 1971 | Parliamentary elections: 4 March 1971 | Reference date changed in accessible areas due to elections |
| 1981 | 11–30 September 1980 + 1–5 October 1980 | 20 April–05 May 1981 + 6–10 May 1981 | Parliamentary elections: 3 January 1980 for Kashmir/ Ladakh; 6 January 1980 for Jammu | Reference date changed in accessible areas 'due to the unfavourable weather conditions … in February 1981' (GoI 1985c: 3) |
| | | 1–10 April 1981 | | For nomadic population in the plains due to bad weather in snowbound areas in September 1980 |
| 1986–87 | 19–30 September 1986 + 1–3 October 1986 | 11–20 June 1987 + 21–23 June 1987 | Assembly elections: 23 March 1987 | Only Ladakh treated as snowbound area |
| 1991 | 11–30 September 1991 + 1–5 October 1991 | | May 1991 | Census cancelled (proposed reference dates inferred from GoI [1991a: 9]); parliamentary elections cancelled as well |

(*Contd*)

**Table 2.3** (*Contd*)

| Census | Enumeration + revisional round* | | Elections | Remarks |
| | Snowbound areas | Accessible areas | | |
| --- | --- | --- | --- | --- |
| 2001 | 11 September–30 September 2000 + 1–5 October 2000 | | | Districts other than those listed below |
| | 11 September–15 November 2000 + 16–20 November 2000 | | | Srinagar, Badgam, Anantnag, Baramula, Kupwara and Doda (Banihal, Ramso, Marwa and Wardwan blocks) |
| | 11 September–15 December 2000 + 16–20 December 2000 | | | Srinagar town and Pulwama district |
| 2011 | 11–30 September 2000 + 1–5 October 2000 | 9–28 February 2011 + 1–5 March 2011 | | Houselisting in accessible areas completed by 30 September 2010 rather than 30 June 2010 due to disturbances |

*Source*: Provisional population totals (PPT), general population totals (GPT) and the Election Commission of India's (ECI) statistical reports for the respective years.

*Note*: (*a*) *The first day of the revisional round is the reference date of a census. The census moment was fixed at sunrise until 1991 and midnight in the later censuses. (*b*) According to the report for the 1987 assembly elections, polling was conducted on 23 March except in Handwara and Langet, where polling was held on 23 May. Both these constituencies are located in Kupwara district and might have been covered later due to inaccessibility on account of being snowbound, but it is not clear why elections were not delayed in other snowbound areas. In neighbouring Himachal Pradesh, too, snowbound constituencies have in the past voted after the rest of the state (A. Bose 2003: 2932–33). (*c*) This table does not cover the houselisting and housing phase.

of villages decreased, is a case in point. Udhampur and Kathua in Jammu and Baramula in Kashmir accounted for the bulk of this increase. There are also disagreements over the count of snowbound villages between the reports of the same census – for example, the general report and the administrative report for 1971 (Table 2.2). The recent censuses have faced the additional problem of belated demarcation of snowbound areas. The list of non-synchronous areas for the 2001 census was not ready as late as February 2000 (GoI 2000c: 1, 17–18).

Further, counting in snowbound areas is plagued with uncertainty due to the weather. Household enumeration of population of snowbound areas had to be conducted in two phases in the 1981 census, which means there were two different reference dates for the same area.

> [A] special time schedule was followed for the enumeration of nomadic segment of population like bakarwals from 1st to 10th April, 1981 under intensive supervision so as to net them in the *villages and pockets of their normal abode* well before their departure for summer pastures. This became necessary because they could not be enumerated in their summer pastures during September–October, 1980 along with the rest of the inaccessible areas as programmed, owing to their advance return to their villages due to sudden worsening of weather conditions in these upper reaches. However, as a precaution against any double enumeration, a special entry was made in their grazing permits about their having been censused with clear instructions to the enumeration agency in the accessible areas to verify the grazing permit in case any of them was found in these areas during the census enumeration period. (Ibid.: 8, emphasis added; see also GoI 1983a: 1)

Reference dates for accessible areas can also vary across censuses. In the 1981 census,

> [u]nlike in the rest of the country … the census in the accessible areas of the Jammu & Kashmir State was conducted from 20th April to 5th May, 1981 followed by a revisional round from 6th to 10th May, 1981 with the sunrise of 6th of May, 1981 as the reference time and date. This special schedule … was decided upon after a careful consideration of the usually adverse weather conditions obtaining in the state during the months of February and March. (GoI 1981a: iii, 7–8; see also GoI 1983a: 1; GoI 1989a: 6)

Census could not be conducted in 1991 due to insurgency. In 2001, census operations were completed in most of Jammu and Ladakh as per schedule (*Kashmir Times* 2011f, 2011g), but there was a delay in Kashmir and some of the adjoining areas of Jammu due to insurgency and

> the enumeration period was extended from 1st October, 2000 to 15th November, 2000 with the reference date of 16th November, 2000 in the six districts viz., Srinagar, Badgam, Anantnag, Baramulla, Kupwara and Pulwama of Kashmir Valley and four blocks namely, Banihal, Ramso, Marwa

and Wardwan of Doda district in Jammu Division. It was further extended in Srinagar town and Pulwama districts until 15th December, 2000 with reference date being 16th December, 2000. (GoI 2009a: 3–4)

In 2011, houselisting was affected by disturbed conditions and

the state government sought extension for completing the house listing operation up to 31.08.2010. As the situation did not show any sign of improvement another extension was sought up to 30.09.2010. (GoI 2011d: 1)

To conclude, reference dates have been contingent on (unpredictable) weather and security conditions in the state. A news report which suggested that 'the deadline [the reference date in 2000] was extended indefinitely' (Rediff 2000a) better captures the uncertainty on the ground than ex-post accounts of orderly changes in the reference date. Changes in reference dates have implications for the regional distribution of population as large-scale migration takes place between the two phases of the census. The descriptive reports of earlier censuses discussed this problem. The 1961 GPT illustrates how a change in migration patterns vis-à-vis the reference date can affect the headcount.

Peasants from Kashmir, particularly from the districts of Anantnag [present Anantnag, Kulgam, Pulwama and Shupiyan districts] and Srinagar [present Srinagar, Badgam and Ganderbal districts], usually move to the plains during winter in search of employment. They stay in open areas where they cook their meals and sleep for the night. Normally, they return home in March but as the month of fasts (Ramzan) had this time started in February and the Id Festival was approaching fast, most of them returned to Kashmir towards the end of February after having been enumerated as houseless, persons. Their original enumeration slips had, therefore, to be cancelled and they were re-enumerated as members of the households in the district to which they actually belonged. (GoI 1964b: viii)

Lapses, if any, in cancellation and re-enumeration would have resulted in double counting or non-enumeration. Concerns about double counting arise because the same report also notes that 'no important fairs and festivals are held in the State in the months of September and February, when the enumeration of non-synchronous and accessible areas was undertaken', and therefore 'it was not felt necessary to make any elaborate arrangements for the Census of pilgrims' (ibid.). The contradiction noted here arises because the earlier reports shared metadata. Such information is not available for the recent censuses because of which the impact of idiosyncratic events is difficult to assess. The 2001 census overlapped with Ramzan, but the reports are silent in this regard. This would have affected the accounting of the non-nomadic population of Srinagar and Pulwama districts in Kashmir as those who usually are in the plains during the winter would have returned for Ramzan.

## Concluding Remarks

This chapter began with a discussion on the inconsistencies in J&K's maps and discrepancies in the estimates of its area. The clumsy treatment of Gilgit–Baltistan in the Jammu and Kashmir Reorganisation Act, 2019, and subsequent maps is a case in point. The chapter then went on to discuss how the census has been conducted amidst several intertwined uncertainties in J&K. Reference dates and the distribution of snowbound villages changed across censuses and even within districts, and political unrest disrupted census activities on several occasions.[64] This vitiates inter-temporal and cross-sectional comparisons – particularly because of substantial internal migration between the two phases of the census. However, information on these changes, let alone their impact on headcount, is not available for post-1981 censuses. In other words, the interpretation of trends in J&K's headcount is difficult because census reports are increasingly deficient even in respect of methodological metadata, let alone quality and conceptual metadata.[65] The decline in the quality of metadata is not confined to census, though. As discussed in this chapter, maps and area estimates of J&K are also deficient vis-à-vis metadata. Likewise, NSSO reports are deficient regarding metadata on non-coverage in J&K (Agrawal and Kumar 2017b).

So, on the one hand, policymakers and researchers alike face difficulties because of the dwindling supply and quality of metadata that are essential to understanding published figures. Even the ORGI inadvertently triggered a political controversy by calculating the growth rate of Muslims in India without accounting for the fact that J&K was not covered in 1991. Interestingly, the debate triggered by the inadvertent miscalculation did not account for the fact that in both 1981 and 2001, the reference date of the census for J&K was different from that of the rest of the country and even varied across districts within the state. On the other hand, as discussed in the later chapters, the deficiency in metadata allows vested interests inside and outside the state to push unsubstantiated claims exacerbating communal tensions as the alleged threat of demographic marginalisation is a key driver of political unrest in Kashmir even as Jammu believes that New Delhi allows Kashmir to cook figures. The decline in the quality of metadata is not limited to politically restive J&K; it is a long-term and pan-India phenomenon (Kumar 2021a). Its impact is, though, most pronounced in J&K, where mobile population groups, uncertain weather, conflicts and a deep politicisation of numbers dynamically shape the headcount.

We have already discussed that earlier reports shared information on the interaction between the festival and census calendars. A few other examples of the richness of information shared by earlier reports are in order. Some of the earlier censuses provided estimates of both the area and the population of non-synchronous villages. In 1931, 621 snowbound or non-synchronous villages accounted for 1,944 square miles and 237,201 persons – that is, 6.5 per cent of the population (GoJK 1933a: 12, 1933c: 17).[66] The reports for the 1961 census give information about the population of the snowbound areas as of 1 October 1960 (GoI 1968a: 39) as well as the adjusted population as of 1 March 1961 – that is, the standard reference date for synchronous areas (GoI 1964b: vii). After the completion of enumeration in snowbound areas, the local administration was asked to provide monthly updates regarding changes in population that were in turn used for adjusting the 1 October 1960 population to arrive at figures as of 1 March 1961.[67] As per the 1961 census, the population of 648 snowbound villages was 292,829 as of 1 October 1960, which increased to 300,657 as of 1 March 1961. The snowbound areas accounted for 8.44 per cent of the state's population.[68] Since Ladakh is the only entirely snowbound region, the share of snowbound areas in the population of the Kashmir and Jammu divisions was 6.11 per cent. Since Gurez and Karnah are also completely snowbound, we can further decompose the headcount and arrive at estimates of population of the rest of the snowbound areas. This decomposition can in turn be used to estimate an upper bound of the nomadic population.[69] Some of the earlier census reports also provided estimates of the size of the water-borne population (GoJK 1923: 5).[70]

There were gaps in earlier reports, too. They did not provide the decomposition of the population of the snowbound areas[71] or clarify if all the census tables used adjusted figures for the snowbound areas.[72] They also did not explain the identification of snowbound villages[73] and the method of accounting of the non-nomadic people who leave the snowbound areas during winters. Instead of filling in such gaps, the post-1981 reports stopped sharing even the elementary details that used to be shared earlier. In the 2001 GPT, the boat population is not even mentioned[74] even as snowbound areas and mobile populations are mentioned in passing. The 2011 GPT report has not yet been released while the administrative reports were not released for the 2001 and 2011 censuses.

# Notes

1. Only three princely states, including J&K, became full-fledged states in independent India. The other two, Manipur and Tripura in the north-east, were initially constituted as Category C states on 26 January 1950 and then as UTs on 1 November 1956 before being elevated to full-fledged states on 21 January 1972. The rest of the princely states were absorbed into larger states.

2. Before 1947, Ladakh was a separate district that included Baltistan and was not part of the Kashmir province (GoJK 1943c: 3). Presently, Baltistan is almost entirely under Pakistan's administration.

3. Until 9 August 2019, the Constitution applied 'to the State of Jammu and Kashmir with certain exceptions and modifications as provided in article 370 ["Temporary provisions with respect to the State of Jammu and Kashmir"] and the Constitution (Application to Jammu and Kashmir) Order, 1954' (GoI 2015b: preface). Article 370 was included in Part 21 (Temporary, Transitional and Special Provisions) of the Constitution. Before being repealed, Article 370 had been diluted over the years and eventually reduced to a symbolic gesture. Sheikh Abdullah seems to have complained against the dilution of Article 370 soon after it was adopted (Noorani 2011: 93). As Rai (2018: 217) puts it, in the end 'its main role is to provoke, on the one hand, the anger of mostly Kashmiri Muslims who see betrayal in its nullity and, on the other, the ire of Hindu nationalists who see in it the "appeasement" of Kashmiri Muslims, read as fundamentalists and separatist traitors'. A political scientist argued that initially Article 370 empowered the state, but later it proved to be inflexible and impeded further progress with its haphazard erosion compounding the problem (interview, Srinagar, 28 May 2022).

4. The general population totals (GPT) offer the most detailed observations on territorial changes between censuses. The reports for Punjab and India do not indicate any change in Punjab's area between 1971 and 1981 (GoI 1982a: 11) even though other sources suggest changes in the external border (Ministry of Defence [MoD] 2014: 430). While Gujarat's area increased between the two censuses (GoI 1983b: 12), it is not commensurate with the estimates provided in other sources (MoD 2014: 438). The lack of change in the area of these states is explained by the fact that under the Simla Agreement both countries eventually withdrew to their side of the international border.

5. 'Population of 41 villages fully and 3 villages partly of Akhnoor Tahsil (District Jammu) falling on the other side of line of control referred to in the Simla Agreement, 1972 has been adjusted in districts Udhampur, Kathua and Jammu on pro data basis' (table A-2, J&K, Census of India 2011). It is not clear what

this adjustment amounts to in terms of the accounting of the population of these villages.

6. '[A]ccording to the 1977 Pakistan White Paper, India gained 340.88 square miles [882.87 kilometres] and lost 58.38 square miles [151.20 kilometres]' (Lamb 1991: 296). The official history of the 1971 war prepared under India's MoD notes the following regarding the loss and gain of territory: 'As a result of the 1971 War, both sides gained and lost some territories.… It is said that India gained 895 sq km and lost 167 sq km territory along the Ceasefire Line (CFL) in Kashmir. In Jammu and Kashmir and Punjab, India gained 1,115 sq km and lost about 192 sq km along the international border. In Rajasthan and Kutch Sector, India gained about 14,272 sq km, and lost 16 sq km. Thus, as a result of the War, India gained about 16,282 sq km territory, while Pakistan gained about 375 sq km territory in the Western theatre' (MoD 2014: 430). The status quo was restored along the international border, but the two sides agreed to respect the ceasefire line of 17 December 1971 in J&K that was finetuned in two areas in November–December 1972, and the two armies took positions on the 'readjusted' border on 17 December 1972. The total gain to India was about 731 square kilometres (ibid.: 430–36), which is close to the Pakistani estimate of India's gain referred to earlier in this note. The census, however, records a gain of 818 square kilometres between 1 April 1971 and 1 March 1981 (Table 2.1). It bears noting that census reports do not seem to adequately capture the actual adjustments on the ground. Gupta (2023: 165, 167), for instance, suggests that half of Badgam village (referred to as Bodagam in census reports) was brought under the Indian administration in 1965, while the other half had to wait until 1971. On the other hand, India lost parts of Drelung to Pakistan in 1965 (ibid.: 167). The census, however, suggests that the whole of Bodagam joined India in 1971 and is silent with regard to Drelung.

7. According to MoD (2011: 299–301), India had captured 1,920 square kilometres and lost 540 square kilometres in the 1965 war. The India–Pakistan Agreement, however, restored the status quo ante by withdrawing forces 'to the positions they held prior to 5 August, 1965' (GoI n.d.1). Das Gupta (1968: 351) provides comparable figures – 740 square miles and 210 square miles – and adds that the 'Pakistan Government claimed that it had occupied 1617 square miles of Indian territory against 446 square miles of Pakistani territory (including Azad Kashmir) occupied by India'.

8. J&K was not the only state whose area statistics have changed inexplicably. For changes in Nagaland's area not explained in census reports, see Agrawal and Kumar (2020a: 62–67).

9. It is not clear if this 'lapse' is explained by the fact that J&K was not covered by the 1991 census, the declining quality of metadata or strategic considerations. Also note that any redistribution of the territory of J&K between Pakistan and India should not affect the overall area of J&K reported in the Indian census that includes all parts of the state irrespective of their present status. It will, however, affect estimates of population density that consider only the territory under India's administration.

10. See Das Gupta (1968: 299) for India's position on the size of the territory ceded by Pakistan to China. Lamb (1991: 40, 51, 246) and Noorani (2006) contest the Indian claim of net transfer of territory to China.

11. Pakistan does not add J&K's area to its own area (Government of Pakistan, n.d.) even though its maps include the occupied territories (Survey of Pakistan 2020). However, it is not clear how Pakistan has estimated the area of the occupied territories. As per the Indian estimate, Pakistan is currently in occupation of 78,114 square kilometres. Pakistan estimates that area to be 85,793 square kilometres. Such discrepancies have puzzled researchers. Snedden (2017) points out, 'In 1951, Azad Kashmir was officially 4,494 sq. miles. Official publications now give the region's area as 5,134 sq. miles. This *inexplicable* additional 640 sq. miles is a 14 per cent increase in Azad Kashmir's area' (emphasis added). Earlier, Lamb (1991: 14–15) noted that both India and Pakistan overestimated their respective shares of the territory:

> An official Pakistani source in 1954 indicated 84,471 square miles. The 1891 Census (of British India) put the area as 80,900 square miles, but the 1911 Census increased the figure to 84,492 square miles. This shrank slightly to 84,258 square miles in the 1921 Census. In 1961 the Government of India suggested that earlier estimates of the area of Jammu and Kashmir were incorrect: the true figure should be 86,023 square miles (no doubt because of the official inclusion in India of the Aksai Chin). The point, of course, was that Jammu and Kashmir possessed vague (to put it mildly) frontiers with both Tibet and China which could be interpreted to give the State a variety of areas.

New Delhi's position remained fluid for quite some time. Several government and private maps depicted the Sino-Indian border inconsistently. The Census of India did not cover J&K in 1951, but its maps showed the state in the national map. At least one census report for 1951 contained two mutually inconsistent maps of J&K (GoI 1957). The report of the first round (1950–51) of the NSS that did not cover J&K did not even include the state in the national map (GoI 1952). As late as 4 January 1962, the president of the ruling Indian National Congress (INC) claimed that India would liberate 42,000 square miles (108,780 square kilometres) 'of Pakistan-held Kashmir as well as the portions under the Chinese occupation' (Das Gupta 1968: 286), which agrees neither with the 1961 census nor the 1971 census (Table 2.1). This is interesting because the aforesaid statement post-dates the third definitive series of postage stamps released in April 1957, which featured the map of India that marked the freezing of the official cartographic imagination of J&K even though the stamps of this series did not include parts of the North-East Frontier Agency (NEFA). A similar diversity of maps is found across the border in Pakistan. The most recent official maps of Pakistan include the territory under India's administration and vaguely exclude the portion claimed by China, but the estimates of the area have not been updated for either the country or the occupied territories.

12.  The area statistics of J&K are widely misunderstood because of multiple estimates in circulation. For instance, Hari Om (1998: 15) suggests that after the Instrument of Accession was signed, the armed forces set out to liberate 83,294 square kilometres under Pakistan. However, as per census publications, this was the area under Pakistan after 1971 including Shaksgam ceded to China in 1963. The correct figure for 1948, following the government estimates, is 84,112 square kilometres.

13.  This is reminiscent of a news report in a leading English daily on a 'surgical strike' along the Indo–Myanmar border (2015). The actual site of the strike was a few hundred kilometres south of the location marked in the map accompanying the report.

14.  The government exempts itself of liabilities in case of errors in maps (see, for instance, the Geospatial Information Regulation Bill, 2016, Section 37). Non-governmental entities that misrepresent India's map can, however, be punished (*Hindustan Times* 2015).

15.  The UT of J&K still mentions its area as 222,236 square kilometres (see, for instance, Government of the Union Territory of Jammu and Kashmir [GoUTJK] 2021, among many other websites). GoUTJK (2021) earlier belonged to the undivided state. The new administration updated the list of languages but not the area (cf. archived pages for 23 October 2020 and 7 February 2021).

16.  Other government sources have followed the nominal changes in administrative atlases. The districts of Ladakh were referred to as Ladakh and Kargil in NSS reports up to the 56th round (2000–01). The reports of the 57th and 58th rounds refer to them as Leh and Kargil, while subsequent reports switched to Leh (Ladakh) and Kargil.

17. The disputed interstate borders of Nagaland generate similar problems as people of some of the border areas are mapped to two jurisdictions (Agrawal and Kumar 2020a).

18. A census report noted that 'it took a good deal of mediation on our part to settle the question of demarcation of jurisdiction between the Forest and Revenue Departments in places like Beerwa, Handwara and Kupwara etc' (GoI 1971b: 6, also 223).

19. British assistance was indispensable for conducting census in princely India. The inability of J&K to conduct a census in 1881 might be explained by political instability and friction with the British towards the end of Maharaja Ranbir Singh's reign.

20. The census continues to report data on sects but clubs it under major religions and presents the details only in an annexure. Numbers reported under sects are very small except in the case of Hindus.

21. Several decades after the aforesaid extension, the applicability of the Census Act continues to be questioned in Kashmir. As Talib (2010: 4) puts it, 'Basically, as per the J&K Constitution, census should have been a state subject.' As recently as 2011, the Committee on Legislative Measures in Statistical Matters suggested that 'the Centre will not be in a position to conduct any census other than human population census in the State of Jammu and Kashmir' as the 'residuary powers are with the State' (GoI 2011i: 12–13). The committee failed to note that even human population censuses were partly constrained because, as discussed later in this chapter, census officers did not enjoy the authority to, say, install additional check posts for enumerating the nomadic population. Also note that J&K was exempt from the supervision of the union government in the case of other statistical exercises as well. It was not covered by the Collection of Statistics Act, 2008, which was amended in 2017 to allow the union government to collect statistics on matters covered in List I (Union List) or List III (Concurrent List) of the Seventh Schedule to the Constitution.

22. Around this time the parliament also extended other laws to the state. For instance, 'regulation requiring Indian citizens to procure permits for entering Kashmir was done away with'. In response, 'Pakistan sent a formal protest note to the Security Council on 9 September 1959, maintaining that the Government of India could not undertake these measures until the question of accession had been settled' (Das Gupta 1968: 266).

23. 'In Jammu & Kashmir, eight communities vide the Constitution (Jammu & Kashmir) Scheduled Tribes Order, 1989, and four communities, namely Gujjar, Bakarwal, Gaddi and Sippi, were notified as the Scheduled Tribes vide the Constitution (Scheduled Tribes) Order (Amendment) Act, 1991.

[The people of] [a]ll the twelve (12) Scheduled Tribes (STs) were enumerated officially [qua STs] for the first time during the 2001 census' (GoI n.d.6: 1).

24. The census reported abnormal changes in the population of Bengal and Punjab in 1941 due to massive over-reporting in the run-up to partition. While the overcount was corrected in 1951 (GoI 1954a: 5), content errors in the language data of Punjab and, to a lesser extent, Assam could not be avoided in 1951.

25. The estimate of the area occupied by Pakistan in this report of the 1961 census is less than the corresponding figure for 1971 in Table 2.1. The two estimates cannot be fully reconciled even if we assume that the former includes Shaksgam Valley transferred by Pakistan to China in 1963. Note that this report was published after the Sino-Indian War (1962).

26. Census reports refer to the floods of 1950 (GoI 1976b: 6) and 1951 (GoI 1974a: 1). The latter would not have affected enumeration (if the government had decided to conduct a census) that would have been completed before the onset of summer. However, the floods of 1950 would have affected the houselisting phase of the census in the accessible areas and both houselisting and household phases in the snowbound areas. The latter areas would not have been directly affected by floods, but census operations would have suffered nevertheless because of the disarray in the administrative machinery based in accessible areas.

27. Census was not the only database affected by the administration's preoccupation with the immediate problems of the state in the early 1950s. A few months after N. Gopalaswami Ayyangar counselled the Constituent Assembly to allow the administration to focus on immediate problems, 'the Ministry of States asked that the CAS [Constituent Assembly Secretariat] should not take up with its [J&K's] government the question of preparation of electoral rolls at present' (Shani 2018: 159). The state was not even covered by the first delimitation commission (Election Commission of India [ECI] 1957). The NSS, too, did not cover the state until the eighth round (July 1954–March 1955) (Suryanarayana and Iyengar 1986: 263). In fact, the map appended to the report of the first round of the NSS did not even show the state as a part of India (GoI 1952).

28. The 1951 census could not be conducted in the NEFA, which includes present-day Arunachal Pradesh, due to administrative incapacity and, possibly, the unsettled international border. The sample verification of the 1951 census could not be 'undertaken in the States of West Bengal, Manipur, Punjab, PEPSU [the Patiala and East Punjab States Union], Himachal Pradesh, Delhi and Bilaspur' due to 'various reasons peculiar to the locality (unconnected with the nature of Census enumeration)' (GoI 1953: 3). Even in 1961, only an abridged household schedule could be administered for 297,853 out of 336,558 persons in the

NEFA, and the all-India schedule was canvased for the remaining 38,705 persons (Chaube 2012: table 6). Assam missed census once (1981) due to civil unrest, while the reference dates were affected by political unrest in two districts of Nagaland (1961) (Agrawal and Kumar 2020a: 186–87, 297). Otherwise, there are only a few instances of non-coverage, but in the absence of adequate metadata the underlying problems are not clear in such cases. Consider, for instance, the 1991 census of Maharashtra. We are blandly informed in some of the reports for that year that '[o]wing to administrative reasons, census could not be conducted in 33 villages of Akrani and Akkalkuwa tahsils of Dhule district. Only their population (persons, males and females) was obtained through secondary source' (GoI 1995a: 3). The 'secondary source' is left unspecified. In any case, the headcount of these villages was 'included in the population of Maharashtra and India. However, their further details are not available' (GoI 1997: 159).

29. When the author enquired about the sparse collection of census publications in the library of DCO, Srinagar, he was told that the office was burnt down in the early 1990s (census official, interview, 30 November 2015).

30. Hari Om (1998: 28) pointed out that '[v]iolence liquidated far less people in Kashmir than in Punjab' and claimed that 'it was not the secessionist militancy but the political game against Jammu which influenced the Registrar General of India in delinking Jammu and Kashmir from the [1991] census operation'.

31. The ORGI organised six conferences of directors of census operations for the 1991 census: 6–9 November 1989 (New Delhi), 5–8 February 1990 (New Delhi), 24–26 October 1990 (Panaji), 8–10 May 1991 (New Delhi), 13–15 May 1992 (Shimla) and 22–24 April 1993 (New Delhi). Various officers of the Directorate of Census Operations in J&K attended at least four conferences, including the last one in 1993. We do not have information on participants of the fifth conference for which we have access to only the agenda notes and not the entire proceedings.

32. Even before militants threatened the exercise, A. Bose (2000: 1433) suggested that there was some uncertainty regarding the census in J&K. Later he noted that 'in spite of frequent threats from militants, the census enumeration did take place in J and K' (Bose 2004: 3595).

33. The Hizbul Mujahideen issued a threat on 3 September that was published in newspapers on the following day (see, for instance, M. Ahmad 2000b). A later census report suggests that the militants were not the only ones opposed to the 2001 census and that the 'threat of boycott calls [came] from various quarters' (GoI 2011d: 1). A researcher suggested that smaller militant outfits, too, were opposed to the census (interview, 23 October 2021). Note that newspapers were used by militant groups in the past, too, to issue warnings. The exodus of

Kashmiri Pandits was triggered, among other things, by warnings in the form of press releases issued in January 1990 (Pandita 2017: 79; F. B. Ahmed 2021).

34. J. K. Banthia's account of the developments in J&K, however, completely sidesteps the seriousness of the political unrest. At one of the conferences of the Directors of Census Operations, Banthia claimed that 'there were some *minor* problems *initially*' in J&K (GoI 2000a, emphasis added). At a United Nations (UN) conference, Banthia mentioned the militant threat in the state but presented the reluctance of the enumerators as an 'administrative problem' unrelated to the political problem. In fact, he bracketed it along with 'strike of Government employees just prior to the commencement of census operations in' other states such as Bihar and Rajasthan (Banthia 2001b: 22) and added that it 'jeopardized the census calendar in the states and much firefighting was done to overcome these obstacles' (Banthia 2001a: 16). He also flagged the long delay in appointing the DCO in J&K (Banthia 2001a: 14) without saying why officers were reluctant to accept the responsibility. This, though, was not the last time J&K troubled Banthia. In September 2004, Banthia presented *The First Report on Religion* and reported the growth rate of Muslims in India during 1991–2001 without accounting for the fact that Muslim-majority J&K was not covered in 1991, which triggered a political controversy (A. Bose 2005: 370; *The Telegraph* 2004). Interestingly, given the obsession with the share of religions, no one, including Ashish Bose, asked any question about the multiple and non-standard reference dates and their impact on the aggregation of headcount and the comparability of different censuses.

35. This is reminiscent of the observation of the superintendent of census operations (SCO) of West Bengal for the 1971 census, who summarised the impact of political disturbance on the quality of census operations as follows: 'One method used to get at least the barest essentials done was a combination of persuasion and compromise: one persuaded one's own staff to do an amount of work that was determined by compromise' (GoI 1972b: Preface).

36. Addressing the Conference of the Directors of Census Operations, the union home minister Advani 'said [that] the current census will be the [first] comprehensive operation in the last thirty years covering all the States and the Union Territories. In the 1981 Census Assam could not be included and in 1991, Census Operation could not be undertaken in Jammu & Kashmir' (GoI 2000a; see also *Kashmir Times* 2000b). Recall that Advani had raised a question in the parliament about the cancellation of the 1991 census in J&K (Lok Sabha 1991: 54).

37. A similar 'sentiment' prevails among international agencies, which collate data, for whom 'data availability is more important than the quality of the data that are supplied' (Jerven 2013: 23).

38. 'Leh witnessed worst Natural Calamity on 5th August 2020 resulting in hundreds of deaths, snapping of road and telecommunication links. It was due to steadfastness and hard labour put in by the local officials and officers dispatched from this directorate that the process was taken to its logical end' (GoI 2011d: 1–2).

39. The state government informed the assembly that '104 persons were killed, 962 persons injured, 4270 persons arrested, 4047 persons released' in Kashmir between June 2010 and March 2011 (*Kashmir Times* 2011e). During this period, 152 persons were detained under the Public Safety Act, 1978.

40. Even the then RGI noted the support for the 2011 census from all sections in Kashmir, including pro-independence groups (interview, 6 August 2021).

41. The 2021 census could not be conducted as per the schedule due to the Covid-19 pandemic. However, the government of the UT of J&K conducted a 'survey covering both nomadic people and under-developed tribal pockets' that 'will be followed by documentation and issuance of smart cards containing each person's details' (*Indian Express* 2021a). The government seems to be planning to replicate the exercise and 'create a digital database of all the families across the Union territory' (*Economic Times* 2022).

42. A lack of familiarity with the geography of non-coverage in the state has meant that researchers end up reporting estimates for parts of the state not covered in surveys. Palmer-Jones and Sen (2010: 10, 12, 13) report poverty ratio for the Jhelum Valley based on the 50th round of the NSS even though the region was not covered in that round. In fact, their map incorrectly identifies NSS regions in J&K. Jha and Sharma (2003) compare 43rd and 50th rounds for the Outer Hills region without noting that the data are not comparable due to sample non-coverage in the latter round. Likewise, the Sachar Committee report relied upon the 43rd to 61st rounds of the NSS (GoI 2006b), which are not comparable in the case of J&K.

43. Non-governmental surveys such as the ones conducted by the Centre for Monitoring Indian Economy (CMIE) have also been affected by political disturbance in the state: 'During the ninth Wave administered during September–December 2016, survey execution was impacted by ... Nagrota and Uri attacks ... [whereas] in the May–August 2017 survey ... the Amarnath attack impacted the survey in Jammu and Kashmir.... During May–August and September–December [2019], internet shutdown in Jammu and Kashmir led to the survey being impacted severely. In September–December the response rate from the state was only 28.5 per cent, which was *largely from the Jammu region*' (CMIE 2020: 14, emphasis added). Interestingly, the CMIE does not mention the massive protests in Kashmir in the aftermath of the killing of the Hizbul Mujahideen's Burhan Wani in 2016.

44. The highest sample non-coverage (about 74 per cent) was reported in the Employment and Unemployment Survey of the 50th round (1993–94), when in addition to the Jhelum Valley (Kashmir), Doda, Punch and Rajauri districts of the Outer Hills region in Jammu could not be surveyed (Agrawal and Kumar 2017b: 37–40). Political uncertainty had affected the coverage of NSSO surveys even before the 1990s. Between 1950 and 1954, the NSSO did not cover J&K at all. In the 38th round (1983), conducted after Sheikh Abdullah's death, the coverage was only 87.5 per cent (report no. 332).

45. A census official suggested that 'the very mention of numbers alarms' the army (interview, Srinagar, 30 November 2015). In fact, the army does not even rebut exaggerated estimates of its deployment in Kashmir (Shukla 2018). The official added that 'the army directly reported to the RGI in 2001, but this time [in 2011 they reported] through the DCO. In 2001, they reported very few [numbers] and we evenly distributed [the figures]. [In 2011,] the army did not give us population statistics until the last moment. We had to go to Udhampur. They gave us district wise [data], and we put them under sub-districts assigning numbers to places with known camps. Media asked us, but we did not give figures for army' (interview, Srinagar, 30 November 2015; see also Bashir 2011). The DCO for the 2001 census confirmed that he had asked 'the army to send [their headcount] directly to the RGI' (interview, Jammu, 4 December 2019).

46. A census official recalled that those not familiar with Ladakh's context were disturbed by the very low sex ratio and as a result '[t]here was enormous pressure on us after the Leh figures came out' (census official, interview, Srinagar, 30 November 2015). A report in the *Indian Express* (2011), for instance, noted that 'Leh district in Ladakh region has a *shocking* sex ratio of just 583, which is a massive drop of 240 from the 2001 Census' (emphasis added). Local newspapers, too, quoted government officials and politicians who were bewildered by Leh's statistics (Ali 2011a, 2011c). The drop in sex ratio can, possibly, be attributed to the better accounting of armed forces and migrant workers, among other things. It bears emphasising that the description of Leh's child sex ratio (CSR) in the PPT, too, contributed to the alarmist perceptions (see GoI 2011d: 81).

47. In Nagaland, the Tikhir Nagas had for long been listed as a separate tribe in the census even though they had not been recognised as a separate indigenous Naga tribe by the state government until January 2022 (Kumar 2023b). The 1951 census identified 'Meithei' as a tribal language even though the speakers of the language were not classified as an ST (GoI 1954d: 8).

48. The suggestion in census reports that Jammu is the permanent residence of the nomadic community agrees with the fact that most members of the community

are enrolled as voters in Jammu (Javaid Rahi, interview, 19 September 2021; Zafar Ali Khatana, interview, 24 September 2021).

49. It is not clear if 'bakerwals and shepherds' includes non-Bakarwal shepherds. Most discussions on the shepherd population pay hardly any attention to other communities such as the Gaddi tribe of Jammu and non-tribal Kashmiri-speaking shepherds. Double counting is relatively less likely in the case of these communities and changes in reference dates do not affect their regional distribution.

50. A census report suggests, 'If the person does not show this token he is likely to be harassed and detained' (GoI 1969: 38).

51. A senior census official pointed out that 'up to 1961–71 [also 1981], Gujjars were issued a slip while crossing checkpoints. This was discontinued and now there is only verbal communication' (interview, Srinagar, 30 November 2015).

52. During fieldwork in Kashmir in May 2016, the author came across settlements of the nomadic community on the outskirts of villages in Kulgam district.

53. The number of houseless households (population) was 1,674 (7,622), 2,123 (12,751) and 3,064 (19,047) in 1981, 2001 and 2011, respectively. The population for 1981 was calculated using the distribution of 'houseless households' by household size, which provides the lower bound because the households with six or more members were combined into one category in table HH-2.

54. In February 2010, in the run-up to the houselisting phase of the 2011 census, Javaid Rahi argued that 'around 3 Lakh nomadic Gujjars who were on upper reaches at that point of time, in connection with annual migration, had not been enumerated and reflected in the census report 2001'. In April 2010, Chaudhary Bashir Ahmed Naz, the vice chairman of Jammu and Kashmir State Advisory Board for the Development of Gujjars and Bakerwals, suggested that 'according to a rough estimate, around 3 to 5 lakh nomadic Gujjars who were on upper reaches at that point of time, in connection with bi-annual migration, had not been enumerated and reflected in the census report 2001'. Later, speaking at a seminar on 'Census and Tribal Development' (5 March 2017), Rahi suggested that 'around 6 to 8 lakh nomadic, semi nomadic and shelterless Gujjars, Bakerwals have not been enumerated' in the 2011 census (see also *Kashmir Times* 2011a).

55. In February 2010, ahead of houselisting for the 2011 census, Rahi (2010) suggested that census should begin 'in the state only after October 2010 to ensure correct enumeration of nomadic population who would be under seasonal migration during the period of houselisting which will be completed during summer months between May to August 2010'. The suggestion is based on a misunderstanding of census because the household phase is anyway

conducted in February. It is only in the snowbound areas that the household phase is completed in September soon after the houselisting, which cannot be postponed due to weather conditions. In principle, the nomadic population of the snowbound areas is counted there as per the de facto method of enumeration. While community activists complain that in practice the upper reaches that support pastures are rarely covered in censuses, they overlook the fact that when the nomads return to plains they are likely to be counted there due to the deterioration of the earlier system of checks to prevent double counting.

56. A comparison of the extensive discussion on boat population in J&K with reports of other states with substantial waterways highlights the salience of this community in Kashmir, where the boat population is not classified as houseless. In Kerala, the census classified 'persons *enumerated* in boats and other inland vessels' along with 'houseless persons' (GoI 1964c: 216, emphasis added). In Assam, the boat population was classified as houseless: 'Houseless persons mean persons like members of the wandering tribes, beggars, boat dwellers, wandering sadhus, etc. who generally do not reside in houses and cannot therefore be enumerated along with the household population' (GoI n.d.11: 53). Interestingly, Assam's administrative report explicitly recognises the 'people who *normally* reside in their big boats in such big rivers as the Brahmaputra, Barak etc' (GoI 1963b: 141, emphasis added). In West Bengal, 'people who live in boats' were classified under 'floating population' along with 'the homeless poor, wandering tribes' and 'coolie gangs' (GoI 1964d: 10), even though the census acknowledged that some people 'live in boats *permanently* and have got no homes on land' (GoI 1963c: 121, also 161, emphasis added). Uttar Pradesh's census did not classify 'population living in boats, *majhis*, *mallahs* and boatmen' along with 'houseless population' (GoI 1964e: iii, GoI 1963d: 67), but the boat population was mostly seen as people in transit.

57. See GoI (2011d: 75) for a brief remark on covering population beyond the fence on the Indo-Pakistan border.

58. During field visits in May 2016, the author found that multiple layers of checkpoints tightly controlled the access to even those villages of Gurez that were at some distance from the border.

59. Even Bangladeshi and Rohingya immigrants have been reported in Jammu (*The Hindu* 2017; *Indian Express* 2017; *Mint* 2021) and have contributed to the growing concern about the city's alleged Islamic encirclement (Hari Om 2017b; Ambale 2020). For estimates of Bangladeshi and Rohingya immigrants in Jammu, see note 69 of Chapter 4.

60. The tables on migration published in 1965 – that is, three years after the 1962 war – classify 1,644 (1,666) individuals, including 1,633 (1,654) in Ladakh

district, as nationals of (persons born in) Tibet and not China for which the figure is zero (GoI 1965: 287, 292). This distinction between Tibet and China was not followed in the later censuses.

61. Snowbound areas (GoI 1963a: 36) are variously referred to as inaccessible (GoI 1981a: 8), non-accessible (GoI 1972c: 8) and non-synchronous (GoI 1964b: viii) areas. Other states with snowbound areas include Himachal Pradesh, Uttar Pradesh (now Uttarakhand) and Sikkim (GoI 1985c: 3).

62. Earlier reports recommended pushing the reference date for all areas into summer to reduce the problems posed by non-synchronous areas handled by census (GoJK 1933a: 12). This will not help as the plain areas of the state that account for the bulk of the population are governed by a different agro-climatic calendar. Moreover, it will not address the substantive question of apportioning the nomadic population between different divisions, which has not been tackled in the 130 years of modern census in the state. Note that enumeration in both synchronous and non-synchronous areas was effectively scheduled around the same time in three decennial censuses during 1981–2001 (excluding the 1986–87 special census but including the cancelled 1991 census). Bad weather and/or political disturbances, though, meant that the actual enumeration did not follow the ex-ante calendar.

63. Several far-flung areas did not have adequate administrative presence in the 1950s and the 1960s (GoI 1963a: 40–41). A civil servant, who happened to be the first sub-divisional magistrate to remain in Tangdhar throughout the winter in the late 1970s, recalled that he had to ask security personnel at the check post at Nastachun Pass to stop non-local government employees from leaving the area ahead of winter (interview, retired civil servant, 13 August 2021).

64. A researcher based in Jammu suggested that conflict not only affected the distribution of population, but also the quality of the institutional set up that collects vital statistics in Kashmir as it was overwhelmed both by the sudden spike in unnatural and unaccounted deaths in the early 1990s and the pressure from all sides to not report certain deaths (interview, 9 May 2022).

65. Metadata can be divided into three strands: conceptual (what and why), methodological (how) and quality (Organisation for Economic Cooperation and Development [OECD] 2007: 658).

66. The method of estimation of the snowbound area followed in the 1931 census is not clear. Ladakh (including Baltistan) alone would have accounted for nearly two-fifths of the state's area.

67. In 1951, enumeration was started in rural Bilaspur a week ahead of the rest of the country most likely in view of administrative constraints and difficult terrain (GoI 1951: 10). In 1961, enumeration began in the Naga Hills long before rest of the country due to disturbed conditions and administrative

incapacity (GoI 1966a: 1). In 1971, the whole of West Bengal was enumerated before the rest of the country due to law-and-order problems. However, the revisional round was synchronised with other states to ensure comparability (GoI 1977: 1).

68. The general report for the 1961 census does not mention the reference date of the snowbound population. We have assumed it to be 1 October 1960, as the GPT reports a higher population total and mentions 1 March 1961 as the corresponding reference date.

69. We can estimate only the upper bound as the nomadic communities account for a part of the snowbound population, and a fraction of non-nomadic population also relocates to accessible areas during winter.

70. In 1921, the boat population was about 12,265 (GoJK 1923: 5). For the past few decades, the estimates of boat population are available in the state government's *The Digest of Statistics* (see, for instance, GoJK 1993: ch. 18) but not in census reports.

71. Earlier reports on Uttar Pradesh provided a break-up of the snowbound population by sex (GoI 1966b: 10).

72. We can try to carry out adjustments using village-level data, but such an exercise will be hampered by the fact that the names of snowbound villages are not reported for the post-1981 censuses. Even in the earlier censuses there were discrepancies in the count of snowbound villages.

73. Earlier reports clearly stated that '[p]roper precautions were, however, taken against *inflation of the list of such places simply to suit the convenience of the Census Agency*' (GoJK 1933a: 12, emphasis added).

74. As per the 2011 census, the boat population was zero. However, *The Digest of Statistics* released by the state government contained a detailed statement on those dependent on boats.

# 3

# Inventing Boys and Miscounting Tribes and Languages[*]

[E]numerators have somehow preferred to differentially over-enumerate Noor (Boy[s]) and Nooristan (Girls).

—Bhat (2011)

## Introduction

More than three decades ago, when Amartya Sen flagged the problem of missing women (Sen 1990), India's overall sex ratio and child sex ratio (CSR) were 927 and 945, respectively. Since then sex-determination technology that 'permits couples to resort to sex selection' has played a major role in skewing sex ratio amidst 'declining fertility and entrenched son preference' (Guilmoto 2009: 524). Son preference also contributes to the neglect of the health of the girl child, which further aggravates the skewed sex ratio by increasing the girl child mortality. The union government introduced the Pre-Conception and Pre-Natal Diagnostic Techniques Act, 1994, to prohibit the misuse of diagnostic techniques to identify the sex of the foetus.[1] Yet, in 2001, 'the child sex ratio (CSR) first dropped below that of the overall sex ratio', and the problem spread beyond the north-western states (John 2011: 11).[2] A decade later, the overall sex ratio increased to 943, even as the CSR further dropped to 918. The CSR has, in fact, steadily declined from 976 to 918 between 1961 and 2011, even as the overall sex ratio fluctuated between 927 and 943 (Figure 3.1).

Very low CSRs were first reported in some of the north-western states, but Jammu and Kashmir (J&K) was not among them. However, defying conventional wisdom, the state, which was above the all-India CSR until 2001, reported a

---

[*] The first part of the title of this chapter is borrowed from Guilmoto and Rajan (2013: 64).

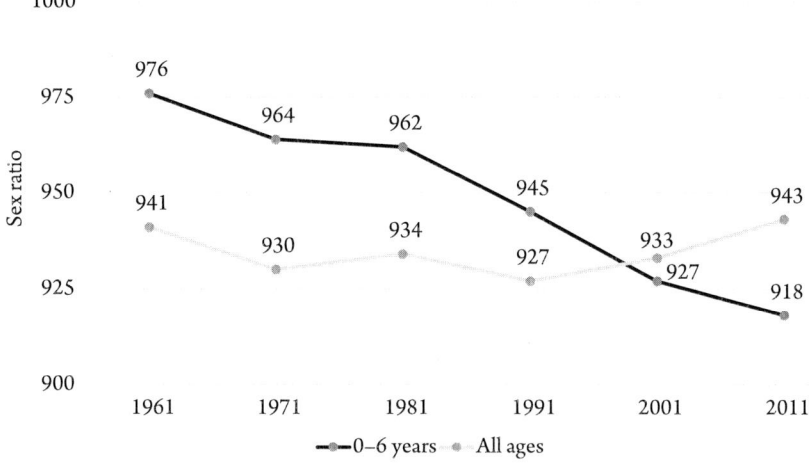

**Figure 3.1**   Sex ratio of India, 1961–2011

*Source*: GoI (2018a: 4).

sharp decline in 2011 that placed it among the worst-performing states in the country (Figure 3.2). This development has not received adequate attention in academic and public debates, possibly because insurgency-hit J&K is widely seen as an exceptional state,[3] and, in any case, it accounts for about 1 per cent of the country's population and less than 1 per cent of the national income. In other words, J&K does not have a large impact on national figures[4] and is not among the major states that are indispensable for national (policy) debates and academic analyses.

Researchers and policymakers have uncritically used the results of the 2011 census for J&K. In fact, census officials, too, did not critically examine the data (see, for instance, Government of India [GoI] 2011d: 81–84).[5] The joint director of census operations (Jt DCO) observed that 'child sex ratio was equally worrisome as it has dropped by 100 points from 963 in 1981 to 863 in 2011. It will have serious effects on our future population and it is time that we take steps to correct it' (*Indian Express* 2011). The union ministry of health expressed concern at the decline of the CSR in the state (M. Ali 2011b), which was followed by the visit of a parliamentary delegation (*Greater Kashmir* 2011a). A week later, the 'embarrassed' state government questioned the 'incomprehensible' and 'unexplainable' data and ordered a survey of births in the previous five years (M. Ali 2011b, 2011c; see also Bhat 2011). Bhat (2018: 170) adds that the 'State Government, particularly the Directorate of Health Services, had certain

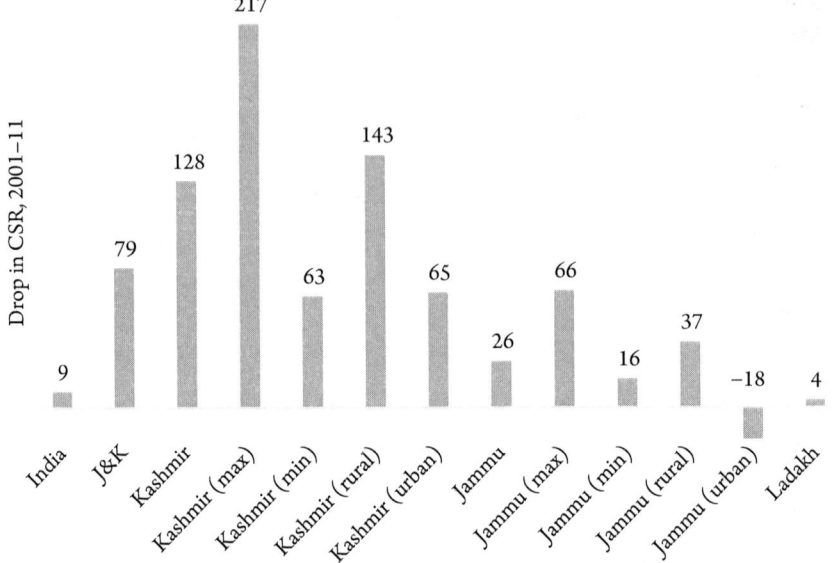

**Figure 3.2**    Drop in the child sex ratio (CSR) of Jammu and Kashmir (J&K), 2001–11

*Source*: Prepared by the author using primary census abstracts (PCA) for 2001 and 2011 and GoI (2014a).

*Note*: 'Drop' is the difference between the 2001 and 2011 figures. A negative drop implies an increase in the CSR between the two censuses.

reservations to accept the Census 2011 CSR figures and ordered a fresh survey to estimate the CSR in the State, but the results of the survey were not made public'.[6] It is not clear if the state government formally conveyed its concerns to the Office of the Registrar General of India (ORGI).[7]

After the provisional census data were released, newspapers carried several reports of raids on illegal ultrasound facilities across Kashmir. Similar reports appeared two years later at the time of release of the primary census abstract (PCA) because illegal facilities had 'survived many of the "newsmaking raids" and crackdowns' (Fayyaz 2013). Once again, the quality of data was left unexamined. Actually, there was hardly any room for critical scrutiny once the data became the basis of comparisons between Kashmir and Jammu. Since the ORGI did not issue any correction, government departments and researchers continue to use the flawed statistics.[8] Most recently, even the Fifteenth Finance Commission used the 2011 data without noting the anomalies. The estimates of total fertility rate (TFR) used by the commission show J&K as the state with the second-highest fertility (GoI 2020b: 162), which is inconsistent with the Sample Registration

System (SRS), the National Family Health Survey (NFHS) and the Health Management Information System (HMIS).

Researchers, too, uncritically use the census data. For instance, commenting on the status of women in the state, Chowdhary (2019: 76–77) observes:

> The state, as per the 2011 census, is among the very few states of India where the sex ratio is going inverse direction. From 892 females per thousand males in 2001, the number has gone down to 883 [889][9] in 2011.... This is really *a worrisome situation* since it is an indicator that despite a constitution with lofty principles of equality, women continue to face a high level of prejudice. However, what is *actually alarming* is the data related to child sex ratio (CSR) which is still lower than the overall sex ratio. (Emphasis added)

An analysis of the state's census data by the Centre for Policy Studies (CPS), Chennai, which was widely covered in the national press, offered a highly misleading reading of the 2011 census. It focused only on the increase in child population share and, curiously, not the connected drop in the CSR. Concerned about the fact that Hindus and Sikhs 'have a significant presence only in the southern parts of Jammu region' (CPS 2016), the CPS alleged:

> There has *obviously* been a great spurt in the fertility of Muslims in the Valley. Such a drastic rise in the fertility measures is unlikely to happen spontaneously. The numbers do indicate a *systematic, concerted and successful effort* among the Muslims of the Valley *to have more children*. (CPS 2017, emphasis added)

Most researchers exclude J&K from their analyses due to the unavailability of good-quality data. Anderson and Ray (2012: 87, 95) note that the phenomenon of missing girls at birth was 'most pervasive in some north-western states', but exclude J&K 'from the analysis due to lack of data'. Others note the abnormal nature of change in the composition of population without discussing possible explanations. Kumar and Sathyanarayana (2012: 68, 71) find that J&K was the only state whose estimates of crude birth rate (CBR) increased between 1994–2000 and 2004–10, and add that while it 'is difficult to conclusively comment on data quality, yet the impression one gets is that the data is suspect and needs further investigation'. John (2011: 11), too, notes the anomalous change in the CSR and fertility in J&K:

> The only state whose figures are so strange that there is every reason to doubt them is Jammu and Kashmir.... According to the provisional figures [of the 2011 census], the CSR has plummeted from 941 to 859 [862][10] – 82 [79] points – along with this, J&K is the only state in the whole country to have

registered a positive increase in its fertility during this period. Whatever the form that the ongoing conflict is taking, such figures are hard to make sense of and require further investigation.

Debroy (2020) noted the unexpected increase in the population share of lower age groups: 'As of 2016, population in the 5–14 age group, which roughly corresponds to the number of elementary school-going children, has already begun declining in India and across all major states except Jammu & Kashmir.'

Bhatt (2011, 2018) and Guilmoto and Rajan (2013) were among the very few who discussed the abnormalities in the state's census. They suggested that only the *reported* CSR declined in 2011 and held the politicisation of census responsible for the over-reporting of children, particularly male children, in Kashmir, which explains the unexpected drop in the CSR and increase in the child population share. However, their discussion is limited to coverage errors in Kashmir's latest census data on the 0–6 age group. An analysis of the larger body of census data, including from earlier rounds, is needed to understand the nature and distribution of over-reporting and its impact on data quality, identify content errors, and uncover the political, economic, legal and administrative contexts of coverage and content errors. This chapter examines the data on child population, Generic Tribes, Gujjar and Bakarwal tribes, and speakers of Bhotia, Gojri, Halam, Ladakhi, Ponchi, Shina and Tibetan languages. It rules out conventional demographic and non-demographic explanations of the abnormal changes in the data on child population, Scheduled Tribes (STs) and non-scheduled languages, and presents estimates of the magnitude of errors.

## Child Population

The Eleventh Five-Year Plan (2007–12) had set the target of 949 for J&K's CSR (GoI 2015a). According to the projections of the Technical Group constituted by the National Commission on Population, the sex ratio of the population aged 0–9 years should have been about 918 in 2011 (GoI 2006a: 144). Moreover, conventional wisdom suggested that the Muslim-majority state would not follow its neighbours known for very low CSRs. Defying plans, projections and expectations,[11] J&K's CSR plummeted sharply from 941 in 2001 to 862 in 2011, compared to the decline in the corresponding national figure from 927 to 918 (Table 3.1).[12] J&K accounted for 6 out of 12 districts with a CSR more than 1,000 in 2001, but in 2011 these districts reported a CSR between 829 and 885.[13] The child population data of J&K stand out when compared to other states, and there are sharp differences even between regions and communities within the state.

**Table 3.1** Child population statistics of Jammu and Kashmir (J&K) and select states, 2011

| State | Sex ratio | | | Child sex ratio | | | Child population share | | |
|---|---|---|---|---|---|---|---|---|---|
| | 1991 | 2001 | 2011 | 1991 | 2001 | 2011 | 1991 | 2001 | 2011 |
| J&K | – | 892 | 889 | – | 941 | 862 | – | 14.65 | 16.10 |
| India (all states) | 927 | 933 | 943 | 945 | 927 | 918 | 17.94 | 15.93 | 13.59 |
| Himachal Pradesh | 976 | 968 | 972 | 951 | 896 | 909 | 16.25 | 13.05 | 11.33 |
| Punjab | 882 | 876 | 895 | 875 | 798 | 846 | 16.30 | 13.02 | 11.09 |
| Haryana | 865 | 861 | 879 | 879 | 819 | 834 | 18.98 | 15.77 | 13.34 |

*Source*: Prepared by the author using primary census abstracts (PCA) for 1991–2011.

## Interstate Comparison

The 2011 census population of J&K exceeded the projected population by 7.03 per cent (Table 3.2). Along with Bihar, Chhattisgarh and Tamil Nadu, it was among the four states and UTs with a population of more than 10 million, whose headcount differed from projected figures by more than 5 per cent. While the actual and projected overall sex ratios of the other three states differ by less than 12 points, the difference is 20 points in the case of J&K. For the lower age groups, the error in projected sex ratios is less than 50 points in the other three states and does not differ much across age groups. In J&K, the error is higher for the lower ages: 46 points (0–9 years) and 69 points (0–4 years). More importantly, unlike the other three states, J&K's reported sex ratios for lower-age groups were lower than the projected figures.

**Table 3.2** Population projections for Jammu and Kashmir (J&K), 2011

| Parameter | Projected (2011) | Reported (2011 census) |
|---|---|---|
| Population: all ages | 11,718,000 | 12,541,302 |
| Sex ratio: overall | 909 | 889 |
| Sex ratio: 0–9 years | 918 | 872 |
| Sex ratio: 0–4 years | 922 | 853 |
| Population share: 0–9 years (per cent) | 19.12 | 22.54 |
| Population share: 0–4 years (per cent) | 9.68 | 11.28 |

*Source*: Prepared by the author using data from GoI (2006a) and table C-13 of Census of India, 2011.

J&K's neighbouring states, Himachal Pradesh and Punjab, reported an improvement in their CSR during 2001–11 (Table 3.1). A district-level comparison shows that there has been a decline only on the J&K side of the respective interstate borders, except a slight decline in one district of Himachal Pradesh but from a much higher level. With the sharp drop in the CSR by 79 points, the highest among all the states, J&K unexpectedly emerged as the third-worst state in terms of the overall CSR (Figure 3.3a) and the second-worst in terms of urban CSR. The magnitude of this decline can be gauged from the fact that it is the first instance of decline in J&K's overall sex ratio after 1931

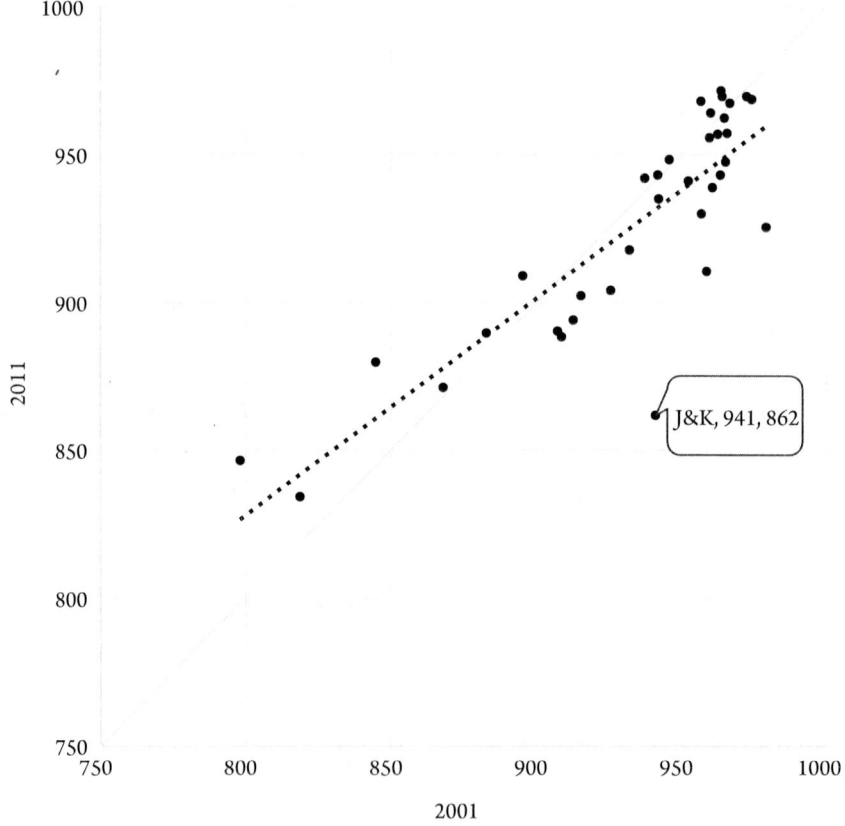

**Figure 3.3a** Child sex ratio (CSR) of the states and union territories (UTs) of India, 2001–11

*Source*: Prepared by the author using primary census abstracts (PCA) for 2001 and 2011.

*Note*: CSR is the sex ratio of the 0–6 age group.

(GoI 2011d: 91). J&K was also among the three states along with Bihar and Gujarat that reported a decline in the overall sex ratio in 2011, and among these it reported the largest decline. Bihar reported a minor decline in its CSR, while Gujarat reported an increase in its CSR, and neither state reported an increase in the child population share.

In fact, J&K is an outlier with respect to child population share as well. It was the only state other than Nagaland that reported an increase in the share of the 0–6 age group in the population (Figure 3.3b). This marked the reversal of the steady decline between 1961 and 2001 (Table 3.3a). J&K's child population

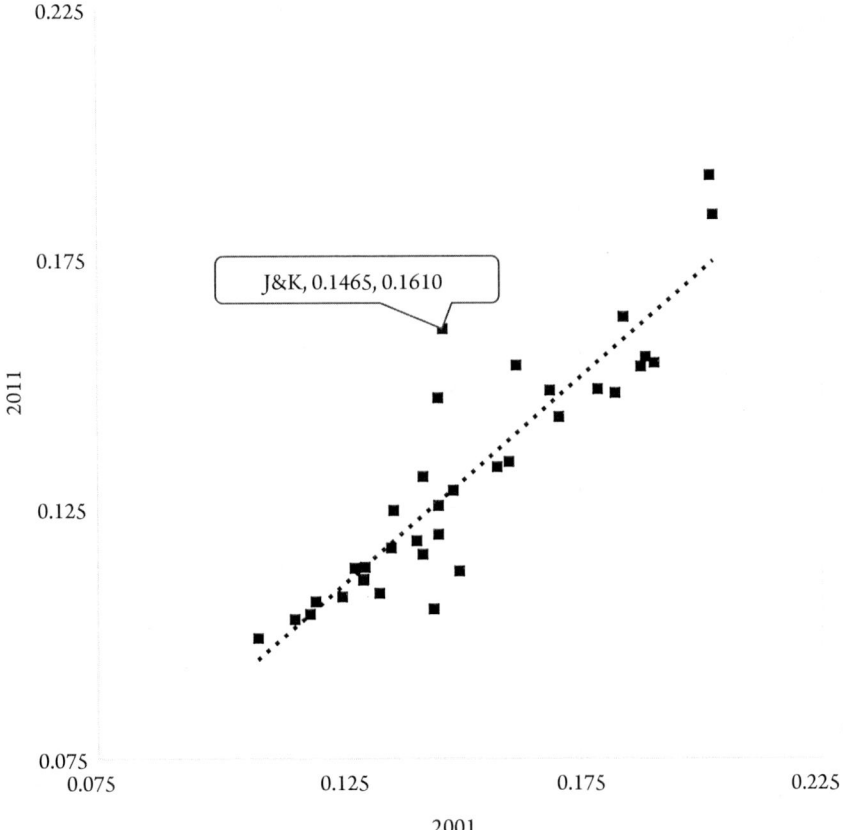

**Figure 3.3b** Child population share of the states and union territories (UTs) of India, 2001–11

*Source*: Prepared by the author using primary census abstracts (PCA) for 2001 and 2011.

*Note*: Child population share is the share of the 0–6 age group in the population.

Table 3.3a   Child population statistics of Jammu and Kashmir (J&K) and its two main divisions, 1961–2011

| Census | J&K | | | | | Jammu | | | | | Kashmir | | | | |
|---|---|---|---|---|---|---|---|---|---|---|---|---|---|---|---|
| | Share | | Sex ratio | | | Share | | Sex ratio | | | Share | | Sex ratio | | |
| | 0–6 | 0–9 | 0–6 | 0–9 | All | 0–6 | 0–9 | 0–6 | 0–9 | All | 0–6 | 0–9 | 0–6 | 0–9 | All |
| 1961 | 0.2139 | 0.2957 | 965 | 957 | 878 | N/A | 0.3172 | N/A | 956 | 898 | N/A | 0.2806 | N/A | 957 | 857 |
| 1971 | 0.2121 | 0.3016 | 959 | 958 | 878 | N/A | 0.3135 | N/A | 952 | 909 | N/A | 0.2933 | N/A | 961 | 848 |
| 1981 | 0.1924 | 0.2774 | 964 | 962 | 892 | 0.1946 | 0.2805 | 968 | 959 | 910 | 0.1915 | 0.2762 | 958 | 964 | 878 |
| 1991 | – | – | – | – | – | – | – | – | – | – | – | – | – | – | – |
| 2001 | 0.1465 | 0.2224 | 941 | 942 | 892 | 0.1506 | 0.2221 | 889 | 893 | 882 | 0.1436 | 0.2235 | 986 | 983 | 904 |
| 2011 | 0.1610 | 0.2254 | 862 | 872 | 889 | 0.1467 | 0.2087 | 863 | 864 | 886 | 0.1739 | 0.2408 | 858 | 874 | 897 |

Source: General reports for 1961–81; primary census abstracts (PCA) and table C-13 for 2001–11.
Note: (a) '–' indicates that census was not conducted. (b) 'N/A' indicates that disaggregated statistics are not available.

share stands out even in comparison to Nagaland because of the magnitude of the increase and the fact that both rural and urban areas of the state were affected, whereas only rural Nagaland reported an increase. In Nagaland, the increase is explained by the correction of accumulated errors in the headcount that led to a relatively larger contraction of the reported adult population in the rural areas (Agrawal and Kumar 2020a). Other than the political and commercial capitals of the state, Kohima and Dimapur, only Phek and Wokha managed to avoid a contraction of population in 2011. Parts of these districts reported anomalous changes in child population statistics. For instance, the CSR of the Kikruma rural development block dropped from 999 in 2001 to 853 in 2011, compared to a smaller drop in the corresponding ratio for Phek district from 926 to 913 (GoI 2004b: 25; GoI 2014g: 24). During this period, the child population share of Kikruma declined slightly from 17.11 to 16.93 per cent, whereas Phek reported a drop from 18.28 to 16.98 per cent. Sharp and counterintuitive changes in the 0–6 age population were also reported in parts of Manipur's Senapati district. Between 1991 and 2001, the CSR of three sub-divisions of the Senapati district of Manipur increased by 28 points, in contrast to the declining trend observed in the district and the state. In the subsequent decade, however, the CSR of the three sub-divisions dropped by 126 points, which swung the district's CSR downwards by more than 100 points. While the population of the 0–6 age group grew in the three sub-divisions and Senapati district in absolute terms between 1991 and 2001, its share in the overall population fell sharply due to relatively higher over-enumeration in older age groups. It bears emphasising that the anomalous changes in the 0–6 age population of both Nagaland and Manipur were driven by the manipulation of headcount governed by political and economic factors (Agrawal and Kumar 2020a, 2020b, 2020e).

Two other points are noteworthy. There is no reason why the child population share should increase with a decrease in the CSR. The experience of Punjab and Haryana suggests the contrary (Figure 3.3c). It is noteworthy that the increase in the population share of the 0–9 age group in J&K is driven by the increase in the 0–6 age group. This is also evident from a comparison between the population shares of single-year ages (0–9) in 2001 and 2011 (Figure 3.4). In J&K, the population share of lower ages is higher in 2011 compared to 2001, and the gap decreases and reverses for the 5–9 age group. In the case of India, the 2011 population share of each single-year age up to 9 years is less than the corresponding 2001 population, except those who had not yet completed the age of one year.

**Table 3.3b** Population share and sex ratio of Jammu and Kashmir (J&K) and its two main divisions by sector, 2001–11

| Population characteristic | J&K | | Kashmir | | Jammu | |
|---|---|---|---|---|---|---|
| | 2001 | 2011 | 2001 | 2011 | 2001 | 2011 |
| Sex ratio (all) | 892 | 889 | 904 | 897 | 882 | 886 |
| 0–6 sex ratio | 941 | 862 | 986 | 858 | 889 | 863 |
| 0–6 population share | 0.1465 (0.0755/0.0710) | 0.1610 (0.0865/0.0745) | 0.1436 (0.0723/0.0713) | 0.1739 (0.0936/0.0803) | 0.1506 (0.0797/0.0709) | 0.1467 (0.0787/0.0679) |
| 0–9 sex ratio | 942 | 872 | 983 | 874 | 893 | 864 |
| 0–9 population share | 0.2224 (0.1145/0.1079) | 0.2254 (0.1204/0.1050) | 0.2235 (0.1127/0.1108) | 0.2408 (0.1285/0.1123) | 0.2221 (0.1173/0.1048) | 0.2087 (0.1119/0.0967) |
| *Rural* | | | | | | |
| Sex ratio (all) | 917 | 908 | 928 | 912 | 907 | 908 |
| 0–6 sex ratio | 957 | 865 | 1002 | 859 | 907 | 870 |
| 0–6 population share | 0.1593 (0.0814/0.0779) | 0.1749 (0.0938/0.0811) | 0.1582 (0.0790/0.0792) | 0.1902 (0.1023/0.0879) | 0.1615 (0.0847/0.0768) | 0.1600 (0.0856/0.0744) |
| 0–9 sex ratio | 958 | 876 | 999 | 876 | 911 | 872 |
| 0–9 population share | 0.2397 (0.1224/0.1173) | 0.2442 (0.1302/0.1140) | 0.2431 (0.1216/0.1215) | 0.2627 (0.1401/0.1226) | 0.2370 (0.1240/0.1130) | 0.2265 (0.1210/0.1055) |

(*Contd*)

**Table 3.3b**  (*Contd*)

| | Urban | | Urban | | Urban | |
|---|---|---|---|---|---|---|
| Sex ratio (all) | 819 | 840 | 839 | 866 | 800 | 812 |
| 0–6 sex ratio | 873 | 850 | 923 | 858 | 808 | 826 |
| 0–6 population share | 0.1076 | 0.1241 | 0.1041 | 0.1388 | 0.1133 | 0.0998 |
| | (0.0575/0.0502) | (0.0671/0.0570) | (0.0541/0.0500) | (0.0747/0.0641) | (0.0626/0.0506) | (0.0546/0.0451) |
| 0–9 sex ratio | 878 | 857 | 923 | 871 | 817 | 820 |
| 0–9 population share | 0.1702 | 0.1755 | 0.1702 | 0.1935 | 0.1712 | 0.1462 |
| | (0.0906/0.0796) | (0.0945/0.0810) | (0.0885/0.0817) | (0.1034/0.0901) | (0.0942/0.0770) | (0.0803/0.0659) |

*Source*: Primary census abstracts (PCA) and table C-13 for 2001 and 2011.
*Note*: Male and female population shares are reported within parentheses.

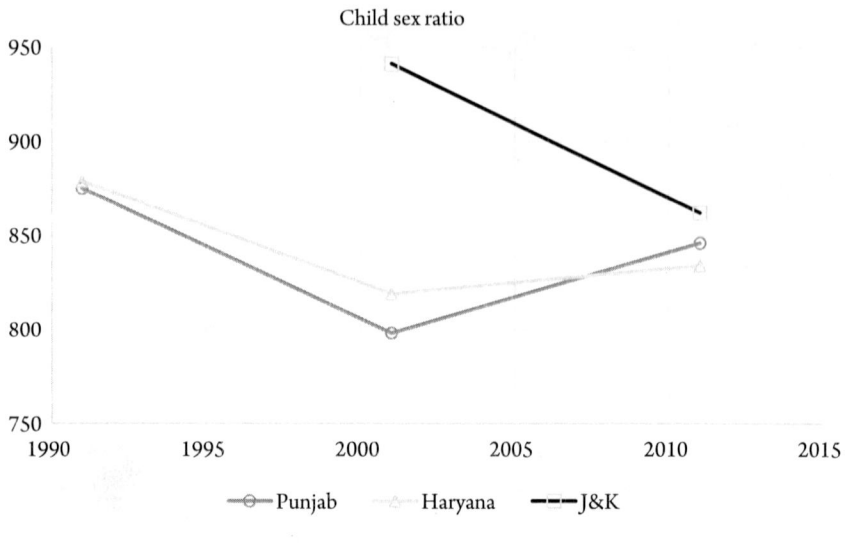

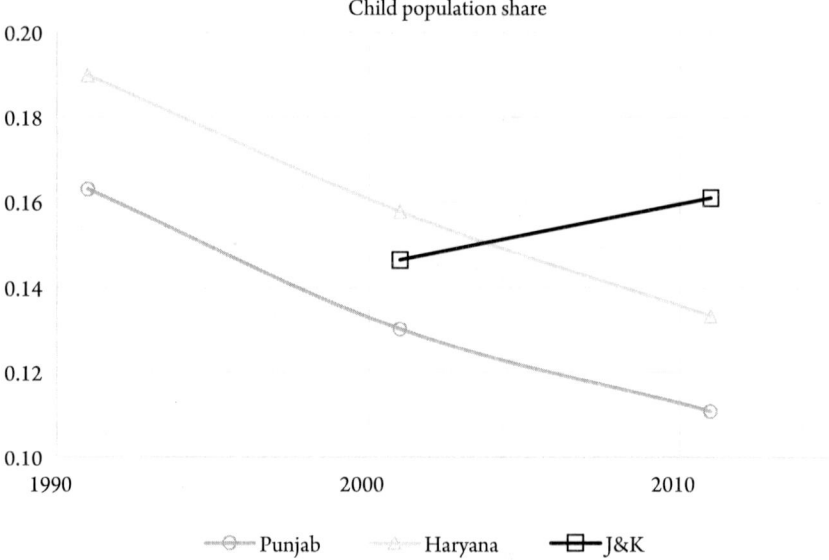

**Figure 3.3c** Jammu and Kashmir (J&K) and other states with low child sex ratios (CSR), 1991–2011

*Source*: Prepared by the author using primary census abstracts (PCA) for the respective years.

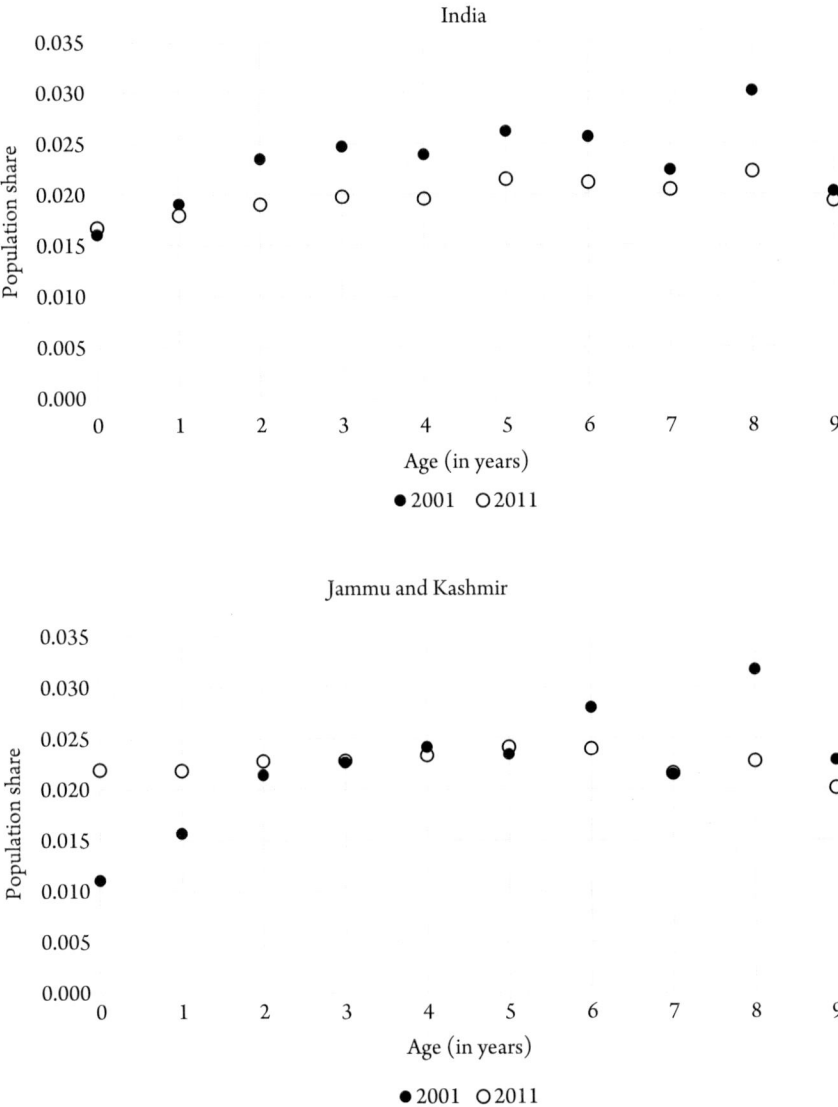

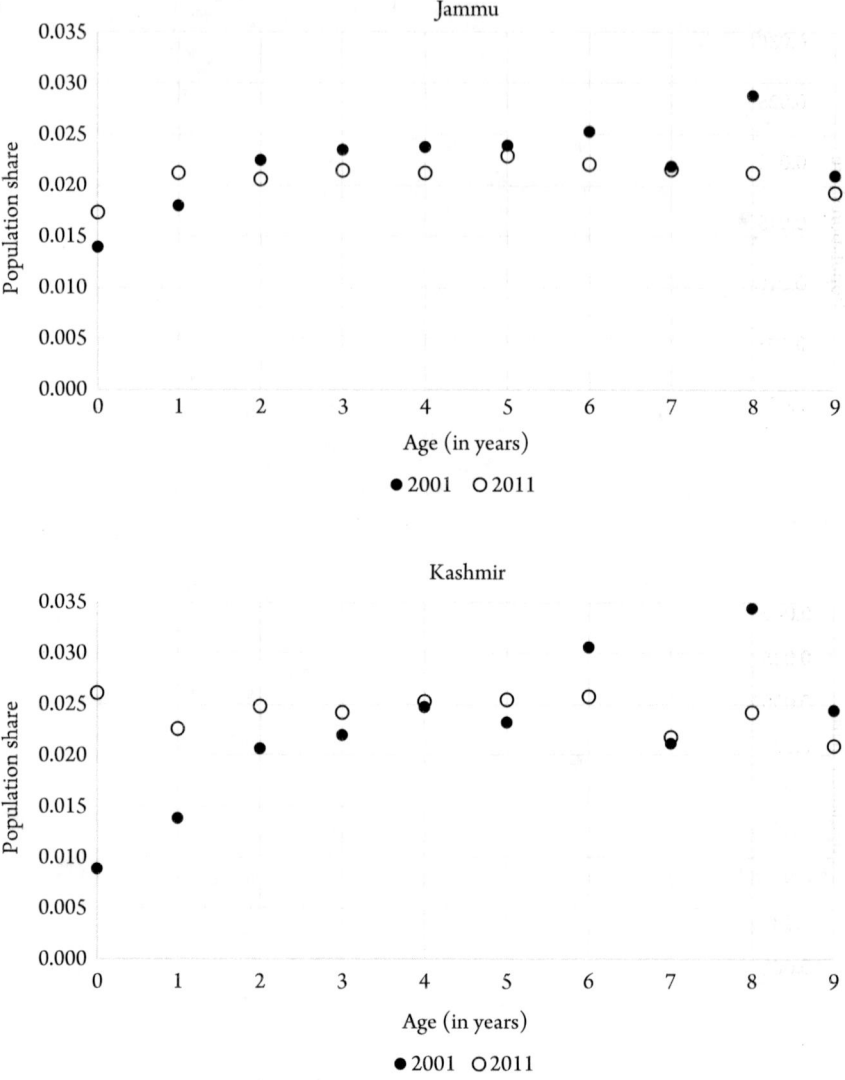

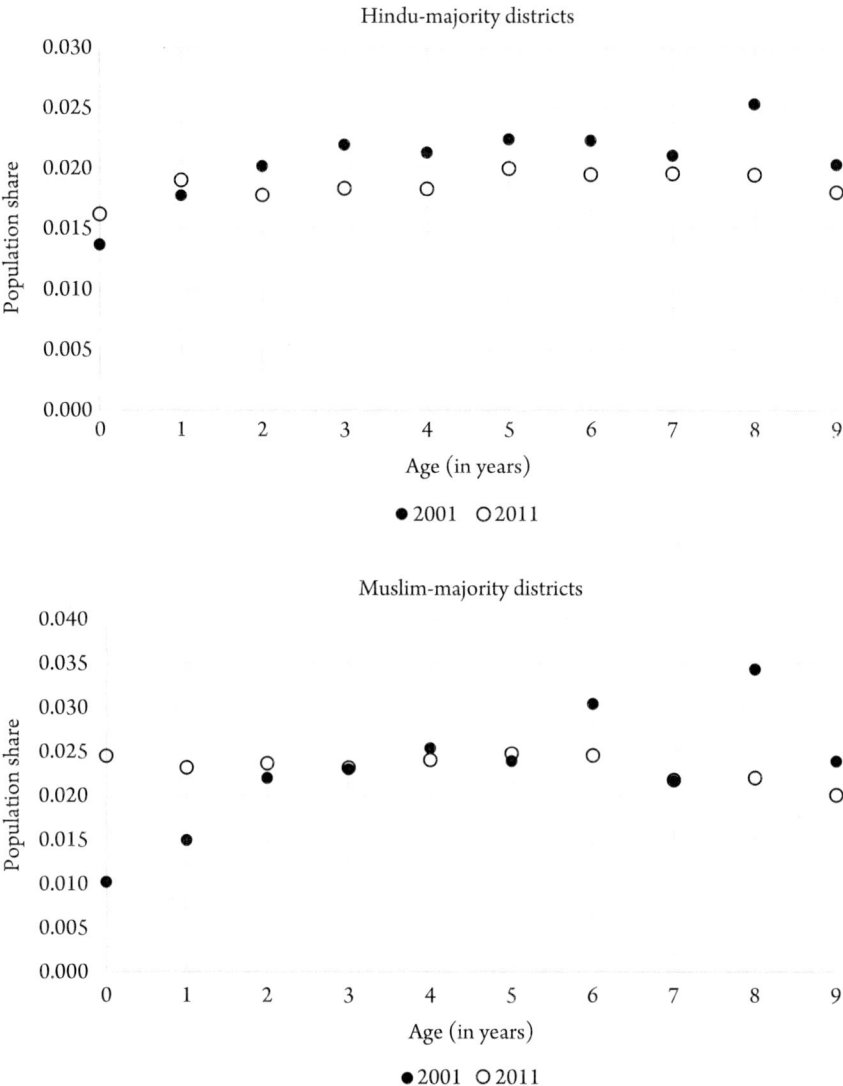

**Figure 3.4** Population share of single-year ages (0–9), 2001–11

*Source*: Prepared by the author using table C-13 of Census of India, 2001 and 2011.

*Note*: (*a*) In 2011, Jammu, Samba, Kathua and Udhampur were Hindu-majority districts of the state, while Leh was a Buddhist-majority district. In all other districts, Muslims constituted the majority. (*b*) For the sake of comparison, Reasi district, where Muslims are the largest community, has been included in Udhampur from which it was carved out after the 2001 census. Excluding it from Udhampur does not affect our conclusions.

The fact that Muslims have reported better sex ratios than Hindus across India makes the drop in the CSR in the overwhelmingly Muslim Kashmir even more puzzling (Tables 3.3–3.4). In J&K's extended neighbourhood, Muslim-majority Mewat reported the highest CSR among districts of Haryana in both 2001 and 2011, and Palwal with 20 per cent Muslim population was a distant second. The overall sex ratios of states with more than 25 per cent Muslim population – Assam, Kerala and West Bengal – have increased steadily since 1951 (GoI 2011b: 84–85). J&K, the only Muslim-majority state of India, broke ranks with these states in 2011. Between 2001 and 2011, the overall sex ratio of Muslims of India increased from 936 to 951, even as the CSR dropped slightly from 950 to 943. Muslims of Kashmir, too, registered a small increase in the overall sex ratio from 933 to 936, even though contrary to the trend for Muslims elsewhere their 0–9 sex ratio registered a much larger decline of 109 points from 984 to 875. Moreover, the population share of the 0–9 age group *decreased* by more than 2 per cent for Muslims of India, while that of Muslims of Kashmir *increased* by about 2 per cent. Figures 3.5a and 3.5b show that the Muslim population of J&K is an outlier in terms of both the CSR and the child population share, with only small union territories (UTs) or north-eastern states, which have very small Muslim populations, reporting comparable figures.

## Intrastate Comparisons

Our discussion so far suggests that the child population statistics of the state diverge substantially from those of the rest of the country, neighbouring states, insurgency-affected north-eastern states and states with large Muslim populations. We will now argue that there are significant differences in this regard within the state. We will examine the differences at the level of region, religion, caste, tribe and gender.

## Region

The anomalies in the data on J&K's child population are distributed unevenly within the state. While the state comprised three divisions – Jammu, Kashmir and Ladakh – we will focus on the first two as Ladakh accounted for barely 2 per cent of the state's population and its CSR registered a very small decline (Figure 3.2).

Kashmir reported much better sex ratios for lower age groups than Jammu in 2001 (Table 3.3a). The CSR of Kashmir was 97 points higher than that of Jammu in 2001, but dropped 5 points below Jammu's CSR even though the latter too declined by 26 points during this period. The sex ratio of the population aged

**Table 3.4** Population share and sex ratio of the two main divisions and religions of Jammu and Kashmir (J&K), 2001–11

| Population characteristics | India | | J&K | | Kashmir | | Jammu | |
|---|---|---|---|---|---|---|---|---|
| | 2001 | 2011 | 2001 | 2011 | 2001 | 2011 | 2001 | 2011 |
| | All | | All | | All | | All | |
| All ages sex ratio | 933 | 943 | 892 | 889 | 904 | 897 | 882 | 886 |
| 0–6 sex ratio | 927 | 918 | 941 | 862 | 986 | 858 | 889 | 863 |
| 0–6 population share | 0.1593 | 0.1359 | 0.1465 | 0.1610 | 0.1436 | 0.1739 | 0.1506 | 0.1467 |
| 0–9 sex ratio | 928 | 919 | 942 | 872 | 983 | 874 | 893 | 864 |
| 0–9 population share | 0.2321 | 0.1980 | 0.2224 | 0.2254 | 0.2235 | 0.2408 | 0.2221 | 0.2087 |
| | Muslim | | Muslim | | Muslim | | Muslim | |
| All ages sex ratio | 936 | 951 | 927 | 935 | 933 | 936 | 908 | 928 |
| 0–6 sex ratio | 950 | 943 | 980 | 871 | NA | 859 | NA | 909 |
| 0–6 population share | 0.187 | 0.1643 | 0.154 | 0.1804 | NA | 0.1783 | NA | 0.1902 |
| 0–9 sex ratio | 947 | 942 | 976 | 885 | 984 | 875 | 950 | 917 |
| 0–9 population share | 0.2724 | 0.2375 | 0.2366 | 0.2504 | 0.2279 | 0.2469 | 0.2696 | 0.2657 |
| | Hindu | | Hindu | | Hindu | | Hindu | |
| Sex ratio of all ages | 931 | 939 | 824 | 795 | 110 | 103 | 872 | 867 |
| 0–6 sex ratio | 925 | 913 | 855 | 834 | NA | 844 | NA | 834 |
| 0–6 population share | 0.156 | 0.1320 | 0.133 | 0.1204 | NA | 0.0295 | NA | 0.1260 |
| 0–9 sex ratio | 926 | 914 | 866 | 831 | 853 | 843 | 866 | 831 |
| 0–9 population share | 0.2277 | 0.1927 | 0.1966 | 0.1735 | 0.0422 | 0.0413 | 0.2027 | 0.1818 |

*Source*: GoI (2004a); table RL-0000 and table C-15 of Census of India, 2011.

0–9 years in Kashmir was 90 points higher than that in Jammu in 2001, but by 2011 Kashmir was barely 10 points ahead, even though the sex ratio of the 0–9 age group of Jammu declined by 29 points. The drop in the CSR was so severe in Kashmir that it dragged down the overall sex ratio of Kashmir as well as the state, even though Jammu reported a marginal increase in the overall

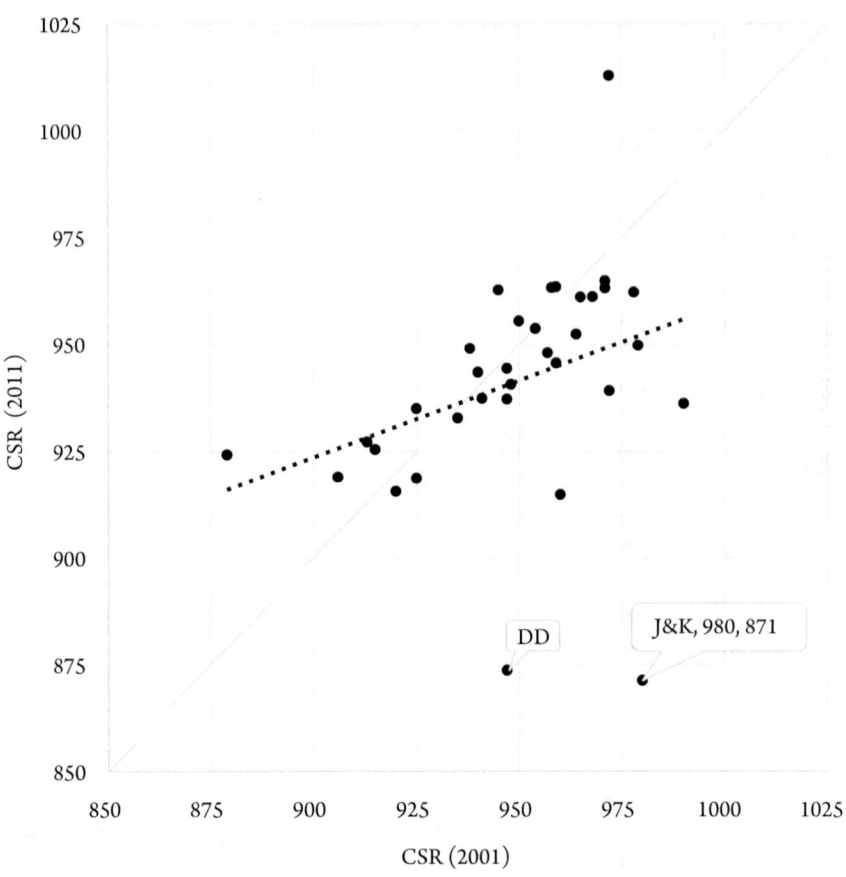

**Figure 3.5a**   Child sex ratio (CSR) (Muslim) of states and union territories (UTs), 2001–11

*Source*: Prepared by the author using GoI (2004a) and table RL-0000 of Census of India, 2011.

*Note*: (*a*) CSR (Muslim) is the sex ratio of the 0–6 age group among Muslims. (*b*) J&K: Jammu and Kashmir; DD: Daman and Diu.

sex ratio (Table 3.3a). It is also noteworthy that in Jammu, the decline in the sex ratio was higher for the 0–9 age group (29 points) than for the 0–6 age group (26 points). In contrast, in Kashmir the decline in the sex ratio was higher in the case of the 0–6 age group (128 points) compared to the 0–9 age group (109 points). A comparison of the distribution of all age groups for Kashmir and the other two divisions of the state shows that the population share of the 0–9 age group in 2011 exceeds the 2001 share only in the case

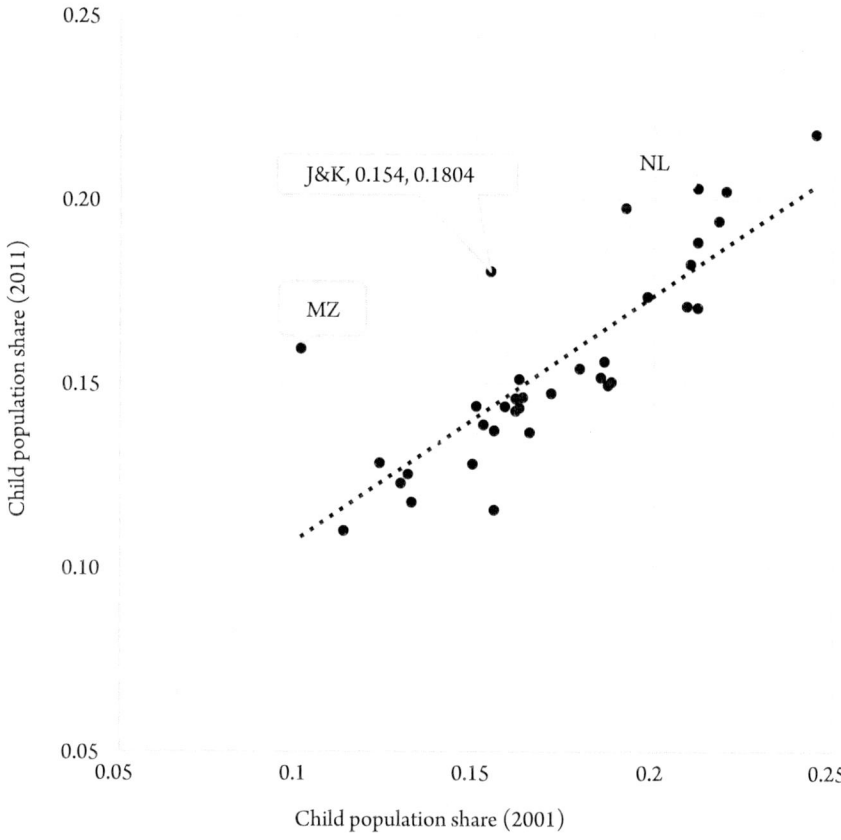

**Figure 3.5b**    Child population share (Muslim) of states and union territories (UTs), 2001–11

*Source*: Prepared by the author using GoI (2004a) and table RL-0000 of Census of India, 2011.

*Note*: (*a*) Child population share (Muslim) is the share of the 0–6 age group in Muslim population. (*b*) J&K: Jammu and Kashmir; MZ: Mizoram; NL: Nagaland.

of Kashmir (Figure 3.6). The 0–9 age group can be further divided into 0–4 and 5–9 age groups. In Kashmir region and the Muslim-majority districts of the state, the population share of the 0–4 age group was higher in 2011 compared to in 2001, and the gap decreased and reversed for the 5–9 age group. In the case of Jammu region and Hindu-majority districts, the 2011 population share of each single-year age up to 9 years is less than the corresponding 2001 population, except those aged one year or less (Figure 3.4).

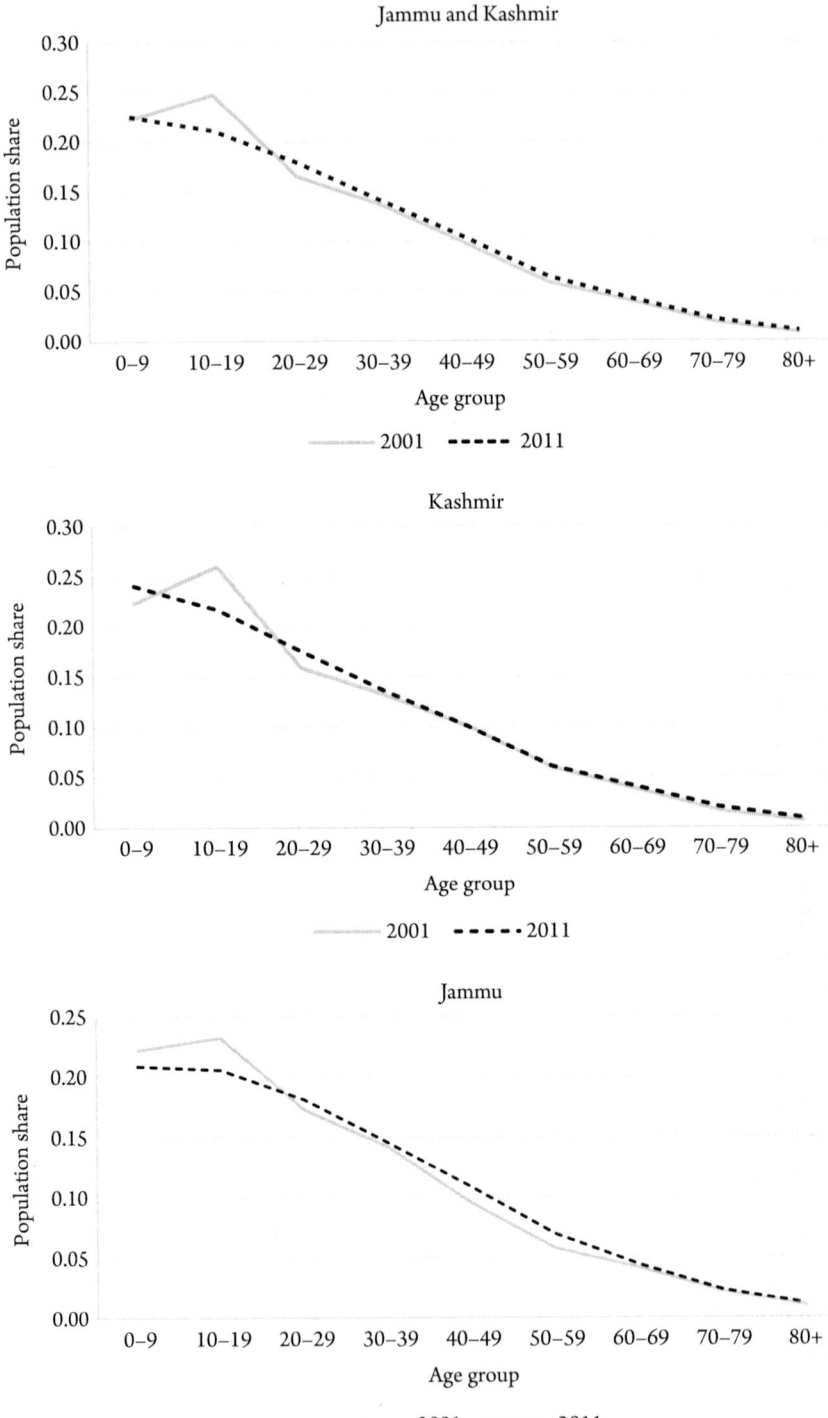

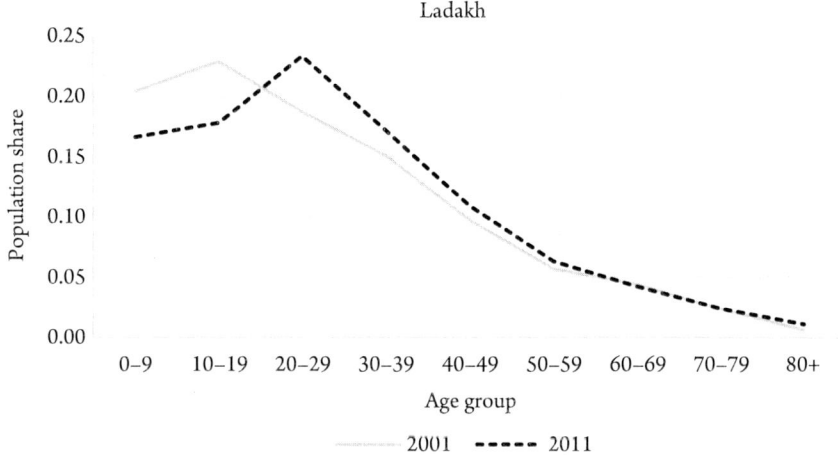

**Figure 3.6** Population distribution of Jammu and Kashmir (J&K) by age groups, 2001–11

*Source*: Prepared by the author using table C-13 of Census of India, 2001 and 2011.

At the district level, the least drop in the CSR reported in Kashmir was 63 points (Srinagar), which is close to the highest drop in the CSR in Jammu that was 66 points (Punch) (Figure 3.2). Several districts of Kashmir reported unusually sharp drops in the CSR between 2001 and 2011: Pulwama (217 points), Badgam (172 points), Ganderbal (151 points), Kupwara (142 points), Anantnag (136 points) and Shopian (133 points). Figure 3.7 captures the regional divide vis-à-vis the CSR within the state.

The drop in the CSR in Kashmir has been associated with an increase in the 0–6 age population share by more than 3 percentage points between 2001 and 2011, whereas the corresponding figure dropped in Jammu (Table 3.3a). The 0–9 age population grew at the rate of 36 per cent in Kashmir compared to 14 per cent in Jammu, whereas the older age groups grew at comparable rates in both the regions. Just as the drop in Kashmir's sex ratio was much steeper in the case of the 0–6 age group compared to the 0–9 age group, the increase in the population share of its 0–6 age group was also higher. In fact, the increase in the population share of the 0–9 age group is driven by the increase in the 0–6 age group because the population share of the 7–9 age group declined even in Kashmir.

Further, both rural and urban CSRs of J&K declined. Kashmir's rural CSR dropped from 1,002 to 859 and its urban CSR dropped from 923 to 858.

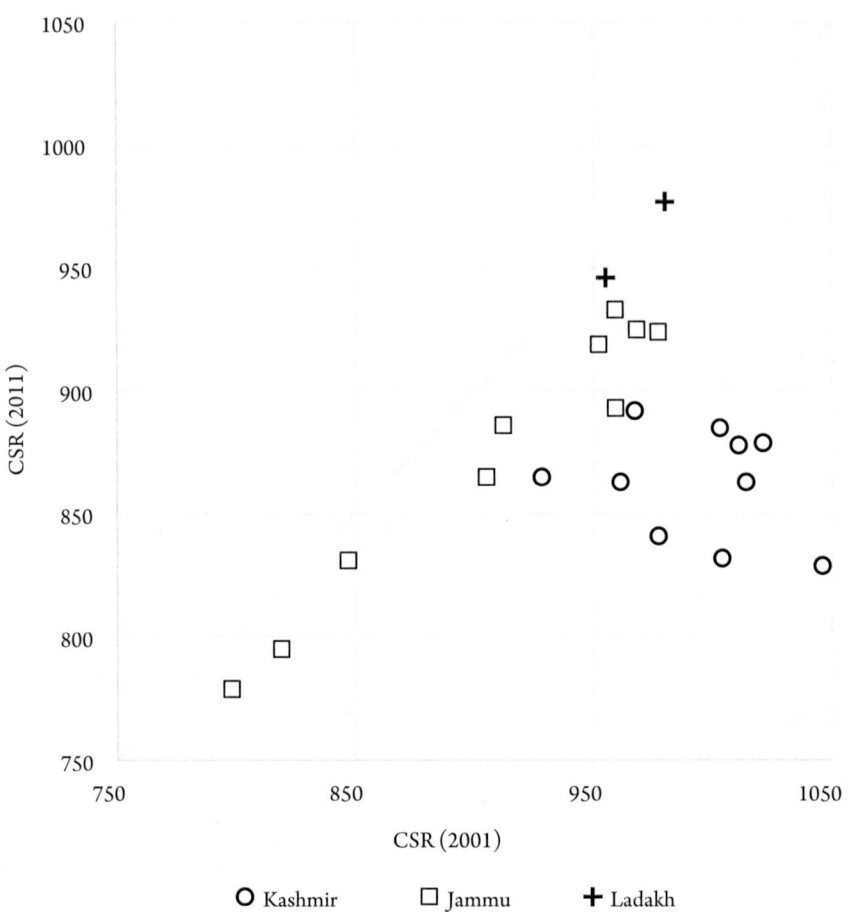

**Figure 3.7**   Child sex ratio (CSR) of the districts of Jammu and Kashmir (J&K), 2001–11

*Source*: Prepared by the author using GoI (2014a: 10–11).

In contrast, Jammu's rural CSR dropped from 907 to 870, while its urban CSR increased from 808 to 826 (Table 3.3b).

## Religion

Between 2001 and 2011, the sex ratio of the 0–6 (0–9) age group dropped by 21 (35) points among Hindus of J&K compared to 109 (91) points among Muslims (Table 3.4). The  sex ratio of the 0–6 age group is lower than that of the 0–9 age group among Muslims, while the opposite is true for Hindus.

At the same time, the population share of the 0–6 and 0–9 age groups decreased among Hindus, even as the population share of both these age groups registered a sharp rise among Muslims.

These observations hold good even when we compare religions within regions. During 2001–11, the population share of the 0–9 age group declined among Hindus in both Jammu and Kashmir. The corresponding figure increased sharply among Muslims only in Kashmir, while it decreased marginally in Jammu. The sex ratio of the 0–9 age group dropped by 109 points among Muslims of Kashmir compared to a 10-point drop among Hindus of Kashmir. The corresponding decline among Muslims of Jammu was 33 points, which is comparable to the 35-point decline among Hindus.

## Tribes and Castes

In 2001 (2011), Muslims accounted for more than 86 (88) per cent of the ST population in J&K (Table 3.5a). The sex ratio of the 0–6 and 0–9 age groups

**Table 3.5a**  Population share and sex ratio of Scheduled Tribes (STs) in Jammu and Kashmir (J&K) and its two main divisions, 2001–11

| Population characteristics | 2001 | 2011 |
|---|---|---|
| J&K | | |
| All STs | 1,105,979 | 1,493,299 |
| Muslim STs | 954,611 | 1,320,408 |
| Population share of Muslim STs | 86.31 | 88.42 |
| Sex ratio of all ages | 910 | 924 |
| Sex ratio of 0–6 age group | 979 | 912 |
| Population share of 0–6 age group | 0.1843 | 0.2030 |
| Sex ratio of 0–9 age group | 969 | 920 |
| Population share of 0–9 age group | 0.2726 | 0.2830 |
| Jammu | | |
| Sex ratio of all ages | 908 | 923 |
| Sex ratio of 0–6 age group | 963 | 907 |
| Population share of 0–6 age group | 0.1955 | 0.2004 |
| Sex ratio of 0–9 age group | 951 | 916 |
| Population share of 0–9 age group | 0.2849 | 0.2797 |

*(Contd)*

**Table 3.5a**   (*Contd*)

| Population characteristics | 2001 | 2011 |
|---|---|---|
| Kashmir | | |
| Sex ratio of all ages | 903 | 899 |
| Sex ratio of 0–6 age group | 1010 | 905 |
| Population share of 0–6 age group | 0.1899 | 0.2394 |
| Sex ratio of 0–9 age group | 1000 | 915 |
| Population share of 0–9 age group | 0.2863 | 0.3312 |

*Source*: Primary census abstracts (PCA), table C-14 (ST) and table ST-14 of 2001 and 2011.

among tribes decreased to a lesser extent than the drop in the sex ratio of the corresponding groups in the state and much less than the drop in the case of Kashmir and Muslims. However, the tribes based in Kashmir reported a sharper decline compared to their Jammu counterparts. Scheduled Castes (SCs) are almost entirely concentrated in Jammu, which accounts for more than 99 per cent of their population. With respect to both the CSR and the child population share, the SC population closely followed the trend of the Jammu division (Table 3.5b). As argued later, the declining share of SCs in the state's population and Hindus among STs of the state are artefacts of the errors in Kashmir's headcount.[14]

## Gender

The share of the 0–6 (0–9) age group in the male population of Kashmir increased by 4 (3) percentage points between 2001 and 2011, while the corresponding female population share increased by less than 2 per cent. The population shares of these age groups contracted in other parts of the state in both male and female categories except for a small increase in the share of the rural male child population in Jammu. The corresponding shares increased sharply across rural and urban Kashmir for both male and female categories, but the increase was larger in urban areas even as the difference between males and females in terms of the increase in the child population share was larger in rural areas.

This discussion shows that the decrease in the sex ratio and the increase in the population share of the 0–6 age group in J&K between 2001 and 2011 are accounted for by the abnormal changes in the child population, with greater anomalies in the case of the male child, particularly among non-tribal Muslims of Kashmir. The anomalies also contributed to the increase in the population shares of Muslims and Kashmir. The next section examines conventional explanations of these anomalous changes and estimates the extent of overcount.

**Table 3.5b** Population share and sex ratio of Scheduled Castes (SCs) in Jammu and Kashmir (J&K), 2001–11

| Population characteristics | 2001 | 2011 |
| --- | --- | --- |
| SC population in J&K | 770,155 | 924,991 |
| SC population share in J&K | 0.0759 | 0.0738 |
| Sex ratio of all ages | 910 | 902 |
| Sex ratio of 0–6 age group | 899 | 861 |
| Population share of 0–6 age group | 0.1520 | 0.1370 |
| Sex ratio of 0–9 age group | 907 | 860 |
| Population share of 0–9 age group | 0.2248 | 0.1978 |

*Source*: Primary census abstracts (PCA) and table C-14 (SC) of 2001 and 2011.

# Conventional Explanations

The temporal and spatial variations in the decline of the CSR have spawned a large literature on son preference and gender discrimination leading to sex-selective abortion or female foeticide, female infanticide and excess girl child mortality,[15] and how these vary across regions[16] and socio-economic categories.[17] We can add migration and conflict as factors that could skew the CSR. Before we discuss these factors, it would bear noting that coverage errors identified by the census cannot explain the anomalies observed in J&K. The census assesses coverage and content errors through post-enumeration checks (PEC) or post-enumeration surveys (PES) conducted after the household phase. While J&K was first covered by a PEC in 1961, it was not covered again until 2001.[18] In 2011, the net omission rate per thousand persons[19] for the northern zone, which included J&K, was 58.13 (40.92) and 56.87 (43.00), respectively, for male and female in the 0–4 (5–9) age group (GoI 2014d: table 3.2). These omission rates are lower than the figures reported in 1961 (GoI 1963a: 38) and 2001 (GoI 2006c: 25). So the anomalous drop in the CSR in J&K cannot be explained by the difference in the net omission rates between sexes identified by the PES.[20]

## Foeticide and Excess Girl Child Mortality

Sex-selective abortion is unlikely to be the driving force behind the sharp drop in J&K's CSR.[21] Kashmir's experience is inconsistent with the literature on the patterns and trends of the CSR generated by sex determination and foeticide, which suggest that the problem begins in urban areas and educationally and economically better-off groups. Guilmoto (2009: 524) suggests that sex determination is

> first adopted by 'pioneer groups' as a response to the fertility predicament common in Asia: reducing the number of children while maximizing the probability of having at least one son. As in many cases of diffusion of innovation, forerunners were initially mainly urban, well-to-do, and better-educated. Urban elites were the first to get information on and access to the new sex selection technology. The fast-rising sex ratio [defined as males per 100 females] at birth observed after a few years followed the diffusion of this new sex-selection strategy to new groups and to neighboring regions. This was made possible by the spread of information; the widening supply of the technology, mostly through private health care facilities; and the declining cost of ultrasound machinery.

In 2001, J&K's CSR was more than the all-India CSR, with the gap being larger in the case of Kashmir. In 2011, Kashmir reported a sharp drop simultaneously across both urban and rural areas, and more so in the latter. Even some of the remote villages in areas where there are hardly any modern health facilities, let alone private clinics, have reported very low CSR in Kashmir. In Jammu, the decline in the CSR was noted in 2001 when it dropped below 900, and, as expected, the decline began in urban areas and was sharper there as well. In 2011, urban Jammu reported an increase in the CSR even as rural areas reported a decline.

Further, it has been suggested that falling fertility rates explain a significant part of the decline in sex ratio (Jayachandran 2017: 134). Most sources, except the 2011 census, suggest that J&K has witnessed a steady decline in fertility. Even if we assume that the drop in the CSR is explained by sex-selective foeticide, it is not clear why this should be associated with a sharp increase in the child population share. In Punjab and Haryana, the child population share declined even when the CSR dropped (Figure 3.3c).

In fact, the sex ratio calculated using single-year age data from the 2011 census falls steadily as age decreases, and the population aged less than 1 has the lowest sex ratio – that is, 779 (Figure 3.8). The pattern revealed by the census is inconsistent with the SRS data. As per the SRS, sex ratio at birth (SRB) steadily increased in J&K between 2005 and 2011. The SRS sex ratio for the 0–4 age group during 2009–11 was higher than the corresponding census figures (SRS 2014: table 17). The difference between sexes in terms of the infant mortality rate (IMR) and the under-5 mortality rate is small and does not increase in the run-up to the 2011 census (GoI n.d.16: tables 8A, 10, 12).[22] In the absence of a widening gap between male and female mortality rates a rising SRB should lead to a rising CSR.

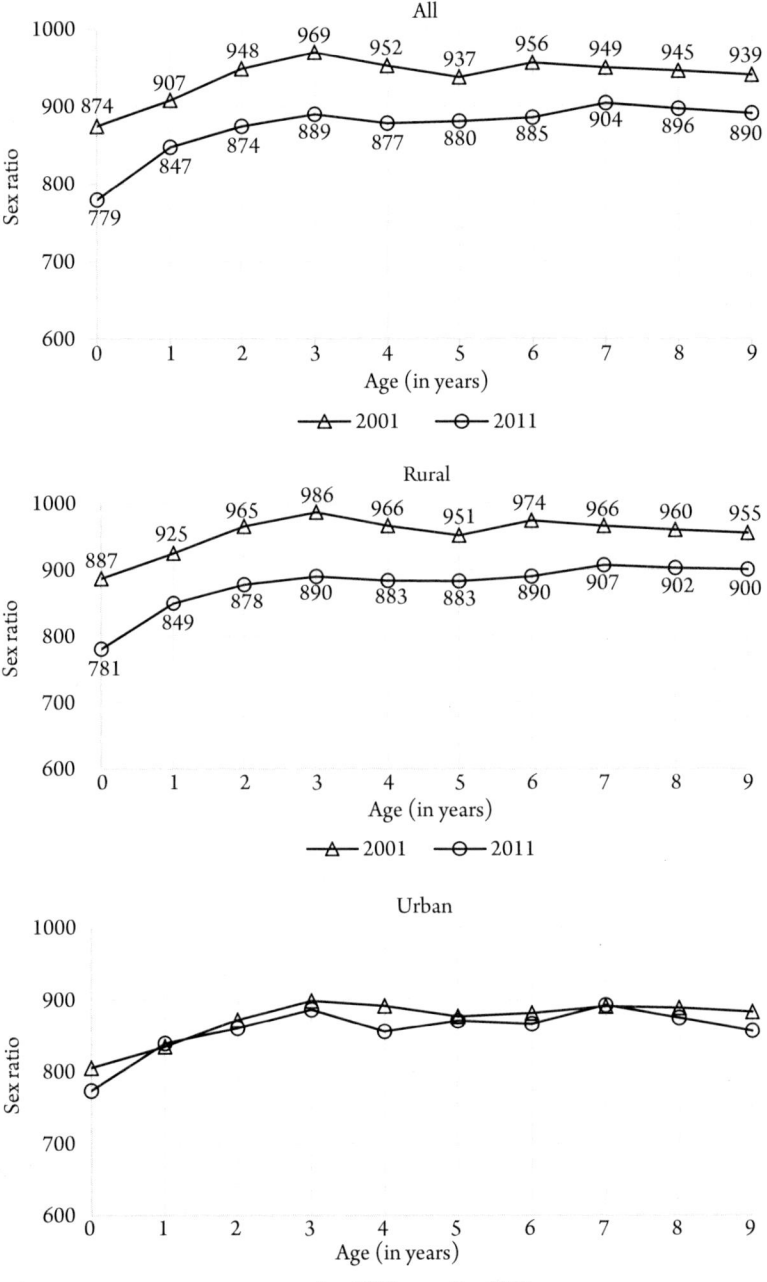

**Figure 3.8**  Age-specific sex ratio of Jammu and Kashmir (J&K), 2001–11

*Source*: Prepared by the author using primary census abstracts (PCA) and table C-13 of Census of India, 2001 and 2011.

Moreover, Bhat (2018: 183–84) points out that according to the state's HMIS, SRB was 909 in 2010–11 compared to 779 for the population aged less than 1, with the difference between the two sources being 200 in Kashmir and 41 in Jammu.[23] He adds that

> if we calculate the correlation between CSR at age 0 and proportion of children at age 0, the correlation coefficient between the two is –0.86 and is significant at 0.0. This indicates with an increase in proportion of children in age 0, there is a decline in CSR. This indicates that in the districts where there is a high proportion of *over reporting* of children in age 0, the sex ratio is in favour of males. (Bhat 2018: 178, emphasis added)

We carried out this exercise for the 0–6 population in urban and rural areas at the level of sub-districts and found that the correlations are negative for Kashmir and positive in the case of Jammu. Moreover, the decrease in the CSR and the increase in child population share are pervasive across *tehsil*s in both urban and rural areas of Kashmir (Figures 3.9a–3.9d).

## Migration

It can perhaps be argued that the drop in the *reported* CSR is explained by in-migration from low-CSR states after violence subsided in the early 2000s. This can be ruled out because migrants from outside the state accounted for less than 0.5 per cent of the *reported* 0–9 age population in 2011 and their population share declined between 2001 and 2011. The sex ratio of the 0–9 age population among migrants from outside the state was only slightly below the state's sex ratio for the 0–9 age population and increased between 2001 and 2011 (Table 3.6). Moreover, migrants from outside the state accounted for only a little more than 1 per cent of the state's reported census population, and their sex ratio improved between 2001 and 2011.[24] The out-migrants had much better sex ratio than the population enumerated within J&K, but this cannot explain the drop in the sex ratio of the state. Out-migrants were barely 1 per cent of the overall population of the state, and the sex ratio of out-migrants was better than the state in earlier censuses too (Table 3.6). So out-migration cannot explain the poor sex ratio reported in the 2011 census.

Since demographic factors such as variations in mortality rates and migration cannot explain the anomalies in child population, we will turn to non-demographic factors. Fertility is discussed alongside conflict in the next section.

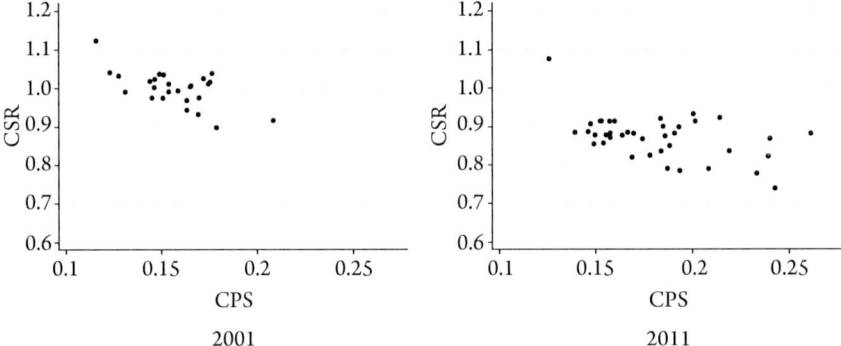

**Figure 3.9a** Child population share and child sex ratio (CSR) of rural Kashmir, 2001–11

*Source*: Prepared by the author using primary census abstracts (PCA) for 2001 and 2011.
*Note*: (*a*) CPS in these graphs stand for child population share. (*b*) The CSR is expressed as the ratio of the respective female to male populations rather than in terms of females per 1,000 males. (*c*) The number of *tehsils* changed between 2001 and 2011.

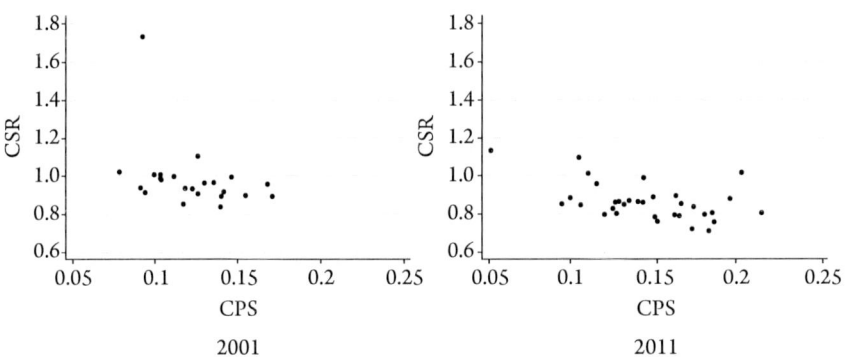

**Figure 3.9b** Child population share and child sex ratio (CSR) of urban Kashmir, 2001–11

*Source*: Prepared by the author using primary census abstracts (PCA) for 2001 and 2011.
*Note*: See the note for Figure 3.9a.

## Conflict

Conflict can affect sex ratio in several different ways, but most of them cannot account for changes in the CSR as their impact is confined to the adult population. First, an increase in the sex ratio is plausible in conflict zones because of, say, the under-reporting of adult males who fear conscription, which is not

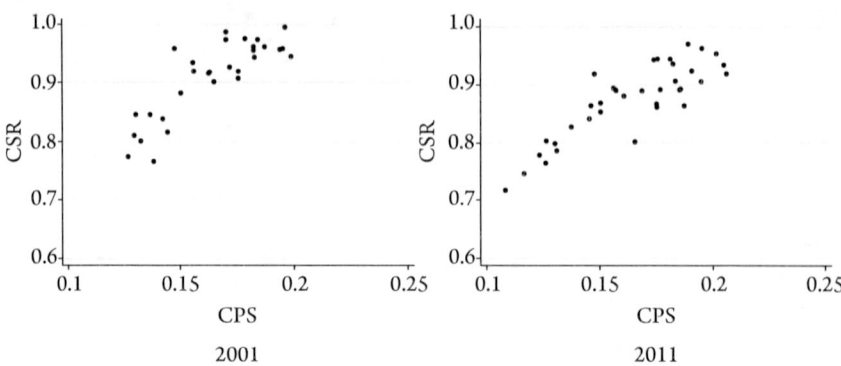

**Figure 3.9c**  Child population share and child sex ratio (CSR) of rural Jammu, 2001–11

*Source*: Prepared by the author using primary census abstracts (PCA) for 2001 and 2011.
*Note*: See the note for Figure 3.9a.

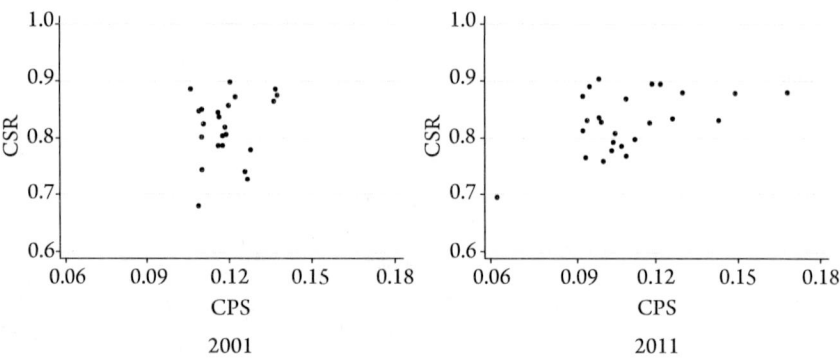

**Figure 3.9d**  Child population share and child sex ratio (CSR) of urban Jammu, 2001–11

*Source*: Prepared by the author using primary census abstracts (PCA) for 2001 and 2011.
*Note*: See the note for Figure 3.9a.

the case in India. Second, the non-enumeration of men living in rebel camps and a relatively higher mortality of men would increase the sex ratio. Third, conflict can also lead to a decrease in the sex ratio if the government deploys armed forces (and labourers) from other parts of the country, who will be counted in the conflict zone under the de facto method of enumeration.[25] The contribution of the deployment of the army to the change in the adult sex ratio between

**Table 3.6**  Migration statistics of Jammu and Kashmir (J&K), 1991–2011

| | 1991* | 2001 | 2011 | Source |
|---|---|---|---|---|
| Total population of J&K | – | 10,143,700 | 12,541,302 | A-2 |
| 0–9 age population of J&K | – | 22,56,366 | 28,26,857 | C-13 |
| 0–9 age population's sex ratio | – | 942 | 872 | |
| *Migration to places of enumeration in J&K* | | | | |
| All migrants | – | 464,229 | 829,109 | |
| Within the state of enumeration but from outside the place of enumeration | – | 374,523 | 748,861 | |
| From states in India beyond the state of enumeration | – | 86,768 | 72,909 | |
| From outside India | – | 2,938 | 7,174 | D-2 |
| Unclassifiable | – | 0 | 165 | |
| All migrants from outside J&K | – | 89,706 | 80,083 | |
| | | (+ 33,917)‡ | (+ 36,796)‡ | |
| Share of migrants from outside J&K in the state's population (per cent) | – | 0.88–1.22 | 0.64–0.93 | |
| Migrants aged 0–9 years from other states and union territories (UTs) | – | 19,501 | 12,223 | |
| Share of migrants from other states and UTs in the 0–9 age population (per cent) | – | 0.86 | 0.43 | D-12 (2001), D-13 (2011) |
| Sex ratio of all (0–9 age population) migrants from other states and UTs† | – | 951 (838) | 1,181 (862) | |
| *Migration out of J&K to other places of enumeration in India* | | | | |
| Out-migration* | 81,254 | 122,175 | 123,104 | |
| | (+ 12,867)‡ | (+ 21,624)‡ | (+ 94,453)‡ | D-2 |
| Sex ratio of out-migrants | 921 | 1,071 | 990 | |

*Source*: Prepared by the author using data from the sources listed in the 'Source' column.
*Note*: (*a*) The reference period for migration is 0–9 years of 'duration of residence in place of enumeration'. (*b*) Figures for out-migration are available for 1991 because only J&K was not enumerated in that year. (*c*) *Migrants for whom the place of last residence – 'rural' or 'urban' *tehsils* – is unclassifiable are excluded. (*d*) † Migrants whose place of last residence is unclassifiable as 'rural' or 'urban' and migrants from outside India are excluded. (*e*) ‡ Figures within parentheses add those who did not state the duration of migration and migrants whose country is unclassifiable.

2001 and 2011 will not be large because the deployment was already close to peak levels in 2001 in the aftermath of the Kargil War.[26] While any additional troop deployment cannot affect the CSR, it should *decrease* rather than *increase* the population share of the 0–6 age group.[27]

The possible impact of conflict-induced migration of the indigenous population on the sex ratio remains to be examined. The exodus of Pandits to Jammu cannot affect the state's sex ratio, but their out-migration from Jammu to other states for employment and education should have affected the sex ratio of the older age groups. This also applies to the out-migration of Kashmiri Muslims from Kashmir. As noted earlier, the sex ratio of out-migrants is better than that of the state, but their population share is very small (Table 3.6).[28] Another type of migration triggered by the conflict involved Kashmiri youth crossing the border into Pakistan. In a press conference after the release of provisional figures, the Joint DCO C. S. Sapru pointed out that 'those who have crossed to other side of LoC [line of control] were not included in the exercise' (M. Ali 2011a). This should have increased the overall sex ratio. These out-migrants would not have affected the CSR and their impact should have reflected in the adult sex ratio in the first census after the onset of conflict – that is, the 2001 census rather than the 2011 census. It has been argued that the migration to Pakistan would have increased the child population as the government encouraged the return and rehabilitation of the cross-border migrants with their Pakistani wives and children. Launched in 2010, the rehabilitation programme covered 'former militants who had crossed the LoC for arms training between 1989 and 2009, but had later given up arms' (M. Ahmad 2018).[29] This is unlikely to have affected the child population. First, circular migration in the inter-censal periods cannot explain abrupt demographic changes reported in censuses. Also, the return of former militants with families adds both adults and children to the population and cannot explain the sharp rise in the child population share. Second, the numbers involved are very small compared to the state's population as only about '15,000 boys crossed into PaK [*sic*]' (A. M. Watali quoted in Habibullah 2020: 23). The number of those returning with Pakistani wives and children was smaller. This agrees with Dulat's (2010: 154) account of his interaction with Irshad Malik, a former secretary general of the United Jihad Council, who suggested that there were '2,400 Kashmiris in Pakistan, most of whom are married and settled. Out of those, about 2,000 are unhappy with their meagre existence and the hostile environment. The rest have done well for themselves and are happy…. We could do very well by bringing that disgruntled 2,000 back.' Third, the scheme was launched barely a year before the 2011 census. In 2017, the state government informed the legislative assembly 'that only 377 ex-militants along with 864 family members have returned from Pakistan via Nepal and Bangladesh since 2010' (S. Yasir 2017; see also Hari Om 2013). It seems many of those who returned did not register with the authorities because they came via Nepal that was not among 'the four identified routes to return to Kashmir' (M. Ahmad 2018). Even the numbers of those who came via Nepal are not very large and spread over a longer time as this route became active as early as 2003. At least 489 men returned with their families via the Nepal route between 2003 and 2016 (ibid.). The population share of migrants from outside India in J&K increased from 2,938 to 7,174 between 2001 and 2011 (Table 3.6). However, only 618 of these international migrants named Pakistan as the place of last residence. Fourth, a census official based in Srinagar who was closely involved with the 2011 census argued that 'families get the cross-border migrants recorded in all official databases' to avoid harassment by authorities that are alarmed by missing youth (interview, Srinagar, 1 December 2015).

Conflict can possibly affect fertility and explain the abnormal increase in the child population share. For a cross section of countries, Janus (2013: 493) found a positive correlation between ethnic diversity and fertility in the context of weak institutions. He suggests that fertility could be a strategic choice in such settings for ethnic groups engaged in conflict because (a) higher fertility increases 'voting power and gains from political office'; (b) 'if ethnic groups allocate society's resources via conflict or bargaining in the shadow of conflict, then fertility might increase their combat strength', and (c) 'larger groups can more easily impose their language or culture on the rest of society'. Alternatively, as Smith (2012: 1512) puts it, 'babies are sites at which geopolitical strategy is animated and made material. The birth of a child contributes to territorial projects: The number of future voters, the number of future soldiers, and the demographic distribution of citizens populating and constituting state territory are determined by this complicated decision.' Morland (2018: 36–38), too, suggests that in ethnically diverse territories communities might deploy 'hard demographic engineering' strategies such as higher fertility. Attane and Courbage (2000: 268, 275) argue that higher fertility among the non-Sinicised minorities could possibly be a defence against the demographic invasion by Han settlers and 'a means of affirming identity'.[30] In the north-eastern states such as Meghalaya and Mizoram, we occasionally come across appeals to 'the tribal people ... to have more children. The appeal is seen as an outcome of the fear that "outsiders", primarily non-tribal migrants, could outnumber the indigenous communities in the near future' (Karmakar 2018). So conflict can possibly result in an increase in the fertility of minorities.[31] Occasional pro-natalist rhetoric notwithstanding, Kashmir has witnessed a steady decline in fertility.

A senior official associated with the 2011 census suggested that the increase in the child population may be explained by the 'baby boom' after the decline in violence in the 2000s (interview, 6 August 2021). It is not clear why this applies only to Kashmir and not even adjoining districts of Jammu, let alone north-eastern states, which, too, witnessed a sharp decline in violence during this period. Moreover, the 'baby boom' hypothesis leaves the large and unexpected drop in the CSR unexplained.[32]

Kumar and Sathyanarayana (2012: 68, 71) note that J&K was the only state for which the CBR estimated using census data increased between 2001 and 2011. The NFHS and the SRS, the two major sources of demographic information, do not agree with the census in J&K (Table 3.7 and Figure 3.10).[33] According to the SRS (NFHS), the CBR fell from 20.9 (23.1) in 1999 to 18.7 (20.9) in 2006 and 15.7 (17.7) in 2016, and the TFR fell from 2.3 (2.38) in 2005–06 to 1.70 (2.01) in 2015–16, which cannot explain the sharp rise in the child population share. In fact, Pulwama district, which reported the largest drop in the CSR during 2001–11 and one of the largest increases in the child population share, had one of the lowest TFRs as per the 2001 census (Guilmoto and Rajan 2002: 668–69). Regional-level estimates, too, confirm a declining CBR in Kashmir. The SRS divided the state into three natural regions: Jhelum Valley (Kashmir and Ladakh), Outer Hills (Doda, Punch, Rajauri, Udhampur, Kishtwar, Ramban and Reasi) and Mountainous (Jammu, Kathua and Samba). The SRS has consistently reported lower CBRs for the Mountainous and Jhelum Valley regions than Outer Hills (SRS 2014, 2017). The rise in the child population share is also inconsistent with the findings of NFHS-3 and NFHS-4 that suggest that the ideal family size did not increase and the use of family planning methods did not decrease in the state. Both NFHS-3 (International Institute for Population Sciences [IIPS] and Macro International 2009: 6) and NFHS-4 (IIPS and ICF International 2017a: 6) suggest that the TFR declines with the years of schooling of women. The state's female literacy rate increased from 43.00 to 56.43 per cent between 2001 and 2011, and the increase was larger in Kashmir than in Jammu. A senior health department official argued that the increase in the child population share is explained by the fact that Kashmir had more women in the 15–49 age group (interview, 12 August 2021). To the contrary, the population share of this group in Kashmir was lower than in Jammu and the rest of the country in both 2001 and 2011, and, in any case, this hypothesis cannot account for the drop in the CSR. So the abrupt rise in the child population share reported in 2011 is inconsistent with a range of covariates of fertility reported in the census as well as other sources.[34] This is also confirmed by the state's HMIS, which recorded about 1.72 lakh births during 2010–11 (census year) (Bhat 2018: 183) compared to the 2.75 lakh population aged less than 1 reported in the census. The aforesaid health official also suggested that the 2011 child population was not overestimated by the census as the department managed to achieve vaccination targets linked to the projected child population. A researcher, however, suggested that the data on vaccination is not reliable as health workers fudge entries to match targets (interview, Srinagar, 23 October 2021).

Further, according to the NFHS, the CSR increased from 903 to 917 between 2005–06 and 2015–16.[35] The sex ratio of the 0–4 age group as per the SRS has been steadily increasing at least since 2009. Likewise, school enrolment data show that the sex ratio of children enrolled in primary classes (classes 1 to 5) has been increasing since 2005 (Figure 3.11a). Children in the 0–6 age group in 2011 would have been in classes 1 through 6 in 2017. We have data on enrolment until 2016, which show that the sex ratio was never less than 893 for any of these classes during 2011–16.[36] Given the relatively lesser importance attached to girl child education and the consequent likelihood of under-enrolment, the actual sex ratio should be higher than the sex ratio of the children enrolled in schools. Further, the sharp rise in the 0–6 age population reported in the census should have translated into a higher enrolment between 2011 and 2017, but enrolment stagnated in line with other sources of demographic information that reveal a decline in fertility and birth rates (Figure 3.11b).[37] This contradicts the claim about meeting vaccination targets linked to projections based on the 2011 census.

In short, the reported drop in the CSR cannot be explained by the change in the intensity of conflict unless it is assumed that conflict accentuates sex-selective foeticide or infanticide.[38] Even if this assumption is maintained despite the absence of any substantial difference in the male and female child mortality rates, the impact should have been more pronounced in the 1990s, rather than the 2000s, which witnessed a semblance of normalcy, at least between 2002 and 2008. Also, the large increase in the child population share would remain unexplained.

## Other Explanations

We have so far seen that both the increase in the child population share and the drop in the CSR reported in the 2011 census are inconsistent with other sources of demographic information. Conventional demographic and non-demographic

**Table 3.7** Census, Sample Registration System (SRS) and National Family Health Survey (NFHS)

| Decade | Year | Source | CBR | CDR | TFR (WTFR) | Decadal population growth rate (per cent) | Population share (per cent) 0–6 | Population share (per cent) 0–9 | Sex ratio 0–6 | Sex ratio 0–9 | Sex ratio All |
|---|---|---|---|---|---|---|---|---|---|---|---|
| 1962–71 | 1971 | Census | | | | 29.65 | 21.21 | 30.16 | 959 | 958 | 878 |
| | 1972 | SRS† | 31.6 | 10.8 | | 22.86 | | | | | |
| 1972–81 | 1981 | Census | | | 3.7§ | 29.69 | 19.24 | 27.74 | 964 | 962 | 892 |
| | 1981 | SRS† | 31.6 | 9.0 | | 25.04 | | | | | |
| 1982–91 | 1991 | Census | | | | | | | | | |
| | 1990 | SRS† | 31.4 | 7.9 | N/A | 26.15 | | | | | |
| | 1998–99 | NFHS-2‡ | 23.1 | N/A | 2.71 (1.74) | 18.13 | | | | | |
| 1992–2001 | 1999 | SRS† | 20.9 | 6.3 | | 15.60 | | | | | |
| | 2001 | Census^ | 24.5 | N/A | 2.98 | 29.43 | 14.65 | 22.24 | 941 | 942 | 892 |
| | 2001 | SRS† | 20.2 | 6.1 | 2.40# | 15.03 | | | | | |
| | 2005–06 | NFHS-3‡ | 20.9 | N/A | 2.38 (1.60) | 16.05 | | 20.7 | 903 | | 976 |
| | 2006 | SRS† | 18.7 | 5.9 | 2.3 | 13.56 | | | | | |
| 2002–11 | 2011 | Census^^ | 25.9 | N/A | 3.0§§ | 23.64 | 16.10 | 22.54 | 862 / 880* / 872** | 872 | 889 |
| | 2011 | SRS† | 17.8 | 5.5 | 1.9 | 13.02 | | | | | |

(Contd)

**Table 3.7** (*Contd*)

| | | | | | | | | |
|---|---|---|---|---|---|---|---|---|
| 2013 | SRS† | 17.5 | 5.3 | 1.9 | 12.81 | | | 972 |
| 2015–16 | NFHS-4‡ | 17.7 | N/A | 2.01 (1.67) | 13.45 | 17.6 | 917 | |
| 2012–21 | | | | | | | | |
| 2016 | SRS†† | 15.7 | 5.0 | 1.7 | 11.23 | | 906* | |

*Source:* IIPS and Macro International (2009: 38–39); IIPS and ICF International (2017a: 45); Census: table A2 for 2011, § GoI (1989b: 23) and §§ GoI (2020b: 6.53); CBR estimates are from ^Guilmoto and Rajan (2002) for the reference period 1994–2001 and ^^Kurrar and Sathyanarayana (2012: 68) for the reference period 2004–10; SRS: † GoI (n.d.16), †† SRS (2017).

*Note:* (*a*) CBR: crude birth rate; CDR: crude death rate; TFR: total fertility rate; WTFR: wanted total fertility rate. (*b*) * The sex ratio at birth (SRB) is the three-year average (2009–11 for 2011 and 2014–16 for 2016). ** Indicates the sex ratio (0–4 years) for 2009–11 . The sex ratio (0–4 years) for the year 2011 is 912. (*c*) ‡ The decadal population growth rates for the NFHS are calculated using the SRS estimate of the CDR. The estimates of the sex ratio in the NFHS are based on the de facto population (the place where each person stayed the night before the survey interview). The NFHS fertility rates are for the period 1–36 months preceding the survey (approximately 1996–98 for NFHS-2, 2003–05 for NFHS-3 and 2013–15 for NFHS-4). (*d*) The TFR is based on births to women in the age group of 15–49 years during the three years preceding the survey. The WTFR excludes unwanted births from the numerators of the age-specific fertility rates on which the TFR is based. # The value of TFR corresponds to the year 2004. (*e*) The sex ratio is the number of females per 1,000 males.

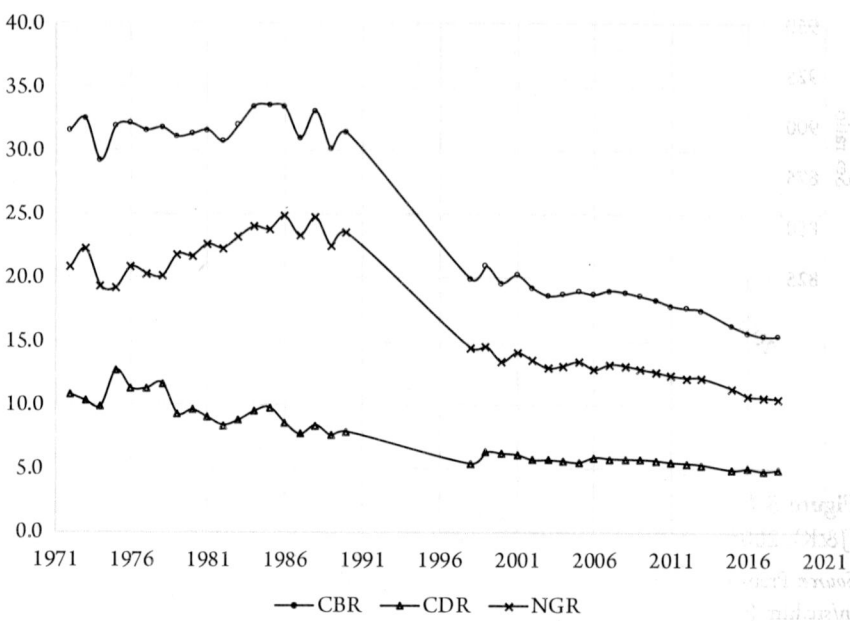

**Figure 3.10** Birth and death rates of Jammu and Kashmir (J&K) as per the Sample Registration System (SRS), 1972–2018

*Source*: GoI (n.d.16) for 1972–2013; 'SRS Bulletins', https://censusindia.gov.in/census. website/data/SRSB (accessed on 9 January 2023), for 2015–18.

*Note*: (*a*) Vital rates are not available for 1991–97 due to the non-receipt of returns. (*b*) Following the SRS, the natural growth rate (NGR) is calculated as the difference between the crude birth rate (CBR, births per 1,000 population in a year) and the crude death rate (CDR, deaths per 1,000 population in a year).

explanations based on fertility, mortality, migration and conflict cannot explain the reported change in the child population during 2001–11. The reported change may, however, be related to coverage and content errors.

Cohn (1987: 236–37) draws attention to the under-enumeration of women irrespective of marital status in nineteenth-century India. It has been argued that the decline in the sex ratio in British India between 1901 and 1946 'was a spurious result of deterioration in the completeness of registration system' (P. Visaria cited in Mari Bhat and Zavier 2007: 2292). More recently, Nagaland's reported sex ratio increased between 1991 and 2001 (Agrawal and Kumar 2020a: 135) and the CSR of three sub-divisions of Manipur varied sharply (Agrawal and Kumar 2020e: 10–11) due to coverage errors. The skewed CSR could also be an artefact of differential misreporting of the age of children

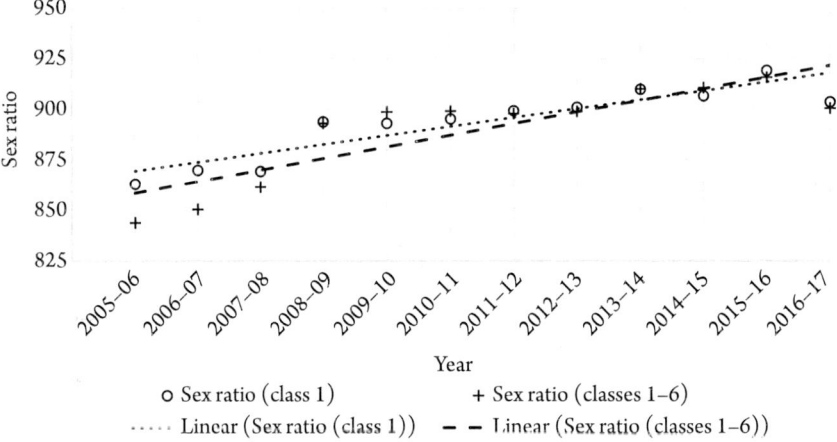

**Figure 3.11a** Sex ratio of children enrolled in schools of Jammu and Kashmir (J&K), 2005–17

*Source*: Prepared by the author with data from 'Elementary State Report Cards', http://udise. in/src.htm (accessed on 9 September 2021).

*Note*: (*a*) Record date for the data on school enrolment is 30 September, which is different from the reference date of the census, 1 March. (*b*) 'Sex ratio (class 1)' signifies the sex ratio for all students enrolled in the class 1 in government and private schools, madrasas and unrecognised schools. 'Sex ratio (classes 1–6)' signifies the corresponding ratio for all those enrolled in classes 1 through 6.

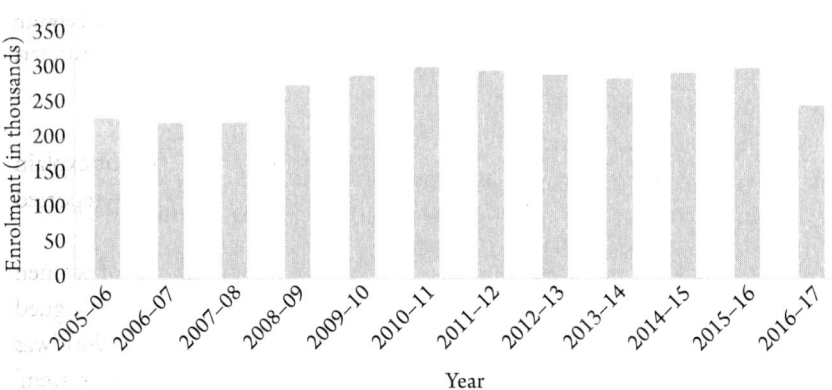

**Figure 3.11b** Children enrolled in class 1 in Jammu and Kashmir (J&K), 2005–06 to 2006–17

*Source*: Prepared by the author with data from 'Elementary State Report Cards', http://udise. in/src.htm (accessed on 9 September 2021).

(Mari Bhat 2002a, 2002b; Mari Bhat and Zavier 2007: 2293),[39] misclassification of the sex of children (*Mint* 2009), under-reporting of girls (Mari Bhat and Zavier 2007: 2302; Goodkind 2011) and over-reporting of boys (Guilmoto and Rajan 2013). J&K's census reports have noted the possibility of under-reporting of children (GoI 1971b: 143) and misreporting of age (GoI 1978b: 3) but not the sex dependence of the reporting errors.[40]

The CSR is also sensitive to government intervention. Delhi's SRB registered a sharp increase from 848 females per 1,000 males in 2007 to 1,004 in 2008 supposedly due to a state-sponsored Ladli Scheme. The sex ratio plunged to 915 the following year, and the 2011 census results raised further doubts about the scheme's success in tackling poor CSR. Financial incentives offered under the scheme encouraged better registration of girls and, possibly, even the registration of some boys as girls, leading to a sharp increase in the reported SRB in 2008 (*Mint* 2009; *Times of India* 2010). China's one-child policy is an example of how government policies can negatively affect the sex ratio (Ebenstein 2010).

The reported child population is particularly susceptible to manipulation as adding non-existent children is one of the easiest ways of inflating headcount. Commenting on the last colonial census, a member of the Constituent Assembly pointed out that 'the census is taken advantage of for bringing into existence a large number of children. That is unfortunately what happened in Bengal and other places' (Maheshwari 1996: 141).[41] An imbalance in the sex ratio of the fictitious children can skew the reported CSR. Changes in the *reported* CSR therefore need not be linked to demographic and other factors that affect the *actual* CSR. So anomalous child population statistics can result from inadvertent errors, which are ruled out in this chapter, and/or deliberate misreporting, which is examined in the later chapters.

Inadvertent errors cannot explain J&K's anomalous child population statistics. First, the state's anomalous census statistics on the child population cannot be attributed to mistaken (or even deliberate) misreporting of real girls as boys. Otherwise, the data will have to be corrected by reclassifying the excess male child population as female. Even without such reclassification the female child population share of Kashmir is anomalously larger in 2011 than 2001. Second, the abnormality cannot be accounted for by assuming that all those for whom information on age is not available ('age not stated') fall in the 0–6 age group because the sex ratio of this category is even more skewed. Moreover, if we maintain this assumption, the child population share that is already very high will increase further. Third, a census official attributed the errors to the double counting of nomadic groups and those who work outside their native places and get counted in both their place of birth and their place of work (interview,

Srinagar, 30 November 2015). Double counting may explain the inflation of (the adult) population but not the drop in the CSR. Fourth, it can perhaps be argued that the adult population was inflated in 2001 but not in 2011. So the child population share was underestimated in 2001 and increased when a proper census was conducted in 2011. This leaves the sharp drop in the CSR unexplained.

There are two other potential explanations linked to coverage errors. As discussed later in this chapter, the explosive growth of Generic Tribes is largely accounted for by ghost population. The population share of Generic Tribes was less than 1 per cent, and their CSR was far above the state average.

Yet another channel through which conflict can possibly affect the CSR is by reducing the reach of census that, unlike other channels discussed earlier, affects the entire headcount rather than just a sub-group. It cannot be argued that the reach decreased in 2011 compared to 2001 because the intensity of armed conflict declined in the 2000s and pro-independence groups also supported the census in 2011. It can also not be argued that the decrease in the reach was more in the case of the girl child. In fact, coverage errors in 2001 cannot even explain the increase in the child population share. Several officials in Srinagar and New Delhi associated with the 2011 census suggested that poor coverage in 2001 could explain the increase in the child population share in 2011 when peaceful conditions allowed better enumeration.[42] This explanation does not hold good for the following reasons. First, this should also apply to other states, say, in the north-east, which emerged out of the violent phase of conflict in the 1990s. But as shown in Figure 3.3b, J&K is the sole outlier. Nagaland, too, reported a small increase in the child population, but that was entirely because of the massive over-reporting in the older age groups in 2001, which was corrected in 2011. Second, Bhat (2018: 179) points out that 'when we look at the percentage of child population among various districts in 2001, it can be seen that there is no indication of any child undercount in any of the districts of Kashmir'. Third, if coverage was poorer in the 2001 census, as argued later, the population share of Kashmir within the state should have decreased relative to 1981. But Kashmir's population share increased in 2001 (Figure 1.1c).

In short, explanations based on demographic factors such as births, deaths and migration and usual content and coverage errors fail to account for the anomalous drop in J&K's CSR and the rise in child population share.[43] These changes are also inconsistent with the corresponding estimates from the SRS and the NFHS. We can conclude that anomalies in J&K's child population statistics are indicative of a significant overcount asymmetrically distributed between sexes and regions (and, by implication, religious groups). We will examine the possible political explanations of the overcount in the subsequent chapters.

The remainder of this chapter will first present estimates of the magnitude of overcount and then examine content errors.

## Corrections

The correction of coverage errors in the overall population of J&K requires certain assumptions about the change of the sex ratio and/or the population share of the child population. The former will not be reliable as sex ratio can move in either direction depending on a variety of factors, but in recent decades the child population share is expected to decrease along with other correlates of fertility. As noted earlier, the child population share has decreased even in states such as Punjab and Haryana that reported sharp drops in the CSR. We will therefore rely on corrections based on the child population share. Since the abnormal growth of Generic Tribes is not explained by content errors, it will have to be accounted as a coverage error and deducted from the total population. Three other points need to be noted. First, our corrections are applied at the level of state even though the data from the Mountainous region are the most reliable and may not require adjustments. This will lead to a slight overestimation of the overcount. Introducing corrections at the lower levels of aggregation will require additional information that is not available. Second, corrections for multiple reference dates in 2000 (see Table 2.3) cannot be carried out for want of metadata. Third, we do not have data for 1991, and the reference dates for the 1971 and 1981 censuses were different from the usual reference dates because of which we do not have access to reliable trends, and the suggested corrections are coarse.

We showed that Muslims of J&K and Muslims of Kashmir reported a sharp increase in the population share of the 0–9 age group. We will assume that contrary to the increase reported in the 2011 census, the actual population share of the 0–9 age group in J&K and among Muslims of the state declined in line with the drop in the population share of this age group in India and among Muslims of India, respectively. We will assume that the population of the older age groups was not inflated. The correction of over-reporting involves two steps: removal of Generic Tribes from the population and scaling down of the 0–9 age population. We scale down the 0–9 age population so that the magnitude of the drop in their population share between 2001 and 2011 is equal to the decline in the corresponding all-India population during this period. If we correct the 0–9 age population, the state's reported population drops by about 672,000 – that is, the census overestimated the state's population by at least 5.67 per cent in 2011 (Table 3.8).[44] Alternatively, if we assume that the position of J&K vis-à-vis the median of the population share of the 0–9 age group for all states

**Table 3.8** Adjusted population of Jammu and Kashmir (J&K), 2011

| | Age group | Year | Population | Population share |
|---|---|---|---|---|
| J&K | All ages | 2001 | 10,143,700 | |
| | All ages | 2011 | 12,541,302 | |
| | 0–9 | 2001 | 2,256,366 | 0.2224 |
| | 0–9 | 2011 | 2,826,857 | 0.2254 |
| All states and UTs | 0–9 (mean) | 2001 | | 0.2321 |
| | 0–9 (mean) | 2011 | | 0.1980 |
| | 0–9 (median) | 2001 | | 0.2168 |
| | 0–9 (median) | 2011 | | 0.1918 |
| | Adjustment for overcounting of Generic Tribes (GTs) | | | |
| J&K | All ages – GTs | 2001 | 10,136,155 | |
| | All ages – GTs | 2011 | 12,429,677 | |
| | 0–9 – GTs | 2001 | 2,254,504 | 0.2224 |
| | 0–9 – GTs | 2011 | 2,796,558 | 0.2250 |
| | 10+ – GTs | 2011 | 9,633,119 | |
| | Adjustment for overcounting of child population (vis-à-vis all states' mean) | | | |
| J&K | Adjusted (0–9 – GTs) | 2011 | 2,235,300 | 0.1883 |
| | Adjusted (all ages – GTs) | 2011 | 11,868,419 | |
| | Overcount | 2011 | 672,883 | |
| | Percentage of overcount | 2011 | 5.67 | |
| | Adjustment for overcounting of child population (vis-à-vis all states' median) | | | |
| J&K | Adjusted (0–9 – GTs) | 2011 | 2,369,687 | 0.1974 |
| | Adjusted (all ages – GTs) | 2011 | 12,002,806 | |
| | Overcount | 2011 | 538,496 | |
| | Percentage of overcount | 2011 | 4.49 | |

*Source*: Author's computations using table C-13 and table C-13 (ST).
*Note*: (*a*) The corrected 0–9 population share of J&K is calculated by assuming that the difference between the state and the corresponding all-India mean or median was same in both 2001 and 2011. Median is calculated for all states excluding UTs except Delhi. (*b*) The adjustments are conservative estimates because of the assumption that the 2001 population and the population aged 10 and above in 2011 were correctly enumerated. (*c*) 'GTs' stands for Generic Tribes.

(including Delhi) did not change between 2001 and 2011, the reported population drops by about 538,000 – that is, the state's population was overestimated by about 4.49 per cent.[45] The aforementioned estimates of overcount imply that the

2011 census overestimated Kashmir's population by 8–10 per cent as inflation of the headcount was largely confined to Kashmir.

The large magnitude of overcount estimated in Table 3.8 is corroborated by an analysis of population change using the fundamental equation of population dynamics.

$$\Delta P(t,t+10) \equiv B(t,t+10) - D(t,t+10) + NI(t,t+10) \qquad \text{(Equation 3.1)}$$

Here, $\Delta P(t, t + 10)$, $B(t, t + 10)$, $D(t, t + 10)$ and $NI(t, t + 10)$ denote the change in population, births, deaths and net in-migrants between '$t$' and '$t + 10$', respectively. Equation 3.1 can be rewritten to account for the possible mismatch between the left- and right-hand sides, where $\varepsilon > 0$ implies overcounting in census.

$$\Delta P(t, t + 10) = B(t, t + 10) - D(t, t + 10) + NI(t, t + 10) + \varepsilon \quad \text{(Equation 3.2)}$$

Following Equation 3.2, Table 3.9 suggests that the overcount ranged between 288,000 and 994,000. The large coverage errors identified in Tables 3.8 and 3.9 are possibly corroborated by very low rates of Aadhaar enrolment in Kashmir, particularly in districts such as Srinagar and Anantnag (Figure 3.12).[46]

**Table 3.9**   Births, deaths and migration

| Base-year population (2001 census) | | 10,143,700 |
|---|---|---|
| Migrants, 2001–11 | In-migrants | 70,628 |
| | Out-migrants | 118,532 |
| Deaths (estimated from the SRS) | | 669,208 |
| Births, 2001–11 | SRS-based estimate | 2,121,048 |
| | 2011 census-based estimate for the 0–9 age population | 2,826,857 |
| Expected-terminal-year population (2011) | SRS-based estimate | 11,547,636 |
| | 2011 census-based estimate for the 0–9 age population | 12,253,445 |
| Terminal-year population (2011 census) | | 12,541,302 |
| Overcount (2011 census) | SRS-based estimate | 993,666 |
| | 2011 census-based estimate for the 0–9 age population | 287,857 |

*Source*: Author's computations using tables C-13, D-12 and D-13 for 2001 and 2011 and SRS (2014).

*Note*: Figures for births are based on the SRS birth rates and the data for the 0–9 age population reported in the 2011 census.

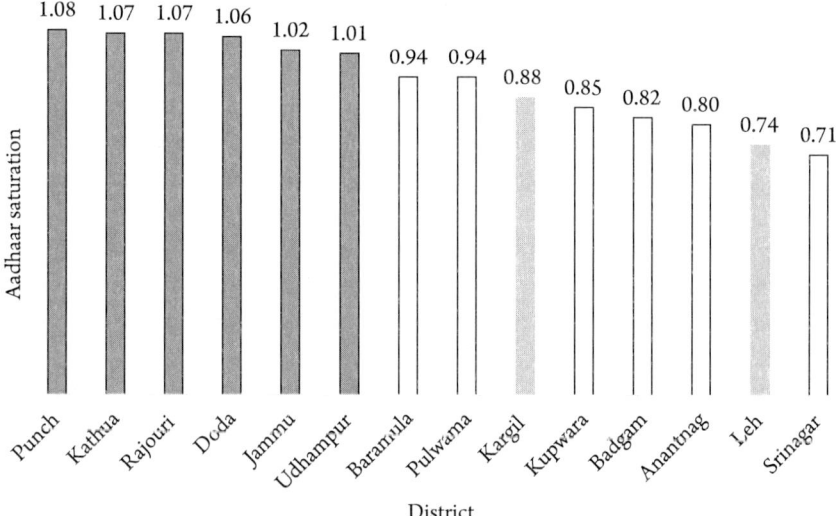

**Figure 3.12**   Aadhaar saturation in Jammu and Kashmir (J&K)

*Source*: Prepared by the author using 'Enrolment Dashboard', https://uidai.gov.in/aadhaar_dashboard/india.php?map_state=Jammu%20and%20Kashmir (accessed on 29 August 21).

*Note*: (*a*) Saturation is equal to the number of assigned Aadhaar IDs divided by the total population (projected). (*b*) This figure shows the 2001 districts because, until recently, Aadhaar used projections based on the 2001 census.

Further, we can also assess the plausibility of the reported change in a cohort's population over time by comparing their expected and actual populations. Let $P_i(t)$ and $P_i(t + 10)$ denote the population of the *i*th cohort reported in the census in year '*t*' and '*t* + 10', respectively. The expected population of this cohort in year '*t* + 10' is given by

$$P^e_i(t + 10) = P_i(t) + B_i(t, t + 10) - D_i(t, t + 10) + I_i(t, t + 10) - O_i(t, t + 10)$$
(Equation 3.3)

Here, $B_i(t + 10)$, $D_i(t, t + 10)$, $I_i(t, t + 10)$ and $O_i(t, t + 10)$ denote births, deaths, in-migration and out-migration between censuses conducted in '*t*' and '*t* + 10'. Our analysis is restricted to the age groups between 10 and 79 years as the SRS reports mortality rate for the 70+ age group as a whole. We have already assessed the discrepancy in the 0–9 age group population in Table 3.8. Since $B_i = 0$ for all cohorts aged 10 and above, the ratio of the reported population of cohort *i* to its expected population can be expressed as

$$R_i(t,t+10) = \frac{P_i(t+10)}{P_i(t) - D_i(t,t+10) + I_i(t,t+10) - O_i(t,t+10)}$$

(Equation 3.4)

If $R_i(t, t+10)$ is much larger than one, it is suggestive of over-enumeration of the population of the $i$th group. For J&K, the ratio is much larger than one for the 10–14 age group even under the stringent assumption of zero deaths and therefore implies overcounting of this age group (Figure 3.13). As a result, the corrections to the 0–9 age group in Table 3.8 are conservative because the 10–14 age group was also overcounted in 2011, whereas we had assumed that the population aged 10 and above was correctly enumerated.

Also, note that we assumed that the 2001 census correctly estimated the population. However, estimates of household size from the census data suggest that the 2001 population was inflated in J&K because its household size increased between 1981 and 2001 (Figure 3.14) even though the CBRs dropped by a third during this period (Figure 3.10). The increase was driven by Kashmir, which reported a sharp increase in household size.

The estimate of overcount in Table 3.8 will also have to take the following into account. On the one hand, relaxing the assumptions that the estimate of the

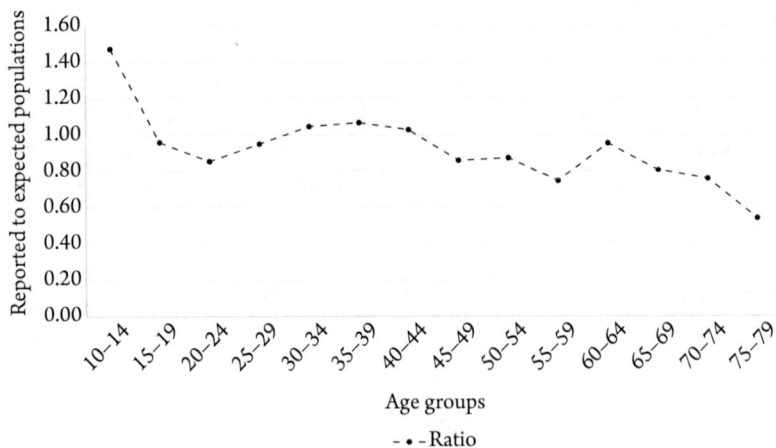

**Figure 3.13** Ratio of reported to expected population of Jammu and Kashmir (J&K), 2001–11

*Source*: Prepared by the author using table C-13 of Census of India for 2001 and 2011.

*Note*: The figure plots the ratio of population reported in the 2011 census to the expected population calculated assuming zero deaths in Equation 3.4. The ratio falls far below one for higher ages due to the highly restrictive assumption of zero mortality.

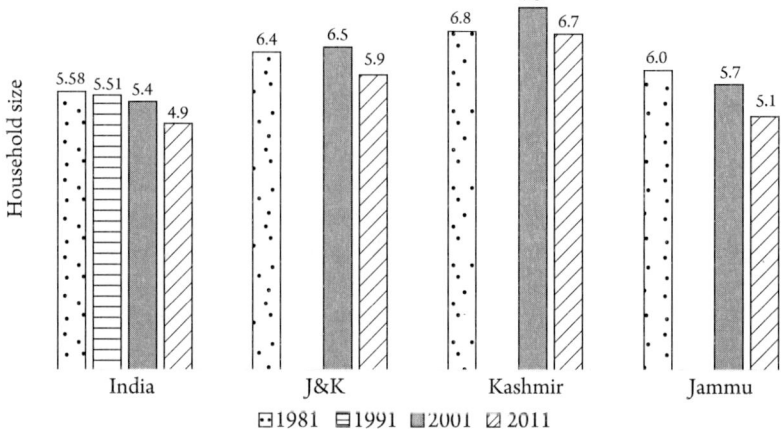

**Figure 3.14**   Household size of India, Jammu and Kashmir (J&K) and the two main divisions of J&K, 1981–2011

*Source*: Prepared by the author using primary census abstracts (PCA) for 1981–2011 and Agrawal and Kumar (2020a: table 4.4a).

*Note*: Figures for India exclude Assam and J&K, where census could not be conducted in one of the decades.

population in 2001 and the 10–14 age group in 2011 are correct will increase the estimates of overcount. On the other hand, the assumption of a fixed difference vis-à-vis national mean and median overlooks the fact that states with lower fertility are likely to register smaller declines in fertility compared to higher-fertility states, and, as a result, we may overestimate the decline in the child population share as well as the overcount. It is difficult to adequately account for this in Table 3.8 and arrive at more accurate corrections because of the erratic changes in the reference dates of post-1961 censuses and the absence of data for 1991, but the degree of overestimation of the overcount would be much less than its underestimation. Therefore, our estimates of the overcount in Table 3.8 should be treated as lower bounds.

## Content Errors

A report of the last colonial-era census in J&K noted the 'political propaganda with regard to questions relating to "Mother-tongue", as "the Urdu-Hindi script" question was then, and had been for some time, the subject of controversy and political discussion' (Government of Jammu and Kashmir [GoJK] 1943a: 12–13). It observed that

> the figures for Hindustani are inflated as the result of the Urdu-Hindi controversy. Propaganda was carried on during the Census by the adherents of both parties to the dispute with the result that many Hindus gave Hindi as their mother tongue and many Muslims gave Urdu quite contrary to the facts in the great majority of cases. The dispute is largely political and so to keep politics out of the Census it was decided to lump Hindi and Urdu together as Hindustani. There are few people in the State who can rightly claim Hindi or Urdu as their mother tongue. The real dispute is in the use of script. (GoJK 1943c: 39)

The Hindi–Urdu controversy does not seem to have affected later censuses.[47] However, given the immense ethnic (Table 3.10) and linguistic (Tables 3.11a and 3.11b) diversity of the state, from the beginning the census stressed the importance of engaging enumerators who spoke local languages. Enumerators were instructed that they were 'not authorised to make any correction' even if

**Table 3.10**  Scheduled Tribes (STs) of Jammu and Kashmir (J&K), 2001–11

| STs | Population in J&K | | | Main (Secondary) | | |
|---|---|---|---|---|---|---|
| | 2001 | 2011 | Change (per cent) | District | Language | Religion |
| All STs | 1,105,979 | 1,493,299 | 35.02 | N/A | N/A | N/A |
| Bakarwal | 60,724 | 113,198 | 86.41 | Rajouri (Reasi, Anantnag) | Gojri (Pahari) | Muslim |
| Balti | 38,818 | 51,918 | 33.75 | Kargil (Leh) | Purkhi (Balti, Bhotia) | Muslim (Buddhist) |
| Beda | 128 | 420 | 228.13 | Leh | Bhotia | Buddhist (Muslim) |
| Bot etc. (Bot, Boto) | 96,698 | 91,495 | –5.38 | Leh (Kargil) | Bhotia (Ladakhi, Tibetan) | Buddhist (Muslim) |
| Brokpa etc. (Brokpa, Drokpa, Dard, Shin) | 51,957 | 48,439 | –6.77 | Kargil (Bandipore) | Shina (Purkhi, Dardi) | Muslim |

*(Contd)*

Table 3.10 (*Contd*)

| STs | Population in J&K | | | Main (Secondary) | | |
|---|---|---|---|---|---|---|
| | 2001 | 2011 | Change (per cent) | District | Language | Religion |
| Changpa | 5,038 | 2,661 | –47.18 | Leh | Bhotia | Buddhist (Muslim) |
| Gaddi | 35,765 | 46,489 | 29.98 | Udhampur (Kathua) | Bharmauri/ Gaddi (Pahari, Dogri) | Hindu |
| Garra | 507 | 504 | –0.59 | Leh (Kargil) | Bhotia (Purkhi) | Buddhist (Muslim) |
| Gujjar | 763,806 | 980,654 | 28.39 | Rajouri (Punch, Anantnag) | Gojri (Pahari) | Muslim |
| Mon | 732 | 829 | 13.25 | Kargil (Leh) | Bhotia (Shina) | Muslim (Buddhist) |
| Purigpa | 37,700 | 39,101 | 3.72 | Kargil | Purkhi | Muslim (Buddhist) |
| Sippi | 6,561 | 5,966 | –9.07 | Kathua (Doda) | Pahari (Bharmauri/ Gaddi, Dogri) | Hindu |
| Generic Tribes | 7,545 | 111,625 | 1,379.46 | Bandipore (Kupwara, Anantnag) | Gojri (Dardi) | Muslim (Hindu) |

*Source*: Table ST-14 and table ST-15 of Census of India, 2011; GoI (2007, 2018b).
*Note*: (*a*) 'Main' and 'secondary' are calculated on the basis of the population share as per the 2011 census. (*b*) Purkhi (spelling as per the census) is also known as Purigi, Purik Bhotia and Purki.

they had 'reason to suspect that in any area due to any organised movement, [say,] mother-tongue was not being truthfully returned'. In such cases, they were asked to 'record the mother-tongue as actually returned by the respondent' and 'report to … superior Census Officers for verification' (GoI 1971b: 153). Precautions notwithstanding, in most censuses, the quality of the data on STs and languages was affected by content errors and large miscellaneous categories.[48]

## Gojri/Gujjari/Gujar

Gujjars and Bakarwals constitute the largest tribal community and the third-largest population group in J&K, next only to Kashmiri Muslims and Dogra

**Table 3.11a**   Major languages of Jammu and Kashmir (J&K), 2001–11

| Mother tongue | 2001 | 2011 |
|---|---|---|
| Kashmiri | 5,425,733 | 6,680,837 |
| Hindi | 1,870,264 | 2,612,631 |
| Gojri | 747,850 | 1,135,196 |
| Pahari | 703,771 | 977,860 |
| Hindi | 192,761 | 304,195 |
| Bhadrawahi | 66,612 | 98,196 |
| Bharmauri, Gaddi | 13,349 | 24,161 |
| Dogri | 2,205,560 | 2,513,712 |
| Punjabi | 190,675 | 219,193 |
| Bhotia | 12,930 | 107,451 |
| Tibetan | 6,390 | 100,499 |
| Khandeshi | 24,767 | 34,862 |
| Shina | 34,206 | 32,027 |
| Urdu | 13,251 | 19,956 |
| Balti | 19,240 | 12,399 |
| Ladakhi | 101,466 | 7,638 |
| Lahnda | 22,224 | 6,102 |
| Others | 92,370 | 16,642 |

*Source*: GoI (2007, 2018b); table C-16 of Census of India, 2011.

Hindus and are also the most widely distributed community of J&K. Gujjars and Bakarwals were belatedly granted ST status in 1991,[49] and Gojri was included as a 'regional language' under the Sixth Schedule of J&K's constitution in 1999 (GoJK 2003). The community alleges that the long-standing neglect reflects, among other things, in its under-enumeration and claims that its population should be close to 20 per cent of the state's population (*Outlook* 2010; Rahi 2019) or 25 lakhs (interviews, Jammu, 13 September 2019 and 24 September 2021), which is roughly two and a half times their 2011 population.[50]

A Gujjar activist suggested that the community does not 'trust the census' and identified several factors that explain its undercount (interview, Jammu, 4 December 2019). First, respondents do not get their identity correctly recorded due to a lack of awareness. Also, a general distrust of government agencies in the (nomadic) community has meant that people avoid sharing information with officials. Second, the community grapples with 'caste stigma' and hostility in

**Table 3.11b**    'Other' languages of Jammu and Kashmir (J&K), 2001–11

| District, region, state | 2001 | District, region, state | 2011 |
|---|---|---|---|
| Baramula | 797 | Baramula | 193 |
| | | Bandipore | 20 |
| Kupwara | 630 | Kupwara | 311 |
| Badgam | 437 | Badgam | 120 |
| Srinagar | 1,659 | Srinagar | 113 |
| | | Ganderbal | 973 |
| Pulwama | 42 | Pulwama | 42 |
| | | Shupiyan | 2 |
| Anantnag | 71 | Anantnag | 15 |
| | | Kulgam | 3 |
| **Kashmir** | **3,636** | **Kashmir** | **1,792** |
| Leh (Ladakh) | 930 | Leh (Ladakh) | 38 |
| Kargil | 73,139 | Kargil | 82 |
| **Ladakh** | **74,069** | **Ladakh** | **120** |
| Punch | 164 | Punch | 120 |
| Rajauri | 633 | Rajouri | 26 |
| Doda | 10,980 | Doda | 4,012 |
| | | Ramban | 1,526 |
| | | Kishtwar | 5 |
| Udhampur | 536 | Udhampur | 89 |
| | | Reasi | 1,277 |
| Kathua | 1,278 | Kathua | 6,482 |
| Jammu | 1,074 | Jammu | 1,109 |
| | | Samba | 84 |
| **Jammu** | **14,665** | **Jammu** | **14,730** |
| **J&K** | **92,370** | **J&K** | **16,642** |

*Source*: GoI (2007, 2018b); table C-16 of Census of India, 2001 and 2011.
*Note*: Non-Scheduled mother tongues with less than 10,000 speakers were clubbed under '123 others' and '124 others' in 2001 and 2011, respectively.

both Kashmir and Jammu.[51] 'We are Gujjar in the Valley and Muslim in Jammu. Pakistanis shells us [along the border] as Indians,'[52] said the activist. He added, 'Kashmiris treat us as Dalits.... [In the Valley,] [w]e are ethnically marked as well.' As a result, 'the upwardly mobile feel ashamed' of their identity.

Third, dominant communities played numbers games, which reflected in the denial of reservation in the legislature[53] and gerrymandering at the village level that divided the Gujjar population among *panchayats* dominated by non-tribal communities. In Jammu, the community increasingly finds itself caught in the Hindu–Muslim crossfire between Kashmir and Jammu. But community leaders and activists maintain their undercount is largely confined to Kashmir, where the dominant community suppresses its numbers fearing electoral competition (interviews, Jammu, 4 December 2019; Baramula, 20 July 2021, 21 July 2021, 18 September 2021; Anantnag, 5 August 2021).[54] An interviewee argued, 'Numbers games begin in the teacher's mind. Kashmiri teachers are not serious about enumerating our community. Because of Kashmiri language consciousness we are misreported [by Kashmiri teachers] as Kashmiri speakers and non-tribals.'[55] In 2019, a petition filed under the Right to Information (RTI) Act, 2005, revealed that there were only 11 Gujjar teachers in more than 100 schools of the Chandanwari educational zone of Baramula district (Zonal Educational Officer 2019).[56] In the past, fewer teachers and administrative officers belonged to the Gujjar community, and that would have translated into a lesser community oversight during enumeration (state government official, interview, Baramula, 20 July 2021).

Fourth, insufficient arrangements for the enumeration of Bakarwals resulted in their undercount. The slip system discussed in Chapter 2 was not practical as there are multiple routes, the number of check posts was insufficient (GoI 1969: 38) and Bakarwals often take shortcuts or avoid check posts, which compounds the problem due to poor coverage of the upper reaches. The coverage was also adversely affected due to disturbed conditions.

In short, Gujjars and Bakarwals maintain that their undercount is explained by the communication gap with the administration, lack of awareness, inadequate arrangements for counting of the mobile population and political and social factors because of which their language and tribal status are misreported and their population is undercounted.[57] We will examine a few inconsistencies in the data on Gujjars and Bakarwals and the Gojri language.

As per the most recent census, the population of 'Gojri/Gujjari/Gujar' speakers grew by 51.79 per cent between 2001 and 2011 (Table 3.11a), whereas the Gujjar tribe grew by about 28 per cent (or 33 per cent if we include Bakarwals) (Table 3.10). The higher growth rate of speakers of 'Gojri/Gujjari/Gujar' can be explained by the growing literacy, identity consciousness and access to government jobs (that translates, among other things, into appointments as enumerators during census) that would have supported better enumeration of their language.

About a week before the household phase of the 2011 census, the secretary of the Tribal Research and Cultural Foundation 'appealed mobile school teachers and local community leaders to cooperate with census officials and get themselves enumerated as ST community with Gojri as their mother tongue' (*Kashmir Times* 2011a).[58] The very high population growth of Gojri speakers in Kashmir too played a role. During 2001–11, the Gojri-speaking population grew by 61.40 per cent in Kashmir compared to 46.42 per cent growth in their population in Jammu. As a result, the share of Jammu (Kashmir) in the Gojri-speaking population of the state decreased (increased) by 2.27 (2.26) per cent. In fact, Generic Tribes alone accounted for nearly 62,000 speakers of Gojri, with a majority being reported in Kashmir.

In 2011, Gojri speakers exceeded the total ST population in Kupwara, Shupiyan and Anantnag districts of Kashmir (Table 3.12a), which is counter-intuitive. First, Kashmiris will not declare Gojri, a language they look down upon, as their mother tongue.[59] In fact, ahead of the 2011 census, Kashmiri leaders such as Syed Ali Shah Geelani exhorted non-Kashmiri-speaking 'Jammu Muslims' including Gujjars to declare Kashmiri as their mother tongue (Ashiq 2010). There have been other attempts to linguistically co-opt Gujjars and Bakarwals. An ethnographic profile of tribes prepared by the state government wrongly suggested that Bakarwals 'speak Kashmiri language and use Arabic script yet their dialect is identified as Gujri which is the main dialect of the Gujjars' (GoJK n.d.1). To the contrary, in 2011, Kashmiri was reported as the first language of barely 1 per cent of Bakarwals and the subsidiary language for another 1.50 per cent. About two-thirds of Bakarwals speak Gojri, and about 27 per cent reported Pahari as their mother tongue. Likewise, Kashmiri was the mother tongue of less than 1 per cent of Gujjars and the subsidiary language of 5.50 per cent.

Second, not all tribes speak Gojri, and the non-Gujjar or Bakarwal tribes are unlikely to report themselves as Gojri speakers. So the tribal population must exceed the Gojri-speaking population, which does not hold in the aforesaid districts and is indicative of misclassification of tribal population as non-tribal leading to a content error.

There is another indication of undercount of Bakarwals in Kashmir. Between 2001 and 2011, the population of Bakarwals contracted in Kashmir, while it grew by 150 per cent in Jammu (Table 3.12d). (The Gujjar population grew faster in Kashmir than in Jammu.) The contraction of the Bakarwal population is inexplicable amidst pervasive overcounting in Kashmir. The census was delayed in Kashmir in 2000 and even extended to December in a few districts by when the Bakarwals had retreated from Kashmir to the plains (Table 2.3). So their

**Table 3.12a**   Scheduled Tribes (STs) and Gojri speakers, 2011

| State/District | ST population | Gojri speakers | ST/Gojri |
|---|---|---|---|
| J&K | 1,493,299 | 1,135,196 | 1.32 |
| Kupwara | 70,352 | 80,163 | 0.88 |
| Baramula | 37,705 | 34,750 | 1.09 |
| Bandipore | 75,374 | 34,586 | 2.18 |
| Badgam | 23,912 | 22,706 | 1.05 |
| Srinagar | 8,935 | 8,202 | 1.09 |
| Ganderbal | 61,070 | 55,570 | 1.10 |
| Pulwama | 22,607 | 22,189 | 1.02 |
| Shupiyan | 21,820 | 23,425 | 0.93 |
| Anantnag | 116,006 | 123,606 | 0.94 |
| Kulgam | 26,525 | 26,827 | 0.99 |
| Punch | 176,101 | 186,658 | 0.94 |
| Rajouri | 232,815 | 221,553 | 1.05 |
| Kishtwar | 38,149 | 33,127 | 1.15 |
| Doda | 39,216 | 30,200 | 1.30 |
| Ramban | 39,772 | 30,654 | 1.30 |
| Reasi | 88,365 | 77,674 | 1.14 |
| Udhampur | 56,309 | 28,770 | 1.96 |
| Jammu | 69,193 | 59,048 | 1.17 |
| Samba | 17,573 | 13,766 | 1.28 |
| Kathua | 53,307 | 21,599 | 2.47 |

*Source*: Primary census abstract (PCA) and table C-16 of Census of India, 2011.

population should have grown in Kashmir in 2011 because in the latter year armed insurgency had subsided, there was no boycott call or threats issued in Kashmir targeting census operations and snowbound areas were enumerated as per the usual schedule. Also, as argued later, the contraction of the Bakarwal population cannot be explained by the abnormally high growth of Generic Tribes. Further, even if the contraction of the Bakarwals in Kashmir is attributed to their misclassification as Gujjars, the disjuncture between the Gujjar and Bakarwal population and the Gojri-speaking population will persist.

We can therefore conclude that a part of the Gojri-speaking Gujjar and Bakarwal tribes was recorded as non-tribal. At the same time, the 150 per cent growth in the Bakarwal population in Jammu cannot be explained unless we

**Table 3.12b**   Speakers of Khandeshi and Gujari, 2001–11

| Language | | 2001 | 2011 |
|---|---|---|---|
| Khandeshi | (1) | 2,075,258 | 1,860,236 |
| Gujari (Khandeshi) | (2) | 48,747 | 57,171 |
| | (3) = (2)/(1) | 0.0235 | 0.0307 |
| Maharashtra | (4) | 1,866,460 | 1,616,730 |
| Madhya Pradesh | (5) | 24,985 | 37,882 |
| Gujarat | (6) | 152,096 | 153,622 |
| | (7) = (4 + 5 + 6)/(1) | 0.9847 | 0.9720 |
| Jammu and Kashmir (J&K) | (8) | 24,767 | 34,862 |
| Gujari (Khandeshi) | (9) | 24,767 | 34,858 |
| | (10) = (9)/(2) | 0.5081 | 0.6097 |

*Source*: Table C-16 of Census of India, 2001 and 2011.

assume massive under-enumeration in the 2001 census. Alternatively, we have to assume that those who earlier reported themselves as belonging to the closely related Gujjar tribe reported themselves as Bakarwals in the 2011 census, but even this explanation cannot account for the contraction of the Bakarwal population in Kashmir and is also inconsistent with the fact that Gujjars, too, reported high growth rates.[60] It can also not be argued that the high growth in Jammu can be explained by the fact that some Bakarwals who were supposed to be enumerated in Kashmir were counted in Jammu because in that case the state-level growth rate of Bakarwals would not have been very high.[61]

In this context, it is also noteworthy that the classification of Gojri has varied across censuses,[62] which may partly explain the erratic accounting of the speakers of the language. The misclassification of 'Gojri/Gujjari/Gujar' speakers has not received attention so far. The language tables list 'Gojri/Gujjari/Gujar' under Hindi, while 'Gujari' is listed as a dialect under Khandeshi, a non-scheduled language.[63] Khandeshi is largely spoken in Maharashtra, which accounted for more than 85 per cent speakers in both 2001 and 2011 (Table 3.12b). Maharashtra and its neighbouring states Gujarat and Madhya Pradesh together accounted for more than 97 per cent of the Khandeshi-speaking population. J&K is home to the fourth largest Khandeshi population in the country. In 2001, all the speakers of Khandeshi in J&K were classified as speakers of the Gujari dialect. In the following census, all but four Khandeshi speakers were classified as speakers of Gujari in J&K. In both years, J&K accounted for, at least, half of the Gujari-speaking population even though the state's share in the Khandeshi-speaking population was less than 2 per cent. Also, there is a high correlation between the number of speakers of 'Gojri/Gujjari/Gujar' and 'Gujari' across the districts of J&K – 0.76 in 2001 and 0.65 in 2011 (Table 3.12c). Add to this the fact that Gujjars and Bakarwals account for more than 83 per cent of Gujari speakers in the state. So Khandeshi speakers are an artefact of content errors introduced at the stage of data processing – that is, misclassification of 'Gojri/Gujjari/Gujar' speakers under Khandeshi as speakers of 'Gujari' due to the similarity in spellings.

We can correct both the content and coverage errors in the headcount of Gojri speakers. The census population of Gujjars and Bakarwals and Gojri speakers can be divided into four categories: both tribe and language are correctly recorded, only tribe is correctly recorded, only language is correctly recorded, and both are incorrectly recorded. We do not have any estimate for the last category – that is, Gujjars and Bakarwals who were reported as neither tribals nor Gojri speakers. The estimates of the other three are reported in Table 3.12e. Three corrections are required to arrive at a more reliable estimate of the population of Gujjars and Bakarwals. First, the Gujjar population was underestimated in Kashmir by at least, 19,016 under the restrictive assumption that Gojri-speaking Gujjars account for the entire tribal population of Kupwara, Shupiyan and Anantnag. The corresponding figure for Jammu is 10,557. These are conservative estimates because we need to account for the following: upwardly mobile Gujjars may report the dominant languages as their mother tongue, the closely related Bakarwal community also speaks Gojri, the tribal population also includes other tribes that do not speak Gojri and the Bakarwal population contracted in Kashmir. Note that Gojri speakers exceed Gujjars and Bakarwals by 41,344. Second, the population of Gojri speakers is overestimated by 59,512 – that is, the speakers of the language belonging to Generic Tribes, which, as will be discussed, largely consist of ghost entries. Third, the misclassification of Gojri speakers as speakers of a dialect of Khandeshi (34,858) needs to be corrected for fictitious Gujari speakers among Generic Tribes (2,179). The corresponding figure for 2001 is 24,767, but this figure cannot be corrected for Generic Tribes as table ST-15 was not released. The population of Generic Tribes was very small in 2001 though.

The corrected population of Gojri speakers grew by 43.46 per cent compared to the uncorrected figure of 51.79 per cent. So the growth rate of the Gojri-speaking population between 2001 and 2011 is more than 10 percentage points higher than the growth of the corresponding tribes even after correcting for coverage and content errors. In the absence of necessary metadata, our corrections do not account for the possible double counting of the nomadic component of the Gujjar and Bakarwal communities in the snowbound areas as well as the plains.

**Table 3.12c** Speakers of 'Gojri/Gujjari/Gujar' and Gujari in Jammu and Kashmir (J&K), 2001–11

| | 2001 | | | 2011 | |
|---|---|---|---|---|---|
| District | Gojri/ Gujjari/ Gujar (Hindi) | Gujari (Khandeshi) | District | Gojri/ Gujjari/ Gujar (Hindi) | Gujari (Khandeshi) |
| **Kashmir** | | | | | |
| Kupwara | 54,501 | 2,852 | Kupwara | 80,163 | 1,110 |
| Baramula | 43,219 | 116 | Badgam | 22,706 | 1 |
| Srinagar | 45,678 | 220 | Baramula | 34,750 | 2,710 |
| Badgam | 9,352 | 1,813 | Bandipore | 34,586 | 969 |
| Pulwama | 21,108 | 1,394 | Srinagar | 8,202 | 1,590 |
| Anantnag | 93,815 | 2,976 | Ganderbal | 55,570 | 4,544 |
| | | | Pulwama | 22,189 | 981 |
| | | | Shupiyan | 23,425 | 116 |
| | | | Anantnag | 123,606 | 255 |
| | | | Kulgam | 26,827 | 488 |
| **Ladakh** | | | | | |
| Leh (Ladakh) | 11 | 1 | Leh (Ladakh) | 123 | 31 |
| Kargil | 19 | 0 | Kargil | 0 | 0 |
| **Jammu** | | | | | |
| Doda | 61,673 | 3,669 | Punch | 186,658 | 5,892 |
| Udhampur | 78,349 | 2,465 | Rajouri | 221,553 | 4,374 |
| Punch | 141,593 | 3,962 | Kathua | 21,599 | 395 |
| Rajauri | 144,765 | 3,003 | Doda | 30,200 | 649 |
| Jammu | 43,540 | 1,272 | Ramban | 30,654 | 3,923 |
| Kathua | 10,227 | 1,024 | Kishtwar | 33,127 | 1,905 |
| | | | Udhampur | 28,770 | 626 |
| | | | Reasi | 77,674 | 2,884 |
| | | | Jammu | 59,048 | 1,090 |
| | | | Samba | 13,766 | 325 |
| **J&K** | **747,850** | **24,767** | **J&K** | **1,135,196** | **34,858** |

*Source*: GoI (2007, 2018b); table C-16 of Census of India, 2001 and 2011.

**Table 3.12d**  Distribution of Gujjars and Bakarwals across divisions of Jammu and Kashmir (J&K)

| Year | State, division | Bakarwals | | | Gujjars | | |
|------|-----------------|-----------|-------|-----------------------|------------|--------|-----------------------|
| | | Population | Share | Decadal growth (per cent) | Population | Share | Decadal growth (per cent) |
| 2001 | J&K | 60,724 | | | 763,806 | | |
| | Kashmir | 24,223 | 0.3989 | | 258,810 | 0.3388 | |
| | Jammu | 35,772 | 0.5891 | | 504,754 | 0.6608 | |
| | Ladakh | 729 | 0.0120 | | 242 | 0.0003 | |
| 2011 | J&K | 113,198 | | 86.41 | 980,654 | | 28.39 |
| | Kashmir | 23,562 | 0.2081 | −02.73 | 345,831 | 0.3527 | 33.62 |
| | Jammu | 89,605 | 0.7916 | 150.49 | 634,707 | 0.6472 | 25.75 |
| | Ladakh | 31 | 0.0003 | −95.75 | 116 | 0.0001 | −52.07 |

*Source*: Respective primary census abstracts (PCA) for Scheduled Tribes (STs).

**Table 3.12e**  Gujjars and Bakarwals and Gojri-speaking population, 2011

| | | Language | |
|------|-----------|----------|-----------|
| | | Correct | Incorrect |
| Tribe | Correct | Gojri-speaking Gujjars or Bakarwals: 979,782 | Non-Gojri-speaking Gujjars or Bakarwals: 114,070 (60,985*) |
| | Incorrect | Gojri speakers recorded as other STs (non-tribals): 3,091** (92,811) | Gujjars or Bakarwals reported as non-tribals and speakers of other languages: (not known) |

*Source*: Primary census abstract (PCA) for Scheduled Tribes (STs); tables C-16 and ST-15 of Census of India, 2011.
*Note*: *Adjusted figure clubs together Pahari and Gojri as the former is spoken by a sizeable section of Gujjars and Bakarwals. ** Adjusted for Gojri speakers among Generic Tribes.

But this correction should not be large for 2011 as the Bakarwal population contracted in Kashmir. Further, if we assume that the Gujjar population was overcounted in Kashmir and the degree of its overcount was comparable to that of Kashmir as a whole, the corrected growth rate of Gojri speakers drops to 38.98 per cent, which is still 6 percentage points higher than the combined growth rate of the Gujjars and the Bakarwals. Note that in the absence of adequate metadata, the aforesaid corrections only indicate the direction and rough magnitude of corrections required. Their impact on Kashmir's headcount will depend on the relative magnitude of the correction of the undercount of the Gujjars and the Bakarwals qua tribes and the overcount of the Gujjars qua de facto residents of Kashmir. The correction of content errors will, however, not affect the overall headcount even as it will affect the population share of the Kashmiri-speaking non-tribal community.

## Gaddi and Sippi

Smaller tribes of Jammu such as Gaddi and Sippi have also complained about under-enumeration and misclassification. These communities claim that their 'population was undercounted due to militancy and negligence of enumerators', which is reflected in the contraction of the population of Sippis (interview, Bhaderwah, 5 December 2019; see also Table 3.10). The non-recognition of Kolis or Kollis as a part of Sippis qua STs seems to have affected the strength of the Sippi community.[64] Kolis claim that 'the tribe was left out as it was inadvertently spelt as "Kohli" instead of "Koli" due to a clerical mistake in the official documents. Koli and Sippi are the same tribe but the Sippis were classified as STs, but Kolis were left out', and since then 'we have been struggling to get "H" removed from the spelling of our tribe but to no avail' (Manhotra 2019). Political parties have in the past supported the demand of Kolis for ST status (*Kashmir Times* 2011g; *Daily News and Analysis* [DNA] 2016). Community leaders claim that the 'population [of Sippis] is decreasing [as per census] but it should have grown as those who had left the community to avoid social stigma are now returning [to access affirmative action benefits] and there has been no out-migration [to other states]' (interview, Bhaderwah, 5 December 2019).[65] While both these communities are primarily concerned about undercount, it seems their headcount has also been affected by misclassification of others as Gaddis – for example, the census reports Gojri-speaking Gaddi Muslims. If we assume that Bharmauri/Gaddi-speaking Hindus have been misclassified as Gojri-speaking Muslims, we will have to posit two content errors to explain the presence of Muslims and Gojri speakers among the Gaddis.[66] It is more likely that Gojri-speaking Muslims were misclassified or misreported as Gaddis.

## Languages of Ladakh

The administrative reports of the first two censuses highlight only Balti and Bodhi, but not Tibetan and Ladakhi languages or dialects like Zanskari and Purkhi or Purigi, even though Ladakhi was recognised as a 'regional language' along with Balti under the Sixth Schedule of the state's constitution (GoJK 2003).[67]

In 1961, the authorities took care 'to appoint Enumerators knowing the mother tongue of the people living in special areas' to ensure that 'the entry regarding mother tongue is correctly recorded' and therefore 'in rural parts of Ladakh district, local Enumerators whose own mother tongue was Balti; Bodhi, etc., were engaged' (GoI 1963a: 37). In the next census, too, 'Bodhi and Balti-knowing staff was recruited locally' (GoI 1971b: 22). While the first two censuses stressed only Balti and Bodhi and not Ladakhi, the language data of 1971 suggest that 'Ladakhi predominates' in the Bhotia group that also includes 'Balti, Tibetan and Budhi' (GoI 1974a: 62).[68] Such swings in language data between censuses are characteristic of Ladakh and, as argued subsequently, can only be explained by the mobilisation of respondents or, at least, enumerators to align linguistic identity with sub-regional and/or religious identities.

In Ladakh (Leh district), the population of Ladakhi speakers declined from 100,371 (94,731) to 5,640 (4,416) between 2001 and 2011 and that of Bhotia speakers increased from 11,801 (65) to 105,186 (90,085) (Figure 3.15a). The people of Leh reported themselves as speakers of the Bauti dialect of Bhotia. Zanskari, a dialect of Bhotia, disappeared completely from the landscape. In Zanskar, nearly two-thirds of the speakers of Zanskari dialect shifted to Bauti dialect in 2011 (Figure 3.15b). During 2001–11, the population of Tibetan speakers in Ladakh (Kargil district) increased from 5,371 (19) to 98,574 (93,786), entirely accounted for by the switch from Balti, Ladakhi and 'Others' to Tibetan. The category 'Others', which was was the largest language of Kargil in 2001 (73,139), disappeared by 2011 (82) (Table 3.10). As per the 2011 census, almost all the speakers of Tibetan in Kargil spoke the Purkhi dialect. Purkhi and Bauti dialects were not reported in the 2001 census.[69] There is not much difference between Tibetan or Purkhi and Bhotia or Bauti spoken in Ladakh. In fact, the 1941 census refers to Balti, Ladakhi and Tibetan as Bhotia of Baltistan, Ladakh and Tibet, respectively (GoJK 1943c: 304–05, 333), while Grierson (1909: 15) calls these languages (including Purik) 'closely related dialects'.[70]

The 'Brokpa, Drokpa, Dard, Shin' tribes that are both Buddhist and Muslim are split among Shina (Bandipore in Kashmir division), Tibetan (Kargil) and Bhotia (Leh) (Table 3.10). About 68 per cent of the 'Brokpa, Drokpa, Dard, Shin' tribes of Leh spoke Bhotia (all Bauti speakers) and 31 per cent spoke Tibetan (only two Purkhi speakers). Nearly 46 per cent of the 'Brokpa, Drokpa, Dard, Shin' tribes spoke Tibetan in Kargil, and 86 per cent of them spoke the Purkhi dialect.

The sharp division within the Kargil district between Kargil and Sanku, on the one hand, and Zanskar, on the other, further establishes that linguistic shifts

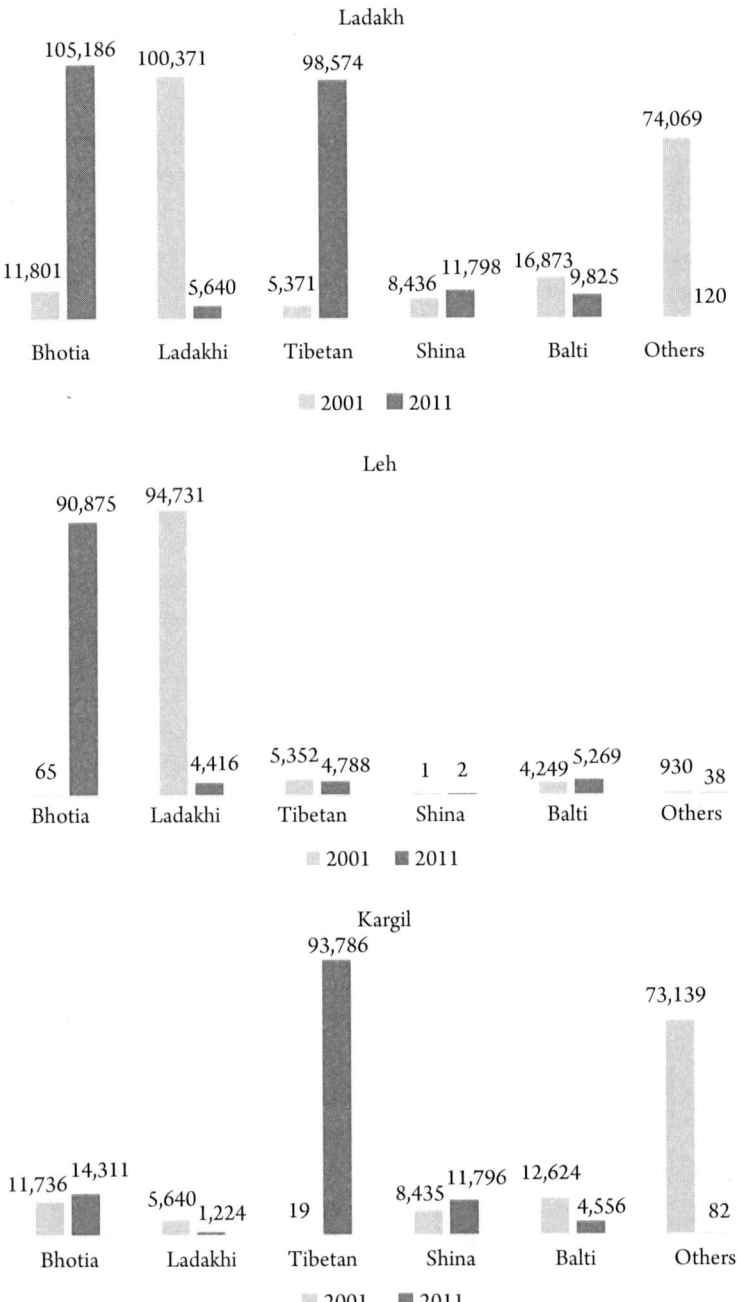

**Figure 3.15a**  Speakers of languages of Ladakh

*Source*: Prepared by the author using data from GoI (2007, 2018b).

Bhotia and its dialects: Leh

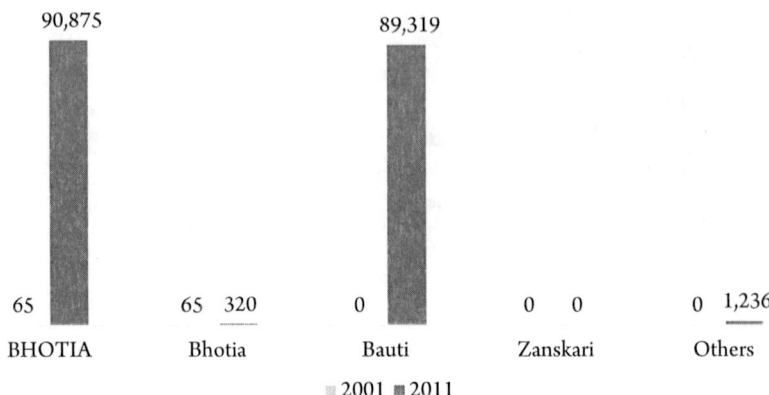

Bhotia and its dialects: Kargil

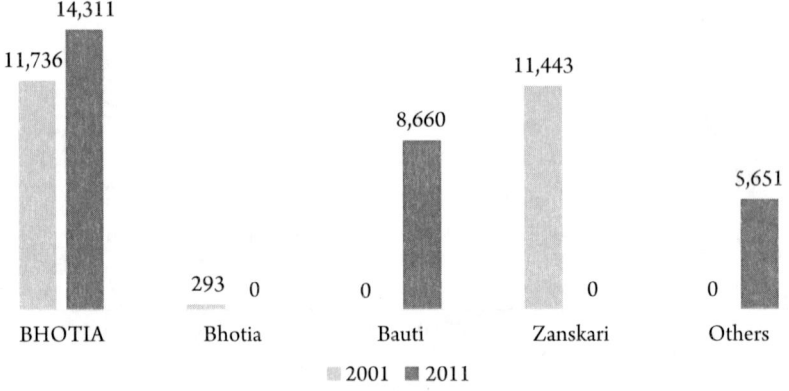

Tibetan and its dialects: Leh

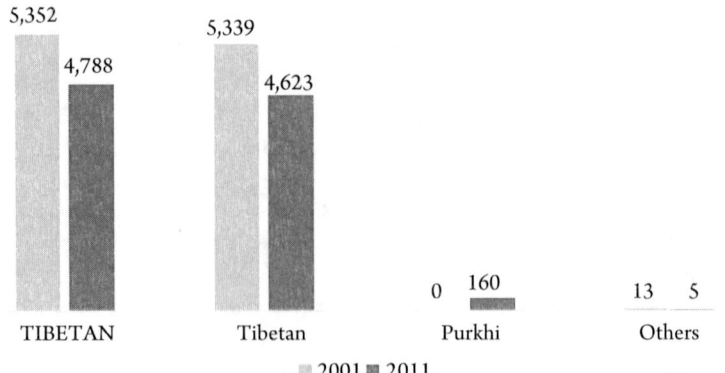

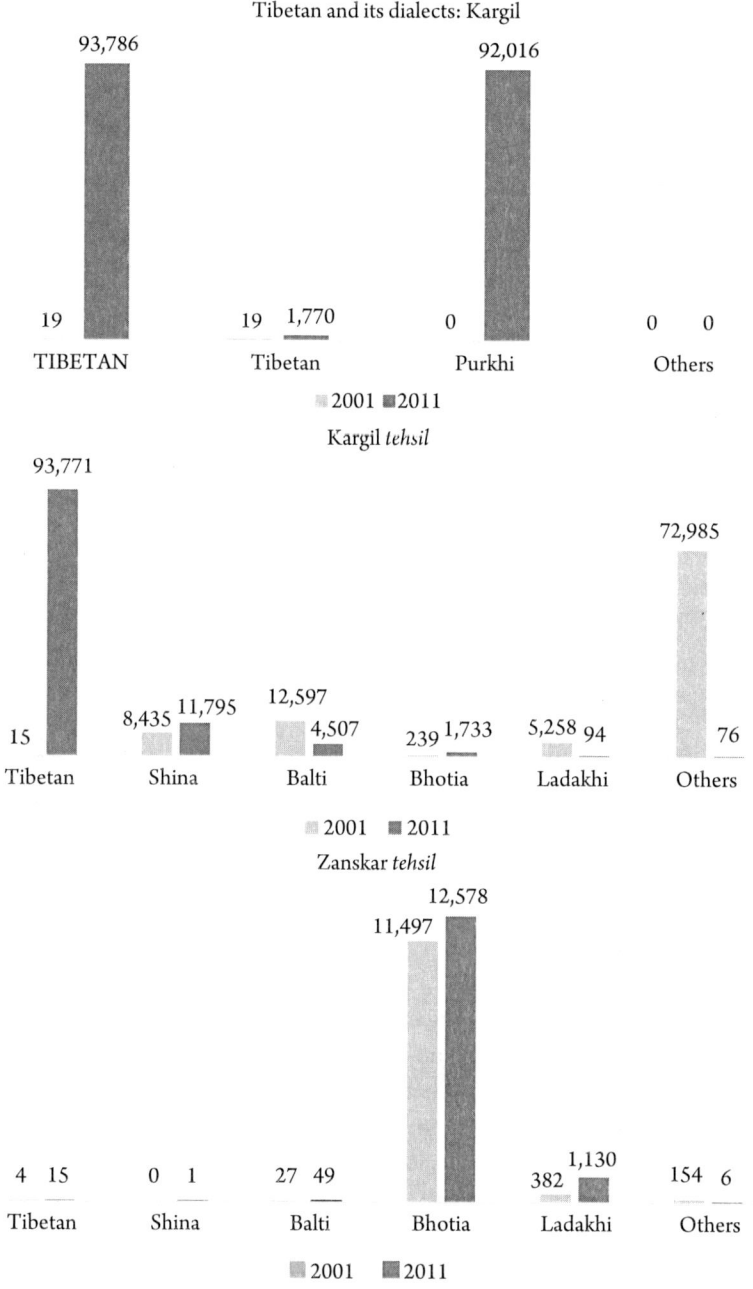

**Figure 3.15b** Speakers of languages and dialects of Ladakh

*Source*: Prepared by the author using data from GoI (2007, 2018b) and table C-16 for 2001 and 2011.
*Note*: (*a*) In graphs with dialects, upper case identifies languages while sentence case identifies dialects. (*b*) Sanku *tehsil* was carved out of Kargil *tehsil*, but it is treated as a part of the latter here.

are aligned with religious identity. Hardly anyone in Buddhist Zanskar was reported as a Tibetan speaker, while neighbouring Shia Sanku is overwhelmingly Tibetan-speaking. Balti, a major Tibetan language of Ladakh, spoken by the predominantly Muslim Balti tribe has ceded ground to Tibetan in Muslim-majority Kargil but not in Buddhist-majority Leh. In 2011, only 11 per cent of Baltis of Kargil reported Balti as their first language, with the rest reporting the Purkhi dialect of Tibetan spoken by the majority in the district.[71] In Leh, nearly 39 per cent of Baltis spoke Balti language even though about 60 per cent spoke Bhotia.

Nearly 94 per cent of the 'Bot, Boto' tribe, the mainstay of Ladakhi Buddhism, spoke the Bauti dialect of Bhotia in Leh. While 76 per cent of 'Bot, Boto' spoke Bhotia in Kargil, only half of them spoke the Bauti dialect. About 97 per cent Purigpa, one of the mainstays of Ladakhi Shiism, spoke the Purkhi dialect in Kargil, and the miniscule population of the tribe in Leh was split between Bauti and Purkhi speakers. In fact, the census fails to capture the true extent of linguistic polarisation between Kargil and Leh as it does not record writing preferences.[72]

The language of Ladakh has ceded ground to Tibetan written in a Perso-Arabic script preferred by Muslim Purigpa and Balti tribes in Kargil and Bhotia written in the Tibetan script preferred by Buddhist Bot (Boto) tribes in Leh.[73] This linguistic shift is also seen in the neighbouring Kishtwar district in Jammu. Atholi (Paddar) *tehsil* of Kishtwar adjoining Kargil's Zanskar *tehsil* was home to the largest indigenous Buddhist population outside Ladakh in the erstwhile J&K. In 2011, the population of Atholi was 21,548 out of which Buddhists, STs, the Boto tribe and Bhotia (Bauti) speakers accounted for 2,038, 2,214, 2,050 and 2,067 (1,922) persons, respectively. Atholi's Boto included 2,031 Buddhists and 2,038 (1,894) Bhotia (Bauti) speakers.[74] In contrast, as per the 2001 census, there were 1,070 (1,070) Bhotia speakers and 570 (365) Tibetan speakers in the Doda district (Kishtwar *tehsil*). Once again we see a shift in linguistic identity from Tibetan to Bauti that is aligned with religious identity.[75] A sharp rise in Bhotia speakers between 2001 and 2011 has also been reported in Arunachal Pradesh and, to a lesser extent, Uttarakhand, even though the Bhotia-speaking population contracted in neighbouring Himachal Pradesh (Table 3.13).

The data on subsidiary languages, too, underscore the deepening linguistic divide in Ladakh. The 2011 census reported Hindi (English) as the subsidiary language for about 21 (23) per cent Bhotia speakers among the Boto tribes, while the corresponding figure for Urdu, the official language of the state, was 16 per cent. In contrast, Urdu was reported as a subsidiary language for about 43 per cent speakers of Tibetan among the Purigpa tribe.[76] Most recently, the linguistic divide was highlighted by the contrasting responses of Buddhists and Muslims to

**Table 3.13** Distribution of Bhotia and Tibetan speakers, 2001–11

| State | Bhotia | | Tibetan | |
|---|---|---|---|---|
| | **2001** | **2011** | **2001** | **2011** |
| India | 81,012 | 229,954 | 85,278 | 182,685 |
| Jammu and Kashmir (J&K) | 12,930 | 107,451 | 6,390 | 100,499 |
| Himachal Pradesh | 8,975 | 2,012 | 18,112 | 21,322 |
| Uttaranchal/Uttarakhand* | 6,217 | 9,287 | 6,712 | 10,162 |
| Sikkim | 41,825 | 41,889 | 1,977 | 2,785 |
| West Bengal | 6,508 | 4,293 | 7,107 | 4,557 |
| Arunachal Pradesh | 267 | 62,458 | 9,527 | 8,500 |
| Mizoram | 3,254 | 1 | 113 | 1 |

*Source*: GoI (2007, 2018b); table C-16 of Census of India, 2011.

*Note*: (*a*) *The state was referred to in census reports as Uttaranchal in 2001 and Uttarakhand in 2011. (*b*) The table is restricted to the Himalayan states except Mizoram that reported a high number of Bhotia speakers in 2001.

an amendment to the Ladakh Revenue (Subordinate) Service Recruitment Rule, 2021, that omits knowledge of Urdu as a criterion for recruitment in the revenue departments (*Daily Excelsior* 2022a).

While the growing religious and linguistic polarisation have affected census data, according to former legislators from both Leh and Kargil, the demand for the inclusion of Bhoti in the Eighth Schedule of the Constitution has also played a role (interviews, Leh, 20 November 2019; Kargil, 22 November 2019; Feroze Ahmed, former DCO, Jammu, 4 December 2019).[77] The Ladakhi Buddhists want to ally with the larger body of Himalayan Buddhists to strengthen their case (interview, Leh, 20 November 2019).[78] A former legislator from Leh district claimed that the shift happened due to 'self-consciousness' rather than a concerted campaign (interview, Leh, 20 November 2019). However, a schoolteacher from Zanskar told the author, 'We decided that we all should report Bhoti' (interview, Kargil, 22 November 2019). Discussions with various leaders in Leh suggest that the Buddhist community is capable of effective mobilisation around demographic issues (see also Smith 2009, 2012; van Beek 2000, 2004). Moreover, some of these issues have been debated over a longer period giving the common people sufficient time to internalise the community aspirations and expectations.

Kargil's opposition to Bhotia as a label for the shared language and mobilisation around another linguistic label is motivated by two factors. Shia Kargil wants to avoid a *Buddhist* label for its linguistic identity (that is also being imagined as distinct from Balti). It also wants to avoid a situation in which its acquiescence to the linguistic shift induced by Buddhist Leh might pave the way for the uncontested rise of Bhotia written in Tibetan as Ladakh's official language.

The linguistic–orthographic divide in Ladakh is reminiscent of how the early twentieth-century census category 'Hindustani', which referred to the spoken language of large parts of north India, gave way to Hindi written in Devanagari preferred by (upper-caste) Hindus and Urdu written in a Perso-Arabic script preferred by (upper-caste) Muslims. Elsewhere, the Serbo-Croatian language split along national lines, which led to the emergence of three 'different' languages: Serbian written in Cyrillic script and Croatian and Bosnian written in Latin.

## Shina

The administrative report for the 1961 census noted that in far-flung 'Tilel and Gurez, local school teachers and Patwaris whose mother tongue is Dardi were detailed on enumeration work' (GoI 1963a: 37). Shina is mainly spoken in the *tehsils* of Gurez (Bandipore district, part of Baramula in 2001), Kargil (Kargil district) and Kangan (Ganderbal district, part of Srinagar in 2001). In 2011, the number of Shina speakers contracted for the first time in India. The Shina-speaking population of Gurez decreased from 23,413 to 12,705 between 2001 and 2011. The Kashmiri-speaking population of Gurez increased from 1,033 to 17,865, which includes 0 and 17,189 speakers of Dardi, respectively (Tables 3.14a and 3.14b). As a result, Kashmiri speakers emerged as the single largest community in Gurez. Shina speakers did not merely drop in terms of population share; rather, their population halved.

This shift cannot be accounted for by the possible migration of Kashmiri speakers from the Jhelum Valley into the extremely remote and inhospitable Gurez Valley[79] and a parallel out-migration of Shina speakers. The miniscule population of Shina speakers outside J&K registered hardly any change between 2001 and 2011. So migration out of the state cannot explain the reported decline in the Shina-speaking population. The same is true of migration to the Jammu region. However, during this period Shina speakers registered a 40 per cent growth in Kargil *tehsil*. Even if it is assumed that half of this growth in Kargil is accounted for by an influx of Shina speakers from Gurez, which seems unlikely, and the rest accounted for by native Shina speakers, the contraction of the corresponding population in the Kashmir division will remain unexplained. Similarly, the increase in the Shina-speaking population in Ganderbal district of Kashmir starting from a small base cannot explain the decline in Gurez.

**Table 3.14a** Distribution of Shina speakers, 1971–2011

| Shina speakers | 1971 | 1981 | 2001 | 2011 |
|---|---|---|---|---|
| India | 10,275 | 15,585‡ | 34,390 | 32,247 |
| Jammu and Kashmir (J&K) | 9,901† | 15,017§ | 34,206 | 32,027 |
| Kashmir | | 12,159 | 25,230 | 19,608 |
| Baramula | | 11,640 | 24,364 | 16,413 |
| Bandipore | | | 24,362 | 16,407 |
| Gurez | | | 23,413 | 12,705 |
| (Gurez *tehsil*) | | | (30,144) | (37,992) |
| (Gurez ST) | | | (26,239) | (31,094) |
| Srinagar | | 303 | 862 | 3,108 |
| Ganderbal | | | 756* | 2,917 |
| Kashmir excluding Gurez | | | 1,817 | 6,903 |
| Ladakh | | 5 | 8,436 | 11,798 |
| Kargil | | 2,853 | 8,435 | 11,796 |
| Jammu | | 0 | 540 | 621 |
| Doda | | | 511 | 612 |

*Source*: GoI (1974a: 112, 2018b, 1987: 108, 1990a: 106, 2007, 2018b); primary census abstract (PCA) (ST) for 2001 and 2011.
*Note*: (*a*) Scheduled Tribes (STs) were recognised in 1989 and 1991, but J&K was not covered by the 1991 census. (*b*) Cross-tabulation of tribes and languages was not released after the 2001 census. (*c*) *Kangan and Ganderbal were *tehsils* of Srinagar in 2001. Ganderbal later became a district that includes Kangan. Shina speakers are concentrated in Kangan. (*d*) † 10,274 as per GoI (1978: 35); ‡ 14,858 as per GoI (2007: 17); § 15,344 as per GoI (1990a: 106). (*e*) See also Shina in Figure 1.1g.

A government official who belongs to Gurez questioned the veracity of census data (interview, 10 February 2021). He added that while there is an older Kashmiri-speaking population in Gurez, Kashmiris from Srinagar and Bandipore have bought land in recent times in anticipation of a tourism boom and may have been enumerated there (even when they live elsewhere). And, at the same time, Shina speakers may have migrated to urban areas in the adjoining districts and may be reporting themselves as Kashmiri-speaking in the new setting. He also suggested that the 2001 census was not conducted properly, which might explain the contraction in 2011 relative to the overestimated population. This will hold if we assume that all the earlier censuses, too, overcounted Gurez's population, but the puzzle posed by an 18-fold increase in the Kashmiri-speaking population

**Table 3.14b**    Shina speakers by tribes, 2011

| | Jammu and Kashmir (J&K) | Bandipore | Kargil | Ganderbal |
|---|---|---|---|---|
| **All Shina speakers** | 32,027 | 16,407 | 11,796 | 2,917 |
| Shina speakers among tribes | 29,827 | 15,933 | 11,597 | 2,213 |
| Bakarwal | 2 | | 2 | |
| Balti | 40 | | 35 | 5 |
| Bot, Boto | 9 | 8 | 1 | |
| Brokpa, Drokpa, Dard, Shin | 23,095 | 11,309 | 9,636 | 2,068 |
| Gujjar | 68 | 38 | | 30 |
| Mon | 323 | | 323 | |
| Purigpa | 99 | | 99 | |
| Generic Tribes | 6,191 | 4,578 | 1,501 | 110 |

*Source*: Tables ST-15 and C-16 of Census of India, 2011.
*Note*: See the note for Table 3.14a. See also Shina in Figure 1.1g.

will remain unexplained. The official agreed that the number of new Kashmiris in Gurez cannot be in thousands and certainly not half of the population.

Since conventional demographic factors and coverage errors cannot explain the abnormal change in the population of Shina speakers, we next explore content errors. Kashmiri included Kashmiri, Kishtwari, Siraji, and 'Others' in 2001, which expanded to Kashmiri, Kishtwari, Siraji, Dardi, and 'Others' in the next census. Both Kashmiri and Shina are classified as 'Indo-Aryan' languages in census reports.[80] Dardi along with Kashmiri was recognised as a 'national' language in Sheikh Abdullah's *Naya Kashmir* (National Conference 1950: 18) and a 'regional' language under the Sixth Schedule of J&K's constitution from the beginning (GoJK 2003), which makes it difficult to explain its disappearance from censuses from 1971 to 2001. It is also not clear why 'Dardi' was included under Kashmiri in 2011.

In 2011, about 93 per cent Shina speakers belonged to STs. 'Brokpa, Drokpa, Dard, Shin' and Generic Tribes accounted for about 77 and 21 per cent of Shina speakers among STs, respectively. In Srinagar and Ganderbal, all Shina speakers belonged to 'Brokpa, Drokpa, Dard, Shin' tribe. However, in Kargil and Bandipore, 13 and 28 per cent of Shina speakers, respectively, belonged to Generic Tribes. A decade ago, in 2001, Generic Tribes accounted for 0 and at most[81] 5 per cent population of Shina speakers in Kargil and Bandipore, respectively.

Together, the contraction of the population of Shina speakers, the sharp increase in the speakers of Shina language among Generic Tribes, an 18-fold increase in Dardi (Kashmiri)-speaking population in Gurez and the presence of a large Kashmiri-speaking population among the 'Brokpa, Drokpa, Dard, Shin' tribes possibly suggest that Shina-speaking tribes were misclassified as Dardi (Kashmiri) speakers in 2011. So the reported contraction of the Shina-speaking population could be a content error. It bears noting that the distributions of Dardi and Shina populations are highly correlated. The three *tehsils* that account for 85.60 per cent of the state's Shina population also account for 81.22 per cent of its Dardi population. Also, in 2001 (2011), the census reported Shina as the first subsidiary language for 15 (3,091) Kashmiri speakers even as the percentage of Shina speakers with Kashmiri as the first subsidiary language did not change.

Alternatively, the contraction could be a manifestation of a closer alignment of the Gurezi and Kashmiri identities as youth from Gurez are increasingly going to Kashmir for higher studies, but we do not have data on language by age to confirm this. It is also possible that some (if not a sizeable fraction) of the Shina speakers heeded the appeal of Kashmiri leaders to non-Kashmiri Muslims to report Kashmiri as their language in census (Ashiq 2010; *Kashmir Life* 2010).

Yet another explanation of content error relates to the widespread belief in the state that the people of Gurez were conferred the tribal status due to their strategic location and the effort of their elite.[82] If true, it can be argued that Dardi (Kashmiri) speakers who wrongly declared themselves or were wrongly recorded or classified as Shina speakers in earlier censuses returned their correct language in the 2011 census. However, for this to hold good, the population of Shina speakers should have registered a very high growth between 1981 and 2001 because tribal communities were formally recognised after the 1981 census. During this period the Shina-speaking population of the state grew by 129 per cent, which is nearly twice the growth rate of the state (69 per cent). Later, between 2001 and 2011, the population of speakers of Shina in Gurez should have contracted but not the ST population of the *tehsil*. Between 2001 and 2011, the rate of growth of the population of Gurez (26 per cent) was a little more than that of the state as a whole (24 per cent). This suggests that the scale of net migration out of the *tehsil* was limited. The tribal population of Gurez grew at 19 per cent, which is comparable to the growth of the *tehsil* as a whole. Interestingly, while the tribal population of the *tehsil* grew, the population of the speakers of the dominant tribe's language contracted by nearly 46 per cent. So those wrongly classified as Shina speakers in 2001 were now recorded as Dardi (Kashmiri) speakers. The key problem with this hypothesis is that after excluding the Generic Tribes, the population growth of the Gurez *tehsil* and the tribal population of the *tehsil* could drop to 10 per cent and 1 per cent, respectively, and the Shina-speaking population of the district would drop to half of the 2001 figure.[83] Yet another hypothesis suggests that the population may have been inflated to retrospectively justify the creation of the Gurez assembly constituency in 1995. As a result, the population contracted once there was no immediate pressure to over-report.[84] But the 18-fold growth of the Dardi (Kashmiri) population will remain unexplained. In short, we do not have a clear explanation for the anomalous changes in the population statistics of Gurez and its Shina-speaking population.

## Generic Categories

Missing or illegible entries cannot explain the very large population share of speakers of other languages that are neither scheduled nor non-scheduled languages because, ideally, in such cases instructions suggest that 'the language of the district should be entered, unless the entries relating to other members of the same household provide a different indication specifically' (GoI 1961: 5). Likewise, '[w]here there is no entry' for religion and caste (and we can add tribe in recent censuses of J&K), the return for 'other members of the same household' is entered (ibid.). The instruction manual for the 2011 census mentions the following regarding the identification and classification of the SCs and STs during enumeration:

> If a person insists on calling herself/himself merely 'Harijan' or 'Achhut' or 'Adivasi' 'Girijan' or repeats the synonym or generic name of a caste or tribe not appearing in the lists provided, please tell her/him that this description is not adequate for census purposes and persuade her/him to give the actual name of the Scheduled Caste or the Scheduled Tribe, as the case may be. This may bring out the actual name of the Scheduled Caste or the Scheduled Tribe. If the person merely claims to be a Scheduled Caste or a Scheduled Tribe, but says that she/he does not belong to any of the notified communities applicable to the area, as reflected in the list supplied to you, she/he will not be reckoned as belonging to a Scheduled Caste or a Scheduled Tribe. (GoI n.d.14: 46)

However, most states report miscellaneous categories of tribes due to incomplete and imprecise lists of recognised tribes, fluid tribal identities, deliberate manipulation of headcount and misunderstanding of enumerators who often belong to non-tribal communities (Agrawal and Kumar 2020a: 300–04).

The population of J&K's unclassified tribes grew from 7,545 to 111,625 during 2001–11 (Table 3.15). This growth cannot be explained by the misclassification of recognised tribes because these tribes did not report lower growth rates in the Kashmir and Jammu divisions (Table 3.10). While the population of some of the tribes decreased in Ladakh, the overall population of Ladakh's recognised tribes increased.

The explosive growth of Generic Tribes is implausible for several reasons. First, we must assume that three decades after the recognition of STs, the degree of familiarity of respondents and enumerators with the official appellations decreased, leading to large numbers being misreported/misclassified across *all* districts. This is highly unlikely given the growing mobilisation around identity and the fact that the census provides multiple names for a community known by more than one name – for example, 'Bot, Boto' and 'Brokpa, Drokpa, Dard, Shin' (see also note 14 of this chapter).

Second, the explosive growth of Generic Tribes cannot be attributed to content errors because the largest tribe of the Kashmir and Jammu divisions (Gujjar) reported a growth rate higher than the uncorrected state average, while the second largest tribe (Bakarwal) grew at an explosive rate of 86 per cent. The combined growth rate of these two closely related tribes was 33 per cent, which is still much higher than the state's uncorrected growth rate. As discussed earlier, even the corrected population growth rate of these two tribes exceeds the state's uncorrected growth rate. Further, more than 60 per cent of Generic Tribes speak Gojri. If the Gojri-speakers among Generic Tribes have to be reclassified and added to the Gujjar or Bakarwal community, the decennial population growth of the latter will be twice that of the rest of the state.

Third, the level of urbanisation of Generic Tribes halved between 2001 and 2011, even as their literacy rate grew, which is counterintuitive.[85] Fourth, if Generic Tribes are accounted for by random errors in recording tribe names during enumeration or sorting, their distribution should not be correlated with that of 'all STs', which was the case in 2001 when their population was tiny. However, the distributions of 'all STs' and 'Generic Tribes' share a significant and strong correlation in 2011. Fifth, in the absence of large-scale in-migration of tribes from outside the state, the spike in the population of Generic Tribes cannot be explained by migration either. Sixth, it cannot even be argued that inter-marriage among tribes is generating newer tribal identities not captured by fixed categories because the state's tribes are divided by ethnolinguistic, religious and geographical barriers. Moreover, the Gujjars, the largest tribe, alone account for as much as 65 per cent of the tribal population, which increases to 75 per cent

**Table 3.15**  Generic Tribes of Jammu and Kashmir (J&K), 2001–11

| State, region, sub-group | 2001 | 2011 |
|---|---|---|
| J&K | 7,545 | 111,625 |
| **Division** | | |
| Kashmir | 6,365 | 66,582 |
| Jammu | 981 | 31,907 |
| Ladakh | 199 | 13,136 |
| **Language** | | |
| Dogri | | 4,666 |
| Hindi | | 69,399 |
| Gojri, Gujjari, Gujar | | 59,512 |
| Pahari | | 5,876 |
| Hindi | | 2,427 |
| Kashmiri | | 14,915 |
| Kashmiri | | 4,819 |
| Dardi | | 8,873 |
| Bhotia | | 4,580 |
| Bauti | | 3,833 |
| Khandeshi (Gujari) | | 2,179 |
| Shina | | 6,191 |
| Tibetan | | 6,311 |
| Purkhi | | 6,276 |
| **Religion** | | |
| Muslims | 6,615 | 93,515 |
| Hindus | 738 | 10,817 |
| Christians | 50 | 833 |
| Sikhs | 13 | 491 |
| Buddhists | 128 | 5,749 |
| Jains | 1 | 9 |
| Other religions and persuasions | 0 | 13 |
| Religion not stated | 0 | 198 |
| **Sex ratio** | | |
| J&K | 763 (910) | 871 (924) |
| Kashmir | 752 | 859 |
| Jammu | 883 | 868 |

(*Contd*)

**Table 3.15**  (*Contd*)

| State, region, sub-group | 2001 | 2011 |
|---|---|---|
| Ladakh | 567 | 943 |
| CSR | 942 (979) | 925 (912) |
| **Urbanisation (per cent)** | | |
| J&K | 14.76 | 7.11 |
| | (4.66) | (5.79) |
| **Literacy rate (per cent)** | | |
| J&K | 32.09 | 53.66 |
| | (37.46) | (50.56) |

*Source*: Tables PCA-IND-A-ST, ST-14 and C-16 of Census of India, 2001; tables A-11, ST-14, ST-15 and C-16 of Census of India, 2011.

*Note*: (*a*) Generic Tribes in the censuses include 'those who returned as Anusuchit Janjati, Girijan, Adivasi, etc.' (*b*) Figures within parentheses are for Scheduled Tribes (STs).

if we include the closely related Bakarwals (Table 3.10). We can conclude that the abnormally high rate of growth of Generic Tribes is likely to be largely accounted for by ghost population a la Nagaland in 2001 (Agrawal and Kumar 2020a: 175).

## Religion

The declining share of Muslims in J&K's population was, until recently, the cornerstone of the demographic propaganda associated with the demand for autonomy or independence. We will rule out the possibility that the decline between 1961 and 1981 was driven by content errors.[86]

There is a sizeable Shia population, particularly in Badgam and Kargil, and Sufi orders also enjoy considerable patronage in Kashmir and Jammu. The Muslim population share could decrease if people reported their sectarian identity and their response was not classified under Muslim. This can be ruled out because sects such '[S]hia, Lingayat, Arya Samajist, Anand Margi, Radha Swami etc' are 'clubbed with the main religion at the tabulation stage' (GoI 1978b: 22), and this practice is meticulously followed. In fact, even requests for sub-categorisation within religions are not accepted. During the 1971 census, the All State Kashmiri Pandits' Conference had 'requested that the Kashmir Pandit community should be tabulated and recorded separately under the general head "Hindu"', but 'the representation was not entertained and Kashmiri Pandits [were] shown under the main religion, Hindu' because 'the practice of collecting and tabulating the data by sub-sections had been abandoned since 1951' (ibid.). The 2011 census identified the sect of only 23 Muslims in J&K. Moreover, the list of 'Other Religions and Persuasions' has never accounted for more than 0.01 per cent of the population of the state (Table 3.16) and does not include entries that can be clubbed with 'Muslim'. This also implies that content errors cannot explain the higher growth of the Muslim population between 1981 and 2011.

It is possible that enumerators inadvertently left the religion column blank or entries were illegible in some cases. This, too, cannot explain the declining population share of Muslims because the census generates missing values as follows: 'If by chance, the entry is blank in both the Individual Slip and in Q. 2 of Part I of the Household Schedule look into the Individual Slips of other related members of the household and write the religion' (GoI 1982b: 486). Moreover, the category 'Religion not stated' has never accounted for more than 0.20 per cent of the population (Table 3.16).

The misclassification of Muslims as non-Muslims is another possibility. The 2011 census, for instance, reports that 805 Gujjars followed a religion called 'Gujarat', while another 304 followed a religion called 'Grover Khatri' (Census of India 2011: table ST-14, appendix). Misclassification cannot explain the *reported* changes in the relative shares of religions as the numbers involved are very small.

In fact, the census takes care to avoid misclassification and asks enumerators to not confound religion with caste or language: 'Do not mistake religion for

**Table 3.16**   Religions of Jammu and Kashmir (J&K), 1961–2011

|  | 1961 | 1971 | 1981 | 1991 | 2001 | 2011 |
|---|---|---|---|---|---|---|
| All | 3,560,976 | 4,616,632 | 5,987,389 | – | 10,143,700 | 12,541,302 |
| Hindu | 1,013,193 | 1,404,292 | 1,930,448 | – | 3,005,349 | 3,566,674 |
| Muslim | 2,432,067 | 3,040,129 | 3,843,451 | – | 6,793,240 | 8,567,485 |
| Christian | 2,848 | 7,182 | 8,481 | – | 20,299 | 35,631 |
| Sikh | 63,069 | 105,873 | 133,675 | – | 207,154 | 234,848 |
| Buddhist | 48,360 | 57,956 | 69,706 | – | 113,787 | 112,584 |
| Jain | 1,427 | 1,150 | 1,576 | – | 2,518 | 2,490 |
| Others* | 3 | 8 | 44 | – | 97 | 1,508 |
| Religion not stated | 9 | 42 | 8 | – | 1,256 | 20,082 |

*Source*: GoI (2004: Statement 2) for 1961–2001; tables C-1 and C-01 appendix of Census of India, 2011.

*Note*: *Other religions and persuasions include unclassified sects.

caste which will not be recorded here. You should also not try to establish any relationship between religion and mother tongue' (GoI 1985b: 207; see also GoI 1971b: 151; GoI n.d.14: 6.52). The 1961 census admits that there were a few cases of mistakes in recording answers to the religion question.

> During the course of sorting of Individual Slips, discrepancies of two types came to notice. In some slips, the entry against religion had been left blank whereas in a few others, a religion other than the one professed by the person enumerated was recorded. In the former case, the discrepancy was set right by inserting the name of the religion followed by members of the same household. In so far as wrong entries were concerned, it appeared that the enumerator, while filling up the Individual Slips of a village or block which was predominantly inhabited by members of one community only did not make enquiries from every individual about the religion professed by him. Mistakes of this nature came to notice in Ramban and Kathua tehsils, where 'Hindu' was recorded in some of the Individual Slips of Muslim inhabitants. The mistake could be traced without much effort as the religion entered was at variance with the name of the person enumerated. The corrections were not, however, effected without reverification of the entries. (GoI 1968b: 235)

A 1961 census report also notes an instance of deliberate misreporting of responses to the religion question by the enumerator.

> No difficulty whatever was felt in obtaining correct replies in respect of Question 5 (b) Religion. One of the Enumerators, however, seemed to have deliberately recorded in a number of cases a religion different from the one professed by the persons enumerated, as was clear by their names. Fortunately, this was detected in time and necessary corrections were made. (GoI 1963a: 38–39)

Similar instances related to Muslims were not reported in later censuses. A few instances of misreporting related to Jains were attributed to the ignorance of enumerators. A census report suggested that 'the Jain community is showing signs of steady decrease. This may be partly due to the error of enumerators in returning Jains with Hindu names as Hindus unless specifically told by the respondent in such cases that he or she is a Jain' (GoI 1978b: 32). The 1971 administrative report discusses 'a complaint [in Jammu city] ... made by one of the leaders of the Jain community that Jains were being wrongly classified as Hindus. These allegations, on examination, were found to be based on a misunderstanding' (GoI 1971b: 20). Otherwise, the 1971 census points out,

'No complaint was received however from any quarter about a different religion having been entered by any enumerator than that stated by the respondent' (GoI 1978b: 22). The 1981 census, too, did not report complaints regarding the misclassification of Muslims. Similar notes are not available for 2001 and 2011, but the Muslim population grew at a faster rate than Hindus in these decades.[87]

### Halam and Ponchi

As per the 2001 census, there were 13,049 speakers of Halam in J&K, which meant it was home to the second largest Halam-speaking population in the country behind Tripura but ahead of Assam. This included 13,002 people in the Ramban (2,789) and Doda (9,479) *tehsils* of Doda district. The sex ratio of the Halam-speaking community was 959, nearly 97 per cent lived in rural areas and a third of its members also spoke Urdu, which suggests that it is unlikely to be a migrant community. A decade later, the 2011 census reported barely eight Halam speakers in the state. Halam is a Tibeto-Burman language spoken exclusively in Tripura and Assam. The director of census operations for the 2001 census was surprised to learn about this language (interview, Jammu, 4 December 2019).[88] Unlike Khandeshi, there is no plausible explanation for the emergence and disappearance of Halam in J&K.[89]

Another content error relates to Gujarati that reported a 389 per cent growth in the state during 2001–11 (Table 3.17). This was driven entirely by the Ponchi dialect of Gujarati. The small Gujarati-speaking population of the state grew by 42 per cent excluding Ponchi speakers. There was no Ponchi speaker in J&K or any other state during the 2001 census. In 2011, the census reported 13,689 Ponchi speakers in J&K. J&K alone accounted for 99.12 per cent of Ponchi speakers in India. The sex ratio of Ponchi speakers was 912, which suggests this is unlikely to be a migrant community. It seems that 'Ponchi' in the 2011 census refers to the language spoken by the people of undivided Punch. Since the speakers of 'Ponchi' were reported almost entirely in Jammu and Ranbir Singh Pora tehsils, it is likely they belong to the Hindu and Sikh refugee community from Pakistan-occupied Punch.

## Concluding Remarks

This chapter analysed the whole range of coverage and content errors in J&K's censuses. We saw that conventional explanations such as female foeticide, excess girl child mortality, misreporting of the sex of girl child, and changes in

**Table 3.17**   Speakers of Ponchi in Jammu and Kashmir (J&K) and India, 2001–11

| Language | J&K | | India |
|---|---|---|---|
| | 2001 | 2011 | 2011 |
| Gujarati | 3,936 | 19,261 | 55,492,554 |
| Gujarati | 3,385 | 5,428 | 55,036,204 |
| Gujrao, Gujrau | 19 | – | 15,431 |
| Pattani | – | 47 | 16,510 |
| Ponchi | – | 13,689 | 13,812 |
| (Jammu district) | – | (11,402) | – |
| (Jammu *tehsil*) | – | (3,929) | – |
| (Ranbir Singh Pora *tehsil*) | – | (7,447) | – |
| Saurashtra, Saurashtri | 1 | 1 | 247,702 |
| Others | 531 | 96 | 162,895 |

*Source*: Table C-16 of Census of India, 2001 and 2011.

the reach of the census, patterns and trends of migration and conflict cannot explain the extremely skewed CSR and the sharp increase in the population share of the 0–6 age group reported in Kashmir. Census data on the child population are inconsistent with other sources of demographic information, including the SRS, the NFHS, the HMIS and data on school enrolment. We attributed the steep decline in the CSR and the increase in the child population share to coverage errors. While both male and female child population were inflated in Kashmir, there was higher over-reporting in the case of the former skewing the sex ratio. Our conservative estimates suggest that the state's (Kashmir's) population was inflated by about 4.5 (8.5) per cent.[90] Since the correction needs to be applied mostly to Kashmir, the corrected population shares of the Jammu region and communities largely based in Jammu such as Hindus, SCs and (Hindu) STs will increase.[91]

We also examined content errors in the data on non-scheduled languages, dialects of scheduled languages, STs and miscellaneous categories of tribes and languages. We saw that the population of Generic Tribes was grossly over-reported in the 2011 census and most likely consists of fraudulent entries. The share of the Gujjar and Bakarwal population was underestimated in some cases and overestimated in others. Even after correcting for errors, the population of Gojri speakers grew much faster than the corresponding tribes, which is indicative of growing language consciousness, among other things. The data for the languages of tribal communities in the Ladakh division show very sharp changes driven by communal politics. Since conventional demographic and non-demographic factors cannot account for the anomalies identified in this chapter, we will have to turn to other sources for an explanation. The following discussion is organised in three parts. In the next chapter we will examine the political context of enumeration. This is followed by a discussion of the legal and administrative context in Chapter 5 to understand why the government fails to check the manipulation of the census. Chapter 6 examines how the union government's prior expectations could affect the census. The last chapter concludes with a discussion on how we can reform the census.

# Notes

1. While abortion was legalised in India under the Medical Termination of Pregnancy (MTP) Act, 1971, 'it is only recently that prenatal diagnostic techniques became widely available' (Mari Bhat and Zavier 2007: 2296). In fact, 'information on the use of these techniques was not collected in NFHS-1 [National Family Health Survey-1] [1992–93]' because 'of its relative rarity' (ibid.). Note that the MTP Act, 1971, and the Pre-conception and Prenatal Diagnostic Techniques (Prohibition of Sex Selection) Act, 1994, did not extend to Jammu and Kashmir (J&K).

2. The Indian experience is not unique as 'almost simultaneous shifts in SRB [sex ratio at birth]' have been 'observed in several Asian countries' (Guilmoto 2009: 526).

3. The British Broadcasting Corporation (BBC) reported on the steep drop in the CSR in a rather sensational manner that confirmed Kashmir's exceptional status: 'The Kashmir Valley, which has been in the grip of an armed insurgency against Indian rule for the past two decades, has now turned on its girls, *killing them ruthlessly*, in most cases even before they are born' (BBC 2011, emphasis added). NDTV (2011), too, used a sensational headline in its report on this matter.

4. There have been a few crucial exceptions to this though. In 2004, while estimating the growth rate of religious groups between 1991 and 2001, the Office of the Registrar General of India (ORGI) forgot to account for the fact that J&K was not covered in 1991. This resulted in the overestimation of the growth rate of Muslims in India (A. Bose 2005: 370).

5. Bhat (2014) attributes the lack of attention to the quality of census data within Kashmir to 'write ups in the local dailies complimenting the Census Directorate to have done a splendid job of completing Census in a sensitive State of Jammu and Kashmir' that 'made people to believe [*sic*] in the authenticity of census figures'. But he also notes that his own article on 'the quality of Census 2011

data' received 'many comments from the public supporting my concerns' (Bhat 2018: 170).

6. A senior health official posted in Kashmir indicated that the department had questioned the validity of census data (interview, 12 August 2021). This was confirmed by a researcher (interview, 21 July 2021) and a journalist (interview, 28 August 2021) who had spoken to the aforesaid official after the results of the 2011 census were released.

7. The state government, civil society and political parties of Nagaland (Agrawal and Kumar 2020a) and Manipur (Agrawal and Kumar 2020b) formally informed the union government and the ORGI that the 2001 census was flawed. However, both the then Registrar General of India (RGI) (interview, 6 August 2021) and the union home secretary (interview, New Delhi, 11 September 2019) told the author that they were not aware of any malpractice or errors in the 2011 census of J&K and do not recall having received any formal complaint against the process or its outcome. A retired senior official of the Ministry of Home Affairs observed, 'Because of persistent law-and-order issues just about that period, there were difficulties faced in the organization of the census exercise [in 2011]', but added that the ministry did not receive any formal complaint (interview, 24 April 2023).

8. Habibullah (2020: 27), in fact, read the high growth rate of districts of Kashmir as a measure of success of Sheikh Abdullah's policies.

9. Chowdhary's figure is based on the provisional population totals (PPT). The figures reported within square brackets are based on later publications.

10. John's (2011) figures are based on the PPT. The figures reported within the square brackets are based on later publications.

11. After the results of the 2011 census were published, Yashpal Sharma, the head of the National Rural Health Mission in Kashmir, observed that they had 'never expected such a drop' (A. Khan 2011).

12. Mari Bhat (2002b: 5258) suggests that 'the sex ratio of the age group 0–14 years is least affected by age errors'. The sex ratio of the 0–14 age population, too, declined in J&K from 940 to 880 between 2001 and 2011, while the corresponding figure for India dropped from 919 to 916. The state's 0–14 age population share decreased from 0.3566 to 0.3381, compared to the much larger decline in the all-India share from 0.3535 to 0.3076.

13. The aforesaid comparison is based on GoI (2014a) that provides data for both 2001 and 2011, following the latest – that is, 2011 – district borders.

14. Content errors due to the growing disjuncture between exonyms and endonyms cannot account for the declining population share of SCs. Census officials have been aware of the disjuncture since the 1961 census, when 'all the Charge

Superintendents ... were requested to submit, after proper enquiries, the names of such [Scheduled] castes as were known by different names than those mentioned in the list circulated to them and about which they were satisfied that they were synonymous with their corresponding original names' (GoI 1963a: 27). Deputy commissioners 'confirmed that there were certain people known as Mahashas, Ramdasias, etc., whose social structure was identical with those of Doom, Chamar, etc. The matter was eventually referred to the State Government and with their approval, a notification (Bidayat No. 10) was issued on 5th January, 1961 directing the Census staff to use both the original and the generic names in the various schedules to be filled up by them during the course of enumeration' (ibid.).

15. See Mari Bhat (2002a, 2002b); Mari Bhat and Zavier (2007); Chakraborty and Kim (2010); Anderson and Ray (2012); Perwez, Jeffery and Jeffery (2012); Mohanty et al. (2016); Jayachandran (2017).

16. Anderson and Ray (2012: 85) point out that 'there is significant variation in the distribution of missing women by age across different states. Missing girls at birth are most pervasive in some north-western states, but excess female mortality at older ages is relatively low. In contrast, some north-eastern states have the highest excess female mortality in adulthood but the lowest number of missing women at birth.' Further, Mari Bhat and Zavier (2007: 2300) add that 'eastern, north-central and north-western parts of India show lower use of prenatal diagnostic techniques (PNDT) than southern states. In other words, the geographical pattern seen in the PNDT use cannot fully be explained by the observed socio-economic variations. This indicates the influence of neighbourhood on the use of technology.'

17. According to Mari Bhat and Zavier (2007: 2299), 'the use of PNDT is certainly much higher in urban areas than in rural areas, but it is not altogether clear as to where the misuse is higher'. Further, the impact of PNDT on the CSR is mediated by a combination of factors: 'While income and education are found to increase the use of PNDT, their misuse is governed more by cultural factors and the sex composition of children already born' (ibid.).

18. PES was known as PEC until 1991. Except for 1961, the results of the PES are available at the level of five zones until 1991 – northern, eastern, southern, western and central. A separate north-eastern zone was created in 2001 to better cover the region where the PEC was limited to Assam in 1971 and Assam and Tripura in 1991. The north-east was left entirely uncovered in 1981.

19. The net omission rate refers to 'the ratio of the number of omitted persons (net of duplication) per 1000 persons enumerated in census' (GoI 2014d: 2.24).

20. PES results are not disaggregated by state. So we cannot conclusively rule out the possibility that the PES results for J&K might have revealed over-enumeration.

21. There were hardly any cases of foeticide registered in J&K around the year 2011 (Lok Sabha 2014). J&K is, in fact, a clear outlier in terms of the availability of prenatal diagnostic facilities (Subramanian and Daniel 2011). It is the only state with both a very low CSR and low access to prenatal diagnostics.

22. Bhat (2018: 175) argues that 'the CSR has shown a decline in [the] most prosperous districts of Kashmir valley where we cannot expect excess female mortality at very young ages'.

23. The HMIS data of Pulwama, Srinagar and Badgam for the years around 2011 accessed by this author too revealed sex ratios above 900. For a comment on the quality of the HMIS data in J&K, see Bhat (2016). *The Digest of Statistics* for 2013–14 reports the number of live births registered by district, sector and sex for each year in the period 2002–11. Kashmir's sex ratio was not less than 900 for any year during this period except in 2008 (885) and 2011 (895). (GoJK n.d.2).

24. Mehbooba Mufti observed that 'there has been an influx of Bengali, Bihari and other women in the state, who marry Kashmiri boys belonging to poor classes' (B. Puri 2004: 1457; see also Peer 2009: 130).

25. A former legislator pointed out that certain government officials readily attributed the sudden drop in the CSR to the inclusion of the armed forces in the headcount (interview, Jammu, 3 December 2019). See also note 46 of Chapter 2.

26. The government does not release statistics on the deployment of the armed forces. The lower bound of the author's indirect estimate of deployment of the central forces in Kashmir based on changes in the distribution of population by age and sex is about 91,000 in 2001 and 136,000 in 2011. Given the ambiguity around the accounting of the central forces, the aforesaid estimates may include deployments in Ladakh that was still part of the Kashmir division in 2011 and also non-local civilians. See Shukla (2018) for an estimate for the period before the reorganisation. Philip (2019) suggests that as of 2019 there were '[o]nly 3.43 lakh Army soldiers and personnel from the central armed police forces' in J&K as a whole including the borders and 'a significant number of which was deployed just ahead of the Narendra Modi government's decision to scrap Article 370 in J&K in August'. In the debate on the reorganisation of J&K, a few parliamentarians suggested that 35,000 additional troops were deployed in the run-up to the decision (Rajya Sabha 2019a: 63, 107).

27. The deployment of the armed forces has a level effect on the aggregate population but no impact on the fertility rates and the child population as most soldiers deployed in combat zones are not accompanied by their families.

28. The quality of the 2011 data on out-migration for J&K is questionable. Unlike in earlier years, the number of out-migrants who cannot be classified

by the duration of stay and place of last residence is almost equal to those who migrated out of the state between 2001 and 2011 (Table 3.6). Our conclusions hold even if we assume that all the unclassified migrants had migrated during the last decade.

29. Wives of two of persons, who had 'crossed over to Pakistan-controlled Kashmir for arms training' and returned through Nepal with their families, won *panchayat* elections in Kashmir (M. Ahmad 2018).

30. The higher fertility rates of non-Sinicised communities might also be explained by the relaxation in the one-child policy in the case of ethnic minorities (see Zhao 2004: 195–99).

31. The link between ethnic tension and pronatalism depends on the context. The Basque nationalist concern over the decrease in both the number and the geographical spread of the Basque-speaking community triggered campaigns to boost the intensity and diversity of the use of the Basque language in public, but it did not trigger pronatalism (Urla 1993). Similar pressures in a developing country may have evoked a pronatalist rhetoric.

32. Anecdotal evidence suggests that the conflict would have suppressed the fertility rate because of delayed marriages, among other things. As Peer (2009: 99) puts it, 'Delays in education and difficulties in finding work kept pushing the wedding dates back, much to the annoyance of impatient parents and grandparents. It had become especially worrisome for girls' parents; most of the dead in the fighting were young men, and further, thousands of young men had deforming injuries, depressions, and non-existent careers.' See also Puri (2004: 1457).

33. Agrawal and Kumar (2020a: 139), too, found inconsistencies between the NFHS and the SRS, on the one hand, and the census, on the other, in Nagaland, where the headcount had been inflated. Also note that the Civil Registration System (CRS) is not reliable for J&K for the period of interest to us as the level of registration was among the lowest in the country (less than 70 per cent in 2011) (GoI 2014c).

34. An interviewee asked, 'If Kashmiri Muslims are affluent and educated and are family-planning conscious, then how is [their] population growing?' (interview, Jammu, 2 June 2016). All Kashmiri Muslim interviewees invariably agreed that the large families are no longer the norm in Kashmir, and the author, too, did not come across large families during his field visits to villages in Kashmir.

35. Anderson and Ray (2012: 89) point out that 'the NFHS does not provide us with a very large sample.... This means that we have to contend with significant variation in the sample SRB'. We do not use district-level estimates of the CSR from the NFHS. For the difficulty in interpreting NFHS data due to the poor quality of metadata, see Kumar (2023d). However, note that the NFHS estimates of the CSR are likely to be closer to the corresponding figures from the census.

The NFHS definition biases the sex ratio upwards vis-à-vis the census by undercounting migrants, often male. The de jure population accounted for by the NFHS includes '[a]ll persons who are usual residents of the selected households, whether or not they stayed in the household the night before the interview' (IIPS and ICF International 2017b: 17). On the other hand, the de facto population includes '[a]ll persons who stayed in the selected households the night before the interview (whether usual residents or visitors)' (ibid.). As a result, the de facto sex ratio should be higher than the de jure sex ratio as per the NFHS (see, for instance, IIPS 1995: 35). In principle, the overall sex ratio should be the highest for the de facto NFHS population followed by the extended de facto census population and the de jure NFHS population. This is corroborated by two facts. First, according to NFHS-1, the de facto and de jure sex ratios of India were 957 and 944, respectively. The corresponding figures were 960 and 949 as per NFHS-2. Second, the CSRs estimated by the NFHS compare favourably with the census as the de facto and de jure populations of children are almost entirely coterminous.

36. Field visits to villages and interaction with healthcare workers across all the districts of Kashmir in 2015 and 2016 suggested that the actual CSRs were higher than the reported figures. Shumriyal in Kupwara was the only village in the sample where the sub-centre or primary health centre maintained a complete date-wise record of births and deaths for the pre-2011 period that can be compared directly with the 0–6 age population of the 2011 census. The data for the period between 1 January 2006 and 1 March 2011 suggest that Shumriyal's CSR was 905 compared to 426 reported by the census. Other villages that allowed assessment for a few years before 2015–16 did not reveal a poor CSR either. However, girls accounted for a larger share of child deaths in Shumriyal, which compares with a study by Kaur and Vasudev (2019) in two villages of Jammu where they found that girls accounted for most of the child deaths.

37. Note that we have referred to school enrolment data only for cross-checking the direction of change of child population statistics. The absolute figures of school enrolment need further investigation.

38. An academic based in Kashmir 'claimed that the Indian Army killed boys and thus people resorted to female foeticide' (journalist, interview, Srinagar, 23 May 2016).

39. Mari Bhat (2002b: 5258) highlights the dynamic impact of differential error in the reporting of age on the CSR: 'In a situation where parents have no clear knowledge of their children's age but have a strong preference for sons, and live in a cultural milieu where daughters are to be married off early, reported ages of boys are more exaggerated than girls. Consequently, to begin with, juvenile sex

ratios are typically biased upward, and as age awareness improves, female-male ratios at childhood could show a spurious drop. Therefore, observed trends of sex ratios for age groups such as 0–6 should be interpreted with caution.'

40. Enumerators were asked to 'make repeated enquiries about infants and very young children for they are often liable to be left out of count' (GoI 1971b: 143).

41. During the 2011 census, in Nagaland, enumerators found families with bogus children, families that insisted on enumerating children studying elsewhere in their native places and families that overstated the age of children 'to show them as eligible voters' (DCO, Nagaland, interview, 19 September 2012, Kohima; GoI n.d.10: 17; GoI 2011h: 2). A Manipuri civil servant told the author that in the earlier censuses people in his jurisdiction had reported fictitious male children to boost the headcount (Senapati, interview, 8 October 2019). Manipulation of household size was also found in the 1991 census of Nigeria (Yin 2007) and the 2009 census of Kenya (Jerven 2013: 73). The latest census data for China seem to 'show 14 million more children in the 0–14 age group than the underlying births for the corresponding years' without any notable addition to this group due to immigration (Kawate 2021). This anomaly is noteworthy because the relaxation of the one-child policy does not seem to have delivered the desired results (Zhao and Zhang 2021).

42. A few interviewees suggested that the 2001 data were possibly obtained through extrapolation or projection wherever it was difficult to access the field (interview, former senior census official, Srinagar, 30 May 2022). In the absence of any further information, it is not clear why extrapolation should alter the trend. However, the allegedly extrapolated population of Kashmir deviated from its long-standing trend that must have been the basis for projections. But see note 40 of Chapter 6 for the possible extrapolation from the houselisting phase to the household phase of the same census.

43. It can also be argued that changes in census-data processing techniques might account for the anomalies in J&K's headcount. Such changes can possibly account for content errors but not the coverage errors in the overall headcount, the 0–6 age population and other such parameters that are covered in the PPT. Moreover, the change in technology and logistics between 2001 and 2011 was not as steep as the change between 1991 and 2001, and their impact on data quality, if any, should have affected other states as well.

44. Bhat (2018: 182) suggests that in many cases third-trimester pregnancies and recently dead persons were included in the headcount, but together these can account for less than a tenth of the overcount because over-reporting affected all pre-school ages and, as shown later, even the higher age groups.

45. The corresponding correction for the Muslim population of the state suggests an overestimation of 7.67 per cent.

46. Other sources, too, suggest low enrolment in Aadhaar. According to NFHS-4 (2015–2016), only 66 per cent of people had an Aadhaar card (IIPS and ICF International 2017a: 3). A senior census official pointed out that there was large under-enumeration in the National Population Register as enumerators had achieved only 70–80 per cent coverage by late 2015 even though field visits suggested that such a large fraction of population did not seem to remain uncovered in Kashmir (census official, interview, Srinagar, 30 November 2015). Interviewees in Jammu alleged that there was opposition to Aadhaar in Kashmir to protect the inflated headcount (Hari Om, interview, Jammu, 30 June 2018; lawyer, interview, New Delhi, 1 July 2018).

47. The population of those who reported Urdu as their mother tongue was 13,251 and 19,956 in 2001 and 2011, respectively. A large fraction of Muslims reported Urdu as their second language though.

48. While we focus on tribes and their languages, it bears noting that there have been complaints about data on other communities too. The 1961 administrative report (GoI 1963a: 27, 37) provides a detailed note on the concerns of Dogri organisations that alleged that 'statistics relating to languages spoken in Jammu and Kashmir would not be collected during the Census'. But the authorities 'pointed out that not only all the statistics regarding mother tongue would be collected in Jammu and Kashmir ... in Jammu province the large majority of Enumerators consisted of those who spoke Dogri'. The 1971 administrative report notes that 'a complaint was received from Kishtwar that contrary to instructions, enumerators in a few cases had not entered the mother-tongue as returned by the respondents' (GoI 1971b: 20). For concerns over the undercounting of SCs, see notes 14 and 57 of Chapter 3.

49. Choudhary (2011: 3) points out that 'the Gujjars were declared as STs in April 1991 after a prolonged struggle of the community which began in 1960s'.

50. See note 54 of Chapter 2 for how the community explains the gap between the headcount reported in the census and its self-estimate. The author cross-checked village-wise estimates of the ST population share prepared by the community activists. In Uri town and 57 villages of Uri and Boniyar, the 2011 census underestimated the population by 28.43 per cent vis-à-vis the community's self-estimate. This cannot be explained by under-enumeration due to the annual migration of nomadic communities. In fact, even in southern Kashmir, the non-enumeration of nomadic population in the snowbound areas can in the best-case scenario explain only a part of the alleged under-enumeration.

51. As Choudhary (2008: 13) puts it, 'The Geelanis have regrets that the Jammu Muslim is not a part of the Kashmir movement as much as the Kashmiri Muslim is. The Khajurias have a grouse that their Muslim neighbours are not as much Indians as they are. This is exactly where their fault lies. Both have a poor

idea of the sentiments of Jammu Muslims.' Nazeer (2022) succinctly sums up how Gujjars and Bakarwals are caught in the crossfire between Kashmir and Jammu: 'The denial of political reservation and the absence of the Forest Rights Act had essentially become a bargaining tool. Kashmir-based parties were not in favour of political reservation and those Jammu based were against the Forest Rights Act. If one side argued for forest and land rights, the other side would oppose it with the political reservation issue. The end result was the continued subjugation and disempowerment of STs in J&K…. The dilution of Article 370 on August 5, 2019, and the subsequent Jammu and Kashmir Reorganisation Act, 2019 changed political dynamics in the region. As far as tribals are concerned, it promised them political reservation under Article 332 and led to the extension of the Forest Rights Act, 2006.'

52. An interviewee (Jammu, 4 December 2019) reminisced that before his 'father went to Pakistan', his paternal aunt 'wrote that he should not tell [anyone there] that he is Gujjar'. Choudhary (2007) writes that the community feels 'isolated in POK [Pakistan-occupied Kashmir] because Gujjars feel inhibited about speaking in their mother tongue and practicing their customs. Unlike in Jammu and Kashmir, Gujjars in POK do not use their surname "Choudhary" as it earns the derision of others. The ethnic, cultural and linguistic identity of Gujjars has increasingly been diluted, and this can be attributed to a state-sponsored hatred against the Gujjar tribe.'

53. A Gujjar interviewee argued that 'Article 370 has resulted in the discrimination of STs. In 1989, we did not get ST status. 8 communities from Ladakh were declared STs. We were excluded at the instance of Farooq Abdullah because *we constitute one-third of the population* and would have captured power from Kashmiris. We finally got reservation under Chandrashekhar…. Their excuse to deny political reservation was that Article 332 has not been extended to the state' (lawyer, interview, Jammu, 13 September 2019, emphasis added).

54. Gujjar interviewees maintained that in Jammu the bulk of their population is concentrated in Rajouri and Punch, where undercounting is highly unlikely.

55. While denying allegations of undercount, a retired civil servant pointed out that it was 'an old refrain among Gujjars. But because this is a migratory community of which there is a majority in areas of Punch and Rajouri district where I was DC [deputy commissioner] as mentioned there have been errors in double counting of populations in home villages in the plains in the Jammu division and *dhoks* [highland pastures] in Kashmir, or vice versa. These were *corrected when noticed*' (interview, 12 January 2021, emphasis added). The DCO for the 2001 census told the author that his office did not receive any complaint from the Gujjar and Bakarwal communities (interview, Jammu, 4 December 2019), but see GoI (2001b: xi).

56. As per the response to the RTI application, there were 12 ST teachers as of 2021. One of them, though, belonged to the 'Brokpa, Drokpa, Dard, Shin' tribal category. Almost a third of these teachers joined service after the 2011 census. Uri and Boniyar *tehsils* are covered by four educational zones with nearly 360 teachers: Uri, Boniyar, Chandanwari and Jullah. In 2011, STs accounted for 16 per cent of the population of these *tehsils*.

57. SC activists claim that their population is under-reported for similar reasons (interviews, Jammu, 14 September 2019, 7 May 2022, 8 May 2022, 10 May 2022).

58. Improved enumeration and growing complaints against under-enumeration are not mutually inconsistent. The community is more vocal now due to growing access to education and awareness about under-enumeration in the past and reacts strongly to any new instance of neglect or marginalisation.

59. Agrawal and Kumar (2020a: 302–03, 309–10) discuss similar instances of disjuncture between data on tribes and languages in Assam and Maharashtra. People not belonging to tribal communities rarely declare tribal languages as their second language, let alone their mother tongue. So the data on tribal languages can be used to cross-check the quality of the data on tribes.

60. The possible contribution of the interaction of Bakarwal qua ethnic identity and Pahari qua linguistic identity to the anomalies in the 2011 data on Bakarwals cannot be examined as table ST-15 was not released for the 2001 census. This interaction can, however, at best account for only a small part of the anomaly in the data on Bakarwals because the Pahari-speaking population grew at the rate of 39 per cent during 2001–11 compared to the state's growth rate of 24 per cent.

61. Most Gujjar and Bakarwal leaders insist that their population is equally distributed between Kashmir and Jammu, with the former likely to account for a slightly larger share (interviews, 26 July 2021, 24 September 2021). Bakarwal leaders also claim that their community's population was close to 6–8 lakhs (interview, 24 September 2021). These claims are a means of both challenging the hegemony of Kashmiris and achieving intra-community harmony between Gujjars and Bakarwals as well as between the community in Kashmir and Jammu. However, the anomalies in statistics on Bakarwals suggest that a sharpening of identities within the larger Gujjar and Bakarwal community and the sedentarisation of the latter maybe underway.

62. 'In the 1941 Census, Gojri, the language of Gujjars and Bakarwals (now declared as Scheduled Tribes), was included as a dialect under Rajasthani due to its close affinities with that language. But Pahari which is closely connected with Gojri and continues to be spoken in much the same areas, was enumerated separately.... The subsequent Census Reports of 1961, 1971 and 1981 have

removed this anomaly of enumerating Gojri and Pahari separately. However, the Census reports of 1971 and 1981 [and later censuses] have followed a new anomalous practice of including Gojri (Rajasthani), Bhadrawahi, Padri [Padari] with Hindi' (Warikoo 2000: 10). Incidentally, the 1961 census reported a small Bakerwali-speaking population entirely based in Jammu (GoI 1965: 209).

63. Grierson (1907: 203) lists Dhed Gujari as a dialect of Khandeshi.

64. Note that Himachal Pradesh recognises Gaddi as an ST and Sippi as an SC. This is not exceptional though. 'The absence of clear guidelines for identifying tribes' and 'the absence of a common national list of the STs' have meant that 'the tribal status of communities varies across states, districts of the same state and even seasons' (Agrawal and Kumar 2020a: 301). Further, 'ambiguous criteria used to identify tribes are conducive to manipulation' (ibid.), which partly explains how political parties can promise ST status to the Paharis of J&K.

65. The Bodos of Assam guard against the attempt of Koch-Rajbhanshis to 'return' to their original community, even as they claim a larger territory and population (Prabhakara 2012: 88, 215–20). They have already secured considerable concessions from the state, and the maintenance of the same is not dependent on the community's size. So keeping the group size small is beneficial. In contrast, Sippis do not seem to have any objection against the return of Kolis as it will boost their miniscule population share and hopefully help them press the government to translate the nominal affirmative action assurances into real benefits.

66. A Gaddi leader alleged that in some parts of the state, Muslims and, to a lesser extent, upper-caste Hindus have also claimed tribal status under the category of Gaddi (interview, 7 September 2021).

67. Balti was included in the Sixth Schedule of J&K's constitution because Baltistan was part of Ladakh Wazarat before 1947, and Balti was a major language of the undivided princely state. Sheikh Abdullah's *Naya Kashmir* manifesto included Baltistani as one of the 'national languages' but did not include Tibetan, Ladakhi or Bhotia (National Conference 1950: 18). Early twentieth-century travellers such as Rahul Sankrityayan referred to the land beyond Zojila as Baltistan (Sankrityayan 1939: 63). Even Bakula (1953: v) refers to 'the Balti (Muslim) area of Kargil'. Presently, Baltis are a minority in Ladakh and seem to increasingly identify with Purigi or Purkhi linguistic identity in Kargil. The Baltis of Baltistan, though, view Purkhi as a dialect of Balti (Kazmi 1996). In fact, Gupta (2023: 132, 181) suggests that even in Kargil partisans of Balti language maintain that Purigi is 'not a proper language', 'an impure, bastard dialect' or 'language of the bazaar'.

68. In the 1970s, the Central Institute of Indian Languages published phonetic readers for Balti (1975), Brokstat (1975) and Ladakhi (1976) but not for Purkhi, Bhotia and Tibetan.

69. The DCO for the 2001 census, Feroze Ahmed, told the author that the issue of classification of Purkhi did not come up in discussions after enumeration (interview, Jammu, 4 December 2019).

70. Purig, Purigi, Purik or Purkhi is also known as Purik Bhotia. Most interviewees, including the DCO for the 2001 census who belongs to Ladakh, suggested that Balti, Bhotia, Purkhi and others refer to closely related languages. As Lama Chosphel Zotpa put it, 'Ek hi bhasha hai, ek hi mula hai' (These languages are one, their roots are the same) (interview, New Delhi, 2 December 2019).

71. The last member of the legislative assembly of Kargil, Asgar Ali Karbalaie, is a Balti but identifies as a Purkhi speaker (interview, Kargil, 22 November 2019).

72. The website of the Ladakh Autonomous Hill Development Council (LAHDC) of Kargil mentions the name of the body in both Tibetan and Perso-Arabic scripts. The LAHDC of Leh uses only the Tibetan script. There have been experiments to revive or at least engage with the Tibetan script among Shia Muslims. A prominent private school in Kargil, Suru Valley Public Higher Secondary School (SVPHSS), has introduced an optional course in which students learn the Tibetan script (Nasir Shabani, interview, Kargil, 23 Nobember 2019). The cover page of the annual magazine carries the school's names in both Latin and Tibetan scripts (Suru Valley Public School [SVPS] 2019). There have also been attempts in Baltistan to revive the Tibetan script (Kazmi 1996). When the author referred to these experiments in a conversation with a leading public figure in Kargil, he strongly disapproved of these because changing the script would amount to abandoning history and dismissed those involved as motivated by outside (Western) influence with no support on the ground (interview, Kargil, 20 September 2019 and 22 November 2019). However, a Buddhist school teacher from Zanskar told the author that the educated youth among Shias are showing interest in their heritage, including the Tibetan script (interview, Kargil, 22 November 2019). Gupta (2023: 188, 190) notes that the Baltistan Cultural Foundation's 'promotion of the Tibetan script' has inspired 'cultural activists in Kargil' who are 'trying to reclaim an identification between the Balti language and Tibetan script'.

73. Interestingly, Shia Muslims have gravitated to the label 'Tibetan' or have been clubbed under it by the census, even though presently there is hardly any direct Tibetan influence in Kargil. Several Shia interviewees pointed out that their spoken language is closer to Tibetan than the language of Leh. While Buddhist interviewees agree, they add that this makes Kargil's quest for a separate linguistic identity even more artificial. The linguistic choice of Leh is driven by

the need to remain close to Buddhism and bond with other Himalayan Buddhist communities within India, while maintaining Ladakhi Buddhist identity and not succumbing to the influence of the globally dominant Tibetan Buddhism of refugees from Tibet. There is a strong resentment among sections of Leh against globally connected Tibetan settlers who refuse to assimilate (interview, Leh, 18 September 2019).

74. Buddhists did not account for a large share of Paddar's population in the past either. There were 439 Buddhists in four villages of Paddar *ilaqa* in 1911 (GoJK 1912: 89).

75. Buddhist STs of Atholi in Kishtwar have called for merger with neighbouring Zanskar 'with whom we are religiously, culturally and mentally attached since our existence and with whom we exchange our trade and culture too', if 'genuine demands' for a reserved assembly constituency and proper implementation of reservation in jobs in UT of J&K cannot be fulfilled (Himalayan Buddhist Cultural Society and Paddar 2021; see also *Daily Excelsior* 2021). While community leaders agree that the tribal Buddhist population is very small and the only pass connecting Atholi to Zanskar is snowbound for an extended period during winters, they lament that the reorganisation has left them with no real options (interview, 23 September 2021).

76. Kargil-based leaders claim a much larger footprint for Urdu. After the amendment of the Ladakh Revenue (Subordinate) Service Recruitment Rule, 2021, one of them even suggested, 'Around 70% of the population in Ladakh can understand Urdu, which is part of our rich culture and heritage' (J. Ali 2022). While we do not have district-wise statistics on second and third languages, anecdotal evidence suggests that this does not hold good even in Kargil (see, for instance, Gupta 2023: 66, 80, 112 for the limited intelligibility of Urdu among illiterate and older people). In any case, the ability to understand spoken Hindustani does not mean one can also read and write Urdu. Several Buddhist interviewees attributed their under-representation in certain government departments to the growing lack of familiarity with Urdu among the younger generation.

77. For the controversy over 'Bhoti' as a name of the language spoken in Ladakh, see Wahid (2022a). 'Bhoti' relates to the census category 'Bhotia'.

78. There is more to the campaign of Buddhists of Leh to secure constitutional recognition than meets the eye. Faced with Srinagar's numbers games, Bakula (1953) reminded the government that his community was part of the much larger Tibetan Buddhist world. Such claims would have been politically incorrect after 1962, which possibly explains the turn to the pursuit of solidarity with other Himalayan Buddhist communities in India. It bears emphasising that Buddhists of Kishtwar are the only Indian Buddhist community outside Ladakh referred to in Bakula (1953).

79. As discussed in Chapter 4, the share of Kashmiri speakers in the population of the relatively accessible and hospitable Karnah and Uri has declined over the past four decades. There is no reason why Kashmiri speakers would migrate in large numbers to the far more inhospitable Gurez. In fact, Kashmiri speakers, who do not speak Dardi that is treated as a dialect of Kashmiri by the census, halved in Gurez between 2001 and 2011. Note that the proportion of Shina speakers with Kashmiri as the second language decreased slightly during 2001 and 2011 (even as those with Urdu as their second language grew by six percentage points). This is not inconsistent with a shift to reporting Kashmiri as the first language, but cannot be confirmed in the absence of disaggregated bilingualism data.

80. Both Shina and Kashmiri are Dardic languages (Grierson 1919: 2), but they are not mutually intelligible. The Kashmiri gentleman who drove the author to Gurez felt utterly out of place and constantly complained about his inability to understand the local language and the strange accent and was not comfortable with the food either.

81. Table ST-15 was not released in 2001, and therefore we can only identify upper limits using tables C-16 and the primary census abstract (PCA) for individual STs.

82. A non-Shina-speaking tribal interviewee suggested that a bureaucrat who belongs to Gurez 'inserted Shina [-speaking communities into the ST list] and a member of the legislative assembly called for giving ST status to all Gurez. This is how many Kashmiris got ST status [in Gurez]' (interview, Jammu, 4 December 2019). Most Gujjar and other interviewees argued that 'sociologically' speaking the Shina speakers of Gurez were not tribes (as they are not nomadic) but were included in the category because of their remoteness and idiosyncratic political factors (interviews, Jammu, 13 September 2019; Jammu, 4 December 2019; Baramula, 20 July 2021). Those opposed to the dominance of Gujjars likewise argue that they got into the list by associating with nomadic Bakarwals (interview, 7 September 2021). A former legislator alleged that Srinagar gave ST status to Gujjars and Bakarwals to balance the rise of Buddhists (interview, Leh, 21 September 2019), while Gujjars claim that Srinagar blocked their case (lawyer, interview, Jammu, 13 September 2019).

83. The revised figures are lower bounds as we do not have disaggregated tehsil-level language data.

84. In several sub-districts of Nagaland that were demanding district status, population contracted sharply after the creation of new districts (Agrawal and Kumar 2020a: 200, 243). While this is a possibility, the undercounting or misclassification of Shina-speaking settlers in other districts of Kashmir in 2011 cannot be ruled out.

85. Similar discrepancies in data on urbanisation and literacy of tribes have also been reported in other states (Agrawal and Kumar 2020a: 302).

86. Conversions out of Islam cannot explain the decline in Muslim population share as there have not been any complaints from Muslims in this regard, and most cases of alleged conversion were reported by the Hindu and the Buddhist communities. One of the very few references to conversion in census reports focuses on Christianity, though: 'The presence of large concentrations of Christians in the urban areas of Jammu and Kathua districts is chiefly attributable to conversions of untouchable Hindus to Christianity. They consist principally of Churas and Chamars connected by marriage and otherwise with their co-religionists in adjoining districts of Punjab, where missionary work has been carried on for many decades in the past' (GoI 1978b: 27). The report adds that 'Christian migration from the Gurdaspur district of Punjab is largest' in Kathua and Jammu. More recently, the SCs have complained that the upper-caste government employees from Jammu do not want to serve in Kashmir and push SC employees into the Valley. The killing of a Dalit government servant in Kashmir in May 2022 highlighted this issue (Raina 2022). According to a Dalit interviewee, in the absence of any social support some of these employees face subtle pressures to convert in the Valley (interview, Jammu, 4 December 2019).

87. The Muslim population share declined between 1971 and 1981 even as the population growth rate of Hindus decreased and that of Muslims increased.

88. Census officials in Nagaland who supervised the 2011 census were surprised when the author pointed out the existence of a hitherto unheard of tribe called 'Viswerna' in the religion tables (interview, Kohima, 11 December 2018). When the author raised this with the ORGI, officials attributed it to coding errors (interview, New Delhi, 21 February 2019), which were left unattended for nearly half a decade and, in fact, remain uncorrected as of writing this.

89. Between 2001 and 2011, the Halam-speaking population of Uttar Pradesh and Uttarakhand increased from 3 to 6,001 (with a very healthy sex ratio of 1,021). We will have to wait until the next census to see if the category 'Halam' persists in Uttarakhand.

90. The estimated magnitude of coverage error in the 2011 census of J&K is high but not unheard of. The 2001 census overestimated Nagaland's population by more than 25 per cent (Agrawal and Kumar 2020a).

91. Since the administrative report has not been released, certain aspects of the census operations that could affect our results remain obscure. For instance, 12 years after the 2011 census, it came to light that '22 boxes with filled-in Schedules [of Phase I from J&K] were not delivered and reported missing' and '[i]nspite of special efforts by DoP, these could not be located' (GoI 2023: 208). We still do not have an official estimate of the number of filled-in schedules lost, the geography of this loss and its impact on coverage errors.

# Part III

# Context

# 4

# Anxious Majorities

Muslims of Kashmir are required to share power as a minority at the national level and a majority at the state level. This is perhaps an unusual experience for a Muslim community anywhere in the world.

—B. Puri (1983b: 231)

## Introduction

The coverage and content errors identified in Chapter 3 are not randomly distributed across Jammu and Kashmir (J&K). Content errors are mostly confined to the data on non-scheduled languages and Scheduled Tribes (STs). The nature of content errors, though, changes across the three regions of the erstwhile state. Coverage errors, on the other hand, mostly affect the headcount of Kashmir and the STs. Since conventional explanations fail to account for the errors, we will have to explore the larger context, including the politicisation of census, that may have affected the *reported* headcount.

Panandiker and Umashankar (1994: 96–97) argue that in states where Hindus are a minority – for example, Muslim-majority J&K and Sikh-majority Punjab – the majority views population control as a measure to reduce their political strength. They add that this is not true of Kerala because of the higher literacy rate and per capita income. However, it is not true that all such provinces are suspicious of population control. Christian-majority states of the north-east are cases in point. Further, in Kerala, the population is divided among Muslims, Hindus and Christians of various castes and sects, with only Muslims having a clear geographical concentration in the north. The fractionalisation and geographical dispersion of communities and strong linguistic ties transcending religions check persistent communal polarisation in Kerala. Even in Punjab there is no region exclusively populated by only one religious community, and despite militancy both the communities share a linguistic and regional identity. Polarisation is further mitigated by the presence of heterodox sects and Scheduled Castes (SCs) among both Hindus and Sikhs in Punjab.

J&K is different. It has two major geographically separable religious groups (Tables 4.1a–4.1c). The exodus of Hindus from Kashmir in the 1990s further hardened this divide. The strongholds of Kashmiri-speaking Muslims and Dogri-speaking Hindus do not overlap even though they share the old Doda district of the Jammu division with other communities. STs are spread across all districts of J&K. They are almost entirely Muslim except in Ladakh and parts of Jammu. On the other hand, SCs are entirely Hindu and confined to Jammu. While the communal polarisation and demographic contest are intense, the resultant pronatalist rhetoric has not affected fertility choices.[1]

This chapter examines the demographic politics involving three anxious majorities: Kashmiri Muslims, who constitute a majority in J&K; Dogra Hindus, who dominate Jammu; and Hindus of 'mainland' India, who constitute the single largest religious group in the country. Each of them believes in its historical right to dominate its chosen territory, which makes conflict unavoidable as these territories are overlapping. Dogra Hindus view themselves as an oppressed minority in the Kashmiri Muslim-dominated state, but Kashmiri Muslims view them as a subversive extension of India's Hindu majority. Hindu nationalists see Hindus of India as a minority in the world dominated by Muslims and Christians and treat Kashmiri Muslims as a subversive extension of the larger Muslim world. We will also discuss the case of Ladakh, where Buddhists were the largest religious community until a few decades ago. In Leh, Shias are seen as an extension of Shia Kargil and Muslim Kashmir. The Buddhists of Kargil are likewise viewed as partisans of Leh.[2]

The predicament of Kashmiri Muslims is unique, though. First, as mentioned in the chapter's epigraph, '[the] Muslims of Kashmir are required to share power as a minority at the national level and a majority at the state level. This is perhaps an unusual experience for a Muslim community anywhere in the world' (B. Puri 1983b: 231, also 233). Second, the Kashmiri Muslim position is reminiscent of the Sinhalese, who believe that in Sri Lanka 'the problem of the Tamils is not a minority problem. The Sinhalese are the minority in Dravidistan. We are carrying on a struggle for our national existence against the Dravidian majority [in southern India]' (Kearney 1985: 903). The case of Punjab, where Hindus are a minority in a Sikh-dominated state and Sikhs are in turn a small minority in the country, is different as Sikhs do not outnumber Hindus globally. Also, unlike in Sri Lanka and J&K, the majority and minority communities of

**Table 4.1a**  Population share of major religions and regions of Jammu and Kashmir (J&K), 1961–2011

| Year | Total population | Population growth rate | Population | | Population share | | Population | | | Population share | | |
|---|---|---|---|---|---|---|---|---|---|---|---|---|
| | | | Muslim | Hindu | Muslim | Hindu | Kashmir | Jammu | Ladakh | Kashmir | Jammu | Ladakh |
| 1961 | 3,560,976 | N/A | 2,432,067 | 1,013,193 | 68.30 | 28.45 | 1,899,438 | 1,572,887 | 88,651 | 53.34 | 44.17 | 2.49 |
| 1971 | 4,616,632 | 29.65 | 3,040,129 | 1,404,292 | 65.85 | 30.42 | 2,435,701 | 2,075,640 | 105,291 | 52.76 | 44.96 | 2.28 |
| 1981 | 5,987,389 | 29.69 | 3,843,451 | 1,930,448 | 64.19 | 32.24 | 3,134,904 | 2,718,113 | 134,372 | 52.36 | 45.40 | 2.24 |
| 1991* | 76.7 lakh | 28.10 | – | – | – | – | 40 lakh† | 35 lakh† | 1.7 lakh† | 52.15 | 45.63 | 2.21 |
| 1991** | 7,718,700 | 28.92 | – | – | – | – | 4,010,202 | 3,537,957 | 170,541 | 51.95 | 45.84 | 2.21 |
| 1991*** | 7,837,051 | 30.89 | – | – | – | – | 4,168,608 | 3,489,033 | 179,410 | 53.19 | 44.52 | 2.29 |
| 2001 | 10,143,700 | 29.43 | 6,793,240 | 3,005,349 | 66.97 | 29.63 | 5,476,970 | 4,430,191 | 236,539 | 53.99 | 43.67 | 2.33 |
| 2011 | 12,541,302 | 23.64 | 8,567,485 | 3,566,674 | 68.31 | 28.44 | 6,888,475 | 5,378,538 | 274,289 | 54.93 | 42.89 | 2.19 |

*Source:* * GoI (1995b: 5); † GoJK (1993: 3); ‡ table A-2 (Census of India, 2011); table C-1 (religion); GoI (2004a).

*Note:* (*a*) Census could not be conducted in 1991. (*b*) * and ** indicate projected figures; *** 1991 population interpolated after the 2001 census. (*c*) The figures are not adjusted for changes in the reference dates noted in Table 2.3. (*d*) † The state government reported rounded-off figures for the headcount in lakhs, indicating the unavailability of precise estimates. (*e*) – indicates the unavailability of data disaggregated at the level of communities.

**Table 4.1b** Share of districts in various sub-groups of population of Jammu and Kashmir (J&K), 2011

| State/District | Population | Scheduled Castes | | Scheduled Tribes | | Kashmiri | | Dogri | | Gojri | | Pahari | | Muslim | | Hindu | |
|---|---|---|---|---|---|---|---|---|---|---|---|---|---|---|---|---|---|
| | | Population | Share | Population | Share | Population | Share | Population | Share | Population | Share | Population | Share | Population | Share | Population | Share |
| J&K | 12,541,302 | 924,991 | 1.00 | 1,493,299 | 1.00 | 6,680,837 | 1.00 | 2,513,712 | 1.00 | 1,135,196 | 1.00 | 977,860 | 1.00 | 8,567,485 | 1.00 | 3,566,674 | 1.00 |
| Kashmir | | | | | | | | | | | | | | | | | |
| Kupwara | 870,354 | 1,048 | 0.00 | 70,352 | 0.05 | 619,592 | 0.09 | 976 | 0.00 | 80,163 | 0.07 | 122,927 | 0.13 | 823,286 | 0.10 | 37,128 | 0.01 |
| Baramula | 1,008,039 | 1,476 | 0.00 | 37,705 | 0.03 | 827,677 | 0.12 | 1,527 | 0.00 | 34,750 | 0.03 | 99,563 | 0.10 | 959,185 | 0.11 | 30,621 | 0.01 |
| Bandipore | 392,232 | 392 | 0.00 | 75,374 | 0.05 | 323,161 | 0.05 | 141 | 0.00 | 34,586 | 0.03 | 7,478 | 0.01 | 382,006 | 0.04 | 8,439 | 0.00 |
| Badgam | 753,745 | 368 | 0.00 | 23,912 | 0.02 | 714,431 | 0.11 | 592 | 0.00 | 22,706 | 0.02 | 1,732 | 0.00 | 736,054 | 0.09 | 10,110 | 0.00 |
| Srinagar | 1,236,829 | 1,068 | 0.00 | 8,935 | 0.01 | 1,164,293 | 0.17 | 1,917 | 0.00 | 8,202 | 0.01 | 2,967 | 0.00 | 1,177,342 | 0.14 | 42,540 | 0.01 |
| Ganderbal | 297,446 | 117 | 0.00 | 61,070 | 0.04 | 206,032 | 0.03 | 181 | 0.00 | 55,570 | 0.05 | 7,750 | 0.01 | 290,581 | 0.03 | 5,592 | 0.00 |
| Pulwama | 560,440 | 402 | 0.00 | 22,607 | 0.02 | 511,687 | 0.08 | 343 | 0.00 | 22,189 | 0.02 | 3,811 | 0.00 | 535,159 | 0.06 | 13,840 | 0.00 |
| Shupiyan | 266,215 | 43 | 0.00 | 21,820 | 0.01 | 234,254 | 0.04 | 243 | 0.00 | 23,425 | 0.02 | 4,749 | 0.00 | 262,263 | 0.03 | 3,116 | 0.00 |
| Anantnag | 1,078,692 | 1,826 | 0.00 | 116,006 | 0.08 | 917,964 | 0.14 | 855 | 0.00 | 123,606 | 0.11 | 13,097 | 0.01 | 1,057,005 | 0.12 | 13,180 | 0.00 |
| Kulgam | 424,483 | 21 | 0.00 | 26,525 | 0.02 | 390,128 | 0.06 | 229 | 0.00 | 26,827 | 0.02 | 2,031 | 0.00 | 418,076 | 0.05 | 4,267 | 0.00 |
| Ladakh | | | | | | | | | | | | | | | | | |
| Leh | 133,487 | 488 | 0.00 | 95,857 | 0.06 | 559 | 0.00 | 896 | 0.00 | 123 | 0.00 | 168 | 0.00 | 19,057 | 0.00 | 22,882 | 0.01 |
| Kargil | 140,802 | 18 | 0.00 | 122,336 | 0.08 | 2,552 | 0.00 | 135 | 0.00 | 0 | 0.00 | 0 | 0.00 | 108,239 | 0.01 | 10,341 | 0.00 |
| Jammu | | | | | | | | | | | | | | | | | |
| Punch | 476,835 | 556 | 0.00 | 176,101 | 0.12 | 25,732 | 0.00 | 443 | 0.00 | 186,658 | 0.16 | 239,402 | 0.24 | 431,279 | 0.05 | 32,604 | 0.01 |
| Rajouri | 642,415 | 48,157 | 0.05 | 232,815 | 0.16 | 14,347 | 0.00 | 10,875 | 0.00 | 221,553 | 0.20 | 356,057 | 0.36 | 402,879 | 0.05 | 221,880 | 0.06 |
| Kishtwar | 230,696 | 14,307 | 0.02 | 38,149 | 0.03 | 158,228 | 0.02 | 1,212 | 0.00 | 33,127 | 0.03 | 1,019 | 0.00 | 133,225 | 0.02 | 93,931 | 0.03 |

(Contd)

**Table 4.1b** (*Contd*)

| | | | | | | | | | | | | | | | | | |
|---|---|---|---|---|---|---|---|---|---|---|---|---|---|---|---|---|---|
| Doda | 409,936 | 53,408 | 0.06 | 39,216 | 0.03 | 227,543 | 0.03 | 16,217 | 0.01 | 30,200 | 0.03 | 9,971 | 0.01 | 220,614 | 0.03 | 187,621 | 0.05 |
| Ramban | 283,713 | 13,920 | 0.02 | 39,772 | 0.03 | 164,103 | 0.02 | 29,078 | 0.01 | 30,654 | 0.03 | 26,237 | 0.03 | 200,516 | 0.02 | 81,026 | 0.02 |
| Reasi | 314,667 | 37,757 | 0.04 | 88,365 | 0.06 | 59,039 | 0.01 | 137,710 | 0.05 | 77,674 | 0.07 | 20,889 | 0.02 | 156,275 | 0.02 | 153,898 | 0.04 |
| Udhampur | 554,985 | 138,569 | 0.15 | 56,309 | 0.04 | 10,394 | 0.00 | 451,530 | 0.18 | 28,770 | 0.03 | 7,518 | 0.01 | 59,771 | 0.01 | 489,044 | 0.14 |
| Jammu | 1,529,958 | 377,991 | 0.41 | 69,193 | 0.05 | 94,649 | 0.01 | 1,084,040 | 0.43 | 59,048 | 0.05 | 5,738 | 0.01 | 107,489 | 0.01 | 1,289,240 | 0.36 |
| Samba | 318,898 | 91,835 | 0.10 | 17,573 | 0.01 | 993 | 0.00 | 269,559 | 0.11 | 13,766 | 0.01 | 1,980 | 0.00 | 22,950 | 0.00 | 275,311 | 0.08 |
| Kathua | 616,435 | 141,224 | 0.15 | 53,307 | 0.04 | 13,479 | 0.00 | 505,013 | 0.20 | 21,599 | 0.02 | 42,776 | 0.04 | 64,234 | 0.01 | 540,063 | 0.15 |

*Source:* Table C-1 (religion), primary census abstracts (PCA) and table C-16 (language) of Census of India, 2011.

*Note:* Share of district $j$ in a sub-group of population, say, religion $i = \dfrac{\text{population of religion } i \text{ in district } j}{\text{population of religion } i \text{ in the state}}$.

Table 4.1c  Share of sub-groups in the population of districts of Jammu and Kashmir (J&K), 2011

| State/District | Population | Scheduled Castes | | Scheduled Tribes | | Kashmiri | | Dogri | | Gojri | | Pahari | | Muslim | | Hindu | |
|---|---|---|---|---|---|---|---|---|---|---|---|---|---|---|---|---|---|
| | | Population | Share | Population | Share | Population | Share | Population | Share | Population | Share | Population | Share | Population | Share | Population | Share |
| J&K | 12,541,302 | 924,991 | 0.07 | 1,493,299 | 0.12 | 6,680,837 | 0.53 | 2,513,712 | 0.20 | 1,135,196 | 0.09 | 977,860 | 0.08 | 8,567,485 | 0.68 | 3,566,674 | 0.28 |
| | | | | | | Kashmir | | | | | | | | | | | |
| Kupwara | 870,354 | 1,048 | 0.00 | 70,352 | 0.08 | 619,592 | 0.71 | 976 | 0.00 | 80,163 | 0.09 | 122,927 | 0.14 | 823,286 | 0.95 | 37,128 | 0.04 |
| Baramula | 1,008,039 | 1,476 | 0.00 | 37,705 | 0.04 | 827,677 | 0.82 | 1,527 | 0.00 | 34,750 | 0.03 | 99,563 | 0.10 | 959,185 | 0.95 | 30,621 | 0.03 |
| Bandipore | 392,232 | 392 | 0.00 | 75,374 | 0.19 | 323,161 | 0.82 | 141 | 0.00 | 34,586 | 0.09 | 7,478 | 0.02 | 382,006 | 0.97 | 8,439 | 0.02 |
| Badgam | 753,745 | 368 | 0.00 | 23,912 | 0.03 | 714,431 | 0.95 | 592 | 0.00 | 22,706 | 0.03 | 1,732 | 0.00 | 736,054 | 0.98 | 10,110 | 0.01 |
| Srinagar | 1,236,829 | 1,068 | 0.00 | 8,935 | 0.01 | 1,164,293 | 0.94 | 1,917 | 0.00 | 8,202 | 0.01 | 2,967 | 0.00 | 1,177,342 | 0.95 | 42,540 | 0.03 |
| Ganderbal | 297,446 | 117 | 0.00 | 61,070 | 0.21 | 206,032 | 0.69 | 181 | 0.00 | 55,570 | 0.19 | 7,750 | 0.03 | 290,581 | 0.98 | 5,592 | 0.02 |
| Pulwama | 560,440 | 402 | 0.00 | 22,607 | 0.04 | 511,687 | 0.91 | 343 | 0.00 | 22,189 | 0.04 | 3,811 | 0.01 | 535,159 | 0.95 | 13,840 | 0.02 |
| Shupiyan | 266,215 | 43 | 0.00 | 21,820 | 0.08 | 234,254 | 0.88 | 243 | 0.00 | 23,425 | 0.09 | 4,749 | 0.02 | 262,263 | 0.99 | 3,116 | 0.01 |
| Anantnag | 1,078,692 | 1,826 | 0.00 | 116,006 | 0.11 | 917,964 | 0.85 | 855 | 0.00 | 123,606 | 0.11 | 13,097 | 0.01 | 1,057,005 | 0.98 | 13,180 | 0.01 |
| Kulgam | 424,483 | 21 | 0.00 | 26,525 | 0.06 | 390,128 | 0.92 | 229 | 0.00 | 26,827 | 0.06 | 2,031 | 0.00 | 418,076 | 0.98 | 4,267 | 0.01 |
| | | | | | | Ladakh | | | | | | | | | | | |
| Leh | 133,487 | 488 | 0.00 | 95,857 | 0.72 | 559 | 0.00 | 896 | 0.01 | 123 | 0.00 | 168 | 0.00 | 19,057 | 0.14 | 22,882 | 0.17 |
| Kargil | 140,802 | 18 | 0.00 | 122,336 | 0.87 | 2,552 | 0.02 | 135 | 0.00 | 0 | – | 0 | – | 108,239 | 0.77 | 10,341 | 0.07 |
| | | | | | | Jammu | | | | | | | | | | | |
| Punch | 476,835 | 556 | 0.00 | 176,101 | 0.37 | 25,732 | 0.05 | 443 | 0.00 | 186,658 | 0.39 | 239,402 | 0.50 | 431,279 | 0.90 | 32,604 | 0.07 |
| Rajouri | 642,415 | 48,157 | 0.07 | 232,815 | 0.36 | 14,347 | 0.02 | 10,875 | 0.02 | 221,553 | 0.34 | 356,057 | 0.55 | 402,879 | 0.63 | 221,880 | 0.35 |

(Contd)

**Table 4.1c** *(Contd)*

| | | | | | | | | | | | | | | | | |
|---|---|---|---|---|---|---|---|---|---|---|---|---|---|---|---|---|
| Kishtwar | 230,696 | 14,307 | 0.06 | 38,149 | 0.17 | 158,228 | 0.69 | 1,212 | 0.01 | 33,127 | 0.14 | 1,019 | 0.00 | 133,225 | 0.58 | 93,931 | 0.41 |
| Doda | 409,936 | 53,408 | 0.13 | 39,216 | 0.10 | 227,543 | 0.56 | 16,217 | 0.04 | 30,200 | 0.07 | 9,971 | 0.02 | 220,614 | 0.54 | 187,621 | 0.46 |
| Ramban | 283,713 | 13,920 | 0.05 | 39,772 | 0.14 | 164,103 | 0.58 | 29,078 | 0.10 | 30,654 | 0.11 | 26,237 | 0.09 | 200,516 | 0.71 | 81,026 | 0.29 |
| Reasi | 314,667 | 37,757 | 0.12 | 88,365 | 0.28 | 59,039 | 0.19 | 137,710 | 0.44 | 77,674 | 0.25 | 20,889 | 0.07 | 156,275 | 0.50 | 153,898 | 0.49 |
| Udhampur | 554,985 | 138,569 | 0.25 | 56,309 | 0.10 | 10,394 | 0.02 | 451,530 | 0.81 | 28,770 | 0.05 | 7,518 | 0.01 | 59,771 | 0.11 | 489,044 | 0.88 |
| Jammu | 1,529,958 | 377,991 | 0.25 | 69,193 | 0.05 | 94,649 | 0.06 | 1,084,040 | 0.71 | 59,048 | 0.34 | 5,738 | 0.00 | 107,489 | 0.07 | 1,289,240 | 0.84 |
| Samba | 318,898 | 91,835 | 0.29 | 17,573 | 0.06 | 993 | 0.00 | 269,559 | 0.85 | 13,766 | 0.34 | 1,980 | 0.01 | 22,950 | 0.07 | 275,311 | 0.86 |
| Kathua | 616,435 | 141,224 | 0.23 | 53,307 | 0.09 | 13,479 | 0.02 | 505,013 | 0.82 | 21,599 | 0.04 | 42,776 | 0.07 | 64,234 | 0.10 | 540,063 | 0.88 |

*Source:* Table C-1 (religion), primary census abstracts (PCA) and table C-16 (language) of Census of India, 2011.

*Note:* Share of a sub-group, say, religion $i$ in the population of district $j = \dfrac{\text{population of religion } i \text{ in district } j}{\text{population of district } j}$.

Punjab share linguistic and cultural bonds and are not geographically separable. Unlike J&K, the demographic anxieties of north-eastern states are centred around the fear of illegal international immigration from Bangladesh and, to a lesser extent, immigration from the 'mainland' and Myanmar. Third, as discussed in Chapter 1, a section of Kashmiri Muslims view themselves and/or are seen by others as sine qua non of a secular India. Likewise, Kashmiri Pandits maintain that their presence *was* key to the secular *image* of Kashmir.

While we will focus mostly on conflicts between the larger communities, there are other conflict dyads in the state as well. The Gaddi and Sippi tribes allege that Gujjars and Bakarwals, the largest tribal community, capture all the benefits meant for tribes as Srinagar favours Muslims (interview, Bhaderwah, 5 December 2019).[3] Gujjars and Bakarwals in turn complain that Hindu Jammu views them as an extension of Muslim Kashmir (interview, Jammu, 4 December 2019).[4] Not all conflicts are structured around religious identity though. For instance, Kashmiri Pandits are occasionally viewed by Jammu's Hindus as part of the dominant Kashmiri-speaking community of Kashmir. Elsewhere, Shia Kargil has narrowly defined its identity in sectarian–linguistic terms around Purkhi or Purigi language and Twelver Shiism, which has limited room for Sunnis, Noorbakshias and Balti-speaking Shias, let alone Buddhists.[5]

# Kashmir

Sheikh Mohammed Abdullah, the best-known Kashmiri leader, who played a key role in steering the princely state towards the Union of India, suggested that Kashmiri Muslims were apprehensive of being driven out of their homeland (Abdullah 2016: 281). In a press statement issued on 21 October 1947, when tribal raiders had reached Muzaffarabad, Abdullah pointed out that there was 'not one single Muslim to be seen now in the Muslim majority state of Kapurthalla [Kapurthala]. The same is true of Alwar, Bharatpur and similar other states' (ibid.). He added that 'if certain people in Kashmir have apprehensions that they might be treated similarly, they deserve sympathetic understanding' (ibid.).[6]

During the Emergency (1975–77), when state governments were out-competing each other to achieve higher sterilisation targets, the Abdullah-led government 'chose to ignore the central government's population policies, being, for example, the only state government not to accept the high cash incentives for vasectomy acceptors made available by the Central Ministry of Health and Family Planning' (Gwatkin 1979: 44). It was also one of the three major states that did not raise their family planning targets, with the other two being

Assam and Kerala (ibid.: 40), which also happen to be states with large Muslim populations. In fact, 'Jammu and Kashmir and Kerala were the only two major states that failed to achieve their originally assigned 1976–77 targets' (ibid.: 56). However, Abdullah did not oppose family planning on religious or communal grounds. His 'government authorized payment of higher incentives only on 18 January 1977, the day Mrs. Gandhi announced the forthcoming elections and reined in the aggressive family planning program' (ibid.: 57).[7] It is also noteworthy that the 1981 census conducted under Abdullah was the most reliable among the recent censuses. That census reported a decline in the proportion of Muslims in the state's population with respect to the 1971 census (Table 4.1a).

Sheikh Abdullah's last term, though, was a turning point in the public discourse on demography. In the 1977 assembly elections, the Sheikh Abdullah-led National Conference (NC) failed to win any seat in Hindu-dominated constituencies of Jammu, while the Indian National Congress (INC) lost all seats in Kashmir (B. Puri 1983b: 233). Faced with competition from Jamaat-e-Islami, among others, Sheikh Abdullah seems to have reinvented 'himself as a defender of the rights of Muslims' (Swami 2008a) and that deepened the communal and regional divide.[8] In fact, in one of his last public speeches, Sheikh Abdullah suggested that giving state subject status to West Pakistan refugees would alter the state's demography, as the return of Kashmiri Muslims from Pakistan-occupied areas was being resisted (S. Hussain 2021: 221). Around this time, Dogra Hindus began voicing concerns about demographic engineering, even as their Kashmiri Muslim counterparts' obsession with the declining population share of Muslims was growing.[9]

It has been argued that after his return to power, Sheikh Abdullah dedicated himself to the task of consolidating the Muslim population of the state and creating a 'Greater Kashmir'. Hari Om (2017a), for instance, suggests that Sheikh Abdullah

> was a known protagonist of Greater Kashmir comprising Kashmir province and the Muslim-majority areas of Jammu and Ladakh regions. One of his plans was to make the medieval and dysfunctional Mughal Road connecting the Muslim-majority Poonch-Rajouri belt with the Muslim-majority Kashmir Valley fully functional ... [to delink] the Muslim-majority Poonch-Rajouri belt from Hindu-majority Jammu province.[10]

Hari Om added this move was in line with how Sheikh Abdullah 'de-linked Muslim-majority Doda area from Hindu-majority Udhampur District and conferred the status of district on the Doda area' in 1948 and 'bifurcated Ladakh into

two districts on purely communal lines' to create a 'Muslim-majority Kargil area
in Buddhist-majority Ladakh' in 1979 (ibid.).[11] In both cases, the 'purpose was to
communalise the situation in the newly-created districts, weaken the movement
in Ladakh and Jammu for political empowerment, and ensure their merger with
Kashmir at an appropriate time' (ibid.).

A retired bureaucrat who served under Sheikh Abdullah rejected the alleged
communal design behind the formation of new districts. He argued that the
decision was driven by administrative compulsions. Commenting on the
formation of Kargil, he observed:

> Sheikh Abdullah did reorganize district administration, redesignation [of] DCs
> as District Development Commissioners, setting up District Development
> Boards (now the elected DDCs), which called for smaller districts and new
> districts that emerged like Kargil, the second largest town in Ladakh, had
> Muslim majorities. On the other hand the Nubra Vall[e]y, which had a
> majority Muslim population[,] was given to Leh despite gre[a]ter distance to
> keep the Muslim majority down in Kargil as was the addition of Buddhist
> majority Zanskar to Kargil. The Muslim majority Jammu districts also were
> carved out not [for] community but for accessibility and development needs.
> (Interview, 12 January 2021)

When the author pointed out that Nubra and Zanskar are even now inaccessible
from Kargil and Leh, respectively, the bureaucrat added, 'As the crow flies,
it [Turtuk] is also closer to Kargil and the terrain is less precipitous than the
connection through the Nubra valley to Leh. Yet, no direct road has been
attempted from Kargil to Turtuk, although this was a demand of the residents
of Turtuk' (interview, 12 January 2021). The people of Zanskar have also
been demanding a direct road to Leh (Buddhist leader, interview, New Delhi,
2 December 2019). In fact, in 1952, long before Turtuk joined India, Zanskar
had demanded merger with Leh, and the people were 'almost on the point of
migrating en masse' otherwise (Alam 2006: 59). This was followed a year later
by a demand for a compact autonomous unit for Buddhist areas (Bakula 1953).

Most observers, though, point out that the creation of Kargil was not entirely
driven by administrative considerations. B. Puri (1983a: 190), for instance,
suggested that the NC 'sympathised' with the Shias of Kargil who feared the
domination of Buddhists 'and lent its support for recognition of their identity'
and 'bifurcated the district of Ladakh in 1979'. While the division of the unwieldy
districts was inevitable, observers have noted the tendency of Kashmiri leaders
starting with Sheikh Abdullah to dominate the rest of the state and relate to

Jammu and Ladakh from a predominantly communal perspective. Chowdhary (2016: 194), for instance, points out that 'Mahore tehsil, which is predominantly Muslim, was carved out of the rest of predominantly Hindu areas to fit in the Muslim belt of Doda' (see also Swami 2008a). Devadas (2007: 37) argues:

> [Sheikh] Abdullah's rhetoric often harked back to Kashmir's independent history but not to its historical territory – which was just the valley and a small area around Muzaffarabad to the west. He did not want to separate from other parts of the vast state that the Dogras had cobbled together. Kashmir's deeply ingrained conviction of superiority was almost unconsciously shaping an imperial template upon which to base its polity, one that would allow the Kashmiris at the centre of the valley to dominate the rest of the state. It was an unstable template, for it was not backed by military might or economic clout. Indeed, it was not even consciously drawn out.[12]

van Beek (2004: 203–04) likewise points out that 'Sheikh Abdullah, after his return to power, pitched his vision for the future in terms of a Greater Kashmir, claiming Muslim-majority Doda and Kargil district (carved out of Ladakh district in 1979) as part of this Kashmir, which shared nothing but religion'.

There is a widespread tendency in Kashmir to view the adjoining Muslim-inhabited parts of Jammu and Ladakh as Kashmir's 'traditional periphery'. An academic added that the armed insurgency impeded the 'historical process' of the integration of the aforesaid periphery with Kashmir (interview, Srinagar, 28 May 2022). Noorani (2008) sums up the consensus in Kashmir regarding the desirable boundaries of Kashmir in the event of exit from the union:

> Three of its [Jammu's] six districts, now broken up into 10, have a Muslim majority – Poonch (91.92 per cent), Rajouri (60.23 per cent) and Doda (57.92 per cent). Two tehsils in Udhampur, Gul Arnas and Gulab Garh, have a Muslim majority. Farooq Abdullah *realistically* warned that these areas would not live with Jammu; the massacres would be worse than those of 1947; and India will be left with two and a half districts while the so-called Greater Kashmir will go on a platter to Pakistan eventually. Mirwaiz Maulvi Umer Farooq [of the All Parties Hurriyat Conference] also said if the Dogras of Jammu's two and a half districts want to secede from the rest of the State we won't oppose it either.[13] (Emphasis added)

It is not clear, though, if Tibetan-speaking Shias of Kargil and Gojri- and Pahari-speaking Muslims of Jammu will want to join Kashmir if it separates from India. Writing after Abdullah's death, B. Puri (1983a: 190) pointed out that Shias of

Kargil were 'not willing to ... merge their identity with Sunni Muslims of Kashmir merely on the basis of a common religion'. B. Puri (1983b: 230) added that the Jammu Muslims who had hitherto avoided the public sphere due to '[the] scars of partition massacres and scorn of Kashmiri leadership for having failed to follow it' began to 'insist on their share in the politics and economy of the state as a distinct entity within Jammu and within the Muslim community' (see also Choudhary 2008). This, though, may not be inconsistent with the willingness of sections of Muslims of Jammu and Kargil to align with Kashmir as long as it remains within India.

The consensus in Kashmir is tempered by an awareness that the facts on the ground do not support the claim to 'Greater Kashmir'. For instance, in 2001, Kashmiri speakers accounted for about 1.48 and 27.80 per cent of the population of the strategically important Karnah and Uri *tehsils* in Kashmir. In the run-up to the 2011 census, the partisans of independence including Syed Ali Shah Geelani (Ashiq 2010) exhorted the non-Kashmiri-speaking Muslims to report Kashmiri as their first language.[14] The Islamic Students League's founder, Shakeel Bakshi, too, appealed to 'the people of Doda and Rajouri to register themselves ... as Kashmiri-speaking Muslims' to 'help the larger cause' (Tantry 2010; see also *Kashmir Life* 2010). Interestingly, before Shakeel Bakshi's statement, the secretary of the Tribal Research and Cultural Foundation had appealed to Gujjars and Bakarwals to 'get themselves enumerated as ST community with Gojri as their mother tongue' (*Kashmir Times* 2011a). Chaudhary Bashir Ahmed Naz, vice chairman of the J&K State Advisory Board for the Development of Gujjars and Bakerwals, reiterated this appeal a few days after Bakshi's statement.[15] A few weeks later, ahead of the prime minister's visit to Kashmir, the community demanded 'a comprehensive socio-economic package', including 'a tribal university, inclusion of Gojri language in Eighth Schedule of Constitution and a special census of nomadic Gujjars and Bakerwals' (*Outlook* 2010). In Karnah and undivided Uri, the population share of Kashmiri speakers dropped to 0.23 and 23.44 per cent, respectively, as per the 2011 census, even as the Kashmiri-speaking population of the state grew by more than 23 per cent. The share of Kashmiri speakers in the Uri *tehsil* was 4.87 per cent in 2011. The population share of Kashmiri speakers has been decreasing in Karnah and Uri, at least, since 1981.

Those who champion a greater Kashmiri realm and call for linguistic assimilation of their coreligionists believe that Kashmir and Muslims are facing the threat of demographic marginalisation. These concerns were fuelled by the decline in the *reported* share of Muslims in the state's population during 1961–81. Some of the 'suspicions' in this regard seem to have emerged as early as the late 1960s when it was feared 'that a grand plan was underway to transform the demographic balance of the state' (Swami 2007: 90).[16] The general report of the next census conducted in 1971 clarified that 'the proportion of Muslim population in the State has been steadily going down except for two decades 1911–31', when the 'fall in the proportion of Hindus during 1911–31 is noticeable in almost every district of the State in varying degrees' (Government of India [GoI] 1978b: 30–31). The report also explained the difficulties in comparing pre- and post-1947 censuses, which will be examined at length later in this section and in Chapter 6.

By the mid-1980s, positions had hardened considerably with even mainstream Kashmiri politicians endorsing communal propaganda on demography. Sheikh Abdullah's death left behind a vacuum amidst a rapidly changing regional situation across the subcontinent and West Asia (Lamb 1991: 322–26), which triggered an intense political competition involving Sheikh's son, Farooq Abdullah, his son-in-law, G. M. Shah, and a variety of other ambitious players, including Mufti Mohammad Sayeed, with the INC allying with different factions at different times.[17] Balraj Puri highlights another significant development largely overlooked in debates on Kashmir. He points out that by the early 1980s the formerly emotional mobilisations had lost steam, with the 1975 accord and 1977 elections being the turning points, and

> people feel involved in all the affairs of the state and are taking elections, administration, and development seriously. Regional and communal urges are being articulated and translates into more militant claims over the *share in the economic cake*. The controversies about their respective share in recruitment and promotions in government services, admissions to technical and higher institutions and development outlays are becoming live. Above all the question of sense of participation and an *equitable share in political power* has started agitating various sections of the people. (Puri 1983b: 230, emphasis added; see also Devadas 2018, 2019a)

So by the mid-1980s communities were increasingly conscious of *shares,* and political parties were trying to adjust to both the growing socio-economic consciousness along regional and communal lines and the departure of Abdullah, who dominated Kashmiri politics for half a century.

## Post-1990 Valley

The communalisation of public debate on demography has continued unabated since the 1980s and interfered with censuses. The 1991 census could not be conducted due to the outbreak of armed insurgency, while boycott calls marred the next census. The provisional population totals (PPT) report for 2001 notes the following in this regard:

> During the course of census, an abnormal situation was created by misinterpretations and misgivings about the entire exercise, without any reason. It was stretched to beyond its purview on the negative side, to gain some political mileage out of it, on communal, linguistic and regional lines. Followed by death threat from the militant outfits to implement the ban imposed by them against the conduct of census. (GoI 2001b: xi)

Once again the falling population share of Muslims was the point of departure. The PPT pointed out that the Hizbul Mujahideen's threat against census operations, which 'appeared in all the newspapers including national dailies [on 4 September 2000]',

> was followed by a press statement by Mr. Saif-ud-Din Soz, former Union Minister on September 6, 2000[18] when he came out with the theory of reducing Muslim majority character of Jammu & Kashmir, as the main intention of census-taking in 2001.... Prevailing situation in the State turned the census into clear cut politics of demography. The state of affairs in Jammu & Kashmir is such, as even a normal routine activity is pounced upon to give political overtones to it. (GoI 2001b)

It is very unusual for a census document to directly refer to a politician who is not associated with the operation in any official capacity – that too for fanning communalism.[19] The then director of census operations (DCO) argued that 'education, healthcare, and economic wellbeing define fertility rates, not religion' and that 'Soz seems to agree with Hindu communalists who say Muslims have more children because of their religion' (Swami 2000a).

Contemporary news reports too singled out Soz,[20] who 'charged that census figures had been manipulated to show that the Muslim population of Jammu and Kashmir had been declining as a percentage of the total population, and that its growth in the Jammu region had been dramatically lower than that of Hindus' (ibid.).[21] A week later, on 13 September, Soz elaborated his concerns at another press conference. He said:

> The lone Muslim majority State had been characterised as one of the strongest elements of India's secular edifice and any manoeuvring and dishonest effort to tamper with this character of the population ratio must be ruthlessly discouraged and condemned. (Ibid.)

Masood Tantray, a Hizbul Mujahideen commander who issued warnings against census,[22] claimed that enumeration was pointless at that juncture when 'so many had migrated and thousands of youth had either gone underground or were reported missing' and warned the state government against 'adventurism' (*The Tribune* 2000b).[23] Under the de facto method of enumeration followed in India, those in jails would have been accounted for under the institutional population, while the displaced and migrants would have been counted in the new places of residence. Kashmiri Muslims are not alone in suspecting the reliability of census data amidst migration and displacement. The Sikhs of Kashmir (*Kashmir Times* 2011b) and tribes of Jharkhand (Agrawal and Kumar 2020a: 310–12) question the validity of headcount as economic out-migrants are not counted in the population of their respective states under the de facto method of enumeration, while tribes across the north-east demand the exclusion of outsiders from headcounts (ibid.: 238). In all such cases, communities are effectively expressing dissatisfaction with the de facto method of enumeration.

Tantray was also concerned that 'demographic changes in the state were being brought about on the advice of the Israelis who did the same in the occupied areas in Palestine'[24] and added that '[i]f the government does not adhere to the warning we will not desist from opening our guns at them' (*The Tribune* 2000b).[25] Abu Ubaid of the Lashkar-e-Taiba, another militant outfit, too issued a threat: 'If any one of the 22,000 government employees is seen participating in the census operation he or she will be killed without warning' (ibid.; see also Rediff 2000b).[26] It bears noting that about a month and a half ago, the Hizbul Mujahideen's field commander, Abdul Majid Dar, had warned against the trifurcation of the state (M. Ahmad 2000a). These threats must have been taken seriously by both enumerators and respondents because by then the state had already witnessed three major instances of killings – Chittisinghpora (March 20), Rangdum (July 11) and Pahalgam (August 1) – in what was one of the most violent years of the conflict (Figure 2.1).

The long run-up to the 2011 census was also marred by a series of communal controversies. In 2004, the state government introduced the Jammu and Kashmir Permanent Resident Status (Disqualification) Bill in response to a 2002 ruling of the High Court, which mandated that women would not lose permanent resident status after marrying a non-permanent resident. This 'blatantly gender discriminatory' bill (B. Puri 2004: 1456) triggered a controversy that dragged for more than a decade and a half and was used, among other things, to justify the repeal of Article 370. Interestingly, the bill was also seen as 'anti-Jammu as most of the girls here marry outside the state with persons who are non-state subject' (Chowdhary 2010a: 17). Two years later, in 2006, Syed Ali Shah Geelani claimed that '[l]akhs of non-state subjects had been pushed into the Valley under a long-term plan to crush the Kashmiris' (Swami 2021). He added that the migrants who were largely 'professional criminals … should be driven out' as the government was using them to 'promote immorality and obscenity' (ibid.).[27]

The 2008 controversy over the grant of land-use rights to the Shri Amarnathji Shrine Board (SASB) further politicised demography. Addressing a Friday congregation at Srinagar's historic Jamia Masjid, the chairman of a faction of the All Parties Hurriyat Conference, Mirwaiz Umar Farooq, warned 'New Delhi that we won't allow anybody to occupy our land and we will fight it tooth and nail. India is trying to change the demographic composition of the state' (Majid 2008). On the same day, Syed Ali Shah Geelani argued that 'New Delhi with the help of pro-India parties like NC and PDP [People's Democratic Party] was changing the demography of the state and the transfer of land to Shrine Board and the setting up of Sadbhavana schools are steps towards that direction' (Gul 2008).

For Geelani, attempts to change the demography were meant to deny the inevitability of Kashmir's secession from India and perpetuate 'Hindu slavery' (Sikand 2010c). The inevitability follows from the fact that it would be difficult 'for a fish [Islam] to live in a desert [Hindu India]' (Sikand 2010d).[28] Others, too, claimed that the government was trying to alter the demography to subvert the plebiscite (Swami 2008b).

While it is 'absurd to say that land for the SASB will affect demography' (Noorani 2008) and 'fantastic to believe that outsiders can settle at high altitude near the shrine' (B. Puri 2008: 8), the context of the demographic anxiety is noteworthy:

> [P]eople overreact when they are told that they are not in a majority in their own State. Witness: Assam. In Srinagar I was told by the 15 Corps Commander in 1995, while working on an article for Frontline, that Kashmiri Muslims were not in a majority in the valley itself. On August 16, 2008, S.K. Sinha said the same thing in the Manekshaw Memorial Lecture. (Noorani 2008)

Roy (2008) highlights numbers that seem to have motivated the protests in Kashmir:

> Until 1989 the Amarnath pilgrimage used to attract about 20,000 people…. In 1990, when the overtly Islamist militant uprising in the valley coincided with the spread of virulent Hindu nationalism (Hindutva) in the Indian plains, the number of pilgrims began to increase exponentially. By 2008 more than

500,000 pilgrims visited the Amarnath cave.... To many people in the valley
this dramatic increase in numbers was seen as an aggressive political statement
by an increasingly Hindu-fundamentalist Indian state. Rightly or wrongly,
the land transfer was viewed as the thin edge of the wedge. It triggered an
apprehension that it was the beginning of an elaborate plan to build Israeli-
style settlements, and change the demography of the valley. (See also Dulat
2010: 340)

Two years later, on the eve of the 2011 census, Mirwaiz Umar Farooq asked
'the people to remain cautious during the forthcoming so-called census' (*Express
Tribune* 2010). Geelani too asked the people to be 'vigilant' and suggested that
they 'must object and resist' if 'the officials try to include a non-Kashmiri in
the population register' (Jameel 2010b; see also Hamid 2011). He suggested
that there was 'a planned conspiracy to change the Muslim majority of the state'
through census (Bhat 2018: 181). He argued that a 'change in the religious
composition of the State may help the Government of India in case India is
compelled to hold plebiscite in J&K' and 'warned the government to desist from
making any change in the Muslim majority character of the State and asked the
Muslims of the State to defeat this conspiracy' (ibid.).[29] Geelani even alleged
that census 'is being carried out by the employees from outside the state and
they have been tasked to erode the Muslim majority status of the state' (Masood
2010; see also Ashiq 2010), which is blatantly misleading as field enumerators
are always entirely from within the state. Given the importance of census for
the 'future political status', 'Kashmiri civil society activists … formed a group to
create awareness at the grassroots' (Jameel 2010b).

In a related development, the Islamic Students League's founder, Shakeel
Bakshi, appealed to 'militant brethren' to refrain from calling for a boycott of
census and asked 'the intelligentsia of Kashmir to make people aware about the
consequences of not taking part in such exercise' (Tantry 2010) because the
results of the exercise will have 'a bearing on a political solution, too' (Hamid
2011). So, unlike in 2001, 'militant organizations … requested all Muslims of
the State to get enumerated and the Census staff of the Valley was asked to fulfill
all obligations towards its community, so that the Muslim majority character
of the State gets reflected in the Census 2011' (Bhat 2018: 181). Note the dual
emphasis on 'the Valley' and 'Muslims of the State'.

Bhat (2018: 182) pointed out that 'many enumerators and the public in the
Kashmir region particularly in South Kashmir, which is still considered to be a
stronghold of separatists and militancy, have resorted to a systematic exaggeration
of their household population in order to boost the overall share of Muslims'.[30]

He added that there were rumours in Kashmir that 'Hindu enumerators in Jammu division in a bid to exaggerate the share of Hindus population are enumerating migrants and non-state subjects and even refugees from Pakistan' (ibid.). As a matter of fact, the community of West Pakistan refugees had reached the state before it adopted a constitution. While the community was denied voting rights in the state assembly elections, it has always been counted in the state since 1961 because of the de facto method of enumeration. Mirwaiz Umar Farooq questioned the enumeration of migrant workers around the Dal Lake and suggested that the government was 'trying to change the region's demographics' (Al Jazeera 2010).[31] The enumeration of migrant workers was not a conspiracy but rather a requirement under the method followed by the census across the country because of which Kashmiri traders based in, say, Cochin's Jew Town are also counted where they are during the period of enumeration.

Guilmoto and Rajan (2013: 63) note that '[i]t was held that there was a plan to exaggerate the share of the Jammu region.... This obviously did not happen since the population growth was lowest in Jammu.' However, 'this rumour may also have, on the contrary, encouraged people in the rest of the state [Kashmir] to react'.[32] Bhat (2014) throws more light on this. He notes:

> We were told by a number of enumerators and supervisors that during the current Census rumor spread in Kashmir Valley that Hindu enumerators in Jammu region are over reporting the Hindu population with a view *to get more funds under State plan*. The enumerators in Kashmir also reacted by over enumeration of population. As it is easy to over enumerate 0–6 population than adult population, because enumerator has to just fill 3–4 columns for a child as compared to more than 15 columns for an adult. So enumerators generally prefer to report more young people than adults in case they resorted to over reporting of population. In this process of over enumeration, enumerators have somehow preferred to differentially over-enumerate boys and girls.[33] (Emphasis added)

Guilmoto and Rajan (2013: 63) agree that there was 'a deliberate over-reporting of children in Jammu and Kashmir' and that 'sex ratio levels suggest that in trying to inflate their child population, many households seem to have *invented boys* rather than girls – as if reporting non-existent boys was easier or more spontaneous' (ibid.: 63–64, emphasis added).[34]

Rumours about 'the use of Census population for issuing new ration cards' may have played a key role in encouraging manipulation of headcount as 'people generally over reported the number of household members' (Bhat 2018: 181). A census official, too, suggested that people exaggerated numbers to increase their access to ration (interview, Srinagar, 30 November 2015). The rumours, whose interaction with the discourse of demographic marginalisation will be discussed in Chapter 6, were not wrong after all. In April 2015, the state government declared that it would update ration cards based on the 2011 census (*Business Standard* 2015; *The Tribune* 2015a).[35]

## Settling Other Scores

In 2000, the political struggle between armed groups and their overground competitors, the Hurriyat and mainstream political parties, fuelled the communalisation of demography. After the Hizbul Mujahideen declared a ceasefire and began engaging the union government, Abdul Ghani Bhat of the Hurriyat claimed that his organisation 'was the only recognised body of Kashmiris' and 'cannot be marginalised' (Noorani 2000: 3956). The Hizbul Mujahideen offered a sharp response:

> If our elders believe that only an armed struggle will liberate Kashmir from the occupation and an honourable solution is possible through militancy, then they should come in the forefront and command the struggle. If not they should at least send their wards to join militancy.... These leaders changed their stance thrice, during the two-week long ceasefire ... the people of Jammu and Kashmir have the right to ask these leaders to muster courage and keep off from the leadership of the movement. (Ibid.: 3957)

In the first week of September, the Hizbul Mujahideen's spokesperson, Masood Tantray, attacked the Hurriyat on the issue of census: 'The leaders claiming to be the representatives of the people should instead of exploiting the public sentiments rise to the occasion and raise their voice against the conspiracies being hatched against the Muslims' (*The Tribune* 2000b). The Hizbul Mujahideen and the Hurriyat 'resolved' their differences a week later (Noorani 2000: 3957), but by then the damage had been done as all sides used the census to score brownie points. The internal politics of the Hurriyat, too, seems to have played a role in souring the relationship of the Hurriyat and the Hizbul Mujahideen as 'Geelani asked Majid [Dar] to wait [for declaring the ceasefire] until elections had been held for both the Hurriyat chair and the Jamaat chief' (Devadas 2019a: 368). In short, the fluid political situation created by the outreach to non-mainstream groups in Kashmir such as the Hizbul Mujahideen and the Hurriyat and their internal divisions helped amplify the anti-census sentiment as a variety of actors used the alleged threat to the Muslim-majority character of the state to secure their position.

Around this time, the NC chief minister, Farooq Abdullah, resurrected the autonomy card in the summer of 2000 (Baweja 2000a). Dulat (2010: 219–20) suggests that Abdullah

> had contested the 1996 election because the government of India wanted him to contest, and at that time he had told the government that he needed a plank to fight an election. Autonomy had been that plank. His committee had in fact been discussing the matter with Governor Gary Saxena while preparing the report; and now another election was coming up, so his government had to do something about autonomy.... He figured the NDA government would just kick the can down the road, giving him both an escape route and an election plank.

While it is true that Abdullah needed an election plank because by 1999 his government began to lose public support (Devadas 2019a: 341–42), immediate developments might better explain his autonomy move. He found himself doubly vulnerable. On the one hand, the outreach to the Hurriyat and armed groups rendered him politically vulnerable. As T. Singh (2000) puts it, '[T]ry and see it for a moment through Farooq's eyes: if Delhi can discuss azadi with the Hurriyat, why not autonomy with a legally elected state government?' Or, as Noorani (2000: 3949) puts it, 'Abdullah becomes dispensable once the "separatist threat" vanishes. His demand for greater autonomy can be rejected with less ado.' On the other hand, Abdullah's government was 'almost completely broke for more than a year. The Fifth Pay Commission added to its problems an annual burden of INR 550 crore and this, as in most other states, literally broke the Government's back', leaving it with hardly any 'money to build schools or provide drinking water and power' (T. Singh 2000). Unable to offer anything tangible to its constituency, Abdullah resurrected as a champion of the autonomy of Muslim Kashmir in the long and messy run-up to the 2002 assembly elections that he lost eventually.[36] It bears noting that the assembly debate on autonomy

> was liberally interspersed with attacks on counter-terrorist forces like the Jammu and Kashmir Police's Special Operations Group. The Chief Minister followed this up with a visit to the family of a victim of an alleged police execution, a privilege most of the dozens of N.C. workers targeted by terrorists had not been accorded. Then, on June 28, he announced State funding for the founding of an Islamic university and *criticised regimes such as those in Saudi Arabia for their failure to promote the faith.* (Swami 2000b, emphasis added)

The union government rejected the partisan autonomy report adopted by the assembly dominated by the NC (Baweja 2000b).

This is the setting in which Saifuddin Soz, 'whose secular credentials have been impeccable', metamorphosed into 'a born-again Islamic chauvinist' (Swami 2000a). Soz was expelled from the ruling NC in 1999, 'when he defied the whip ... to vote against the motion of confidence moved by the Vajpayee government in the Lok Sabha' and played a crucial role in defeating the vote of trust sought by the Atal Bihari Vajpayee government (*The Tribune* 2006). Soz was toying with the idea of launching a political outfit 'for highlighting the problems of the people' and fill in the political space that Mufti Mohammad Sayeed was also trying to occupy (*The Tribune* 2000c). Sayeed does not seem to have openly taken a stand on census.

In other words, while the propaganda about the demographic marginalisation of Muslims was already in place, entirely unrelated political clashes within the ruling NC and the pro-independence camp coalesced and aggravated communal passions (Timeline 4.1). This is reminiscent of how the Jamaat-e-Islami's electoral forays in the late 1970s pushed mainstream politicians to resort to communal rhetoric. Several iterations followed in the 1980s. In the 1983 assembly elections, the first to be held after Abdullah's death, Indira Gandhi campaigned in Jammu as a defender of the state's Hindu minority (Kashmiri journalist, interview, Srinagar, 9 July 2021; retired civil servant, interview, 3 September 2021), which aggravated the growing communal polarisation in the state. This complemented the competition over the Muslim vote in Kashmir triggered by the death of Abdullah. G. M. Shah, Abdullah's eldest son-in-law, served as the chief minister between 1984 and 1986 after defecting from the NC and toppling his brother-in-law Farooq Abdullah's government. Once out of office, Shah 'rediscovered the virtues of Islam' (Devadas 2019a: 173) and used demography to portray his successor, Farooq Abdullah, as a stooge of New Delhi.[37] In 1989, the year of outbreak of armed insurgency, Shah claimed:

> The government has hatched a conspiracy to reduce the Kashmiri Muslim population. [Chief minister] Farooq Abdullah is an instrument of this plot. Our State had an 82 percent Muslim population in 1947; it is now a mere 54 percent as the 1981 census figures reveal. We should reject the government's family planning program. This is aimed at further reducing the Muslim population in Kashmir. Every Kashmiri Muslim should have four wives to produce at least one dozen children. (Panandiker and Umashankar 1994: 96)

G. M. Shah had only one wife and three children including two sons and a daughter (Abdullah 2016: 149; Majid 2009). None of his children seems to have followed his advice either. In any case, his statistical claims were blatantly

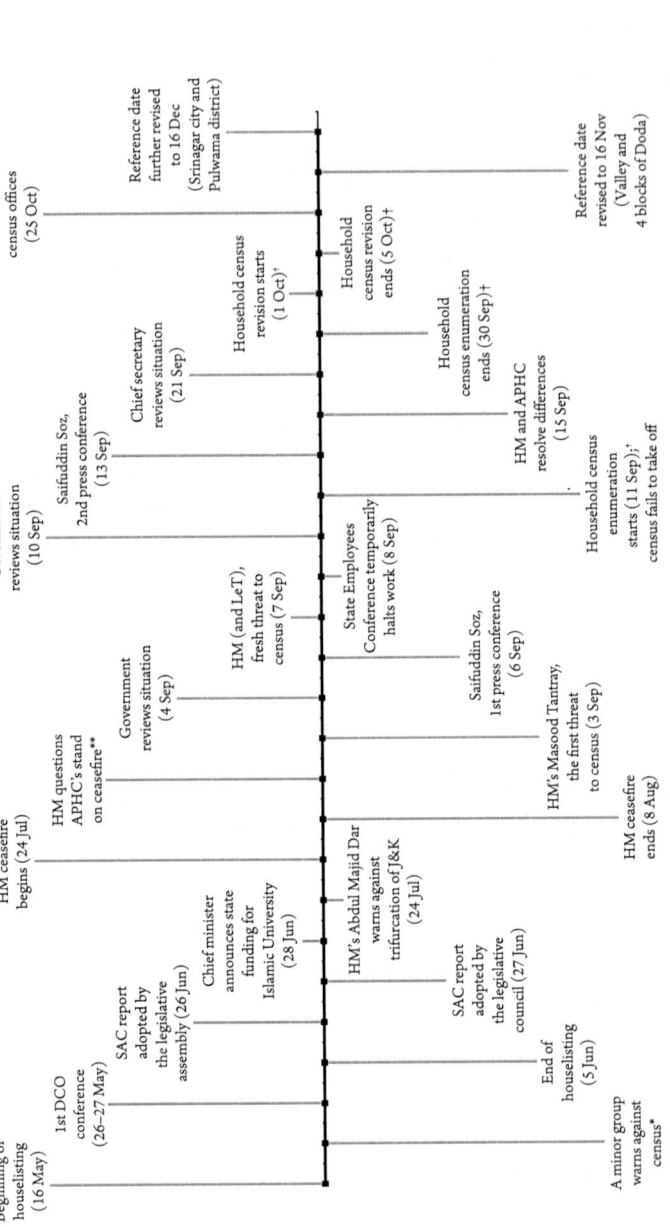

**Timeline 4.1**  Political contests in Jammu and Kashmir (J&K) around the 2001 census, May–December 2000

*Source:* Table 2.3; GoI (2001b); *Hindustan Times* (2000); Noorani (2000); Rediff (2000a, 2000b); Swami (2000a); *Times of India* (2000a, 2000b); *The Tribune* (2000b).

*Note:* (*a*) The timeline focuses on political developments in the second half of the year 2000 and the disruption of census. (*b*) * A minor terrorist outfit issued a threat even before 5 June (GoI 2001b: 16; Rediff 2000a) but the exact date is unavailable. ** Exact date not available. (*c*) † identifies the normal census calendar. (*d*) SAC: State Autonomy Committee; APHC: All Parties Hurriyat Conference; HM: Hizbul Mujahideen. LeT: Lashkar-e-Toiba. (*e*) The visit of the registrar general of India (RGI) to J&K has not been marked due to a lack of clarity about the dates.

Labels within the timeline:

Beginning of houselisting (16 May)

1st DCO conference (26–27 May)

SAC report adopted by the legislative assembly (26 Jun)

Chief minister announces state funding for Islamic University (28 Jun)

HM ceasefire begins (24 Jul)

HM questions APHC's stand on ceasefire**

Government reviews situation (4 Sep)

HM (and LeT), fresh threat to census (7 Sep)

Government reviews situation (10 Sep)

Saifuddin Soz, 2nd press conference (13 Sep)

Chief secretary reviews situation (21 Sep)

Household census revision starts (1 Oct)†

HM raids census offices (25 Oct)

Reference date further revised to 16 Dec (Srinagar city and Pulwama district)

A minor group warns against census*

End of houselisting (5 Jun)

SAC report adopted by the legislative council (27 Jun)

HM's Abdul Majid Dar warns against trifurcation of J&K (24 Jul)

HM ceasefire ends (8 Aug)

HM's Masood Tantray, the first threat to census (3 Sep)

Saifuddin Soz, 1st press conference (6 Sep)

State Employees Conference temporarily halts work (8 Sep)

Household census enumeration starts (11 Sep);† census fails to take off

HM and APHC resolve differences (15 Sep)

Household census enumeration ends (30 Sep)†

Household census revision ends (5 Oct)†

Reference date revised to 16 Nov (Valley and 4 blocks of Doda)

misleading as there are no census data for 1947[38] and the Muslim population share was 64.19 per cent in 1981 (Table 4.1a).

Another example of opportunistic politics around demographic fears is offered by the 'incendiary' intervention of the PDP in the Amarnath controversy to undermine the growing political base of Ghulam Nabi Azad, who headed the INC–PDP coalition government (Devadas 2019a: 411). More recently, after losing all three parliamentary seats in Kashmir to the PDP in the run-up to the 2014 assembly elections, the NC vowed to 'spill its blood to maintain the identity of the state' and not allow 'demographic change' (Ganai 2014).

The local media in Kashmir, too, played a role in communalising the debate. During the 2001 census, Swami (2000a) pointed out that the local media 'frequently makes reference to the 1987 Census, which is asserted to have shown that the numbers of Muslims as a percentage of State subjects was also declining. The document is in fact a record of the percentage of State subjects who belonged to specified Scheduled Castes and Scheduled Tribes.' Others claimed that the 1986–87 headcount showed the 1981 census had under-reported the Muslim population share by more than 1 percentage point (*Kashmir Life* 2010; Hamid 2011). Since the data were not made public, they are particularly amenable to partisan use. Later, during the Amarnath controversy, without checking the actual data, let alone understanding the changing contexts of data collection in the state,

> many local newspapers such as *Greater Kashmir*, the *Daily Etalaat* (Daily Bulletin), *Rising Kashmir*, the *Srinagar Times*, and the *Daily Aftab* (Daily Sun) carried reports and presented data from the Indian census documenting a consistent decline of the Muslim population from 72.4% in 1941, to 68.3% in 1961 and 66.9% in 2001. (Tremblay 2009: 941)

It was almost as if Kashmir was preparing for 'a Hindu invasion' (ibid.: 946).

Researchers have also contributed to the communalisation of statistics. Dabla (2012: 189–90) argued that '[t]he Indian leadership, to the dismay of those who espoused the cause of secularism in the state, was successful in reducing the majority character of the Muslim population from 80 percent in 1941 to 67 percent in 2001' and that the 'downward trend in the Muslim population of the state was matched with an upward trend of 8 percent increase in the Hindu-dominated region of Jammu during the same period, that is, 1951–2001'. This is outrightly misleading. In 1941, Muslims accounted for 72.4 per cent of the population in the territory currently under Indian administration. Census could not be conducted in 1951. Between 1961 and 1981, the share of Muslims and Kashmir in the state's population decreased before increasing in the subsequent decades (Figure 1.1). During 1961–2001, Jammu region's share in the population of the state decreased by half a per cent, and there was an increase in the population share of Muslims in Jammu.

## The 'Outsider' Bogey

Kashmiri leaders fanned conspiracy theories regarding the settlement of outsiders to alter the demography. The DCO, Feroze Ahmed, argued that 'if the government of India was engaged in a sinister plot to flood the State with migrants, it wouldn't have advertised it in the census figures [that is, the increasing share of Hindus between 1961 and 1981]. The fact is, we just record who lives where' (Swami 2000a). A retired civil servant, too, ruled out conspiracy theories. He pointed out:

> Punjabis and other North Indians have obtained State Subject certificates fraudulently in Jammu Division. There were well organised rackets in the offices of the Deputy Commissioners in Jammu and Kathua for this. This had nothing to do with the GoI and was just another instance of smart people taking advantage of the widespread corruption in the State. In fact, some very prominent people in Jammu are reputed to have obtained State Subject certificates in this manner. In another case, when there was a shortage of sanitation workers in Jammu Municipality, the then district administration brought in sweepers from Punjab who were settled in Jammu and later surreptitiously given State Subject certificates. (Interview, 22 June 2016)

He added that these cases were 'statistically insignificant' from the perspective of relative population shares of communities or regions.

It is also alleged that staff from outside the state manipulate the data. This claim has three layers. One, outsiders serve as enumerators (Masood 2010; Ashiq 2010). This is blatantly false. Feroze Ahmed pointed out that 'each census form is signed by local employees' (Swami 2000a).[39] Two, '[w]hile there is no doubt that all primary data is collected by the local members of the local government administration, there are many missing links between the Census Department and government manpower in terms of co-ordination, training, geographical coverage, logistics and data consolidation' (Talib 2010: 4). Even a cursory familiarity with census publication for the state would clarify that both Kashmiris and Muslims have been adequately represented in each census. In particular, five out of six DCOs of J&K during 1961–2011 were Muslims. A Kashmiri Pandit was the only non-Muslim DCO, while a Ladakhi Muslim was the only non-Kashmiri DCO. Three, it has been suggested that '[a]fter data collection, all compilation and analysis takes place at the central level.... It is quite surprising that there is not a single Muslim member in the Task Force on Quality Assurance [for the 2001 census], which is responsible for the final clearance of census data for J&K' (ibid.). This, too, is incorrect. Two task forces were constituted for the 2001 census: Task Force on Quality Assurance (10 members) and Special Task Force for Religion and Scheduled Castes and Scheduled Tribes (12 members) (GoI 2003: v). The members of these task forces included 'the Heads and senior officers of the Census Division, Data Processing Division, Map Division, Demography Division and Social Studies Division' in the ORGI (ibid.: xiii). Eighteen officials were members of these two task forces, with four officials including the RG&CC and the Deputy Registrar General (Census & Tabulation) being members of both taskforces. Seventeen of these officials were regular employees of the ORGI, while the registrar general of India and the census commissioner (RGI&CC) was an IAS officer. All DCOs were ex officio members of the Task Force on Quality Assurance, which included Feroze Ahmed from J&K. Explaining the functioning of the task force the final population totals report of J&K notes that '[t]he Directors [of Census Operations] and their senior officers were required to make detailed presentations of data for their own state both in respect to the quality and the coverage and only after the full possible satisfaction of the TFQA, the population data was cleared' (ibid.: xiii). Feroze Ahmed was also a member of the 'Working Group on Religion' (GoI 2004a: 123).

## Misleading Comparisons

The fear of demographic marginalisation is grounded in misleading comparisons. Some of the older census reports discussed concerns regarding the declining population share of Muslims. They stress that pre- and post- 1947 census figures for J&K cannot be compared without adjusting for the loss of Muslim-majority areas and cross-border transfer of population. The 1971 general report points out that

> the decadal growth for Muslims is 25% as compared to 38.6% for Hindus [during 1961–71]. Since no census could be taken in 1951, it is not possible to compare the rates of their growth during the decade 1951–61.... The areas of the State seized by Pakistan became almost denuded of the Hindu population bulk of whom migrated to this side of the line of control [LoC]. In return a large number of Muslims crossed over to the so-called Azad Kashmir territory. It is not possible to arrive at an accurate assessment of the impact of this large scale movement and gauge the extent to which it tilted the proportion of communities as it stood prior to 1947. (GoI 1978b: 26)

The report adds that the decline of Muslim share was a longer-term phenomenon except for 1911–31, when the proportion of Hindus fell across 'almost every district of the State in varying degrees' (ibid.: 31). It points out:

> The decline in the growth-rate of Muslims was reflected in 1941 census also in which their overall proportion had come down to 72.4% from 73.0% as it stood in 1931. These were normal years and the effects of the calamitous events of 1947 were not yet felt by the population. (Ibid.: 26)

The report divides the puzzle posed by the declining proportion of Muslims into three geographical parts.

> Examining the district-wise position, it is evident that the decrease in the proportion of Muslim population during 1931–61 is confined to 5 out of 10 districts of the State namely Jammu, Rajauri, Kathua, Ladakh and Srinagar.... The small decline in the proportion of Muslims in the case of Srinagar district is insignificant. The fall in their proportion in Ladakh district seems to be due to a large exodus of Muslims from Kargil tehsil in 1947 and afterwards to areas now under the illegal occupation of Pakistan. The occupation by Pakistan of one of its tehsils, Skardu, with a cent per cent Muslim population is yet another reason for depressing the overall proportion of Muslims in Ladakh district. (Ibid.: 31)

The report goes on to explain the drop in Muslim population in Jammu. Migration triggered by partition

> enabled the Hindu population to add to its size by new arrivals from Pakistan held areas while the proportion of Muslims inhabiting parts of the State particularly the districts of Jammu, Kathua and Punch-Rajauri on this side of the line of control [LoC] became reduced. The 1965 Indo-Pak confrontation added further to their losses. (Ibid.: 26)[40]

The last point about the exodus of Muslims in 1965, which perhaps contributed to the declining share of Muslims in the state's population, is not discussed further in census reports. There was some migration in 1971 as well. It seems some Muslims from Rajouri and Punch relocated into Pakistan-occupied territories during the wars. The magnitude of the migration is not clear though. Robinson (2013: 52) suggests that as many as 50,000 people may have migrated during the 1965 and 1971 wars. The accounting of this population is difficult for several reasons. One, many came back after the war as the state government encouraged their return (interviews, Jammu, 19 September 2021; Punch, 21 September 21).

Moreover, interviewees from these areas suggest that the numbers were much smaller than 50,000 (interview, Jammu, 19 September 2021). Even in Rajouri where the intrusion of Pakistani forces was more substantial and 'parts of the district remained in an unsettled condition' (GoI 1974b: iv), the population grew at a rate comparable to the rest of the state as per the 1971 census. In fact, Rajouri's population growth rate during 1961–81 was higher than that of the state. Two, Robinson (2013: 52) suggests that the displacement of people was largely 'from areas where the front lines of warfare altered the boundaries of military control'. This means that at least some of the migrants were settled in the territory under Pakistan's administration before the war and left after it fell under India's control. People in areas formerly under Pakistan were not counted in the 1961 census and their relocation cannot explain the drop in the state's Muslim population share. Two such villages are identified in census reports – Nakarkote and Titri – that were added to Punch after 1971. Three, Punch reported a growth rate lower than the state during 1961–71 but grew faster than the state in the following decade. This, however, does not amount to evidence of migration from Punch to Pakistan-occupied territories because in 1961 enumeration was carried out in February, while it was shifted to March in 1971 (Table 2.3). This may have affected the headcount as the nomads would have begun migrating to snowbound areas within and outside Punch in 1971.

As discussed at length in Chapter 2, changing reference dates and weather conditions have also influenced headcounts and, by implication, relative population shares. While commentators in Kashmir allege that an 'engineered demographic change' was being carried out 'to change Kashmir's majority Muslim character' (Talib 2010: 2), census officials suggest that changes in the intrastate migration and the timing of census operations affect the headcount. Consider, for instance, Jammu city that reported a phenomenal growth of 103.93 per cent during 1941–61. The district census handbook attributes the growth to

> the settlement of displaced persons in Roulki colony and other areas from which large scale migrations have taken place and partly to the fact that the Census was taken in the month of February, 1961 when the city was inhabited, among others, by:- 1. Kashmiri Government officials employed in Secretariat and administrative departments which move with the Government to Jammu during winter, 2. Kashmiri labour which usually comes down to Jammu during winter for employment, 3. Members of the two Houses of Legislature and their families (the Census was taken while the budget session of the Legislature was on), 4. Kashmiri shop-keepers and their families, such as, butchers, bakers, tarugars, tailors, shoemakers, etc. (GoI 1966d: 32)

The point about Kashmiri labour is though partly contradicted by the 1961
GPT, which noted that 'they return home in March but as the month of fasts
[Ramzan] had this time started in February and the Id Festival was approaching
fast, most of them returned to Kashmir towards the end of February' and explained
the procedure followed to re-enumerate them in Kashmir (GoI 1964b: viii).
In principle this also applies to other categories of seasonal Kashmiri migrants
in Jammu. The contradiction between the two census reports highlights the fact
that even government officials in charge of enumeration are not conversant with
the complex accounting of population in the state.

Responding to allegations of forced demographic change, Feroze Ahmed, the
DCO for the 2001 census, highlighted the implications of changing reference
dates. He said:

> In 1961, the census operations were conducted in February, and in 1971, in
> mid-March. Census operations in 1981 were, however, conducted in April.
> Large populations of government employees move between Jammu and
> Srinagar for six months a year. During the 1961 and 1971 census operations,
> over 15,000 employees and their families would have been present in Jammu,
> but in 1981, they would have begun the process of moving back to Srinagar
> and been enumerated there ... [also] in 1981, a large number of these people
> [nomadic communities] would have moved on to pastures up in the mountains
> in 1981. (Swami 2000a)

The impact of the migration and reference date on headcount is nicely illustrated
by Kathua district in Jammu. Its Muslim population dropped from 29,812 in
1971 to 25,699 in 1981. Saifuddin Soz asked:

> How could the Hindu community mark a steep rise in its population growth,
> and Muslims not only fall but fall below the State's average population
> growth? Did Muslims adopt family planning and score a march over other
> communities? Did any natural calamity occur during these decades especially
> to deal with Muslims? (Ibid.)

Feroze Ahmed pointed out that the Kathua's Muslim population was concentrated
in a few areas and consisted of large numbers of Gujjars and Bakarwals and
that 'a large number of these people would have moved on to pastures up in
the mountains' during the 1981 census (ibid.). The nomadic population that
moved into snowbound areas would have been accounted for in the household
census conducted in inaccessible areas in September 1980 (wherever weather
conditions had not forced early departure to the plains) or in the special round

of enumeration conducted for the community in accessible areas of Jammu in April 1981 before they left for the mountains (and added to the headcount of inaccessible areas) (Table 2.3). We cannot, however, rule out that a section of the nomadic people may have been counted in Kashmir as the reference date was shifted to May. So, in 1981, the nomadic population was accounted in three parts and then added to the population of Kashmir. The DCO's remarks wrongly suggest that the nomadic population is counted in the plains of Jammu and should once again alert us to the fact that avoiding double counting is a difficult task and even senior officials may lack clarity amidst multiple reference dates changing in an ad hoc manner.

Administrative reorganisation too plays a role in stoking fears when unadjusted numbers are used in public debate. Feroze Ahmed illustrates this with an example.

> The population growth of Muslims massively outstripped that of Hindus in five of six districts in Kashmir Valley between 1971 and 1981. In Badgam, the one district where the growth of the Hindu population was higher than that of Muslims, it was because the Kashmiri-Pandit dominated areas of Barzulla, Rawalpora and Hyderpora, on the outskirts of Srinagar city, had been transferred to the revenue district of Badgam. (Ibid.)

## Nature of the Argument

The fear of demographic change has been a fixed point of Kashmiri politics, but the conceptualisation of the alleged threat and responses to it have shifted over time. The Hizbul Mujahideen and the Lashkar-e-Taiba did not call for a boycott of the 2001 census only because participation in the exercise would have amounted to an acceptance of the territorial order of the Indian state. They also argued that under the prevailing circumstances, Kashmir would have suffered a loss because of conflict-induced displacement of its population. More specifically, the Hizbul Mujahideen spokesperson alleged that 'the government's plans were to change the demographic pattern of the state so that avenues of development for Muslims were restricted' (*The Tribune* 2000b). After the conclusion of the census, it was again alleged that the exercise was 'a deliberate attempt to show the Muslim population of the state less than actual' (Masood 2010; see also S. Yasir 2011). In hindsight Kashmir was relieved as 'even the 2001 census that militants opposed tooth and nail was not such a bad exercise. It, in fact, *negated a trend* that earlier censuses had build. It showed the percentage of Muslims across J&K growing and Hindus exhibiting a decline' (*Kashmir Life* 2010, emphasis added).

In the next census, the position of supporters of independence shifted further, and they exhorted people to participate in the census to protect the Muslim-majority character of the state and Kashmir's share in development funds.[41] Mohammad Yasin Malik, the chairman of Jammu and Kashmir Liberation Front (JKLF), said, 'Last time, we had asked people to stay away from the process but now we feel that it was not a wise decision.... We don't want it to happen again' (Masood 2010; see also *Kashmir Life* 2010). Around this time the partisans of independence also reconsidered their opposition to elections (Swami 2008c). These developments in Kashmir are reminiscent of the north-east where insurgent groups fighting for independence, which used to oppose state-sponsored census as an instrument of subjugation and even tried to conduct census on their own, began to support their respective community's numerical claims in various government-sponsored exercises including census, map-making and delimitation (Agrawal and Kumar 2020a: 36, 245).

A decade later, the repeal of Articles 370 and 35A in 2019, followed by the introduction of new domicile rules and regulations governing transfer of land, accentuated the demographic anxiety. Three aspects of the changing engagement with census in Kashmir would bear elaboration.

First, there is a key difference between the pre- and post-2001 demographic propaganda. Published census data were used before 2001 as evidence of conspiracy to change the state's demography. However, after the 2001 census showed a marked increase in the Muslim population share, the propaganda has quietly shed its older statistical moorings and now stresses the potential threat to the *natural* numerical preponderance of Kashmir. In a related development, boycotts of statistical exercises gave way to active engagement to secure favourable numbers.

Second, the rationale for engagement with census has changed over time. We need to distinguish between different motivations for manipulation. *Pre-emptive* manipulation aims at avoiding a potential loss. It is different from both *defensive* manipulation aimed at avoiding detection of past manipulation as well as *prospective* manipulation aimed at securing hitherto unavailable benefits. *Altruistic* manipulation, on the other hand, overlaps with the aforementioned categories but differs from them insofar as it is meant to potentially benefit the next generation.[42] In the 2001 census, the manipulation was defensive in nature as it involved aligning the latest figures with the larger allocation of electoral seats secured in the past. The manipulation of headcount in Kashmir in 2011 was pre-emptive in nature insofar as the aim was to avoid a potential loss – namely, an unfavourable outcome in a possible plebiscite. Manipulation linked to plebiscite, which by all accounts is unlikely to be held in the foreseeable future, can also be treated as altruistic manipulation insofar as it will potentially benefit Kashmir in the distant future. However, in 2011, at the level of households, manipulation was defensive in character and was aimed at retaining past benefits obtained through the overstatement of family size in ration cards.[43] As discussed in the next chapter, contrary to popular perception, responses to census questions do not determine one's access to welfare schemes.

Third, the interference in census resulting in content errors reflects differently in the data depending on the context.[44] Before 1991, tribes of J&K suffered *statistical erasure* due to the denial of ST status. This continued in Kashmir to a lesser extent even later insofar as Gujjars and Bakarwals were not recorded as Gojri-speakers and/or tribes in certain areas. After the grant of ST status, however, there were also attempts at *statistical assimilation* of Pahari-, Gojri- and Shina-speaking Muslims into the Kashmiri fold. The re-engagement of the partisans of independence with census referred to previously did not just stop at ensuring Kashmir was correctly enumerated; it also involved appeals to non-Kashmiri Muslims to report themselves as Kashmiri speakers to strengthen the cause, but there was no claim or attempt to forge real sociocultural and linguistic bonds. In Gurez, we are possibly witnessing the *statistical marginalisation* or *submergence* of a language. In Ladakh, the census data revealed *fission* and *fusion* of communities. Finally, Halam and Generic Tribes are examples of *ex nihilo genesis* of categories with no real counterpart on the ground.

## The Nowhere People

Kashmiri Muslims highlight the decline in their population share in J&K between 1961 and 1981 to substantiate their demographic anxiety. Kashmiri Pandits express a reciprocal concern about the sharp decline in their numbers in Kashmir. A section of Pandits maintains that 'in 1947 the Pandits constituted 15 percent of Kashmir's population, which fell to 5 percent by 1981, and after the exodus to 0.1 percent' (Behera 2006: 125), and that

> 1941 marks the beginning of a statistical assault on the Pandit numbers by the junior local Muslim officials, who underestimated the strength of the Pandits by nearly 10–15 percent. The 1981 census had put the Pandit numbers at a little over 124,000 in a total population of 3.1 million, which stood exposed in 1990 when 300,000 Pandits fled the Valley. (Ibid.: 312)[45]

As per the 1941 census, Hindus (including Pandits, Dogras and Punjabis) accounted for only 4.95 per cent of the population of Kashmir province

(Government of Jammu and Kashmir [GoJK] 1943c: 341). Even if we accept the allegation of under-enumeration by 15 per cent, which is unlikely under the Dogra rulers, the population share of Hindus in Kashmir province would have been 5.77 per cent in 1941. So, for Hindus to account for 15 per cent of the population of Kashmir in 1947, about a million Muslims – that is, two-thirds of the Muslim population – must have migrated out of Kashmir, which is not true.

Participating in the parliamentary debate on the repeal of Article 370, Subramanian Swamy and Shamsher Singh Rana claimed that 500,000 Pandits and Sikhs were forced to leave Kashmir (Rajya Sabha 2019a: 56, 66).[46] It is not clear if the figure corresponds to the current population of those who left Kashmir over the decades[47] or the population of those living there in 1989–90. Writing a month after those debates, Thapar (2019) pointed out that while it is widely claimed that 'nearly 300,000 were thrown out of the Valley',[48] the official estimates suggest that '150,000 Pandits, 24,202 families migrated in January 1990…. The number of Pandits killed between 1989 and 2004 is put at 219. [Wajahat] Habibullah [who was divisional commissioner (Kashmir) in 1990] … says the total number of Pandits in 1990 was 200,000. After the migration that year, 20,000 were left. Over the next quarter century, that's come down to around 3,500 people.'

Assuming the Hindu child populations reported in the 1981 and 2001 censuses are correct and the child population share among Hindus was same across J&K in 2001, we can estimate that at least 135,000 Hindus, *including* Pandits, Punjabis and Dogras, would have left Kashmir in the 1990s for Jammu and other parts of the country.[49] Note that these estimates are based on the assumption that armed forces were either not deployed in the Valley in 1980–81 (and, also, 1990–91) or were not enumerated.[50] It also bears noting that the unusual reference date of the 1981 census meant that enumeration happened when the Darbar was in Srinagar rather than Jammu, which would have inflated the Hindu population of Kashmir in that year. So a simple extrapolation of the 1981 census, without accounting for the reference dates of the 1981 and 2001 censuses, will overestimate the January 1990 population of Hindus in Kashmir.

Kashmiri Pandits argue that their exodus was preceded by long-standing political marginalisation. For instance, Habbakadal, the only assembly constituency in Kashmir that was dominated by the community, was gerrymandered to deny it any representation in the legislative assembly (Panun Kashmir n.d.2: 7–8; Hari Om 1998: 117–118). As a result, after 1967, only two non-Muslims successfully contested assembly elections, and, significantly, both were fielded by the INC. In fact, even without gerrymandering the Pandits were electorally superfluous as they found it difficult to exercise their franchise amidst growing religious polarisation. For instance, Devadas (2019a: 154) points out that '[t]he communal antagonism that [1983] assembly campaign spewed had so crystallised by the time Parliament elections were held the next year, 1984, that many Pandits were not allowed to vote'. Around this time, Pandits also began to question the reliability of population estimates (GoI 2001b: 29; *Kashmir Times* 2000c).

Unfortunately, the plight of Kashmiri Pandits is lost somewhere in the statistical tug of war between Srinagar and Jammu and between the Hindu nationalists and the secular nationalists. Despite ostensibly being the darling of Hindu nationalists across India, the Pandits are perhaps the only community uniformly disliked *across* the union territory (UT) of J&K. Several interviewees based in the Valley told the author that the Pandits left Kashmir as part of a conspiracy to allow the army to indiscriminately kill Kashmiri Muslims. Even those who admitted that Pandits left due to communally motivated attacks hastened to add that Muslims left behind suffered more casualties or that Muslims migrated out of the state in larger numbers in the late nineteenth century. Pandits, however, were not quite welcome in Jammu (see, for instance, Panun Kashmir n.d.2: 33), which otherwise imagines itself as the bastion of Hinduism in the region. As Pandita (2017: 100) puts it, 'Initially, like us, the Jammuites thought our exodus was temporary. Though they benefitted economically because of us, they developed an antipathy towards us. For them, we were outsiders.' There are several reasons for the antipathy.

First, the sudden influx of a highly educated community increased competition in Jammu's job market.[51] The permanent transfer of Pandits employed in state and union government departments from Kashmir to Jammu, too, affected job opportunities, especially when the government was even otherwise avoiding new recruitments in the aftermath of liberalisation. While Jammu's complaint against Pandits in the job market seems to predate the exodus of the latter from Kashmir (see, for instance, Abdullah 2016: 575), the 'unwritten policy [of the Public Service Commission] in fact favoured Muslims – and even Jammu Hindus' (Devadas 2019a: 151; see also Panun Kashmir n.d.2: 13, 33). The forced migration of Pandits must have worked in favour of Kashmiri Muslims and Dogra and other Jammu-based Hindus in public service examinations as most Pandit families were in a complete disarray in the first half of the 1990s, and later many migrated to other states and countries.

Second, the union government seems to have offered more support to Pandits compared to the West Pakistan refugees and refugees from Pakistan-occupied territories, who came in the late 1940s. The Group of Interlocutors for J&K, for instance, noted, 'Migrants from PoK who have settled in Jammu from 1947–48 onwards also suffer from a feeling of neglect. Their entitlements are nowhere on par with those enjoyed by the Kashmiri Pandits' (Kumar, Ansari and Padgaonkar n.d.: 138). This has been a source of very bitter differences in Jammu district. The refugees from Pakistan-occupied territories of J&K feel that compared to 'valley migrants [Pandits] – a[n] elite', they have been treated as 'a delete class' (SOS International n.d.2). The West Pakistan refugees, too, complain that they have been neglected (West Pakistan Refugees Action Committee [WPRAC] 2015: 9). Pandits point out that they, too, were treated as 'unwanted' and 'untouchables' in Jammu (Panun Kashmir n.d.2: 13, 33) but agree that after the initial disarray the administration eventually offered some support to the members of the community who stayed back in Jammu.[52] They lament that the government's failure to address the grievances of other refugee communities has triggered a vicious 'politics of competitive victimhood' (interview, Jammu, 13 July 2021), which has been exploited by vested interests including the Kashmiri Muslim leadership and sections of the Rashtriya Swayamsevak Sangh (RSS) (interview, 14 July 2021).

Third, Kashmiri Pandits enjoy greater political visibility at the national level, which compounded the sense of marginalisation of Dogra Hindus who were already grappling with New Delhi's tilt towards Srinagar. Fourth, occasionally some Pandits side with Kashmir in public debates. The 'betrayal' evokes very strong responses in Jammu (see, for instance, Hari Om 2019a). Last but not least, a few Jammu-based Muslim interviewees complained that Kashmiri Pandits, who 'were half-Muslims in the hills … became Hindus here [in Jammu]' (interview, Jammu, 13 September 2019)[53] and accentuated the Hindu dimension of Jammu's identity to make themselves more acceptable in the plains (interview, Jammu, 13 August 2021).

## Jammu

The demographic anxiety of Jammu (city) has two layers. While the larger anxiety is about the Jammu division and, occasionally, the territory of the erstwhile Dogra kingdom, the immediate anxiety is about the region that borders upon Punjab – that is, Jammu, Samba and Kathua districts – which is also referred to as the Mountainous region, and Udhampur. In 2011, the Mountainous region (including Udhampur) accounted for 46 (56) per cent of population and 22 (32) per cent of the area of Jammu division. Further, the Mountainous region (including Udhampur) accounted for more than 74 (92) per cent of Dogri speakers and 59 (73) per cent of Hindus of the entire state (Table 4.3). Likewise, Dogri speakers accounted for more than 75 (76) per cent of the population of

**Table 4.2** Population of major religions in the Mountainous region of Jammu and Kashmir (J&K), 1961–2011

| Census | District/Region | All religions | Hindu | | Muslim | | Sikh | |
|---|---|---|---|---|---|---|---|---|
| | | | Population | Share | Population | Share | Population | Share |
| 1961 | Jammu | 516,932 | 428,835 | 0.8296 | 51,847 | 0.1003 | 32,788 | 0.0634 |
| | Kathua | 207,430 | 177,666 | 0.8565 | 27,005 | 0.1302 | 2,553 | 0.0123 |
| | *Mountainous* | 724,362 | 606,501 | 0.8373 | 78,852 | 0.1089 | 35,341 | 0.0488 |
| 1971 | Jammu | 731,743 | 611,164 | 0.8352 | 58,182 | 0.0795 | 55,614 | 0.0760 |
| | Kathua | 274,671 | 239,010 | 0.8702 | 29,812 | 0.1085 | 5,097 | 0.0186 |
| | *Mountainous* | 1,006,414 | 850,174 | 0.8448 | 87,994 | 0.0874 | 60,711 | 0.0603 |
| 1981 | Jammu | 943,395 | 824,209 | 0.8737 | 40,309 | 0.0427 | 72,034 | 0.0764 |
| | Kathua | 369,123 | 336,503 | 0.9116 | 25,699 | 0.0696 | 6,082 | 0.0165 |
| | *Mountainous* | 1,312,518 | 1,160,712 | 0.8843 | 66,003 | 0.0503 | 78,116 | 0.0595 |
| 1991 | *Mountainous* | – | – | – | – | – | – | – |
| 2001 | Jammu | 1,588,772 | 1,366,711 | 0.8602 | 90,272 | 0.0568 | 117,490 | 0.0740 |
| | Kathua | 550,084 | 493,966 | 0.8980 | 44,793 | 0.0814 | 9,152 | 0.0166 |
| | *Mountainous* | 2,138,856 | 1,860,677 | 0.8699 | 135,065 | 0.0631 | 126,642 | 0.0592 |
| 2011 | Jammu | 1,529,958 | 1,289,240 | 0.8427 | 107,489 | 0.0703 | 114,272 | 0.0747 |
| | Samba | 318,898 | 275,311 | 0.8633 | 22,950 | 0.0720 | 17,961 | 0.0563 |
| | Kathua | 616,435 | 540,063 | 0.8761 | 64,234 | 0.1042 | 9,551 | 0.0155 |
| | *Mountainous* | 2,465,291 | 2,104,614 | 0.8537 | 194,673 | 0.0790 | 141,784 | 0.0575 |
| Growth | 1961–2011 | 2.40 | 2.47 | – | 1.47 | – | 3.01 | – |
| *(Mountainous)* | 1981–2011 | 0.88 | 0.81 | – | 1.55 | – | 0.82 | – |

*Source:* Table C-1 of Census of India, 2001 and 2011; General Reports of the Census of India for the earlier years.

*Note:* (*a*) The Mountainous region includes Jammu, Samba and Kathua districts.

(*b*) $Growth_{1961-2011} = \dfrac{Pop2011 - Pop1961}{Pop1961}$ and $Growth_{1981-2011} = \dfrac{Pop2011 - Pop1981}{Pop1981}$.

**Table 4.3** Population of speakers of major languages in the Mountainous region of Jammu and Kashmir (J&K), 1961–2011

| Census | District/Region | All languages | Dogri Population | Dogri Share | Hindi Population | Hindi Share | Kashmiri Population | Kashmiri Share | Punjabi Population | Punjabi Share |
|---|---|---|---|---|---|---|---|---|---|---|
| 1961 | Jammu | 516,932 | 416,807 | 0.8063 | 17,066 | 0.0330 | 11,175 | 0.0216 | 59,889 | 0.1159 |
| | Kathua | 207,430 | 191,538 | 0.9234 | 980 | 0.0047 | 3,712 | 0.0179 | 3,962 | 0.0191 |
| | *Mountainous* | 724,362 | 608,345 | 0.8398 | 18,046 | 0.0249 | 14,887 | 0.0206 | 63,851 | 0.0881 |
| 1971 | Jammu | 731,743 | 593,254 | 0.8107 | 29,913 | 0.0409 | 15,789 | 0.0216 | 80,831 | 0.0415 |
| | Kathua | 274,671 | 236,399 | 0.8607 | 21,773 | 0.0793 | 6,439 | 0.0234 | 7,329 | 0.0267 |
| | *Mountainous* | 1,006,414 | 829,653 | 0.8244 | 51,686 | 0.0514 | 22,228 | 0.0221 | 88,160 | 0.0876 |
| 1981 | Jammu | 934,577 | 764,426 | 0.8179 | 53,836 | 0.0576 | 8,207 | 0.0088 | 95,379 | 0.1021 |
| | Kathua | 366,221 | 319,506 | 0.8724 | 30,595 | 0.0835 | 6,819 | 0.0186 | 7,337 | 0.0200 |
| | *Mountainous* | 1,300,798 | 1,083,932 | 0.8333 | 84,431 | 0.0649 | 15,026 | 0.0116 | 102,716 | 0.0790 |
| 1991 | *Mountainous* | – | – | – | – | – | – | – | – | – |
| 2001 | Jammu | 1,588,772 | 1,162,897 | 0.7319 | 157,028 | 0.0988 | 88,508 | 0.0557 | 127,727 | 0.0804 |
| | Kathua | 550,084 | 468,007 | 0.8508 | 53,628 | 0.0975 | 11,931 | 0.0217 | 9,858 | 0.0179 |
| | *Mountainous* | 2,138,856 | 1,630,904 | 0.7625 | 210,656 | 0.0985 | 100,439 | 0.0470 | 137,585 | 0.0643 |
| 2011 | Jammu | 1,529,958 | 1,084,040 | 0.7085 | 178,064 | 0.1164 | 94,649 | 0.0619 | 123,874 | 0.0810 |
| | Samba | 318,898 | 269,559 | 0.8453 | 27,766 | 0.0871 | 993 | 0.0031 | 15,799 | 0.0495 |

*(Contd)*

**Table 4.3**  (*Contd*)

| | | | | | | | | | |
|---|---|---|---|---|---|---|---|---|---|
| Kathua | 616,435 | 505,013 | 0.8192 | 75,739 | 0.1229 | 13,479 | 0.0219 | 11,471 | 0.0186 |
| *Mountainous* | 2,465,291 | 1,858,612 | 0.7539 | 281,569 | 0.1142 | 109,121 | 0.0443 | 151,144 | 0.0613 |
| Growth 1961–2011 | 2.40 | 2.06 | — | 14.60 | — | 6.33 | — | 1.37 | |
| Growth 1981–2011 | 0.90 | 0.71 | — | 2.33 | — | 6.26 | — | 0.47 | — |

*Source*: General Reports of the Census of India and sociocultural tables for 1961–81; table C-16 of Census of India, 2001 and 2011.

*Note*: (*a*) The Mountainous region includes Jammu, Samba and Kathua districts. (*b*) Hindi includes Gojri and Bhadrawahi. Kashmiri includes Kishtwari. (*c*) Dogri was reported under Punjabi in 1961.

(*d*) $Growth_{1961-2011} = \dfrac{Pop2011 - Pop1961}{Pop1961}$ and $Growth_{1981-2011} = \dfrac{Pop2011 - Pop1981}{Pop1981}$.

the Mountainous region (including Udhampur) (Table 4.3). Dogri speakers, almost entirely Hindu, constituted more than 46 per cent of the Jammu division's population. These four districts constitute what can be called the Dogra, Dogri, and Hindu heartland in the state. However, Dogras do not dominate Jammu the way Kashmiris dominate Kashmir. In 2011, Kashmir accounted for 78 per cent of the state's Muslim and 88 per cent of its Kashmiri-speaking population. On the other hand, Muslims and Kashmiri speakers accounted for 96 and 86 per cent of Kashmir's population, respectively.

Several factors have shaped the perspective of Jammu (city). First, the union government and national political parties prevented the emergence of a strong and independent political identity and leadership in Jammu.[54] To begin with, '[i]t was not easy for most people in Jammu to reconcile themselves to the transfer of power in 1947 from the Jammu-based Maharaja to the Srinagar-based leadership' (B. Puri 1974: 185).[55] The complete dominance of Kashmir over representation in both the national and the state constituent assemblies and the exclusion from several rounds of talks between New Delhi and Srinagar compounded Jammu's sense of loss.[56] B. Puri (1983a: 188) points out that 'as champion of Kashmiri nationalism, the National Conference could not strike roots in Jammu', yet 'the national leadership recognised only Sheikh Abdullah and his party as representative of the state. The attempts to organise Congress party in Jammu were discouraged by its High Command' (see also B. Puri 1983b: 230). In fact, as late as the early 1960s, Jawaharlal Nehru seems to have refused to 'open a Congress unit in the state' as he treated the NC 'as a sister party' whose leaders were permanent special invitees at the Congress Working Committee (Devadas 2019a: 118).[57] The obstruction of the growth of leadership in Jammu to ensure that Kashmir does not feel threatened within the state resulted in a situation where Jammu and Hindus were effectively barred from the chief minister's office.[58] Before independence this meant that '[t]he Hindu Sabha and the Muslim League leaders moved in to fill the vacuum' (B. Puri 1983a: 188). Later, it manifested as Jammu's exclusion from the dialogue between New Delhi and Sheikh Abdullah and pushed sections of the Hindus of Jammu (including those opposed to land reforms) towards the Hindu nationalists (B. Puri 1974: 185), who were interested in Jammu from a pan-Indian Hindu perspective rather than in Jammu's regional concerns.

In the absence of a strong regional political voice, Jammu was reduced to 'a prop in the hands of New Delhi to counter the Kashmir narrative or to strengthen integrationist politics' (Jamwal 2020a). The 'perpetual neglect' has reduced Jammu, 'the epicentre of the princely state of J&K', to 'a forgotten footnote in the national perspective' (B. Singh 2018; see also Hari Om 2022a). Now 'with the administrative project of integration achieved and Kashmir's politics stepping down into virtual servitude, Jammu's political worth has further diminished' (Jamwal 2020a). It is not the case that the nationalist stance has not yielded any returns to Jammu. The city has gradually emerged as the de facto economic and administrative headquarter of the state, but that has not alleviated its sense of political marginalisation.

Second, symbolic marginalisation compounded the political marginalisation. It reflected in, say, the names of the state's administrative and police services that were called Kashmir Administrative Service and Kashmir Police Service.[59] Kashmir first featured on a postage stamp in 1955, while Jammu had to wait until 1999. The tourism department foregrounded Kashmir as the pre-eminent destination in the state. Jammu often found no mention in the tourist brochures issued by the government (see, for instance, GoI 1966e).[60] Dogri was not recognised in the Eighth Schedule as late as 2003 even though Kashmiri was included from the beginning. The inclusion of Kashmiri did not translate into its promotion though, as the state government was committed to Urdu (as the official language). Kashmiri leadership and intelligentsia enjoyed greater visibility at the national and international levels, even as non-Dogri refugees from Pakistan-occupied territories dominated the intellectual space within Jammu (journalist, interview, 20 August 2021).

Third, Jammu received several waves of refugees and internally displaced communities from West Pakistan, Pakistan-occupied territory of J&K, the international border, the LoC, Kashmir, Muslim-dominated districts of the Jammu division, and even Bangladesh and the Rakhine province of Myanmar, which turned it into 'the City of Refugees' (Hari Om 2022c). These complement the educational and economic migrants from across the state.

The less than satisfactory treatment of Hindu refugees in the state has aggravated Jammu's demographic concerns more than anything else. The economic and political competition between Kashmiri and Dogri Hindus notwithstanding, the helplessness of the state in face of the near complete exodus of indigenous Hindus from Kashmir seriously dented the sense of security of Hindus of Jammu (and even Buddhists of Ladakh). We have already discussed the experience of Pandits earlier. The case of refugees from West Pakistan and Pakistan-occupied territories would bear elaboration here.

The West Pakistan refugees believe that they suffered discrimination due to their religious (non-Muslim) and ethnic (non-Kashmiri) background. They viewed themselves as 'slaves', 'aliens' and 'second-class citizens' within the undivided state. The community was until recently not allowed to vote in assembly elections, deprived of government jobs and denied access to affirmative action benefits meant for SCs, as they were not state subjects.[61] A memorandum pointed out that 'J&K Constitution was implemented in the State [in] 1957 i.e. after ten years of our migration. But while determining the citizenship of the State under Part-III, Section 6 of J&K Constitution, we people were deliberately ignored' (WPRAC 2015: 3). As a result, 'unfortunately we are not even eligible to became a class IV employee in the J&K State' (ibid.: 2). They were also denied 'Scheduled Caste/OBC [Other Backward Classes] Certificates ... depriving them of even Central Schemes' (ibid.: 5). The memorandum added that 'West Pakistan Refugees comprises of only 18428 families and such a meagre and marginalised society/number cannot change the demographic structure of a 1.25 crores soul populated state' (ibid.: 4). It added that the state's 'demography was never threatened' when Uyghur Muslims and Tibetan Muslims were, respectively, granted state subject status in 1952 and 1959, while the West Pakistan refugees were 'till date living as slaves in independent India, the only reasons that these families are Hindus' (ibid.: 4). A community leader complained that Kashmir does not feel threatened by the presence of illegal Bangladeshi and Rohingya settlers even as longstanding residents were kept out of the polity (Labha Ram Gandhi, interview, Jammu, 3 June 2016; see also Hari Om 2017).[62]

In Jammu, the fact that 'the Kashmiri leadership would settle in Srinagar the Uyghur and Tibetan Muslims with full citizenship rights and make unjust, invidious and humiliating distinctions between them and the non-Muslim refugees from West Pakistan' was widely seen as 'communally motivated' (Hari Om 2013).[63] The Bharatiya Janata Party (BJP) used the treatment of the West Pakistan refugees, among other things, to justify the decision to repeal Article 370.[64] It is noteworthy that while the West Pakistan refugees were not allowed basic citizenship rights in J&K, they have always been covered in censuses because India follows a de facto method of enumeration. This has been deliberately misinterpreted by Kashmiri leaders to suggest that the union government is using census to change the demography. Sheikh Abdullah, too, viewed the demand to grant the status of state subject to West Pakistan refugees as detrimental to the state's demographic composition but linked his stand to the resistance to the return of Kashmiri Muslims from Pakistan-occupied areas (S. Hussain 2021: 221).[65]

The community of refugees from Pakistan-occupied parts of J&K, who were state subjects even before they relocated to Jammu, is much larger than West Pakistan refugees. Despite being state subjects, the community feels neglected as it was 'warehoused in ... refugee camps or segregated settlements' after reaching

Jammu and were not considered as refugees as they were displaced within their state (SOS International 2004).[66] Starting with a population of about 300,000 the community seems to have grown to 'over one million' or '12 lakhs' (SOS International n.d.2, 2006, 2014a). The community has been demanding reservation of eight seats in the state legislative assembly out of the 24 seats reserved for Pakistan-occupied territories, where they accounted for a third of the population at the time of independence (interview, Jammu, 2 June 2016).[67]

Fourth, the Amarnath land controversy of 2008, which we have discussed earlier, was second only to the exodus of Hindus from Kashmir in the 1990s in terms of the impact on Jammu's perception of Kashmir. Kashmir opposed the 'grant of land-use rights to the Amarnath Shrine Board' that would allegedly 'alter the demographic character of the state' (Swami 2013). Jammu responded with a massive (economic) blockade. This incident deepened the chasm between Kashmir and Jammu. The Amarnath controversy was not the only source of 'demographic' discontent in the run-up to the 2011 census, though. The debate on whether women married to non-permanent residents can access the benefits of permanent resident status was revived in 2010 (Chowdhary 2010a). Jammu was apprehensive due to the high incidence of interstate marriages. Around this time, inter-district appointments emerged as another bone of contention as Kashmiris resented the appointment of 'outsiders' from Jammu in Kashmir, while Jammu viewed this as another instance of denial of opportunity within the state.[68]

Fifth, partisans of Hindus of Jammu fear Kashmir's alleged 'Land Jihad'. It is argued that Kashmir has been expanding its footprint outside Kashmir by (*a*) co-opting Muslims in the districts of Jammu bordering Kashmir, (*b*) facilitating the settlement of Muslims in and around Jammu city and (*c*) transferring government land in Jammu, primarily to Muslims under the J&K State Land (Vesting of Ownership Rights to the Occupants) Act, 2001. The last two are allegedly facilitated by the Kashmiri Muslim-dominated revenue and police departments that also allegedly aided the settlement of Bangladeshis and Rohingyas.[69] In short, Jammu allegedly suffered both funded colonisation and induced migration aided by the Kashmiri Muslim-dominated administration[70] that offered legal immunity to encroachers, who have been encouraged since the mid-1990s to strategically settle along highways and on state land including forests (journalist, interview, Jammu, 2 June 2016; lawyer, interview, New Delhi, 1 July 2018; former union home secretary, interview, 27 June 2023).[71]

Commenting on an earlier iteration of Jammu's concerns, Swami (2000a) pointed out that 'the Hindu Right in Jammu has sought to play on exactly the kind of fears raised by [Saifuddin] Soz' as 'BJP members of the Legislative Assembly Piara Singh and Ashok Khajuria announced the existence of a State government-run plot to settle Muslims in Jammu and alter its demographic profile. The MLAs claimed that State support had led to the construction of 35,000 homes in Jammu by Muslim migrants over the past seven years.' A retired bureaucrat rejected claims of conspiracy to encircle Jammu. He pointed out:

> Given the superiority of the quality of life in Jammu city compared to any other district [or] town in the division, people from all over Jammu Division aspiring for a better life, both Hindu and Muslim, have been migrating to and settling down in Jammu city and surrounding areas for ages. There may be a few Muslims from Kashmir who have done this, but the vast majority of Muslims who have migrated to Jammu city are from Jammu Division itself, with a significant proportion being Kashmiri speaking Muslims from old Doda district. (Interview, 21 June 2016; see also Jamwal 2020a)

Commenting on the nature of land transactions, he added:

> The wholesale encroachment of Khalsa Sarkar or Shamilat land by these migrants [in Jammu] is also a myth, and I am saying this as a former Financial Commissioner (Revenue) ... rent seeking opportunities in the capital cities are huge and I am sure encroachments on State land and common land in connivance with Revenue authorities have occurred. However, the vast majority of these transactions would have involved straightforward sale and purchase of proprietary land. I am not sure that it is impossible for a Jammuwalla to buy land in Srinagar: the point is would anyone want to in the present circumstances. (Interview, 21 June 2016)

He also explained the spatial patterns of settlement in Jammu:

> With the communal divide in Jammu being what it is, most of these migrants have settled in or around areas which already had a Muslim population like Gujjar Nagar.[72] Ghettoisation in these circumstances is not just inevitable, but preferable too from a public order point of view. (Interview, 21 June 2016)

Jammu's experience agrees with that of economic hubs in other parts of the country. Bengaluru, Mumbai and Dimapur attract more migrants from within their respective states than from other states. Yet the political discourse remains sharply focused on 'outsiders'. The rise of Jammu as the commercial hub of the state accounts for a large part of the migration to the district. A former finance minister observed:

Whether one like[s] it or not, much like Article 370, the substantive basis of Srinagar as a capital has been eroded over time. It has been emaciated by the successive governments at the centre. Its administrative importance starting dwindling in the 1980s and was completed in the 1990s. All the Central Government institutions, departments, agencies and undertakings have chosen to locate their regional offices in Jammu. Same is the case with all the national and multi-national private companies or organizations. They too have their offices and business operations from out of Jammu. (Drabu 2020f)

Migration of Muslims to Jammu, whether due to conspiracy or pursuit of education and employment, should reflect in the census. Between 1961 and 2011, the share of Muslims in the population of Jammu (including Samba) changed from 10.03 to 7.06 per cent, with the lowest share being reported in 1981 when Muslims accounted for 4.27 per cent of the district's population (Table 4.2). It is noteworthy that the Muslim population share of Jammu district increased from 4.27 per cent in 1981 to 5.68 in 2001, even though the exodus of more than 150,000 Hindus and Sikhs from Kashmir and the migration of Hindus from the northern districts of the Jammu region added substantially to its non-Muslim population. Also, Srinagar district, which was home to 42 (48) per cent of Kashmir's non-Muslim (Hindu) population before the exodus, reported a growth rate twice that of Jammu district, which received almost the entire refugee population from Kashmir. It cannot be argued that migration from other districts accounts for Srinagar's growth as they too reported very high growth rates. Note that the share of Muslims in Jammu grew between 2001 and 2011 as well, because of the increase in the population of largely Muslim STs (Tables 4.2 and 4.4).

The real and perceived asymmetry in migration has aggravated Jammu's sense of siege. In 1981, the census reported that 7,944 (2,877) persons enumerated in Jammu (Srinagar) district were born in the Kashmir (Jammu) division (Table 4.5). Two decades later, Jammu district's headcount included 72,147 persons who were born in the Kashmir division (most of them must be Pandits and Sikhs migrants), and Srinagar (including Ganderbal) district reported only 5,163 persons born in the Jammu division. The asymmetry was also reflected in the number of employees who travelled with the Darbar to the other division. After the practice of 'Darbar Move' came to an end, 'the officials who were allotted around 2198 flats and quarters, that is, 1737 in Jammu and 461 in Srinagar' were 'directed to vacate the accommodations' (Shah 2021). These figures do not reflect the true extent of the asymmetry because they capture only the movement of those entitled to government accommodation. A much larger

**Table 4.4** Population of Scheduled Tribes (STs) in the Mountainous region of Jammu and Kashmir (J&K), 2001–11

| District | Population (1) | All STs (2) | Gujjar (3) | Bakarwal (4) | Gujjar and Bakarwal (5 = 3 + 4) | Generic Tribes (6) | Gujjar and Bakarwal / (All STs – Generic Tribes) (7 = 5 / [2 – 6]) |
|---|---|---|---|---|---|---|---|
| **2001** | | | | | | | |
| Jammu | 1,588,772 | 53,304 | 50,555 | 1,576 | 52,131 | 491 | 0.9871 |
| Kathua | 550,084 | 34,174 | 15,149 | 2,754 | 17,903 | 57 | 0.5248 |
| *Mountainous* | 2,138,856 | 87,478 | 65,704 | 4,330 | 70,034 | 548 | 0.8056 |
| **2011** | | | | | | | |
| Jammu | 1,529,958 | 69,193 | 58,404 | 6,997 | 65,401 | 3,036 | 0.9886 |
| Samba | 318,898 | 17,573 | 13,183 | 3,758 | 16,941 | 588 | 0.9974 |
| Kathua | 616,435 | 53,307 | 23,437 | 7,946 | 31,383 | 2,784 | 0.6212 |
| *Mountainous* | 2,465,291 | 140,073 | 95,024 | 18,701 | 113,725 | 6,408 | 0.8508 |
| Growth (2001–11) | 0.1526 | 0.6012 | 0.4462 | 3.3189 | 0.6239 | 10.69 | — |

*Source:* Primary census abstracts (PCA) of 2001 and 2011 for STs.

*Note:* (a) The Mountainous region includes Jammu, Samba and Kathua districts. (b) Figures for 2001 have not been adjusted for the changes indicated in the Appendix. (c) $Growth_{2001-2011} = \dfrac{Pop2011 - Pop2001}{Pop2001}$.

**Table 4.5** Migrants of Jammu and Kashmir (J&K) by place of birth, 1981–2011

| 2011 | Born in Jammu division | Born in Kashmir division | Born in Ladakh division | Source |
|---|---|---|---|---|
| Kashmir | 30,308 | 211,158 | 1,573 | |
| Jammu | 257,603 | 79,470 | 896 | |
| Ladakh | 331 | 392 | 2,770 | |
| Jammu district | 98,265 | 71,786 | 552 | |
| Samba district | 46,144 | 778 | 10 | |
| Jammu + Samba district | 144,409 (93,244*) | 72,564 | 562 | D-11 |
| Srinagar district | 4,387 | 53,474 | 825 | |
| Ganderbal district | 6,604 | 14,619 | 527 | |
| Srinagar + Ganderbal district | 10,991 | 68,093 (54,137*) | 1,352 | |

| 2001 | Born in Jammu division | Born in Kashmir division | Born in Ladakh division | |
|---|---|---|---|---|
| Kashmir | 14,700 | 45,466 | 523 | |
| Jammu | 124,699 | 80,979 | 512 | D-11 |
| Ladakh | 1,486 | 1,463 | 1,427 | |
| Jammu district | 61,912 | 72,147 | 377 | |
| Srinagar district | 5,163 | 16,207 | 459 | |

| 1981 | Born in Jammu division | Born in Kashmir division | Born in Ladakh division | |
|---|---|---|---|---|
| Kashmir | 7,714 | 65,304 | 730 | |
| Jammu | 97,727 | 15,785 | 243 | D-13 |
| Ladakh | 284 | 1,016 | 868 | |
| Jammu district | 37,752 | 7,944 | 107 | |
| Srinagar district | 2,877 | 16,239 | 579 | |

*Source*: Compiled by the author using data from the tables of the respective censuses listed in the 'Source' column.

*Note*: (*a*) *The figures reported within parentheses have been adjusted for the reorganisation of district borders and correspond to undivided Jammu and Srinagar. (*b*) For want of information, the adjustments do not account for the transfer of territory from Kathua and Udhampur to Jammu and Baramula and Badgam to Srinagar (see Figure A.2 for details). (*c*) For want of information, the division-level figures have not been adjusted. The adjusted figures for Kashmir and Jammu divisions will be lesser than those reported here. The figures for Ladakh division will remain unaffected in absence of any territorial change during 1981–2011. (*d*) The figures reported here exclude migrants whose district of birth was unclassified. Treating these migrants as persons born in the Jammu division and enumerated in the Kashmir division will not alter our conclusions.

entourage followed politicians and officials to Jammu to escape the harsh winter of Kashmir (GoI 1966d: 32), and for many others the Darbar Move offered an escape from turmoil in Kashmir (Ashiq 2021). In contrast, after 1990, most officials from Jammu went alone to Srinagar.[73]

Several interviewees based in Jammu asked if their district is the single most important destination for both intrastate and interstate migrants, what explains its low growth rate between 2001 and 2011. Srinagar (including Ganderbal) district grew by 20.35 (23.18) per cent between 2001 and 2011. The corresponding figure for Jammu (including Samba) was 12.74 (13.45) per cent. In fact, after 1981, internal migration to Jammu increased, but its population share in the state decreased. This also holds good for the population share of the southern districts (Jammu, Samba, Kathua and Udhampur). Interestingly, between 2001 and 2011, the share of both Jammu district and the southern districts decreased even within the Jammu division.

Srinagar Urban Agglomeration (UA) grew by 27.93 per cent during 2001–11, compared to 7.38 per cent growth in Jammu UA. The urbanisation rate of Jammu (including Samba) grew marginally from 44.14 to 44.28 per cent. The urbanisation rate of Srinagar (including Ganderbal) grew from 78.69 to 82.55 per cent. Interestingly, the slum population of Jammu district contracted by 81 per cent, while that of Srinagar district grew by 150 per cent (Table 4.6a). The slum population of Jammu district should have grown with the inflow of economic migrants and those escaping conflicts. Jammu district accounted

Table 4.6a    Slum population of Jammu and Kashmir (J&K), 2001–11

| State/Division/District | Slum population (share in district in per cent) | |
| --- | --- | --- |
| | 2001 | 2011 |
| J&K | 268,513 | 662,062 |
| Kashmir division | 191,356 | 619,124 |
| Srinagar district | 137,555 (11.04) | 343,125 (22.36) |
| Jammu division | 77,157 | 35,175 |
| Jammu district | 77,157 (4.73) | 14,980 (0.81) |
| Ladakh division | 0 | 7,763 |

*Source*: GoI (2005b: 63–64); primary census abstract (PCA) for slum (2011).

*Note*: Jammu district includes Jammu (municipal corporation + outgrowth) and Bishna (municipal committee) in 2011 and Jammu (municipal committee) in 2001. For the sake of comparability of 2001 and 2011, we have combined Srinagar and Ganderbal districts and Jammu and Samba districts. There was no slum in Ganderbal in 2011, while a slum population was reported in Samba's Bari Brahamana (municipal committee).

for the entire slum population of the Jammu division in 2001, whereas only a third of the division's slum population lived in Jammu district in 2011. In 2001, slums accounted for about 4.73 per cent of Jammu district's population (see Table 4.6a for details of the calculation). A decade later the share of slums in Jammu district's population was barely 0.81 per cent. In the case of Srinagar district this share increased from 11.04 to 22.36 per cent. It cannot be argued that Srinagar's slum population grew at a very high rate due to influx from other districts from Kashmir because almost all districts of Kashmir reported large increases in the overall population as well as the slum population during 2001–11. The districts of Kashmir excluding Srinagar reported a 413 per cent growth in the slum population.[74] Kashmir's slum population grew by 224 per cent, while that of Jammu contracted by 54 per cent. The large growth of the slum population has had a huge impact on the regional distribution of population. If we exclude the slum population, Kashmir's population grew by 19 per cent during 2001–11 compared to Jammu's 23 per cent. The sharp increase in the slum population partly explains the very large increase in Kashmir's urban population by 47.57 per cent, which cannot be accounted for by new towns (Table 4.6b) or migration. The anomalies in slum statistics also partly account for the sharp increase in the urbanisation rate of Kashmir. A researcher suggested that the growth of slum population could be traced to headcount-linked funding received by the government (interview, Srinagar, 30 May 2022), but this cannot explain the sharp contraction of Jammu's slum population.

To conclude, while 'Jammu's Hindu belt ... has been the most cosmopolitan part of J&K' with 'the ability to assimilate many cultures and the influx of people from other parts of the State due to economic reasons and conflict' being its 'greatest strength', it 'appears to have reached a saturation point' (Jamwal 2020a). Pervasive over-reporting in Kashmir has meant that despite bearing a much larger burden of migrants, Jammu district's headcount-linked share in the legislature and government spending decreased between 1981 and 2011.[75] In the post-reorganisation period, 'Jammu may not share Kashmir's anxieties of a demographic change', but it harbours 'fears of a large-scale scavenge[r] hunt for business opportunities, land purchase and jobs by outsiders' because its 'geographical, cultural and religious proximity with the Indian heartland' and 'the absence of a violent conflict' make it a 'lucrative destination' (ibid.). Indeed, two years after the repeal of Article 370, the Jammu Chamber of Commerce and Industries called for a day-long shutdown against people from outside, ruling like the East India Company (Ganai 2021).

**Table 4.6b**  Urbanisation in Jammu and Kashmir (J&K), 2001–11

| State/ Region | Sector | 2001 | 2011 | Growth (per cent) | Urbanisation 2001 | Urbanisation 2011 | New towns [population added] |
|---|---|---|---|---|---|---|---|
| J&K | Total | 10,143,700 | 12,541,302 | 23.64 | | | |
| | Rural | 7,627,062 | 9,108,060 | 19.42 | 0.2481 | 0.2738 | 46 (13 MCs + 33 CTs) [360,321] |
| | Urban | 2,516,638 | 3,433,242 | 36.42 | | | |
| Kashmir | Total | 5,476,970 | 6,888,475 | 25.77 | | | |
| | Rural | 4,001,529 | 4,711,096 | 17.73 | 0.2694 | 0.3161 | 28 (12 MCs + 16 CTs) [231,888] |
| | Urban | 1,475,441 | 2,177,379 | 47.57 | | | |
| Jammu | Total | 4,430,191 | 5,378,538 | 21.41 | | | |
| | Rural | 3,428,290 | 4,184,684 | 22.06 | 0.2262 | 0.2220 | 16 (1 MC + 15 CTs) [113,632] |
| | Urban | 1,001,901 | 1,193,854 | 19.16 | | | |
| Ladakh | Total | 236,539 | 274,289 | 15.96 | | | |
| | Rural | 197,243 | 212,280 | 7.62 | 0.1661 | 0.2261 | 2 (2 CTs) [14,801] |
| | Urban | 39,296 | 62,009 | 57.80 | | | |

*Source*: Primary census abstracts (PCA) for 2001 and 2011; appendix 1 of table A-4 of Census of India, 2011.

*Note*: (*a*) MC: municipal committee; CT: census town. (*b*) The decadal growth rate of the urban population of Kashmir and Jammu will be 31.86 per cent and 7.82 per cent, respectively, if we exclude the new towns.

# Ladakh

The small population share of Ladakh meant that it was entirely inconsequential in the demographic politics of the erstwhile state. But this has not meant an absence of demographic tensions. In fact, the Leh–Kargil and Buddhist–Shia binaries are no less polarising than the Kashmir–Jammu and Muslim–Hindu binaries.

Buddhist organisations in Leh allege that Muslims wage 'love jihad' (forcing demographic change by inducing Buddhist girls to marry Muslims) and even otherwise try to convert Buddhists. As a result, 'Buddhist–Muslim intermarriage, once unremarkable, is now forcefully prevented, and Buddhist leaders claim that marriages of Buddhist women to Muslim men are part of a strategy of deliberate demographic aggression' (Smith 2012: 1512). Shia leaders of Kargil point out that one of the recent cases of conversion that deeply embittered Buddhists involved Sunnis, and Srinagar-based leadership's partisan and triumphalist interference in what was a purely local matter made things worse.

Buddhists, in fact, allege that the Muslim-dominated state government in Srinagar used a variety of instruments, including monetary inducements, to promote conversions (interview, Leh, 18 September 2019).[76] However, Buddhist leaders, including those who referred to the threat of 'love jihad' and conversion linked to employment, admitted that the actual numbers of conversion and inter-religious marriages may not be large enough to explain the change in population shares (interviews, Leh, 19 September 2019; Leh, 21 September 2019).[77] It appears that Leh was angered more by Srinagar's partisan interventions than disputes arising out of actual instances of conversion that could have been resolved locally (interviews, Buddhist leader, Leh, 19 September 2019; Buddhist leader, New Delhi, 2 December 2019).

Fertility differences are the real bone of contention in Ladakh. Fertility rates have fallen faster in Buddhist Leh than in Shia Kargil due to delayed exposure of the latter to modern education and employment.[78] According to Guilmoto and Rajan's (2002: 668) estimates based on the 2001 census, the total fertility rate (TFR) (crude birth rate [CBR]) of Leh and Kargil was 1.3 (10.6) and 3.4 (26.7), respectively.[79] The gap between fertility rates has flipped the shares of the two main religions in Ladakh's population (Figures 4.1a–4.1d). The long-feared reversal was first reported in the 2001 Census. The changing population shares triggered demographic anxiety among Buddhists, who believe that Srinagar favoured Kargil in the game of numbers and played divide-and-rule (Buddhist leader, interview, New Delhi, 2 December 2019; former legislator, interview, Jammu, 3 December 2019; see also Bakula 1953; Lok Sabha 2019: 160).[80]

Smith (2009: 203) traces back some of the tensions in Ladakh to the outbreak of armed insurgency in Kashmir that 'may have created economic concerns among Leh residents for the future of tourism in the region, additional economic competition from the influx of Kashmiri traders fleeing Kashmir, and fears that the Srinagar chaos could spread to Leh'. But the census figures for 1981, 2001 and 2011 suggest that there were only 1,145, 871 and 559 Kashmiri speakers in Leh and 331, 1,178 and 2,552 in Kargil, respectively. The decline in Leh's small Kashmiri population between 1981 and 2001 is steeper than indicated by the aforementioned figures because the 1981 data on households and household population by language(s) mainly spoken in the household 'includes houseless households but excludes institutional households' (GoI 1987: 6). The institutional households include the police and army barracks, among others. However, the census may not have captured the alleged influx from Kashmir if traders returned to Kashmir before the 2001 census was conducted in inaccessible areas. In any case, the real problem is not the numbers but the larger context. Older Buddhist interviewees remembered the exodus of Pandits as a very distressing event. Their worst fears almost came true in the summer of 2000, when terrorists killed three monks at the Rangdum monastery (Swami 2000c).[81] They alleged that the plan was to push Buddhists out of Zanskar but was thwarted due to the timely intervention of the community and the army (interview, former legislators, Leh, 21 September 2019 and 20 November 2019; interview, schoolteacher, Kargil, 22 November 2019).[82]

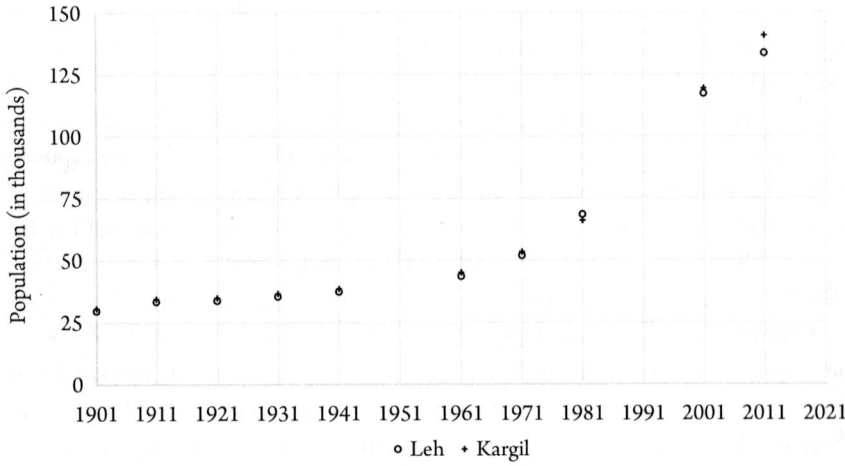

**Figure 4.1a**   Population of Leh and Kargil, 1901–2011

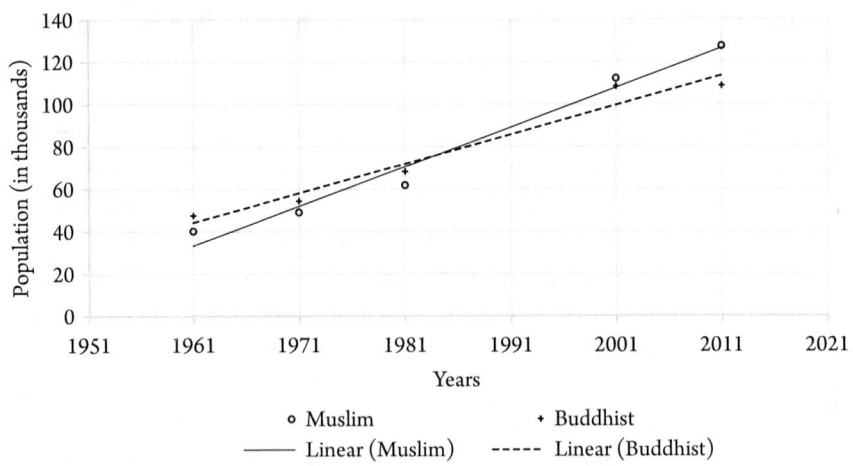

**Figure 4.1b**   Population of Buddhists and Muslims in Ladakh, 1961–2011

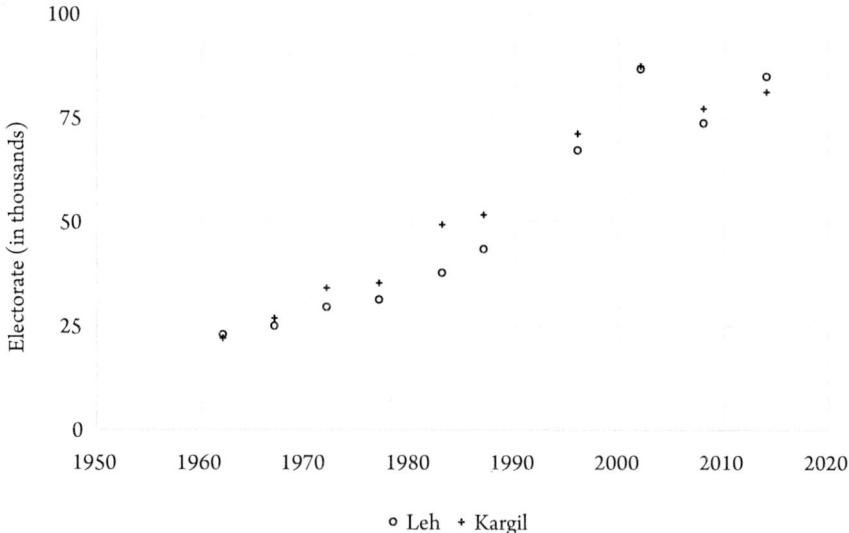

**Figure 4.1c**   Electorate (state assembly) of Leh and Kargil, 1962–2014

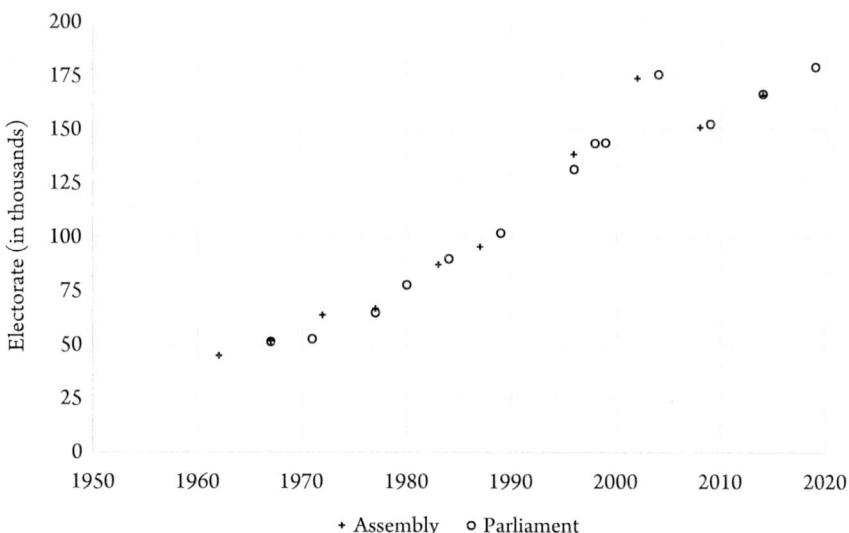

**Figure 4.1d**   Electorate of Ladakh, 1962–2014

*Source*: Table A-2 of Census of India, 2011; Election Commission of India (ECI) reports.

Smith (2012: 1517) suggests that the communal tilt in Leh 'was a strategic decision made after other attempts to secure autonomy [from Srinagar] failed (not the result of long-standing animosity)' with Muslims.[83] In fact, Shia Kargil, too, leaned on Sunni Srinagar to balance Buddhist Leh that seemed to be backed by New Delhi, national parties and Hindu nationalists. The divergent political orientations of the two districts strained their relationship.[84] Leaders on both the sides admit that communal anxieties peaked around elections to the lone parliamentary seat because the two districts have comparable electorates (Figures 4.1a–4.1d). Asgar Ali Karbalaie suggested that the allocation of a parliamentary seat to Ladakh 'increased the competition between the two communities' (Donthi 2019). He argued that otherwise there was 'no deep cause for disputes' that flare up around the parliamentary elections and then subside (interview, Kargil, 20 September 2019; also P. T. Kunzang, interview, Leh, 19 September 2019; Rigzin Zora, interview, Jammu, 3 December 2019). So, as the chief executive councillor of the Ladakh Autonomous Hill Development Council (LAHDC) of Kargil puts it, 'five years of brotherhood disappears during parliamentary elections' (interview, Kargil, 20 September 2019). However, parliamentary elections in Ladakh cannot be reduced to a Buddhist–Muslim binary as there have been several instances of cross-voting (van Beek 2000: 545–46; Swami 2006). An interviewee alleged that Asgar Ali Karbalaie contested against Kargil's consensus candidate in the 2019 Lok Sabha elections and that ensured the victory of the BJP's Jamyang Tsering Namgyal (ex-councillor, interview, 20 November 2019).

Interestingly, some in Leh feel that New Delhi is more comfortable with Muslims rather than Buddhists as the Indian state has a much longer experience of working with the former in the 'mainland' (interview, Leh, 18 September 2019). On the other hand, Kargil believes that it has been neglected because 'Hindu' New Delhi cannot trust Muslims[85] and showers all benefits on Buddhists who follow an 'indigenous' religion and live close to the headquarter of Ladakh that has also benefitted from tourism and better connectivity (interview, Kargil, 22 November 2019).[86] Karbalaie complained, 'The Buddhists of Leh feel that they are dominated by Muslims in the J&K state. But Kargilis feel the same discrimination. The state government thinks we are Muslims, but Shias. The centre thinks we are Ladakhis, but Muslims' (Donthi 2019). This is reminiscent of the complaint of Gujjars and Bakarwals that they are treated as tribes in Kashmir, while in Jammu they are viewed as Muslims. Karbalaie added that they have demanded that the UT should be renamed 'UT of Leh and Kargil instead of Ladakh, because Leh people are misusing Ladakh. Everything that is given to Ladakh, goes to Leh. They show Leh as Ladakh' (ibid.).

We are witnessing a status conflict between ethnolinguistic twins in Ladakh, with neither hesitating to occasionally rely on external help. As an interviewee put it, 'Kargil was a halting place, our equivalent was Skardu. The Valley *raised* Kargil to a higher status' (interview, Leh, 18 September 2019, emphasis added). A Buddhist leader lamented that Leh was not paying attention to the fact that both the population and the economy of Kargil were growing faster (interview, New Delhi, 2 December 2019). That the conflict between the two districts is essentially about parity is reflected in demands such as renaming the UT as 'UT of Leh and Kargil' like Daman and Diu, J&K, Andaman and Nicobar, and Dadra and Nagar Haveli (Asgar Ali Karbalaie, interview, Kargil, 22 November 2019). Kargil is demanding a redefinition of the symbolic space hitherto dominated by Leh, and this battle is being waged in all possible spaces including signboards installed by the Border Roads Organisation – 'Welcome to the land of lamas', 'Don't be a Gama in the land of Lama' – that invariably introduce Ladakh as a Buddhist territory.

## National Politics

Census appears to be 'just a pretext for a religious war, led by the competing elites of Srinagar and Jammu' (Swami 2000a) and has been 'used as a weapon both by the militant organizations as well as by the State Government to show their strength' (Bhat 2018: 181). The 'war' is, however, not confined within the state's border and has a disproportionately large national footprint. At one level, both the partisans of Kashmir and Jammu alike distrust national parties. Kashmir is convinced that national parties, cutting across ideological divides, are consistent in their suspicion of (Muslim) Kashmir and promote the interests of (Hindu) Jammu (Abdullah 2016: 375–76, 556). While alerting Kashmiris towards New Delhi's plans to alter Kashmir's demography, Syed Ali Shah Geelani claimed that 'Congress has always been a collaborator in BJP's scheming against Jammu and Kashmir' (Ashiq 2010). Likewise, Jammu believes that the INC has always appeased the separatist Kashmiri Muslims. Interviewees in Jammu argued that since the mid-1990s the BJP and its mentor, the RSS, have adopted a similar approach (interview, Jammu, 21 September 2020; Jammu, 21 July 2021; note 95 of this chapter).[87] As discussed earlier, in Ladakh, too, both Kargil and Leh believe that the other side is favoured by New Delhi. These, though, are not the only links between communal contests over demography at the local and national levels that were first established in the late nineteenth century.

Some of the earliest questions regarding Kashmir's demography can be traced to the Punjab-based Arya Samaj, which played a major role in the communalisation of demography across colonial north India. Arya Samajis expressed concern about the falling population share of Hindus in the princely state of J&K.[88] Lala Jai Chandra of the Punjab Arya Samaj published the tract *Hamare Bichchre Hue Bhai* (Our Lost Brethren) in 1898, which pointed out that Kashmir that was 'once populated by Hindus' was 'no longer so as the census indicated' (Jones 1981: 88).[89] 'Written in the 1890s, this was one of the first tracts to employ census data as concrete "proof" of a particular argument' (ibid.).[90] This was also the decade when the Arya Samaj reached Kashmir with the maharaja's Punjabi employees, and their activities seem to have eventually included 'converting people of other religions including Muslim to Hinduism' (Kaur 1996: 56).[91] Arya Samajis were among the senior officers of the princely state around the time of independence (Abdullah 2016: 281, 284). While the foregrounding of a particular understanding of 'Hindu' in Arya Samaji propaganda was ideological in nature, there was an instrumental aspect as well because Pandits resented the competition in the job market and claimed Kashmir for Kashmiris, which eventually led to the introduction of the category of 'state subject'.[92]

Leh's demographic anxieties, too, have Arya Samaji linkages. The influence of Rahul Sankrityayan, who visited Ladakh in 1926 and 1933, is noteworthy. He 'had discussed with prominent religious leaders ... the dangers of growing numbers of Muslims and the low birth rate among Buddhists due to monasticism and polyandry' (van Beek 2004: 200; see also Sankrityayan 1939: 70–76, 88, 93). Around this time Kashmiri Muslims began to articulate their claims *qua* 'majority' using statistics (Abdullah 2016: 40, 43, 50, 55, 102, 108). Muslim notables of Kashmir submitted a memorandum to the viceroy in 1924 that called for, among other things, 'grant of Muslim representation in the State Council according to the number of Muslims in the population of the State' (Kaur 1996: 127). In the following decade, the majority began to be visualised in statistical terms when, for instance, the Kashmir Day resolutions sought '70 percent of the appointments in the State services to the Muslims' (ibid.: 153). Muslims accounted for more than 70 per cent of the population of J&K in the late colonial period. Notably, newspapers run by Muslims in Punjab, which were 'patronized by the British to intimidate and pressurize the new Maharaja [Hari Singh]', foregrounded the 'predominantly Muslim' character of the princely state and the oppression of Muslims (ibid.: 134; see also Devadas 2007: 23). Sheikh Abdullah notes that Muslims 'from outside the state' had 'helped a great deal', but he blames key groups such as the All-India Majlis-e-Ahrar and the Ahmadiyas for using Kashmir to further their agenda in the rest of the country (Abdullah 2016: 110–15). This also applies to the activities of Punjabi Hindus in Kashmir. So Lahore served as the site for the initial entanglement of J&K with national debates.

The Hindi–Urdu controversy provided another connection between the census politics in British India and princely J&K (GoJK 1943a: 12–13, 301),[93] which again had roots in Punjab. The interlinkages between the local and national debates are also evident from the colonial administration's concern about the impact of riots in Kashmir on the 'communal situation in British India' (Devadas 2007: 23).

After independence, the belief of the founding fathers that Muslim-majority Kashmir, as discussed in Chapter 1, was essential for Indian secularism facilitated the intertwining of the local and national debates on demography. J&K emerged as a cause celebre among the Hindu nationalists across the country only in the late 1940s, even though by the early 1940s the princely state was already a part of the propaganda on Pakistan. A week after the accession of the princely state, *Organiser*, an English daily published by the RSS, carried a piece by Satyananda Shastri that dwelled upon the princely state's natural resources and pointed out that 'Kashmir has vast space for the expansion of population. If industrialized, it can absorb crores of people' (Shastri 1947). He added, 'Not a thousand years back, it was a seat of Hindu culture and learning. At every step in Kashmir there lie our worship and a sacred place. If it goes, it will injure our feeling of pride and self-respect.' Noorani (2015) pointed out that Shastri 'mentioned everything about the State – except its people' (also K. B. Ahmad 2017: 288). Shastri's counterparts in the Pakistan movement (Dhulipala 2015: 160, 343, 379) were also primarily concerned with natural resources and strategic advantages offered by the princely state and referred to the Muslims of J&K only as a statistically quantifiable, homogenous and additively separable entity that could be freely appended to Pakistan to increase its area, resource base and share of Muslim population.[94]

In 1953, Syama Prasad Mukherjee, the founder of the Bharatiya Jana Sangh, the precursor to the BJP, died in custody while in Srinagar. This event permanently etched Kashmir in the self-imagination of Hindu nationalists. However, irrespective of this history, the Sangh Parivar would have been drawn to the state that was identified by the Nehruvian elite as one of the foremost sites where India's secularism would be validated.[95]

The exodus of Kashmiri Pandits and other Hindu communities from Kashmir in the early 1990s finally catapulted J&K's demography into the national political arena. Ever since Kashmir has been an integral part of the national politics of numbers. The complete ouster of communities belonging to the nationally dominant religion from a province dominated by a minority religion rocked the national politics[96] just when the country was grappling with the fallout of the Shah Bano case, ban on Salman Rushdie's *Satanic Verses* and the increasingly violent mobilisation around the Babri Mosque. Hindu nationalists view J&K as a wobbling domino, whose fall will have catastrophic territorial and demographic consequences for the cradle of Hinduism and the only large Hindu-majority country.

An exchange in the *Indian Express* after the reorganisation of the state illuminates the intertwining of Kashmir's demography with national politics. Writing after the reorganisation, the editor argued that the erstwhile state's 'unique demography [was] protected by the Constitution' (*Indian Express* 2019). A year later, another editorial added that the government has done 'nothing to assure the people of the former state that their concerns on demographic change are being taken into consideration' (*Indian Express* 2020). The *Indian Express*, though, lacked clarity about J&K's demography that its editorial wanted the government to preserve. This is evident from a piece by a staff member aimed at demystifying the numbers. Among other things, it wrongly claimed that '[t]he Census of 2011 showed that the religious make-up of the erstwhile state of Jammu and Kashmir had remained almost entirely unchanged over the previous half century' (Shaikh 2020). An opinion piece in response to the editorials laid out the 'Hindu' perspective:

> For the mainstream discourse, the demography of Jammu is not a matter of concern. It is only Kashmir where this issue is deemed legitimate. When concerns about Kashmir's demography are raised, it is clearly about the percentage of Muslims in the region.... For the sake of the nation's collective memory, there may also be a need to mention that the only forcible change in Kashmir's demography so far has come with the barrel of the gun pointed at the Hindus, making them flee the Valley ... demographic concerns of the Jammu region had rarely been discussed. (Ambale 2020)[97]

Ambale asks, if an almost entirely Muslim Kashmir 'can be concerned about demographic change by the people of a religion that does not even proselytise', can Hindus facing 'proselytising religions' 'not have legitimate demographic concerns'? He then draws attention to the national implications of the liberal 'concern only for one region's demography':

> Are some demographic changes better or worse than others.... If a mainstream editorial is asking for legal safeguards for Kashmir's demographic character, shouldn't it be logical that India, too, may need an anti-conversion law that protects it from demographic change? Then, should foreign-funded missionary activities in India not be restricted, because, clearly, the money comes from somewhere else to change India's demographic character? (Ibid.)

He finally adds the international implications:

> Demographic aggression is a reality, especially in Pakistan and Bangladesh. Since we all accept that the fallout of demographic realities can be disastrous for those at the receiving end, then what is so wrong about making India a safe haven for demographically-threatened minorities from Pakistan and Bangladesh? Also, in the same context, isn't the NRC [National Register of Citizens] a legitimate exercise? Wouldn't Assamese have genuine demographic concerns, especially given that the magnitude of demographic change the state has seen is massive? (Ibid.)

In short, Ambale argues that Hindus are a minority both in J&K and the world, and therefore they require greater degrees of freedom to counteract the dual demographic threat from within and without the borders (see also Prakash 1979: 16 for an earlier iteration of this argument). So while Kashmiri Muslims allege that the Indian state is determined to alter the state's religious demography,[98] their Hindu counterparts feel that the state has aided the champions of 'Greater Kashmir'. The Hindu nationalists argue that while 'Indian religionists' are able to defend themselves in the heartland, 'the vigilance and support of a state committed to protecting and preserving the civilizational identity, pride and genius of the nation' is needed along the borders, but '[w]e have so far failed to fashion such a state for ourselves' (Joshi, Srinivas and Bajaj 2003: 175).[99]

Both the *Indian Express* and Ambale overlook the fact that Kashmir is not the only region of the country concerned about potential change in its ethnic composition. Insider–outsider conflicts and calls for regulating the inflow of ethnolinguistic 'others' are commonplace across the north-east (Ambale restricts himself to one strand of Assam's concerns) as well as economically dynamic southern and western states. More importantly, as discussed at length in Chapter 6, it is not the case that the Indian government does not intervene to preserve the relative population shares of states or the ethnolinguistic composition of minority dominated provinces.[100]

## Delimitation Politics

The interlocked anxieties across the three tiers discussed earlier affect key policies such as delimitation. The belief that New Delhi had offered Kashmir a carte blanche to appease the separatist sentiment was the point of departure for debates on delimitation in Jammu.[101] It was argued that New Delhi allowed Srinagar to manipulate census and delimitation and enhance its grip on political power at the expense of other regions. These concerns go back to the 1950s.

J&K was represented in the Constituent Assembly of India by four members of the NC (Abdullah 2016: 357; Rai 2018). In 1951, Sheikh Abdullah-led NC convened a constituent assembly for J&K. There was no reliable guide for apportioning seats as census could not be conducted in 1951, and the 1941 data were not of much help due to the significant redistribution of population after 1947. The NC manipulated the regional distribution of seats of the Constituent Assembly. 'Constituent Assembly of seventy-five deputies ... comprised forty-three representatives from the Kashmir Valley, thirty from the Jammu region, and two from Ladakh. Twenty-five additional seats were left vacant for the areas of Pakistan-controlled Kashmir' (S. Bose 2003: 55). The NC 'managed to "win" all seventy-five seats' in the Constituent Assembly (S. Bose 2007: 171).[102]

The fairness of the seat allocation can be *retrospectively* checked against the 1961 census, the first post-1947 headcount of the state. As per the 1961 census, 40 out of 75 seats should have been allotted to Kashmir (instead of the 43 allotted to it) and 33 seats should have been allotted to Jammu, with the remaining two going to Ladakh.

Ved Bhasin suggested that Abdullah 'did not tolerate any opposition. He crushed the freedom of press. He and other NC leaders did not tolerate any voice of dissent. He acted as an authoritarian ruler. The constituent assembly elections of 1951 were totally rigged. Personally I believe that if the elections were held in a fair manner NC would have still secured a two third majority. There was no threat to NC. May be 15 seats had gone to opposition – five to Praja Parishad in Jammu, another 10 or 15 to pro Pakistan elements in Kashmir. If that had happened, then the situation in J&K would have been different. That constituent assembly, in that case, would have been representative of the people' (S. Khan 2009; see also B. Puri 1983a: 189; Fayyaz 2021).

Abdullah (2016: 364–65) defended himself by claiming that the Pakistan loyalists and the Praja Parishad either did not field candidates or withdrew them on flimsy grounds. He also deflected the blame, if any, to Ghulam Mohammad Bakshi, a protégé and close confidante, until he replaced Sheikh as the prime minister in 1953, who was 'in charge of the election campaign had Jammu': 'It may be that some electoral irregularities were committed there – a fact not unthinkable, given Bakshi's character' (ibid.: 365). Whatever may have been the factors behind the manipulation, '[t]he manner in which this election was conducted made a mockery of any pretence of a democratic process, and set a grim precedent for future "free and fair elections"' (S. Bose 2003: 55).

In a letter to Nehru written on 27 January 1954, Sadr-e-Riyasat Karan Singh highlighted the 'considerable dissatisfaction at the manner in which the previous [Abdullah] regime carved out constituencies' because '[i]t is alleged that they did so … to keep one community in a majority wherever possible regardless of such factors as geographical, cultural and ethnic affinities' (Alam 2006: 153). He also requested Nehru to entrust delimitation to 'the Delimitation Commission which has been appointed to do the job in the rest of India'. Two years later, in another letter to Nehru, he suggested that at least the delimitation of parliamentary seats could be entrusted to the 'Indian Delimitation Commission' (ibid.: 201).[103]

In the first legislative assembly elections held in 1962, the regional distribution of seats followed the precedent of the Constituent Assembly. This distribution changed after the 1966 delimitation when a seat was transferred from Kashmir to Jammu. In 1976, under a fresh delimitation, 1 out of 25 seats of Pakistan-occupied territories was transferred to Jammu, increasing its seats to 32, which was still less than its entitlement of 34 seats as per the 1971 and 1981 censuses. This new allocation held until 1995 when another delimitation increased the strength of the assembly to 87 seats[104] that were distributed as follows across the regions: Kashmir (46), Jammu (37) and Ladakh (4).

Between 1951 and 1995, Jammu's seat allocation was at least two less than suggested by its population share (Table 4.7).[105] Since census could not be conducted in 1991, the fairness of the 1995 delimitation can be checked against the 1981 or 2001 census. As per the 2001 census, Kashmir's share was 46.97 (47) out of 87 seats. The entitlement of Jammu and Ladakh was 38.00 (38) and 2.03 (2), respectively. Following the 2001 census, seat allocation of both Kashmir and Jammu was one less than warranted by the population share, which they lost to Ladakh. It is interesting that the 2001 census almost offered an *ex-post* justification for the 1995 delimitation. However, in 1995, the Delimitation Commission could *not* have used data published six years later. It would have used data available at the time of delimitation – that is, either the 1981 census data or the projected figures reported in, say, *The Digest of Statistics* of 1991–92 (GoJK 1993). As per the 1981 census, Kashmir's share should have been 45.55 seats (rounded off to 46) with Jammu and Ladakh being entitled to 39.50 (39) seats and 1.95 (2) seats, respectively. Kashmir's share was 45.20 (45) according to the projected figures for 1991. The corresponding entitlements of Jammu and Ladakh were 39.88 (40) and 1.92 (2), respectively. From the perspective of the 1981 census, Jammu lost two seats to Ladakh in the 1995 delimitation. Likewise, following the projected figures for 1991, it lost two seats to Ladakh and one to Kashmir. Actually, both Kashmir and Jammu should have shared the burden of accommodating additional seats for Ladakh. Following this criterion, in 1995 Jammu lost one seat each to Ladakh and Kashmir as per the 1981 figures, and one seat to Ladakh and two to Kashmir if the projected figures for 1991 were used. Since delimitation was not carried out in 2002, the 1995 delimitation

**Table 4.7**  Delimitation of state assembly constituencies, Jammu and Kashmir (J&K)

| Period | Reference year† | Total seats | Kashmir | | Jammu | | Ladakh | | Rest |
|---|---|---|---|---|---|---|---|---|---|
| | | | Seats | Loss | Seats | Loss | Seats | Loss | |
| Constituent Assembly and 1962 | Actual | 100 | 43 | | 30 | | 2 | | 25 |
| | 1961 | | 40.01 (40) | −3 | 33.13 (33) | 3 | 1.87 (2) | 0 | |
| 1966–75 | Actual | 100 | 42 | | 31 | | 2 | | 25 |
| | 1961 | | 40.01 (40) | −2 | 33.13 (33) | 2 | 1.87 (2) | 0 | |
| | 1971* | | 39.57 (39) | −3 | 33.72 (34) | 3 | 1.71 (2) | 0 | |
| 1976–95 | Actual | 100 | 42 | | 32 | | 2 | | 24 |
| | 1971 | | 40.10 (40) | −2 | 34.17 (34) | 2 | 1.73 (2) | 0 | |
| | 1981** | | 39.79 (40) | −2 | 34.50 (34) | 2 | 1.71 (2) | 0 | |
| 1996–2019 | Actual | 111 | 46 | | 37 | | 4 | | 24 |
| | 1981*** | | 45.55 (46) | 0 (−1) | 39.50 (39) | 2 (1) | 1.95 (2) | −2 | |
| | 1991‡ | | 45.20 (45) | −1 (−2) | 39.88 (40) | 3 (2) | 1.92 (2) | −2 | |
| | 2001 | | 46.97 (47) | 1 (0) | 38.00 (38) | 1 (0) | 2.03 (2) | −2 | |
| | 2011 | | 47.79 (48) | 2 (1) | 37.31 (37) | 0 (−1) | 1.90 (2) | −2 | |

*Source*: GoJK (2007); respective Election Commission of India (ECI) and census reports.

*Note*: (*a*) 'Total seats' is the sum of the seats of the three regions. 'Rest' refers to seats nominally assigned to the Pakistan-occupied territories. (*b*) † 'Reference year' indicates the source of the population data used to calculate the allocation of seats. Except in the case of 1991, the census is the source. 'Actual' indicates the seats in the assembly. (*c*) See ECI (2008b: 535–36) for the treatment of fractions. Following the Delimitation Commission, (*i*) *33.72 is rounded off to 34 and 39.57 to 39, (*ii*) **39.79 is rounded off to 40 and 34.50 to 34 and (*iii*) ***45.55 is rounded off here to 46 and 39.50 to 39. In each of these cases, the fraction in the first number is larger. (*d*) 'Loss' is the difference between estimated seats as per the census and the actual seat allocation. A negative loss indicates a gain. (*e*) Ladakh's population share cannot support more than two seats. In 1995, two additional seats were created for Ladakh. Figures within parentheses indicate the loss, assuming both Kashmir and Jammu sacrifice a seat each for Ladakh. (*f*) ‡ We have used projections given in GoJK (1993: 3) for 1991 as the census was not conducted in that year. We can also use the figures given in table A-2 of the 2001 census, but that will not affect our conclusions. (*g*) Constituencies were formally delimited in 1975 and 1995.

governed the assembly elections until 2014. Further, note that as per the flawed 2011 census, Kashmir's share was 47.79 (48) out of 87 seats. The entitlement of Jammu and Ladakh was 37.31 (37) and 1.90 (2), respectively – that is, Kashmir had to sacrifice two seats: one for Ladakh and another for Jammu. Before 2011, Jammu's share in the state assembly was always less than its population share (Table 4.7).

Several interviewees (Jammu, 30 June 2018, 9 May 2022) claimed that Jammu's seat share has not only been less than its reported population share but also less than its share in the electorate (see also Sharma and Sengupta 2008; Swami 2002). Talib (2010: 3–4), though, questions Jammu's electoral rolls:

> Jammu and Kashmir's chief electoral officer BR Sharma recently made a significant statement when he said that the latest revised electoral rolls show that the number of voters in Kashmir is 32 lakh and that in Jammu around 30 lakh. In 11 assembly segments in Jammu, 94,000 bogus voters were found and their names deleted. It was, however, not explained why Kashmir, despite having nearly 1.3 million people more than Jammu, had only about 2 lakh more voters. It is common knowledge that an unspecified number of eligible voters in Kashmir are not registered either due to their disinterest in the democratic process or administrative lethargy. Whatever be the case, by modest estimates, no less than a million voters in Kashmir are missing from the electoral rolls, even if one takes the 2001 census figures as the base line data.

Interviewees based in Jammu turned Talib's argument on its head and asked how Kashmir could have a much larger population if the gap between the electorate size of the two regions was small (interviews, Jammu, 2 June 2016 and 30 June 2018).[106] During 1977–87 (1961–72 and 1996–2019), the largest constituencies of the state were located in Kashmir (Jammu) (Figures 4.2a–4.2c and 4.3). In 2002, Jammu's total electorate was about 8 per cent larger than that of Kashmir.[107] Also, after 1987, the mean constituency size, both assembly and parliamentary, was always larger in Jammu compared to Kashmir. For both parliamentary and assembly seats, the ratio of the electorate (registered voters) to the census population of Kashmir and Jammu has continued to diverge since the early 1980s (Figures 4.4a and 4.4b). The ratio of the electorate per seat of Jammu to Kashmir has also increased (Figures 4.4c and 4.4d). So the relative voting power of Jammu has continuously eroded as the number of registered voters in Jammu per seat has been growing in comparison to Kashmir. Interestingly, the trend persisted even after the delimitation in the mid-1995, which is quite unusual because the exercise was aimed at equalising the ratio of the electorate

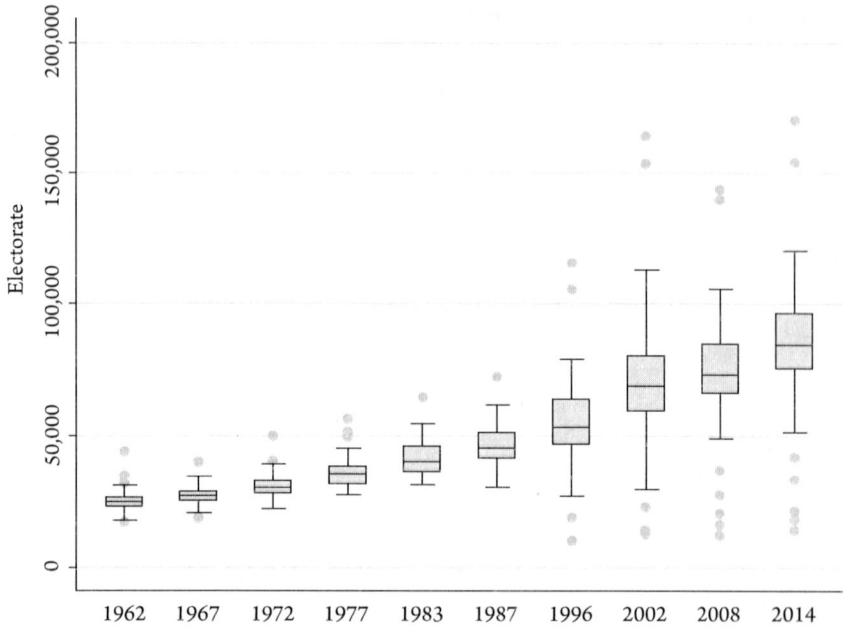

**Figure 4.2a**   Distribution of the electorate size of assembly constituencies, Jammu and Kashmir (J&K), 1962–2014

*Source*: Prepared by the author using Election Commission of India (ECI) reports.

*Note*: (*a*) In 1967 and 1983, the state's electorate mentioned in the ECI reports does not tally with the total calculated using the assembly constituency data. We have used the latter. The difference between the two figures is 592,798 and 72 in 1967 and 1983, respectively. (*b*) The 1983 figure excludes Doda (constituency no. 48) as the result was withheld by the High Court of J&K. (*c*) In 2002, two candidates won unopposed. The electorate reported here includes the respective constituencies.

to the population and should have arrested differences in the short run.[108] Jammu's complaint regarding the delimitation was not limited to assembly constituencies. It also questioned the distribution of *panchayat*s, or local-body constituencies, as there were 'lesser number of Panchayats in Jammu province as compared to Kashmir despite having more electors than it' (*Kashmir Times* 2011e).

It is not the case that flawed delimitations have not adversely affected sections of Kashmir. Devadas (2019a: 380) points out that 'the mirwaiz's support base in Srinagar's inner city had been split into three constituencies' through delimitation. Several Gujjar interviewees complained that their villages in Kashmir are divided and combined with *panchayat*s dominated by Kashmiris. We have already noted the complaint of Pandits about the gerrymandering of Habbakadal constituency

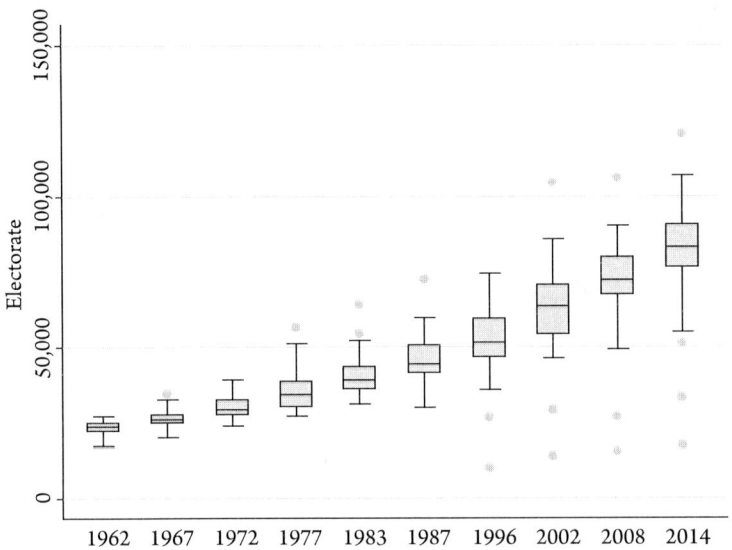

**Figure 4.2b** Distribution of the electorate size of assembly constituencies, Jammu division, 1962–2014

*Source*: Prepared by the author using Election Commission of India (ECI) reports.

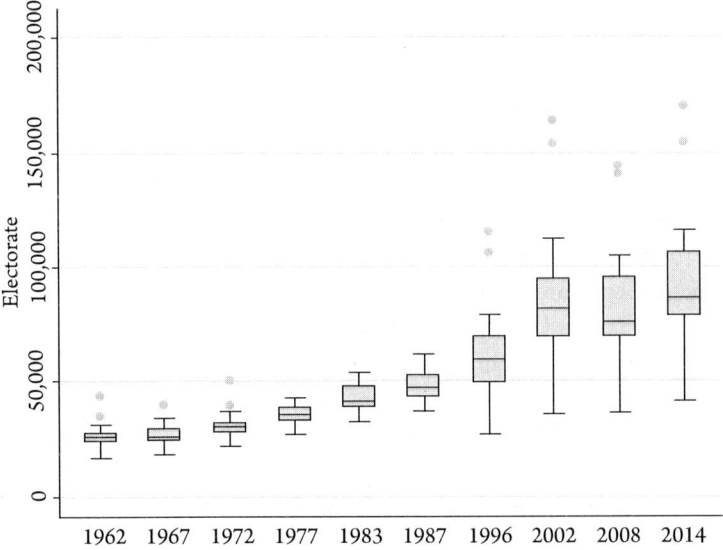

**Figure 4.2c** Distribution of the electorate size of assembly constituencies, Kashmir division, 1962–2014

*Source*: Prepared by the author using Election Commission of India (ECI) reports.

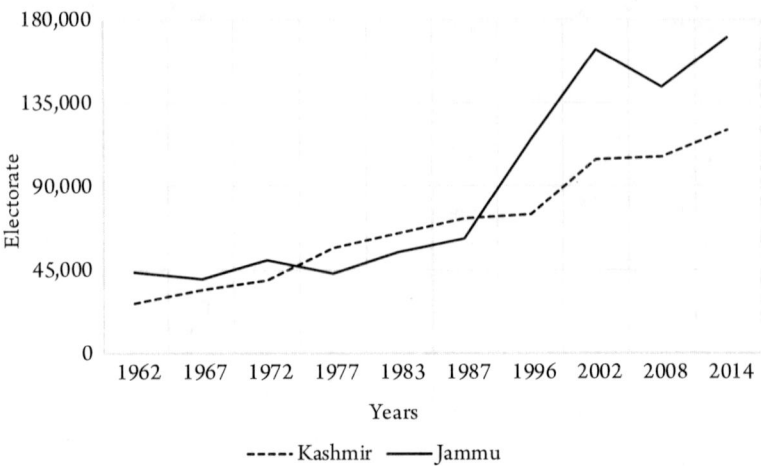

**Figure 4.3**   Largest assembly constituency in Jammu and Kashmir (J&K), 1962–2014
*Source*: Prepared by the author using Election Commission of India (ECI) reports.

in Kashmir. It is also not the case that the dominant groups in other regions have not benefitted from flawed delimitations. The north–south orientation of the Lok Sabha constituencies in the Jammu region has meant that the relatively less dense hilly districts are divided into two parts that are combined with densely populated plains (Map 4.1). This has ensured that only candidates with a strong support base in the plains can win the parliamentary elections in Jammu. Indeed, over the past seven decades, Jammu has never elected a Muslim to the parliament, and Kashmir-based parties have never won any parliamentary election in the division either. Chowdhary Talib Hussain of the NC, who won the 2002 by-election to the Jammu parliamentary seat, is an exception. The latest delimitation has made it even more difficult for Muslims to win elections to the Lok Sabha from the Jammu and Udhampur constituencies.

Ladakh did not complain against under-representation in the state assembly, but that did not mean an absence of conflicts over delimitation. van Beek (2004: 215–16) pointed out that '[t]he fact that the voters' list in Kargil has more people on it than that of Leh, despite the latter's slightly larger population, is taken as further evidence of Muslim manipulations aimed at dominating the Buddhists'.[109] A more enduring conflict is related to the smaller assembly constituencies of Ladakh – Nubra (Leh) and Zanskar (Kargil) – that are also home to the religious minorities of the respective districts. Buddhists complain that the Zanskar constituency was created in 1996 for what is perhaps one of the most remote

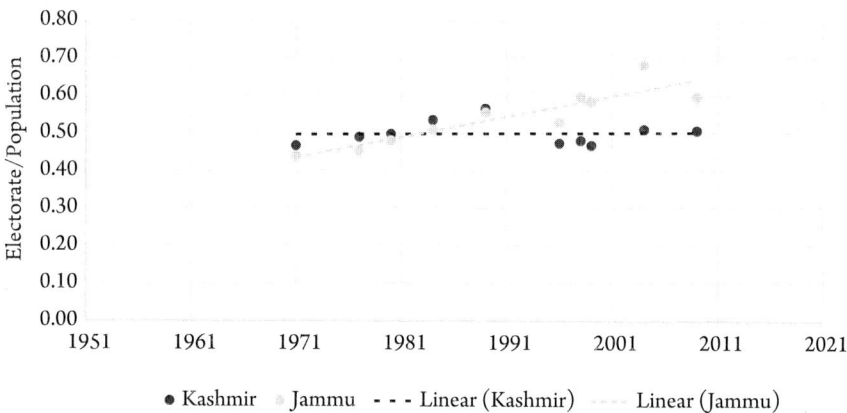

**Figure 4.4a** Ratio of the electorate to the census population, Kashmir and Jammu divisions: parliamentary elections, 1971–2009

*Source*: Prepared by the author using Election Commission of India (ECI) reports and table A-2 of Census of India, 2011.

*Note*: (*a*) The figure depicts the ratio of the electorate for parliamentary elections to census population. (*b*) Unlike other states, in J&K voter eligibility differed between parliamentary and assembly elections. The electoral roll of parliamentary constituencies aggregated the electoral rolls of the constituent-assembly constituencies and added those who were not eligible to vote in the assembly elections. (*c*) We have not included the 1967 parliamentary elections because of the unavailability data for Anantnag constituency. (*d*) For the break-up of the electorate by parliamentary constituencies in 2004, we use the second volume of the ECI report because the first does not have the constituency-wise information. There is a minor difference between the two sources.

regions of the country, but it has never elected a Buddhist because the state's Delimitation Commission deliberately added a large Muslim population to ensure that Buddhists can never win (Dorje 2019; executive councillor, LAHDC [Kargil], interview, Kargil, 20 September 2019; schoolteacher, interview, Kargil, 22 November 2019).[110] Indeed, only once did a Buddhist manage to emerge as the runner-up. In 1996, both the Nubra and Gurez constituencies were half the size of Zanskar in terms of the electorate. In other words, a smaller constituency limited to Zanskar was not inconceivable, and it was not necessary to add the geographically separate Muslim-dominated Sanku. Zanskar Buddhists complain that they are also under-represented in the LAHDC of Kargil. They boycotted the first election to the LAHDC of Kargil in 2003 to protest the allocation of only three seats to Zanskar (*The Tribune* 2003).

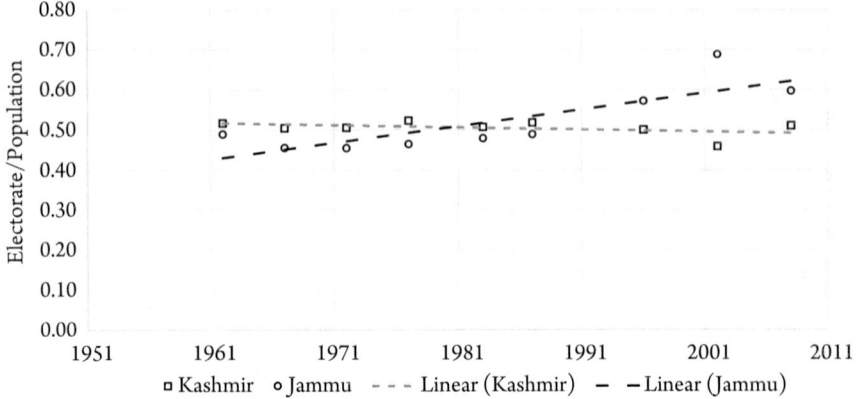

**Figure 4.4b** Ratio of the electorate to the census population, Kashmir and Jammu divisions: assembly elections, 1962–2008

*Source*: Prepared by the author using Election Commission of India (ECI) reports and table A-2 of Census of India, 2011.

*Note*: See the note for Figure 4.2a.

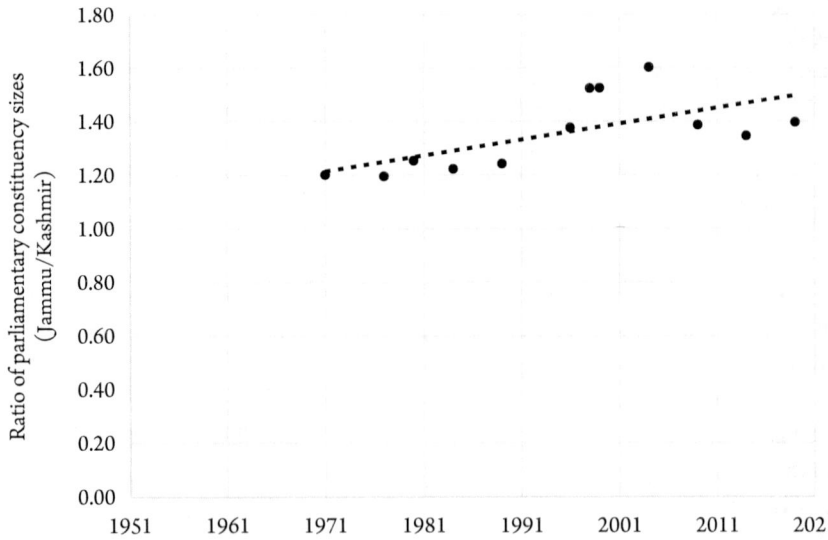

**Figure 4.4c** Ratio of the mean parliamentary constituency size of Jammu to Kashmir, 1971–2019

*Source*: Prepared by the author using Election Commission of India (ECI) reports.

*Note*: (*a*) The figure plots the ratio of the mean sizes of parliamentary constituencies (in terms of electorate) of Jammu division to Kashmir division. Values above 1 are indicative of the mean size of constituencies being larger in Jammu than in Kashmir. (*b*) See also Figure 4.4a.

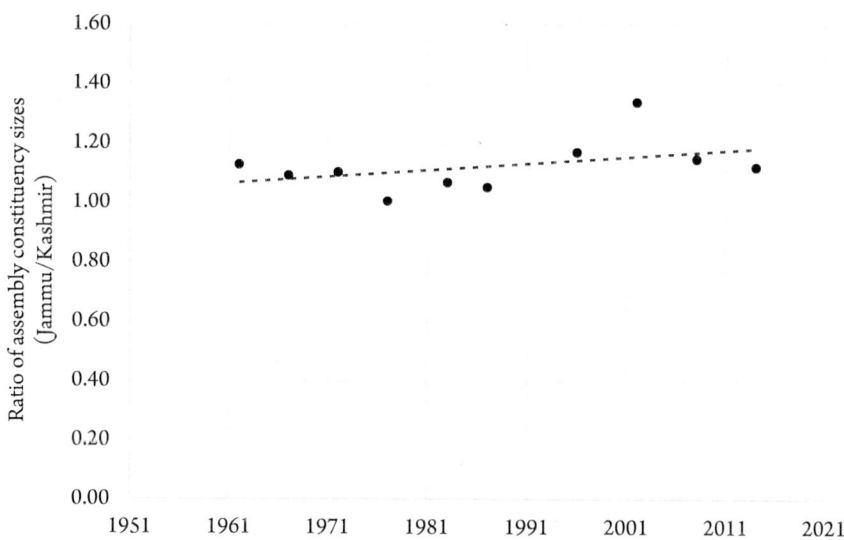

**Figure 4.4d**   Ratio of the mean assembly constituency size of Jammu to Kashmir

*Source*: Prepared by the author using Election Commission of India (ECI) reports.

*Note*: (*a*) The figure plots the ratio of the mean sizes of assembly constituencies (in terms of electorate) of Jammu division to Kashmir division. Values above 1 are indicative of the mean size of assembly constituencies being larger in Jammu than Kashmir. (*b*) See also Figure 4.2a.

The question of delimitation of constituencies of J&K was reopened after the reorganisation. In the past, delimitation acts did not apply to J&K (Election Commission India [ECI] 2008a: 20). In April 2002, the assembly amended the state's constitution and froze delimitation until after the first census conducted after 2026 (Talib 2010: 3). A legal challenge to this amendment was rejected by the court (*National Panthers Party v. The Union of India & Ors. 2010*). The Jammu and Kashmir Reorganisation Act, 2019, however, allowed the government to conduct a fresh delimitation. We will rely on Haseeb Drabu and Hari Om to understand the regional divide over the latest delimitation.

Haseeb Drabu, a former minister of finance of J&K, questioned the decision to delimit constituencies amidst a nationwide freeze on the exercise until after the first post-2026 census (Drabu 2020b). A fresh delimitation was, however, required due to the change in borders, the introduction of reservation for STs in the assembly, the extension of the right to vote in assembly elections to the West Pakistan refugees and the increase in the number of constituencies. Arunachal Pradesh and Mizoram were delimited after they attainted statehood in the

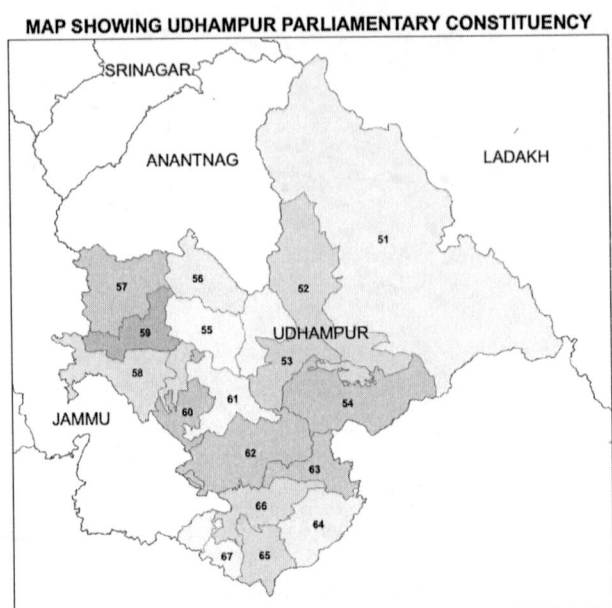

MAP SHOWING UDHAMPUR PARLIAMENTARY CONSTITUENCY

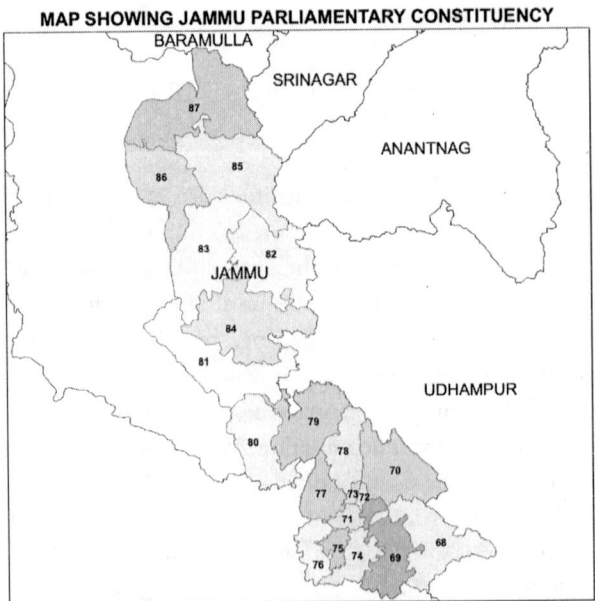

MAP SHOWING JAMMU PARLIAMENTARY CONSTITUENCY

**Map 4.1**   Parliamentary constituencies of the Jammu division, 2019

*Source*: 'Parliamentary Constituency Maps', https://ceojk.nic.in/PC_Maps.htm (accessed on 26 December 2022).

*Note*: Maps not to scale and do not represent authentic international boundaries.

1980s (Kumar 2022b). Likewise, Delhi was delimited after it was recognised as the National Capital Territory, with its own legislative assembly (Sivaramakrishnan 1997: 3280). Parliamentary seats were redistributed on ad hoc basis when three new states were formed in 2000 (McMillan 2001: 1272). In all these cases, delimitation was conducted amidst a nationwide freeze. However, Jammu and Kashmir's reorganisation has not affected the borders of any existing constituency. So it can be argued that it was pointless to delimit seats just before the next census – that too in the middle of a pandemic when it was difficult to organise public hearings. The NC in its note of dissent submitted to the Delimitation Commission added another objection – namely the increase in the number of constituencies was not followed by delimitation when Andhra Pradesh was reorganised into two states (GoI 2022: 24).

Drabu then argued that given the lack of clear guidelines for demarcating constituencies and the absence of judicial remedy, Kashmiris fear that 'constituency redefinition can provide ample scope for manipulation along communal contours in such a manner that the advantages of the majority are diluted' (ibid.).[111] He finally questioned the received wisdom that non-Muslims were under-represented in the erstwhile legislature. He concluded that 'it is Kashmir that has been discriminated against', which contrasts with the under-representation of Muslims in Hindu-majority states like Assam, West Bengal, Kerala and Uttar Pradesh.[112] He added that 'J&K stands out as a shining example of egalitarianism vis a vis its minorities, be they Buddhist or Hindu' (Drabu 2020c). This, as discussed earlier, has been a fixed point of Kashmiri Muslim self-image since the late 1940s.

Drabu extended his claim regarding the absence of discrimination against non-Muslims by incorporating area as a criterion for delimitation. He noted that while 'using area as a criteria violates the basic concept of representation' and 'flies in the face of all established national and international practices', his conclusions held good even if we take into account area (qua population density) (Drabu 2020d, 2020e, 2021b). So 'contrary to the perception ... distribution across the two divisions, no matter how hard one looks at the data, is surprisingly fair and egalitarian' (Drabu 2020e). Drabu used uncorrected population figures of the 2011 census in his analysis, and the weights used to account for area are not self-evident. As discussed earlier, Jammu's representation in the assembly has always been less than its population share except after 2011 due to over-reporting of Kashmir's population (Table 4.7).

Drabu (2021a) summarised his argument in a national daily. Curiously, he did not repeat his claim about the under-representation of Muslims in India.

He did not discuss religion at all and confined himself to the regional distribution. Indeed, his claim about the over-representation of Buddhists in J&K is pointless. Using the over-representation of geographically isolated micro-minorities such as Buddhists of Ladakh, Muslims of Lakshadweep and Christians of Mizoram, New Delhi can likewise claim a moral high ground. But micro-minorities are over-represented in most democracies due to the allocation of a minimum number of seats to them as their population is much smaller than the mean constituency size.

In response to Drabu, Hari Om, the former dean of the Faculty of Social Sciences, University of Jammu, claimed that it is

> not a secret that the number of voters in Jammu province was more as compared with Kashmir till 2002. Hold a fair census or decide the number of voters in the UT on the basis of the Aadhaar Cards issued so far to the people of both the regions and you will know that Kashmir is inferior to Jammu province both in terms of population[113] and voters.[114] It needs to be underlined that Jammu province will get at least 50 out of 90 seats on the basis of population alone and at least 7 to 8 seats more if its land area, nature of its terrain and accessibility also are taken into consideration, which is mandatory but which no commission ever took into consideration to appease Kashmir and jeopardize the legitimate interests of the people of Jammu province, Hindus, Muslims and Sikhs included. (Hari Om 2021c)

Both Drabu and Hari Om paper over internal heterogeneity of their respective regions even as they highlight the fissures in the other region, which is reminiscent of the parliamentary debate discussed in Chapter 1. Concerned about 'efforts ... to establish that Kashmiris are not one homogenous entity; they are fragmented with different, if not conflicting interests and affiliations', Drabu suggested that 'J&K comprises four distinct regions — Jhelum Valley (South Kashmir, Central Kashmir and North Kashmir), Chenab Valley (Kisthwar, Doda, Ramban and Reasi), Pir Panchal (Rajouri and Poonch), and the Tawi basin or the plains (Jammu, Kathua and Udhampur)' (Drabu 2020e). So '[t]he geography of Jammu, unlike that of Kashmir which is one large valley, is a mix of ... distinctive terrains of topography, [that] also overlap with the religious, ethnic and linguistic affiliations of their inhabitants' (Drabu 2020d). In other words, as a Kashmiri academic put it, Jammu is 'doubly artificial – an artificial construct within an artificial state' (interview, Srinagar, 28 May 2022).

Drabu's representation of Jammu's heterogeneity is not novel. In 2000, the ruling NC spelt out its approach to autonomy that called for communal partition of Ladakh and Jammu. The NC's Farooq Abdullah warned that if autonomy was denied 'you will have one border along the Chenab river and another between Kargil and Leh' (Swami 2000b).[115] The NC proposed a radical reconfiguration of the state's internal political geography along communal lines:

> Based on the proposals of a committee in which opposition groups, religious minorities and the Jammu region were *unrepresented*, the state government advocated the creation of six new provinces. Muslim-majority districts Rajouri and Poonch were to be carved out from the Jammu region as a whole, and recast as a new Pir Panjal Province. Udhampur's single Muslim-majority tehsil, Mahore, was to form part of the Chenab province, while the rest of the district was incorporated into Jammu. Even the single districts of Buddhist-majority Leh and Muslim-majority Kargil, were to become separate provinces. (Swami 2008a, emphasis added)

Observers pointed out that the proposal was 'meant to scuttle the demand for regional autonomy within the state' and that '[b]y challenging the regional status of Jammu and Ladakh, the committee seeks to negate the very basis of regional autonomy' (Chowdhary 2000: 2603) and 'divide the state on religious lines' (B. Puri 1999: 1400)

While this confirmed (Hindu) Jammu's fear vis-à-vis an irredentist Kashmir, Muslims of Jammu felt that Kashmir was using them as a bargaining chip. As Choudhary (2008: 13) puts it, 'The Kashmiri leadership never reached out to the Muslims in Jammu; they never explained to them what they stand for and never enquired from them what they want. But the Kashmiri leadership has always maintained that they represent the "majority sentiments in Jammu and Kashmir"' (see also note 51 of Chapter 3). Even the Gujjars of Kashmir feel marginalised in the Kashmiri discourse on autonomy.

Commenting on 'the sinister narrative' that 'presents Kashmir as a homogeneous unit and Jammu province as socially, politically and geographically heterogeneous region' (Hari Om 2021c) and aims at '[p]itting Jammu against Jammu to defeat Jammu' (Hari Om 2021f), Hari Om (2021c) pointed out that 'the length of Kashmir Valley is 132 Km and width just 32 Km and its part, which is inhabited predominantly by ethnic Kashmiris.... The rest of Kashmir is hilly and mountainous like the whole of Jammu province, barring four to five tehsils. The bulk of population which inhabits the mountainous areas of Kashmir is non-Kashmiri (Pathwaris, Gujjars, Bakerwals etc)'. Indeed, while the Valley per se is visibly Kashmiri, 'the mountain slopes and valleys surrounding Kashmir are studded by the Gujjar settlements. These areas are [in] Uri, Baramulla, Kupwara,

Ganderbal, Kangan, Pahalgam, Anantnag, Daksum and Kulgam administrative divisions' (Warikoo 2000: 7; see also Tables 4.1b and 4.1c and Choudhary 2011: 2).

After highlighting the internal divisions of Kashmir, Hari Om (2021c) turned to Jammu and argued that even laymen knew that 'all these districts of Jammu province [Kisthwar, Doda, Ramban, Reasi, Rajouri and Punch] are located in what is called mighty Pir Panjal Mountain Ranges – the natural boundary between Kashmir and Jammu'. A few months later, after the Delimitation Commission concluded its visit to the UT, Hari Om (2021b) added that '[i]f at all Chenab Valley exists in Jammu region, it exists in Reasi district' but not 'anywhere in the erstwhile Doda district'.[116] He suggested that 'religiously and politically motivated' proponents of 'Greater Kashmir' are interfering in Jammu. They deliberately misrepresent 'the erstwhile Doda district' as 'a Muslim-majority area' even though 'the Muslims and the Hindus in the area are almost evenly balanced' because they want to trifurcate Jammu, the historical Duggar land, that is, the people who 'did not allow the separatist movement to engulf their region' (ibid.). The fear that Jammu may be split along Chenab is echoed in Leh where Buddhists fear that a 'Greater Kashmir' could confine them to the north of Indus (former legislator, interview, Leh, 21 September 2019).

While commentators based in Jammu stress that their division has a much larger area than Kashmir (Hari Om 2021c), their Kashmiri counterparts point out that the bulk of the area is dominated by Muslims (Drabu 2020d; Mir 2021). Indeed, while two-thirds of Jammu are Hindu, six out of ten of its districts are dominated by Muslims. Hari Om overlooks this fact, and, like Drabu, he sidesteps the question of fairness of delimitation within the region. A similar cherry-picking is seen in Ladakh. 'Ladakhi demands ... always refer to the large size of the area to counter the argument that per-capita government spending in Ladakh is among the highest in India' (van Beek 2000: 537). Buddhists of Kargil district, too, highlight the fact that they inhabit the larger part of the district's area (Lok Sabha 2019: 161), while the Shias foreground the much smaller population share of Buddhists (interview, Kargil, 20 September 2019).

The people of Jammu add that political marginalisation affects their share in government funds,[117] jobs[118] and creation of new districts.[119] However, a retired bureaucrat suggested that

Jammu Hindus practically opt out [of] selection to certain categories of posts where a knowledge of Urdu is required because they do not know Urdu ... which just happens to be the official language of the state. Land revenue, police and judicial records are maintained in Urdu, and a knowledge of Urdu is rightly specified as a condition for eligibility in the relevant recruitment rules for posts in these departments. (Retired civil servant, interview, 3 January 2016).

But the extensive use of Urdu is confined to a few departments, including the three referred to earlier, and that too at the lower echelons, and even there English is also used alongside Urdu (retired civil servant, interview, 19 June 2016). So not learning Urdu can only partly explain the alleged difference in access to jobs. While a more systematic study is needed to verify the question of representation in jobs, the latest data suggest that at least in one case the distribution is not skewed against Jammu. Presently, Hindus account for at least 48 per cent of the officers of the Kashmir Administrative Service (GoJK 2021c) and 50 per cent of the officers of the Kashmir Police Service (GoJK 2021d), which means Jammu's share is certainly more than half.[120] But Jammu alleges that its candidates are posted in unimportant departments (see note 70 of this chapter). Also note that the partisans of Jammu overlook the complaints of SCs regarding the half-hearted implementation of reservation.

While the contest between Kashmiri-speaking Sunni Muslims of Kashmir and Dogri-speaking upper-caste Hindus of the plains of Jammu dominates the public debate, other groups manoeuvre away from the limelight. A submission to the Delimitation Commission demanded 'a "say" in political-democratic process' for the STs of the Kashmir through reservation of five seats 'which have an ample percentage of ST population': Baramula (Uri), Kupwara (Lolab), Anantnag (Dooru and Tral) and Kulgam (Noorabad) (Badhana 2021).[121] This list does not include Gurez, one of two entirely snowbound *tehsils* of Kashmir, where 82 per cent of the population belongs to STs – mostly 'Brokpa, Drokpa, Dard, Shin' tribes, which are ethnolinguistically different from the Gujjars and the Bakarwals. It does not even include Bandipora, where STs account for 23 per of the population, including the Gujjars and the Bakarwals, or Kangan in Ganderbal, where the Gujjars and the Bakarwals account for as much as 40 per cent of the population.[122] In Jammu, most submissions to the Delimitation Commission overlooked the Hindu Gaddi and Sippi (for an exception, see Kalsotra 2021) and Buddhist Boto (see note 75 of Chapter 3) tribes.

Unlike sections of STs, who were eagerly awaiting delimitation so as to access political reservation for the first time, the All India Confederation of SC/ST/OBC Organisations (hereafter, the Confederation) pleaded against using 'faulty data for such a crucial policy like delimitation' as SCs feared the loss of representation (interview, Jammu, 7 May 2022). It elaborated the concern about faulty data:

Everywhere in India the population share of Scheduled Castes has increased over time as they have higher fertility compared to other social groups due to socio-economic disadvantage. However, the share of Scheduled Castes in Jammu and Kashmir's population has been decreasing. In 1991, census could not be conducted in the state. The population share of Scheduled Castes was 7.6 per cent as per the 2001 Census, which declined to 7.38 per cent in 2011! This is contrary to the trend observed across the country[123] and socio-economic parameters ... if we accept delimitation on the basis of 2011 Census, we will be governed for the next decade and a half by faulty data because the next delimitation will happen along with the rest of the country after the first census conducted after 2026, i.e., after the 2031 Census. (Ibid.)

The Confederation urged the Delimitation Commission to postpone delimitation until 'the provisional population totals of the next census are released' or 'constitute an expert panel to correct the 2011 Census data' (ibid.). Note that a correction of regional population shares (Tables 3.8 and 3.9) will increase the population share of SCs too. In a subsequent representation, the Confederation added that if 'some constituency [can be] named after Shri Mata Vaishno Devi and Bahu Jammu, a constituency should be named on the names of Shahdara Sharief & Dr B R Ambedkar' (Kalsotra 2022). More importantly, the Confederation 'urge[d] the commission to rotate all the reserved seats' because '[b]y repeatedly reserving a constituency, the Commission is not only denying representation to Scheduled Castes in other parts of the state but also disenfranchising citizens belonging to unreserved categories who are denied the right to contest for an extended period' (ibid.). Indeed, non-SC and non-ST communities of Jammu complained that they have to bear a disproportionate burden of political reservation because all but three reserved seats are located in Jammu. In several constituencies, other communities have been denied the right to contest for an extended period in the absence of rotation of reserved seats. The All India Jat Maha Sabha complained that 'all border constituencies from Kathua to Mendhar were reserved by the Delimitation Commission ... leaving no scope for the Jat community in the polls' (*State Times* 2022).

These bigger and smaller contests within J&K over delimitation are nested within the larger national politics. A Jammu-based civil society leader argued that the BJP government wants to tell its core support base in the 'mainland' that by repealing Article 370 it has done for national integration what no other government could, but at the same time it wants to tell Kashmiris that there has been no real change (interview, Jammu, 14 July 2021). This dual purpose is best served by carrying out delimitation using the 2011 census that overcounted

Kashmir's population (ibid.). This is not entirely true as installing a Hindu chief minister in J&K is the ultimate trophy. But given the large gap in the reported populations of the two divisions, a Jammu-centric party cannot achieve this goal unless it can win a few seats in Kashmir. The extension ·of reservation in the assembly to the tribes was viewed as a step in that direction.[124] A near-clean sweep in Jammu coupled with independent candidates and smaller parties in pockets of Kashmir could deliver the trophy. Kashmir based parties can, however, exploit the complex demographic mosaic of the UT and compensate for the loss due to the BJP's incursion into Kashmir with a few wins in the Kashmiri-speaking and Muslim periphery of Jammu. Indeed, the report of the latest round of delimitation confirms that the complex demography of J&K and competing political objectives governed by considerations of religion, region, language, caste and national security circumscribe opportunities for gerrymandering.[125]

# Fortuitous Majorities

This chapter discussed the interlinked demographic anxieties of communities locked in zero-sum contests and how these contests spill over into the realm of delimitation of legislative assembly constituencies, which fuels the politicisation of numbers. The anxiety of Kashmir is despite the absolute majority enjoyed by Muslims in both Kashmir and the state as a whole and the fact that during the last six decades the state has never had a non-Sunni and non-Kashmiri-speaking chief minister. We saw that Jammu's representation in the assembly has always been less than its population share, except after the 2011 census. Ladakh did not have a similar complaint, but that did not imply an absence of conflict. Only the locus of conflict shifted from the inter-regional to intra-regional level and involved ethnolinguistically close communities.[126]

Competing claims to 'majority' need to be understood in their proper perspective. Several interviewees argued that unlike Kashmir, Jammu was heterogeneous. A retired civil servant, for instance, noted that 'it is difficult to think of a Jammu region' because of its enormous heterogeneity that makes it 'very difficult to think of the Jammu region having any common interest at all' (interview, 3 January 2016). He added that 'when Jammu city Hindus talk of Jammu region they are actually talking of a notional area around the city. Their world view does not even extend to Batote or perhaps even beyond Akhnoor and they have consistently opposed the development of the hilly and mountainous areas of Jammu Division' (ibid.). We can, in fact, go a step further and argue that the self-perception of Jammu city's Hindu elite is not only geographically limited but also socially limited to upper castes leaving out lower castes and tribes among Hindus. One could, however, also argue that it would be inappropriate to presuppose a 'common interest' within Kashmir or among Muslims. As Talib (2010: 6) puts it:

> The Muslim Pahari ethnic group[127] is at odds with the Muslim Gujjars and Bakerwals for political and economic considerations. There are divisions between Sunni and Shia Muslims[128] .... Kashmiri-speaking Muslim residents of Doda-Baderwah-Rajouri-Poonch do not necessarily share a common political vision with the Pahari-speaking Muslims of the same areas. Within Kashmir, another divide has evolved over the years: that is the divide between 'well-developed', 'ever-complaining' urban Srinagar and the rest of rural Kashmir.[129]

The Gujjar–Pahari divide is, indeed, a major fault line that cuts across the regional divide with the two communities being locked in a contest over affirmative action benefits (cf. Choudhary 2011 and Maini 2011). Choudhary (2011: 2) outlines the geography of this contest:

> In Rajouri and Poonch districts of the Jammu province and some parts of Baramulla, Kupwara, Bandipore and Badgam districts of the Kashmir province where these communities are predominant, the Gujjar-Pahari divide is a major political and social discourse in everyday life and their divide becomes a dominant factor in all decisions – political and administrative.

The aforesaid divisions are politically salient. S. Bose (2007: 191) suggests that '[t]he Kashmir Valley would be likely to return a strong pro-independence majority, but even there a substantial minority of Muslims would vote for Pakistan while a smaller but not insignificant number of Hindus, Sikhs, *and Muslims* would vote for India' (emphasis added). Likewise, Swami (2001) points out that it is not

> clear what linguistic affiliation the tehsils of Karnah and Uri in Kashmir ... might have with the valley. Similarly, while Ramban and Bhaderwah tehsils in Doda are not Kashmiri-speaking and principally trade with Jammu, the KSG [Kashmir Study Group] proposals make the a priori assumption that they would vote to join the new state. Indeed, these tehsils have recorded some of the highest voter turnout in successive elections from 1996, suggesting their residents have little sympathy for Kashmir valley-centred secessionist politics.

The difference in the degree of statistically assessable heterogeneity of the two regions is then not the primary determinant of the emergence of the Kashmiri 'majority'. The contribution of idiosyncratic historical developments to the political rise of Kashmir is often overlooked. Until 1941 Jammu was the most populous province of the princely state (GoJK 1933b: 4–5, GoJK 1943c: 72–73). Snedden (2017) draws attention to an important detail:

> The process of Kashmir becoming pre-eminent in Indian J&K was assisted by Jammu's geographic and demographic diminution in the latter part of 1947. Jammu went from being the princely state's most populous and influential province to a clear second to the Kashmir Valley in the smaller political entity of Indian J&K that came into existence in late October 1947 ... this changed Jammu from a Muslim-majority region to a Hindu-majority one. Furthermore, after 1947, Kashmir became Indian J&K's most populous region, which also helped it to become politically supreme in India[n] J&K.

In fact, all the anxious majorities of J&K are products of communal politics around the violent partition of the princely state. The Hindus of Jammu complain that Sheikh Abdullah was not interested in recovering the territories occupied by Pakistan where his NC had very limited political influence and interest. Muslims of Jammu add that for the same reason Sheikh did not do much to protect them from partition violence. The alleged inaction turned Jammu into a Hindu-majority region, whereas undivided Jammu was a Muslim-majority region. Interestingly, the Shias of Ladakh, too, believe that the Buddhists of Leh did not want Skardu, their main rival, to be part of India as that would have reduced them to a minority (ex-councillor, interview, Leh, 20 November 2019). Leh eventually lost the majority status in Ladakh, while Kashmir is battling to protect the political recognition of its majority status vis-à-vis Jammu. Hindus are trying to preserve their political and numerical dominance within Jammu and leverage it to ameliorate their political and symbolic marginalisation within J&K.

In the end, we can argue that the political marginalisation of (Hindu) Jammu vis-à-vis (Muslim) Kashmir was not predetermined by numbers because the majorities in the three regions of the present J&K fortuitously emerged during partition. Rather, the ability of the dominant elite to establish its hegemony and, perhaps, equally importantly, getting that hegemony endorsed by outsiders played the key role. Sheikh Abdullah himself noted that as late as the 1920s the Muslims of Kashmir were a divided lot and lacked political consciousness (Abdullah 2016: 73–74, 97, 100–01, 112–15, 134–37, 140–41). The broad convergence between various groups on a few major issues that is seen today in Kashmir is not an inevitable development rooted in faith or numbers.

Rather it is a product of political mobilisation whose outcome was accepted by outsiders too.[130] The geographical, administrative and economic centrality of the area inhabited by the numerically dominant Kashmiri-speaking population and the emergence of a charismatic leader among Kashmiris at a crucial juncture allowed the Kashmiri Sunni elite to dominate their region to the extent that Shia, Pahari, Gujjar and Bakarwal Muslims are entirely absent from the outside world's perception of Kashmir. The ham-handed counter-insurgency and New Delhi's inability to build credible channels of communication in Kashmir allowed the Kashmiri Sunni elite to further consolidate their dominance.[131]

At another level, away from the regional and communal battles within Jammu and Kashmir, the ruling party in New Delhi is in a dilemma after the reorganisation of the state because Kashmir and Muslims have already 'won' the census in 2011, which in turn seals the outcome of the contest over the chief minister's office. New Delhi's hand is therefore constrained, which explains the eleventh-hour manoeuvrings around the otherwise formulaic process of delimitation. The correction of coverage errors in the census would have improved the demographic weight of both Jammu and Hindus, but the reality that the current state of J&K (as well as the larger realm including the area under Pakistan) has been a Muslim-majority area for a very long time cannot be altered. A 1911 census report attributes the demographic preponderance of Muslims to, among other things, 'physiographical conditions': 'The largest part of the [Muslim] community inhabits the temperate Kashmir and Baltistan while the Bud[d]hists live in the severe cold of Laddakh and the Hindus on the dry hillocks and malarious terai lands of Jammu' (GoJK 1912: 94).

The complex ethnolinguistic and religious mosaic of J&K ensures that its political future cannot be determined by a mechanical quest to achieve statistical proportionality commensurate with enumerated identities bounded by administrative borders. The real tension, though, is between a union government that wants to showcase a peaceful, prosperous Muslim-majority sub-national unit, on the one hand, and the Hindu nationalist ruling party that wants to rework the political geography to penetrate the last (electoral) frontier and the Kashmiri elite that wants to sustain the unequal status quo, on the other. As a result, in J&K, delimitation is not merely an optimal mapping of equal-sized constituencies predetermined by the census. Instead, it is an instrument for achieving communal and political goals using census data that are themselves shaped by communal politics involving state and non-state actors at multiple levels in a federal polity. However, given the historical distribution of communities across a complex terrain, every iteration of delimitation generates further dissatisfaction with the political geography and demands for partition of the 'artificial' administrative unit. The latest round of reorganisation and delimitation of J&K has shown that no matter how we partition this complex geography, the historical distribution of population is such that the resultant anxious majorities will continue to feel both suffocated (sharing administrative units with 'others') and dismembered (separated from a larger imagined community beyond the 'artificial' borders),[132] even as the minorities will feel alienated across the successor administrative units.[133] Among the constituent regions of the erstwhile state of J&K, only Ladakh has a capacity to resolve its internal differences without or, even, despite external intervention. Despite deep differences, both Kargil and Leh have repeatedly shown the capacity to wage a united struggle for shared regional goals (see, for instance, Puri 1983a: 190, Wahid 2022b and Gupta 2023: 196–97).

# Notes

1. During one of his visits to a village in Ganderbal, the author was searching for an address when a young sportsman approached him and enquired about the purpose of the visit. When he learnt that the author was studying skewed sex ratio, without any prompting he suggested that every Kashmiri family needs 20 boys to fight the army (interview, December 2015). This is the most egregious but not the only encounter of its kind. Gentle probing on such occasions, however, revealed that the propagandists themselves fell well short of the lofty targets they had set for their people.

2. A Muslim leader of Kargil complained that 'our Buddhist people take cues from Leh' (interview, 20 September 2019). The Zanskaris maintain that even if this were true, Kargil has to blame itself because, as a Buddhist schoolteacher put it, 'we feel like refugees in the district headquarter where we are effectively barred from owning property, building places of worship or even cremating our dead' (interview, Kargil, 22 November 2019). A recent agreement between the two communities, though, has helped resolve some of these concerns (Wahid 2022b).

3. The predominantly Hindu Gaddis of Jammu are the largest tribal community in neighbouring Himachal Pradesh. 'Sippi' that is recognised as an SC in Himachal Pradesh is one of the smaller communities of the state. In 2011, there were 5,133 Sippis in Himachal Pradesh compared to 102,547 Kolis, who are also recognised as an SC. As discussed earlier, in Chapter 3, in J&K, Kolis are recognised neither as an SC nor as an ST.

4. Barely 0.5 per cent of Gujjars and Bakarwals of J&K were reported as Hindus in the 2011 census. As discussed in Chapter 3, this seems to be a content error, but otherwise Gujjars are both Hindus and Muslims in the

rest of the country. Former chief minister Omar Abdullah's brother-in-law, Sachin Pilot, a prominent leader of the Indian National Congress (INC), belongs to Rajasthan's Hindu Gujjar community. The Gujjars of J&K derive support from the larger body of the community spread across the country (Chowdhary 2010a: 19; interview, Baramula, 19 July 2021).

5. A former legislator suggested that Sunnis complain that they face discrimination in Kargil (interview, Jammu, 3 December 2019). Twelver Shias of Kargil are almost entirely Usulis and in turn seem to be polarised around competing *marjis* (living guides who can be emulated).

6. This is perhaps the first instance in Sheikh Abdullah's autobiography where he expresses the fear of demographic change. Interestingly, the last three censuses – 1981 (Sheikh Abdullah), 2001 (Farooq Abdullah) and 2011 (Omar Abdullah) – were conducted under chief ministers belonging to the Abdullah family.

7. In his autobiography, Sheikh Abdullah points out that before he was arrested in 1953 he had 'urged ... people to plan their families for their own sake and for the sake of the whole country' (Abdullah 2016: 346). Abdullah's *Naya Kashmir* (National Conference 1950), too, stressed women's rights.

8. Abdullah (2016: 535), though, was justified in blaming New Delhi for promoting Jamaat-e-Islami against him (see also Devadas 2019a and Fayyaz 2021). However, it bears noting that he was also accused of hobnobbing with Hindu extremists (see, for instance, various issues of *Kashmir Times* from January 1979).

9. The demographic and political tension between the Muslim majority and the Hindu minority of Kashmir predates the 1980s. In his autobiography, Sheikh Abdullah refers to the Pandits as 'a fifth column' (Abdullah 2016: 568), 'moles for central leadership' (ibid.: 576) and 'secret agents for Delhi' (ibid.: 576), who have since the days of Akbar allowed 'themselves to be used as instruments of tyranny against the majority community' (ibid.: 563). Sheikh summed up his position in the following question: 'Is it not possible for Kashmiri Pandits to understand the *logic of history* to consider the numerical strength of their Muslim compatriots and to keep in mind the topography of the state with its concomitant demands and stop looking up to the bureaucracy in Delhi?' (ibid.: 576, emphasis added, also 102, 575). Most Pandit accounts suggest that Sheikh Abdullah barely tolerated the community (see, for instance, Pandita 2017: 21). A Pandit interviewee pointed out that the community was first called an *amaanat* (collateral), then 'fifth column' and finally *mukhbir* (New Delhi's informers) (interview, Jammu, 21 July 2021). Note that *mukhbir* is almost synonymous with the 'other' in Kashmiri nationalist parlance (see, for instance, Wani 2010: 186 and Hari Om 1998: 108). Attacks on Pandits continue to be framed in terms of their alleged role as informers (Zargar 2021; Majid 2022).

More recently, the terrorist groups have expanded the scope of their attacks to cover 'Pandit employees who were recruited under the Prime Minister's Rehabilitation Package' (*Hindustan Times* 2022). The extent to which Pandits have been 'othered' in Kashmir is evident from the following. A Kashmiri journalist suggested that 'Pandits are Indian by faith' and added that 'Kashmir will either remain Muslim or lose its identity [and become Indian]' (interview, Srinagar, 29 May 2022).

10. The Mughal Road was completed three decades later in 2009, but even now it is motorable only during summers. The road between Zanskar and Leh is as contentious as the Mughal Road. Srinagar allegedly went out of its way to block the road as it did not want Zanskar to gravitate toward Leh (former legislator, Interview, Leh, 21 September 2019; Buddhist leader, interview, New Delhi, 2 December 2019). The politicisation of infrastructure in ethnically divided societies has been reported in other states as well. See Ziipao (2020) for the neglect of infrastructure in the hill districts of Manipur.

11. A former legislator claimed that the leaders based in Leh were not even informed that Kargil was being separated (interview, Leh, 21 September 2019).

12. For Sheikh Abdullah, 'Kashmir' and 'Kashmiri' referred to a specific part of J&K – namely, Kashmir that was 'a place inhabited mainly by Muslims' in which he occasionally included the undivided district of Doda (Abdullah 2016: 462). However, he also used 'Kashmir' and 'Kashmiri' to refer to the entire state. In a letter to Nehru in 1951, he referred to himself as 'the protector of the destiny of 40 lakh Kashmiris' (ibid.: 374, also 102, 123, 257, 268). As per the 1961 census, the first to be held after 1947, the state's population was 35.61 lakh (3.56 million). The population of the undivided princely state was 40.21 lakh in 1941. So Abdullah claimed the leadership of Pakistan-occupied territory too. He had very limited following outside Kashmir. Ladakh appears barely twice or thrice in the nearly 600-page long autobiography. There are very few substantive references to Jammu, which figures as a place of incarceration and transit and a place where Hindus and Muslims allow themselves to be used as puppets by external forces. In reality, Kashmir was all that mattered, and it is therefore not surprising that the idea of a greater Kashmiri realm was not seriously fleshed out in his thought. In fact, in his autobiography, even Kashmir appears *only* as the stage on which his destiny as a great leader unfolds, not as a real place. The fact that Kashmir could not put together even a semblance of pan-J&K response that could credibly appeal to people in other regions and challenge the union government's unilateral decision to repeal Article 370 confirms Devadas' observation regarding the foundations of 'Greater Kashmir'. Sikand (2010b) points out that 'Kashmiri Islamist leaders consistently argue that what they want is for the whole of the state of

Jammu and Kashmir ... to secede from India and form part of an independent "Islamic state" of Jammu and Kashmir or else join Pakistan. They curiously ignore the obvious contradiction in their stance in this regard – opposing what they call "Hindu Indian colonialism" in Kashmir while at the same time championing what, from the point of view of the non-Muslims and non-ethnic Kashmiris, would be Muslim Kashmiri colonialism in Jammu and Kashmir'.

13. Other Hurriyat leaders, such as Abdul Ghani Bhat (Oberoi 2004: 184) and Syed Ali Shah Geelani (Sikand 2010e: 132) also occasionally showed willingness to let the non-Muslim areas secede. In fact, before independence, Sheikh Abdullah once 'accepted the principle of self-determination for all communities, interpreting this principle not in its limited implication of religion but in the wider sense of culture' (Abdullah 2016: 238). Even after independence, in the 1950s, in response to the Praja Parishad's agitation, Sheikh Abdullah suggested that Kashmir would not mind the separation of Jammu (Balraj Puri referred to in Bhatia 2020: 10).

14. 'Syed Ali Shah Geelani, reportedly asked the people of Rajouri-Poonch to demand merger with Kashmir valley, a concept better known as Greater Kashmir' (L. Puri 2022).

15. This statement was released on 2 April 2010 (Javaid Rahi, secretary, Tribal Research and Cultural Foundation, Jammu).

16. The controversy around the alleged conversion of Parmeshwari Handoo and her marriage to a Muslim in 1967 had strained inter-community relations (see, for instance, K. B. Ahmad 2017: 168).

17. Between 1982 and 2005 – that is, for almost a generation – Farooq Abdullah, G. M. Shah and Mufti Sayeed served as chief ministers of the state.

18. There is some confusion about the chronology. Swami (2000a) notes that 'Former Union Minister and National Conference rebel Saifuddin Soz's intervention on the issue could not, however, be so easily dismissed. At a September 6 press conference, *which evidently prompted* the Hizb's threats ...' (emphasis added). According to *The Tribune* (2000b), the Hizbul Mujahideen's '*fresh* warning' appeared on 7 September, while Soz held his first press conference on census on 6 September (emphasis added). So the Hizbul Mujahideen's first warning was published on 4 September. Also, the 2001 PPT suggests that 'a militant out-fit' had issued a threat 'against the conduct of census' even before 5 June (GoI 2001b: 16; see also Rediff 2000a).

19. The 1971 administrative report for West Bengal, too, offered a very frank assessment of the disturbed conditions but without directly naming any political actor (GoI 1972b: preface).

20. Saifuddin Soz's book *Kashmir: Glimpses of History and the Story of Struggle* (2018) is silent on demographic change. Two decades after the 2001 census,

the then DCO, Feroze Ahmed, reiterated that 'Saifuddin Soz said that Muslim population was undercounted due to the Government of India' (interview, Jammu, 4 December 2019). Later he added that Soz's intervention was driven by 'political gain' (interview, 22 October 2021).

21. Soz's intervention received approval in Kashmir. See, for instance, Riaz-ud-din (2000) who suggested that census was a ploy to 'Hinduise' the state.

22. Masood Tantray (or Abdul Hamid Tantray) had 'played a leading role in trying to start peace talks with the Indian Government a year ago after Hizbul Mujahideen declared a unilateral ceasefire' (British Broadcasting Corporation [BBC] 2001). The ceasefire did not hold for long as there were differences within the separatist camp. Tantray was allegedly killed in an encounter in Kashmir in late July 2001 (ibid.; *The Tribune* 2001). Majid Dar had reportedly condemned the 'fake encounter' and custodial killing of Tantray (BBC 2001). A retired intelligence official, though, suggested that Tantray was killed by the Hizbul Mujahideen (interview, 13 August 2021). Dulat (2010: 145), too, notes, 'The Salahuddin people began bumping off Majid Dar's people, starting with those who participated in the 3 August talks with the government of India, Hamid Tantray and Farooq Mircha. And, of course, it included the murder of Majid Dar himself.'

23. Schofield (2003: 231), too, notes that '[a]s in 1991, when the militant groups forced the census to be cancelled, a ban was imposed' by the 'Hizbul Mujaheddin' against the 'futile exercise' as 'it was not possible to hold a credible census because thousands of Kashmiris had either been displaced, had migrated or were in jail'.

24. Kashmir (*The Tribune* 2000b; Roy 2008) and parts of it (Peer 2009: 110) are routinely equated with Palestine by civil society activists as well as armed groups. Dabla (2012: 191) alleges that 'the Government of India, during the rule of the Bharatiya Janata Party in the 1990s, *confirmed* that talks had been held with the Israeli Foreign Minister, at the time, about orchestrating change in the existing demographic situation in Kashmir' (emphasis added; see also *Kashmir Life* 2010). Others suggest that New Delhi is pursuing the 'Palestine model' to promote settlements in Kashmir (interview, Srinagar, 28 May 2022; see also Roy 2008). In the 1970s, '[i]t was even alleged that the Government of India had dispatched a team to Andalusia, headed by the Kashmiri Pandit [politician] D.P. Dhar, to investigate how Islam was driven out of Spain' (Swami 2008c). Then there were mundane stories of 'Israeli soldiers ... dressed as ghosts ... to frighten our people' (Sikand 2010a: 224). Most recently, in the parliamentary debates on the reorganisation of J&K, Asaduddin Owaisi (Lok Sabha 2019: 206), T. K. Rangarajan (Rajya Sabha 2019a: 63) and K. K. Ragesh (Rajya Sabha 2019a: 107) equated Kashmir with Palestine.

25. It seems 'Hizb was willing to give one concession to the government that is till all the doubts were removed the census should be suspended' (*The Tribune* 2000b). The then DCO told the author that whatever doubts people had were addressed through a press conference (interview, 22 October 2021). The press conference, though, failed to allay the apprehension. See *Kashmir Times* (2000k) for a letter to the editor questioning the method of accounting of three classes of migrants: those who migrated out of the state (who were not being enumerated in J&K), those who came to Jammu with the Darbar (who were being enumerated in Jammu but accounted in Kashmir's headcount) and the nomadic community (which was being enumerated in Jammu even as counting was in progress in Kashmir). This is just one example of utter confusion spawned by ad hoc and piecemeal shifting of reference dates amidst a lack of faith in public institutions and a widespread misunderstanding of the extended de facto (synchronous) method of counting.

26. Ibrar Ahmed, alias Abu Ubaid of the Lashkar-e-Toiba, was killed in an encounter in Rajouri on 4 October 2008 (South Asia Terrorism Portal n.d.).

27. A Kashmiri journalist pointed out that 'if the Government of India wants [to settle outsiders] they will do it in Kathua and give state subject' rather than bring them all the way to Kashmir where they can be easily detected and would even otherwise find it difficult to survive due to the difficult climate and political unrest (interview, Srinagar, 23 May 2016).

28. In the late 1980s, the Muslim United Front (MUF) 'declared that Islam could not survive [in India] under the authority of a secular state' that allowed the sale of liquor and banned cow slaughter (Swami 2008a).

29. When several groups called for a boycott of the 2001 census, the then DCO, Feroze Ahmed, remarked that everyone needed census, even the Hurriyat. The media presented this as an appeal to militants (interview, 4 December 2019). In his preface to the 2001 PPT, Ahmed wrote, 'I am confident after the publication of this book, they [militant outfits that issued threats against census] will realise their mistake and will rue for it, but then it will be too late, as census has no chance for retake' (GoI 2001b: xi). A decade later, Geelani's appeal seems to have vindicated him. See also note 41 of this chapter.

30. Elsewhere, Bhat (2018: 175) suggests that 'South Kashmir which is considered to be the hot bed of militancy seems also to be the hot bed of CSR decline'. The singling out of South Kashmir is somewhat problematic because both the districts with the lowest (Kulgam) and highest (Anantnag) decadal growth rates were from this region. The other two districts reported the fifth (Pulwama) and the eleventh (Shupiyan) highest growth rates. Moreover, Kupwara, Ganderbal and Badgam reported steeper declines in the child sex ratio (CSR) than all but one district of south Kashmir.

31. During the Amarnath controversy, Geelani had asked Kashmiris to refrain from employing or providing accommodation to outsiders and asked migrant workers to 'leave Kashmir peacefully' (Swami 2008b). A non-Kashmiri journalist who was living in a rented accommodation in Srinagar during the 2011 census recalled that his neighbours asked the enumerator to not count him as he was not from Kashmir (interview, 20 August 2021).

32. Jammu's claim to more seats in the legislature supported by its larger electorate must have contributed to the politicisation of the 2011 census in Kashmir. See, for instance, Sharma and Sengupta (2008) for one of the most extensive statements of this claim. Coming after the divisive protests and counter-protests around the Amarnath issue, such claims must have added to the communal anxiety of Kashmir.

33. See Maheshwari (1996: 141) and note 41 of Chapter 3 for further examples of inflation of child population.

34. The author came across a family in Kashmir that gets a male relative, who stays elsewhere in the same town, recorded as a member in all documents including the ration card because it was not safe to tell enumerators or surveyors – that is, strangers – that the family had no adult male.

35. Bhat (2018: 181) points out that 'J&K is a food importing State, and almost [all] food items like Rice, Atta [wheat flour], Sugar, Kerosene is distributed through public distribution system.... These ration cards have not been updated for so many years and there are large scale fake ration cards and even some families have more than one ration card and some families had none. So government had a plan to use the Census 2011 population as the basis for issuing new ration cards. Therefore there was a rumor that Census 2011 will be used by the Government to include or exclude persons from ration cards.' Economic incentives are known to have resulted in the manipulation of records elsewhere too. In 2012, more than a fifth of all domestic cooking gas connections in Karnataka were found to be illegal (*The Hindu* 2012). Likewise, the Andhra Pradesh government found several hundred thousand duplicate ration cards (*Indian Express* 2012). A 2014–15 study commissioned by the union government suggested that there were more than 12 million bogus ration cards (GoI 2015d). More recent reports suggest that the more than 30 million ration cards have been cancelled for failing to fulfil documentary requirements (Rajagopal 2021). In 1980, in some parts of Manipur 'people believed that houselisting [phase of the 1981 census] had something to do with permits for distribution of sugar, the figures tended to inflate' (GoI n.d.13: 35). Sugar rationing seems to have pushed people in parts of Sudan to overstate headcount in the 1993 census (A. H. Ahmed n.d.). In erstwhile Yugoslavia, housing rationing pushed people to inflate headcounts

in the 1953 census (Zarkovich 1989). International donors and aid agencies struggle because of the lack of 'incentive compatibility between data systems and funding rules' (Sandefur and Glassman 2015: 129). Even post-war western Europe manipulated figures to access greater benefits through the Marshall Plan (Morgenstern 1973: 20–21).

36. This echoes developments in Nagaland. The S. C. Jamir-led Congress government was cornered both by its exclusion from peace talks with the Isak–Muivah faction of the National Socialist Council of Nagaland (NSCN) and severe resource crunch. Both Jamir and Farooq Abdullah lost the following elections as their parties failed to secure majorities of their own. In both cases, observers suspected that the then union government supported the hastily put together opposition parties, Nagaland People's Front (NPF) in Nagaland and the People's Democratic Party (PDP) in Kashmir. (Dulat [2010] argues that the union government had no role in the PDP's success except that the rejection of the autonomy report undermined Abdullah's appeal.) There is a key distinction between Jamir and Abdullah though. Jamir did not resort to communal appeals in the run-up to the assembly elections despite the enormous backlash generated by his frank assessment of the Naga political problem and later refused to form government after the elections even though he was the leader of the largest party.

37. See also Saifuddin Soz's detailed statement in the Lok Sabha on G. M. Shah's hobnobbing with Islamic extremists when he was the chief minister (Lok Sabha 1986a: 326–27; see also Thakur 1985). Farooq Abdullah, too, leaned on demography to clarify his credentials. After the 1987 elections, he questioned the 'patently manipulated' census on various platforms including the state assembly (*Kashmir Life* 2010). Also note that in the 1987 assembly elections, the Muslim United Front had suggested that 'the authentic Muslim character of Kashmir' was inherently insecure in a secular India (Swami 2008a).

38. Sheikh Abdullah claimed that Muslims accounted for 'over eighty-five per cent of the population' of pre-partition J&K (Abdullah 2016: 219, 222, 342, 348). The source of this figure is not clear, but it is widely accepted in Kashmir (see, for instance, Syed Ali Shah Geelani quoted in Sikand 2010d). On other occasions Abdullah suggested that 80 per cent of the population was Muslim (Abdullah 2016: 170, 175). In the 1941 census, Muslims accounted for 77.11 per cent of the population of the entire princely state (GoJK 1943c: 338). The corresponding figure for the areas currently under the Indian administration was 72.4 per cent (GoI 1978b: 26), which reduced after 1947 due to the out-migration of Muslims and the in-migration of Hindus and Sikhs.

39. After the 2001 census, certain tribes in Nagaland that benefitted from the anomalies in the headcount argued that earlier censuses were flawed due to the involvement of staff from outside the state, but this baseless claim was rejected by other tribes (Agrawal and Kumar 2020a: 160–61, 189).

40. Interestingly, a census report suggests that in addition to in-migration from Pakistan-occupied territories, 'higher sex ratio among Hindus', too, contributed to their relatively higher population growth rate (GoI 1978b: 26).

41. In his outreach to various sections of the Valley, Farooq Ahmad Factoo, the DCO for the 2011 census, pointed out that a boycott would affect the headcount and, eventually, development funds (interview, Srinagar, 30 May 2022).

42. Our typology of the rationale for manipulation follows Agrawal and Kumar (2020a: 194), but we add the clarification that altruistic manipulation can potentially overlap with other types of manipulation.

43. Evidence from Nagaland, too, shows that often individuals first inflate numbers for welfare schemes and electoral rolls followed by inflation of headcount in the next census. Agrawal and Kumar (2020a: 230) show that if the electorate of a constituency in Nagaland is inflated in a given year, the headcount is also overestimated in the following census.

44. See Agrawal and Kumar (2020a: 301–03) for the categories used in the following discussion.

45. A retired civil servant who served in Kashmir noted, 'This has been an old refrain by the Pandit community, but no instances could be cited although I specifically sought as DC Srinagar in preparing for elections in 1977 to verify any such complaint' (interview, 12 January 2021).

46. Panun Kashmir (n.d.1), too, refers to the figure of 500,000: 'In the recent past, that is after 1947 and prior to the present mass exodus, more than 2.5 lacs of Kashmiri Pandits have been forced to bid farewell to their homeland and find settlements in different parts of India and foreign countries. If all those Kashmiri Pandit State Subjects who at present are away from Kashmir are put together and called back to the Valley, they shall form a sizeable proportion, over five lacs in number.' Elsewhere, Panun Kashmir (n.d.2: 3, 27, 30) demanded a homeland for 700,000 Kashmiri Pandits (see also *Kashmir Times* 2000c). Anupam Kher pointed out that as the lead actor of *The Kashmir Files*, he 'was representing all those 5 lakh people whose exodus took place *on* January 19, 1990' (Zee News 2022, emphasis added). Our analysis suggested a lower bound of 135,000 for the scale of out-migration of Hindus. Since the community maintains that more than 200,000 escaped the Valley that winter, a more detailed analysis of administrative statistics is required to arrive at an

appropriate upper bound (see also note 50 of this chapter). Sections of the Sikh community, too, claim numbers way above the census estimates. A Kashmiri Sikh delegation told the union home minister, 'We are only 1.5 lakhs left with 60,000 registered voters from Sikh community which is spread to 6 districts and 135 villages' (J. K. Singh 2021). Kumar, Ansari and Padgaonkar (n.d.: 137) suggest that there were 73,000 Sikhs in Kashmir in 2010–11. As per the 2011 census, there were 55,950 Sikhs in Kashmir. See also note 49 of this chapter.

47. Vivek Tankha, a Rajya Sabha member and a Kashmiri Pandit, stressed the need to locate '[t]he last big exodus [that] started after 1990' in the long process of their out-migration 'over a period of 100 years' (Rajya Sabha 2019a: 121).

48. See, for instance, Hari Om (1998: 15) and Behera (2006: 125).

49. Dulat (2010: 157) points out that 'there was a business community [in Srinagar] of Punjabis – not just Sikhs, but Hindus as well – who left along with the Kashmiri Pandits in the exodus of 1990'. According to contemporary news reports, 'By the middle of 1990, the total number of Sikhs and Kashmiri Muslims, who had registered themselves for relief in Jammu, had gone upto 18,000 and 800, respectively' (Hari Om 1998: 109). A retired bureaucrat, though, suggested that the magnitude of out-migration of Sikhs was not large as the community had decided against leaving Kashmir, and that during his visit to camps he rarely came across Sikh refugees (see also Kaur 2010: 241). He added that they might have left due to the dwindling economic and educational opportunities in Kashmir (interview, 13 August 2021). Kashmir accounted for 24.82 per cent of J&K's reported Sikh population in 1981, which dropped to 23.28 per cent in 2001 before marginally increasing to 23.82 per cent in 2011. Kashmir's share of Sikh population of J&K had peaked in 1971 (27.63 per cent). Note that largely stable share of Kashmir in J&K's Sikh population does not rule out out-migration because under the de facto method of counting Sikhs in the armed forces could make up for the shortfall in numbers due to out-migration of the local Sikh community. It is also noteworthy that while the share of Kashmir in the state's Sikh population is stable, the share of Sikhs in Kashmir's population has been steadily declining since 1971.

50. Evans (2002) arrives at a similar figure, but his estimate is based on extrapolation of the population. He suggests that figures based on the number of families in camps in Jammu overestimate the exodus. This is because '[o]fficials sympathetic to the KP plight, worked to allow for the smallest possible family groups to register as family units, often leading to elderly couples registering separately from their children' (ibid.: 25). Pandita (2017: 100) notes the following in this regard: 'Every ration card had to include a photograph of the male head of the family, along with his wife. So in some cases, husbands and wives made

separate ration cards to ensure that more money came in. To do this, many got their pictures taken with migrant labourers from Bihar and elsewhere. It was shameful, but there was little else one could do in those treacherous times.' This inflated the number of migrant families. There were also outrightly fraudulent registrations. While Evans is right about the overestimation of camp population, we also need to account for the fact that a sizeable part of the community left for other states.

51. The appointment of a Pandit as the youngest vice chancellor of Jammu University in 2002 is a case in point. This academic, who would normally be counted as a Hindu in Jammu, was identified by his competitors as a Kashmiri in Jammu's job market. It is noteworthy that sections of the Pandit community, too, did not like their fellow Pandit's appointment. They argue that vested interests in New Delhi occasionally need Pandit faces to mask their pro-Kashmiri Muslim policies (interview, Jammu, 14 July 2021).

52. Pandits, in turn, demand parity with Sikhs in terms of the investigation of crimes against the community. As the counsel for Roots in Kashmir put it, 'If the court can scrutinise each case of anti-Sikh riots which happened 33 years ago and order re-opening of the closed ones, why can they not order a probe into the killings of Kashmiri Pandits which took place 27 years ago?' (Nair 2017). But Pandits who did not leave Kashmir, too, demand parity with those who migrated (*Kashmir Times* 2011i; interviews in Srinagar).

53. Dulat (2010: 16, 294) suggests that Punjabi Muslims of Pakistan view Kashmiri Muslims as 'Brahmin ke aulaad' (offspring of Brahmins).

54. Abdullah (2016: 181–82) offered a three-layered explanation of the absence of a strong leadership in Jammu among both Hindus and Muslims. First, people respected the rulers and could access the Dogra court for employment and other perks. So 'interest in public maters was only nominal'. Second, demographic heterogeneity thwarted the emergence of leadership. Third, the Hindus and Muslims of Jammu relied upon co-religionists in Punjab to solve their problems.

55. A retired bureaucrat pointed out that Jammu city's Hindus 'feel deeply aggrieved by their perceived fall in stature from being rulers of the state to *mere citizens* of the state' (interview, 3 January 2016, emphasis added). Before 1947, Kashmiri Muslims used to complain against their marginalisation within quasi-decision-making bodies dominated by Hindus of Jammu and Kashmiri Pandits (Abdullah 2016: 43; Kaur 1996: 127, 153; Hari Om 1998: 114).

56. It bears emphasising that even the Muslims of Jammu continue to be excluded from New Delhi–Srinagar dialogues (see, for instance, *Kashmir Times* 2000l).

57. Writing months after the outbreak of armed insurgency and the exodus of Pandits, Hari Om (1990) lamented that Jammu was always asked to sacrifice

its legitimate interests in order appease Kashmir and integrate with the country, but in the end it did not help. An interviewee noted that New Delhi uses Jammu as a 'sandbag', which is abandoned as soon as conditions improve in the Valley (interview, Jammu, 19 April 2023). But Jammu did not challenge its mistreatment 'thinking that any such thing would club them with Delhi-bashers [in the Valley]' (Joshi 2020).

58. Ghulam Nabi Azad, an ethnic Kashmiri, was the only chief minister of the state from Jammu. Kashmiri Muslims viewed him as someone from Jammu, and for the Hindus of Jammu he was a (Kashmiri-speaking) Muslim. In the 2002 assembly elections, the Congress won twenty seats, including five in Kashmir, while the PDP won 16 all in Kashmir. They had to come together since neither was willing to forge a post-poll alliance with the NC, the largest party with seats distributed across all three regions of the state. Ideally, with more seats and representation from both Kashmir and Jammu (and even Ladakh if we include Nawang Rigzin Jora, a Congressman, who was elected from Leh but had contested as an independent candidate due to the UT movement), the Congress should have led the Congress–PDP alliance government. However, '[d]efying the logic of numbers, Mufti [Mohammad Sayeed] made public the argument by which Kashmir lived: Kashmir was the centre of the world. Despite its smaller number, his party must lead the government since it had more seats from Kashmir, he argued. It did not matter that the Congress' nominee was an ethnic Kashmiri from Doda, just across the Pir Panjal' (Devadas 2019a: 381–82). The parties agreed to share the office of the chief minister. Azad become the chief minister in 2005, but the government was destabilised by the PDP because, among other things, it feared that Azad wanted to 'reduce Kashmir to an appendage of Jammu' (ibid.: 411).

59. The princely state of Jammu and Kashmir was often referred to as 'Kashmir' in official publications even during the Dogra period – for example, in the census reports of 1911 (GoJK 1912) and 1921 (GoJK 1923). The administrative report of the 1941 census refers to the state as both 'Jammu and Kashmir' and 'Kashmir state' (GoJK 1943a). During the Dogra period, even Jammu-based traders referred to the princely state as Kashmir in their correspondence (author's personal collection). The question of an appropriate appellation was also debated in the Constituent Assembly of India, where a few members questioned Jammu's omission from the state's name in a motion tabled by N. Gopalaswami Ayyangar. Nehru proposed the following as a compromise: 'State of Kashmir (otherwise known as the State of Jammu and Kashmir)' (Rai 2018: 209–11).

60. Hari Om (1990) suggested that '95 per cent of the State's tourism budget is spent in the valley' (see also Behera 2002). Within Jammu division, other

districts complain that Jammu receives disproportionate attention. Kargil complains that Leh corners most of the tourism budget meant for Ladakh (see also GoJK 2001). Shia villages of Leh complain that they, too, are neglected by the tourism department (ex-councillor, interview, Leh, 20 November 2019).

61. A community leader lamented, 'Manmohan Singh served as the prime minister and Advani as deputy prime minister, but we [West Pakistan refugees] are denied even the right to contest local body elections' (interview, Jammu, 3 June 2016; see also WPRAC 2015: 2).

62. There are very few Uyghur Muslims left in Srinagar, while Tibetan Muslims do not seem to have been granted state subject status (interview, Srinagar, 30 May 2022). We should also note that it is not the case that the West Pakistan refugees did not have access to permanent resident certificates. In fact, after the Article 370 was repealed, not many from the community stepped forward to claim benefits under schemes meant to address their longstanding disadvantages. It seems 'members of almost all of these refugee families managed to get their names included in voter lists for J&K elections and procure permanent residency certificates', which made them ineligible to access schemes meant for West Pakistan refugees who were not permanent residents under the earlier regime (A. Sharma 2020).

63. Abdullah (2016: 212) complained that Central Asian refugees were mistreated under the Dogra regime.

64. Retired bureaucrats (various interviewees) told the author that the union government, which had committed itself to Article 370, failed to resettle the West Pakistan refugee community elsewhere in the country. However, this community reached J&K before Article 370 was enacted and the rights of communities living in the state on the date of enactment *should* have been protected a la Nagaland and several other states that are covered by special constitutional provisions. Nagaland is, though, trying to repeal some of its foundational commitments to longstanding non-Naga indigenous inhabitants.

65. The return of Kashmiri Muslims from Pakistan-occupied territories and the enfranchisement of the West Pakistan refugees appear as separate issues in the more recent discourse in Kashmir (see, for instance, A. Hussain 2016).

66. Devadas (2007: 63) suggests that Sheikh Abdullah was not keen on settling refugees from Pakistan-occupied parts of the state in Kashmir as he 'was intent on Kashmir's ethnic exclusivity'.

67. The claims on the 24 vacant seats ignore the fact that if refugees are excluded, Jammu's present allocation of seats within the UT will drop as its population share will decrease. Also, note that the relevant population is dispersed across the Jammu division and even outside the state, making it difficult to map electors onto such constituencies. The Hindu and Sikh refugees from the

Pakistan-occupied territories are not the only ones demanding a share of the 24 vacant seats though. Turtuk, which became part of India in 1971, has also demanded a seat from that quota (interview, Leh, 20 November 2019).

68. See Chowdhary (2010a) for a nuanced analysis of how in the course of this controversy the concerns of 'backward' districts were recast in terms of the usual Jammu-versus-Kashmir and Hindu-versus-Muslim binaries.

69. The census reported 40 migrants from Burma (present-day Myanmar) in 1961. Five decades later the number increased to 508. The population of migrants from Bangladesh increased from 87 in 1981 to 309 in 2011. It is not clear if the pre-1981 figures for Pakistan-origin migrants included those from East Pakistan. These numbers do not rule out the presence of large numbers of migrants from Bangladesh and Myanmar in Jammu as it is possible that their numbers swelled after the 2011 census, and the popular debates are responding to this more recent development. Indeed, the last chief minister of the state informed the legislative assembly that '5,743 Burmese (Rohingyas) are staying in ... the districts of Jammu and Samba' (*The Tribune* 2017). Moreover, it is likely that fearing deportation, illegal immigrants evade enumeration, report themselves as non-migrants or report themselves as migrants from eastern or north-eastern states (see Agrawal and Kumar 2020a: 25, 171, for international migrants in the north-east).

70. An interviewee based in Jammu argued that Kashmiri Muslims keep others out of key departments such as law, revenue, police and general administration (interviews, Jammu, 2 June 2016; Jammu, 14 July 2021). After killing a Kashmir Pandit posted in the Revenue Department in Badgam, a little-known terrorist outfit, Kashmir Tigers, threatened to target more 'migrant Pandits who are getting government jobs especially in departments which are important for demographic changes' (Ashiq 2022b). Devadas (2022b) adds that 'recent spate of killings indicates that those engaged in sensitive work within the government, pertaining to the potential transfer of land, are being particularly targeted'. Terror groups have since then extended their threat to all Pandits employed in Kashmir under the Prime Minister's Reconstruction Plan (*Hindustan Times* 2022).

71. The conspiracy to divide Jammu and settle more Muslims in the Hindu-dominated areas allegedly followed the setback Pakistan received in Geneva in 1994 at the United Nations Commission on Human Rights (UNCHR) session and the failure to precipitate a Kashmir-style exodus of the indigenous Hindu population of Jammu's northern districts (Hari Om 1995a; Hari Om, interview, Jammu, 30 June 2018; see also Baweja 1994 for an insightful assessment of the post-Geneva mood in Kashmir). Sikand (2001: 223) suggests that '1996 witnessed a major shift in the LIT's [Lashkar-i-Taiba's] Kashmir strategy, with

greater attention being paid to the border districts of Jammu, particularly Rajouri and Poonch, as well as Doda ... with the aim of causing an exodus of the Hindu population of these areas' (see also Swami 2021). Even in the absence of exodus, as Oberoi (2004: 173–74) points out, the attacks altered the distribution of population within the districts and polarised the communities: 'After nearly achieving their goal of changing the demographic structure in the [Kashmir] valley, militants are extending their activities across the Pir Panchal mountain range in the Jammu division. A series of Hindu massacres in the Doda, Poonch and Rajouri districts ... has led to the migration of minorities from the upper ranges. Militants in these mountainous regions are ruling the roost in the absence of security positions, as difficult terrain makes it impossible for troops to station themselves on these hilly ridges. As a coup d'état, the government has come up with Village Defense Committees (VDCs). Most VDCs consist of Hindus who have been issued arms by the government to protect the villagers. This situation widens the split between Muslims and Hindus.... Minor incidents lead to scuffles, which then lead to communal tensions.' A Gaddi leader suggested that the community lost access to pastures in the upper reaches due to terrorism, which translated into an inability to support large flocks (interview, Bhaderwah, 5 December 2019). The most recent killing of Hindus in Rajouri has revived the demographic anxiety of the Hindus of Jammu (N. Kumar 2023).

72. The Hindus of Jammu city are not alone in resenting Gujjar Nagar. Paharis feel that Gujjar Nagar reflects the state's tilt toward Gujjars and have demanded a 'Pahari Nagar at Jammu and Srinagar on the pattern of Gujjar Nagar' to 'provide avenues to those Paharis living in remote areas to own houses in big cities, thereby giving some access to a better life' (Maini 2011: 4). The allocation of land for settlement in urban areas has been a contentious issue in J&K for a much longer time. Hari Om (1998: 38), for instance, complained that 'while more than 20 per cent of the plots in government colonies in Jammu have been allotted in favour of the people of the Valley, the people from Jammu have at no point of time have been allotted any plot in the officially developed colonies in Kashmir. That is one of the reasons why the people of Jammu and Ladakh are conspicuous by their absence in Kashmir' (see also Swami 2000a for related allegations levelled by BJP leaders).

73. While the asymmetric flow of population along with the Darbar was viewed as a demographic assault by certain sections in Jammu, for the city's traders it was a major source of demand. In fact, the perception of assault post-dates the outbreak of armed insurgency that effectively put an end to Jammu's annual summer forays into Kashmir.

74. Observers noted the sharp increase in the reported rate of urbanisation (see, for instance, Swami 2012) but did not ask if this might be an artefact of the poor quality of census data. The urban and slum statistics of J&K merit closer scrutiny. A few interviewees attributed the abnormal increase in slum population between 2001 and 2011 to the influx of migrants working for the army and various infrastructural projects. This is unlikely to be true as migrant workers are mostly male. But, in 2011, the sex ratio and the CSR of the slum population was 933 and 860, respectively, and its child population share was 14.23 per cent. The growth in slum population outside Srinagar, Baramula, Sopore and Anantnag districts in the Kashmir division and Jammu district in the Jammu division can partly be explained by a definitional change. In 2001, the census data on slums was 'restricted to 640 cities and towns having 50,000 or more population in the 1991 Census and reporting slums' (GoI 2005b: v). However, in the 2011 census, slum areas were 'identified in all statutory towns irrespective of their population size' (GoI 2010b: 3–4). The definitional change cannot account for the change in Srinagar's slum population. As per the 2011 census, 28 (22) per cent of the population of Srinagar district (including Ganderbal) – that is, one in every four persons – lived in slums. But, as all Srinagar-based interviewees including academics, bureaucrats and politicians admitted, this is difficult to reconcile with the ground reality. Even if the definitional change explains the change in Srinagar's slum population, the contraction of Jammu's slum population will remain unexplained.

75. By inflating census and eroding Jammu's power within the state, Kashmir paradoxically hurt its own cause by further undermining faith in public institutions and thereby pushing Jammu to support the repeal of Article 370.

76. The youth wing of the LBA alleged that '[a]ny Muslim government servant marrying a Buddhist girl is immediately rewarded with juicy postings' (Smith 2009: 204; also interview, Leh, 20 November 2019). This echoes the concern in Jammu that subtle pressures nudge the SCs posted in the Valley to convert (see note 86 of Chapter 3).

77. Concerns about conversion from Buddhism to Islam have been around for a long time. The 1941 census noted, 'The Buddhist community shows a far smaller increase in numbers than any other. This has been the case since the beginning of Census taking. In 1911 it was attributed partly to the practice of polyandry and *partly to conversions to Islam*. These conversions are not so numerous as to be noticeable and most occur as a result of intermarriage' (GoJK 1943c: 13, emphasis added). These concerns were part of a larger demographic anxiety that led to the ban on polyandry in 1941 followed by a ban on primogeniture in 1943 (van Beek 2000: 533). There was also a concern about Christian proselytisation in the late 1980s (Devadas 1988).

78. There has been a persistent gap between the literacy rates of Kargil and Leh: 18.86 and 25.17 per cent in 1981, 60.85 and 65.34 per cent in 2001, and 71.34 and 77.20 per cent in 2011, respectively. Smith (2009, 2012) suggests that growing access to literacy and modern technologies of family planning amidst rising opportunity cost of raising children explain the decline in fertility in Leh. P. T. Kunzang of the Ladakh Buddhist Association (LBA) attributed the drop in fertility among Buddhists to education and greater exposure to the rest of the world due to tourist inflows (interview, Leh, 19 September 2019). However, a former legislator claimed that family planning was selectively promoted in non-Muslim districts of J&K as part of a conspiracy (interview, Leh, 20 November 2019). In Leh, the LBA 'actively campaigns against birth control measures (officially for violating Buddhist precepts regarding the sanctity of life)' (van Beek 2004: 215–16).

79. We do not have district-level TFR estimates based on the 2011 census. The corresponding CBR estimates for Leh and Kargil are 13.4 and 25.6, respectively (Kumar and Sathyanarayana 2012: appendix, table A1). The 2011 census suggested a marginal increase in the CBR of Leh district (ibid.). Interviewees in Kargil suggested that there has been a further decline in birth rates of Shias since 2011 as the educated youth increasingly prefer only two children.

80. Leh raised questions about the impartiality of Kashmiri establishment as early as 1953 when Kushok Bakula rebutted a statement of Maulana Masudi in the parliament that Muslims accounted for a majority of Ladakh's population. He asked why Kashmir was '*anxious* to pass off the District [Ladakh] as a Muslim majority one' and added that '[a]ll this has given rise to apprehensions … that the Buddhists would be officially relegated to the position of a minority in the Census of 1961' (Bakula 1953, emphasis added). This was perhaps the first allegation regarding manipulation of data levelled against Kashmir.

81. This was not the first instance of violence related to Rangdum that stirred communal passions. In 1951, 'the Muslims of Suru Kartse murdered 3 innocent Lamas of Rangdum Gumpa (Zanskar) in cold blood appropriating the supplies worth 2500/- which they were carrying from Kargil for their monastery' (Bakula 1953).

82. Around this time Leh reacted very strongly to Farooq Abdullah's autonomy report. The Ladakh Autonomous Hill Development Council (LAHDC) expressed concern about the 'gradual secession of the State from the Union of India' (Swami 2000c). The LBA's youth wing threatened 'an armed struggle against the government in Srinagar', while its head warned of a 'mass appeal by the Buddhists of Ladakh for asylum in a Buddhist country' (ibid.).

83. Narendra Jadhav, a nominated member of the Rajya Sabha, pointed out that 'most of the financial allocations, which were recommended in my report ['as a

member of the then Planning Commission'] and accepted by the then Central Government, were usurped by the Government for Kashmir Valley and not left for Laddakh for whom it was intended to be' (Rajya Sabha 2019b: 60). For similar observations about the neglect of Ladakh during 1951–65, see the report of Jammu and Kashmir Commission of Inquiry, headed by P. B. Gajendragadkar (GoJK 1968: 24–29; see also van Beek 2000: 536–37).

84. Leh and Kargil are reluctant to support each other's political initiatives. Asgar Ali Karbalaie complains that 'Leh was opposed to the divisional status, but Kargil fought for it. But when the order came, headquarters was given to Leh' (Donthi 2019). Leh, too, complains that while it fought for ST status and hill council, Kargil opposed the demands to protect its relations with Kashmir even as it accessed both the benefits later (community leader, interview, Leh, 19 September 2019; Buddhist leader, interview, New Delhi, 2 December 2019). The difference in approaches of the two districts is also evident from their choice of symbols. The emblem of the LAHDC of Kargil included J&K's state emblem, while Leh's emblem included the national emblem. This difference was also reflected in their divergent responses to the reorganisation of the state. Later misgivings notwithstanding, Buddhist Leh's first response to the government's decision to repeal Article 370 and separate the Ladakh region was to celebrate 'freedom' from Kashmir (A. Mishra 2019). Commenting on the reorganisation, former legislator Tsering Samphel said, 'Hum sattar saal rote rahe' (We kept weeping for seven decades) (interview, Leh, 21 September 2019). Lama Chosphel Zotpa added, 'Kashmiri ne pure Ladakh ke logon ko daba kar rakha tha. Humein kuch din pehle hi azaadi mili' (Kashmiris had suppressed people throughout Ladakh. We got freedom only a few days ago) (interview, New Delhi, 2 December 2019). A Lok Sabha member from Leh, Jamyang Tsering Namgyal, too, expressed similar sentiments in the parliamentary debate on Article 370 (Lok Sabha 2019). An interviewee reminisced that Kargil had celebrated its separation from Leh as the day of deliverance from slavery (Tsering Samphel, interview, Leh, 21 September 2019). Kargil, on the other hand, felt let down and protested. As Karbalaie put it, 'In J&K's reorganisation, Kargil is the biggest loser' (Donthi 2019). It is not the case that Karbalaie is rosy eyed about Kashmir. He argued that the reorganisation disempowered Ladakh completely by throwing it to the mercy of a physically and emotionally distant New Delhi represented by bureaucrats (interview, Kargil, 20 September 2019). The Buddhists of Kishtwar, too, complain that reorganisation has isolated them from Ladakh and reduced them to a micro-minority in the UT of J&K without any safeguards (see note 75 of Chapter 3).

85. To illustrate New Delhi's lack of trust in Muslims of Kargil, Asgar Ali Karbalaie points out that 'till 2002, not even 0.1 percent of the Muslims of Ladakh were recruited in the Ladakh Scouts [an infantry regiment of the Indian Army]. It's the same with the SSB [Sashastra Seema Bal] and the ITBP [Indo-Tibetan Border Police] as well. They have trained only Buddhist people in villages with mixed population. They openly say that they do not want to train Muslims' (Donthi 2019). He further asked, 'What does it [a Buddhist flag right in the middle of 121 brigade headquarters] convey in the Muslim-majority district?' (ibid.). Sheikh Abdullah, too, seems to have complained to General K. M. Cariappa about the lack of recruitment of 'the Muslims of Kargil in the Ladakh militia' (Abdullah 2016: 389).

86. Even Zanskar that is far more remote than Kargil is a major tourist hub. Leh's greater success in attracting tourists is therefore largely explained by the global appeal of its rich Buddhist heritage. Leh, on the other hand, complains that while the 1999 war between India and Pakistan was fought in both Leh and Kargil, New Delhi wrongly called it 'Kargil War' because of which development funding and war-related tourism has gone to Kargil (interviews, Leh, 18 September 2019 and 21 September 2019). Swami (2000c) suggests, 'During the Kargil war, elements in the LBA campaigned against the temporary resettlement of Muslim refugees in Leh' (see also van Beek 2000: 544).

87. See also several opinion pieces Hari Om wrote for the *State Times*, a daily published from Jammu, around the 2020 District Development Council elections and delimitation (for instance, Hari Om 2022a).

88. In J&K, Arya Samajis grew from about 79 in 1901 to 1,065 in 1911 (1,047 as per the 1931 census reports, which is incorrect because it does not include the figure for Ladakh), 23,116 in 1921, 93,944 in 1931 and 87,356 in 1941 (GoJK 1912: 87, 1933a: 296, 1943c: 11, 337). Three points are noteworthy. First, in 1931 (1941), there were 92,725 (85,656) Arya Samajis in Jammu and 1,219 (1,675) in Kashmir. Second, 43 per cent of the Arya Samaji population in 1911 belonged to the Megh caste (recognised as an SC after independence) but perhaps none among Pandits (GoJK 1912: 90). In 1941, 45 per cent Arya Samajis were from the lower castes (GoJK 1943c: 337). Third, the drop in the Arya Samaji population between 1931 and 1941 might be explained by the introduction of a law reserving public employment and land grants to state subjects.

89. While the role of the Arya Samaj in imagining Kashmir as a Hindu territory is known, the possible contribution of Jains in this regard has not received scholarly attention.

90. This is not an exceptional case of use of census data as evidence. Jalal (2008: 118) points out that the government's population data were used by the *ulema*

worried about Christian proselytisation: 'Population censuses showing an increase in the number of Christians in India were distributed as *proof* of the mala fide intentions of the British' (emphasis added).

91. Balraj Madhok, who introduced the RSS in Kashmir, belonged to a Punjabi Arya Samaji family.

92. Kashmiri Pandits still resent Punjabi Arya Samajis. An interviewee recalled how Balraj Madhok's writings targeted the community (interview, Jammu, 14 July 2021).

93. See Abdullah (2016: 51, also 211) on the Hindi–Sanskrit versus Urdu–Persian–Arabic controversy during the late Dogra period.

94. According to Dulat (2010: 108), in a conversation with a Kashmiri militant, Majid Nizami – the chief editor of the Urdu newspaper *Nawa-i-Waqt*, published from Lahore – seems to have suggested that 'an independent Kashmir would reduce Pakistan's strategic depth, as the international border would only be 10 km from the arterial Grand Trunk Road that connected Lahore to Rawalpindi; and an independent Kashmir would deprive Pakistan of water, since all of the water that irrigated the breadbasket in Punjab came from the Mangla dam in Mirpur'.

95. A Kashmiri academic (interview, Srinagar, 28 May 2022) noted that the Dogras feared becoming 'a perpetual minority' in Muslim-majority J&K. So, from the perspective of the RSS, 'Jammu had to be rescued, liberated, separated'. He added that in the late 1990s, the then home minister, L. K Advani, convinced the RSS against the separation of Jammu. But Swami (2002) suggests that the RSS continued to demand separation.

96. Interviewees in both Kashmir and Jammu pointed out that the national debate rarely, if ever, refers to Dogra and Punjabi Hindus and Sikhs, let alone Muslims, who, too, left Kashmir around this time. Focusing on a single indigenous community perhaps helped build and sustain a campaign, which would have been difficult if a cluster of communities cutting across religious lines, including non-indigenous ones, were referred to. In fact, the Dogra exodus from the Valley in the late 1940s, too, has not received attention in the national debate on Kashmir.

97. In the Hindu nationalist discourse 'Kashmir' is widely used as a label for places where followers of 'indigenous' religions allegedly face imminent demographic extinction in the absence of an urgent intervention (S. Sharma 2016; *India Today* 2017a; *Times of India* 2021; NDTV 2022; *The Wire* 2023). The most recent iteration of this argument is due to Assam's chief minister who said, 'People ask me if Assamese people will face the same fate as Kashmiri Pandits. It is the duty of Muslims [who constitute 35 per cent of the state's population] to allay our fear. Muslims must behave like a majority and give us assurance that there will be no repeat of Kashmir here' (NDTV 2022b).

98. Soon after the exodus of Pandits, the RSS had suggested 'balancing' Hindu–Muslim population in Kashmir 'to check the separatist and anti-national activities' (*Kashmir Times* 1990). After the repeal of Article 370, the union minister Jitendra Singh, who belongs to Jammu, seems to have claimed that demographic change will be good for the UT as it would boost the economy (Sandhu 2020). He was echoing the long-standing position of the Hindu nationalists that the Kashmir issue cannot be solved in the absence of demographic engineering.

99. Some of these views have seeped into the administrative apparatus as evident from, say, an unusually blunt op-ed, 'Kashmir's Future Is Tied to India's, It Is for All of Us to Decide What That Will Be', in the *Indian Express* (11 August 2020) by a serving police officer that called out the 'hypocrisy' of those worried about demographic change in Kashmir and advocated deep educational and other reforms.

100. Even governments in the West have intervened, even if erratically, in order to control immigration to maintain the ethnic composition of the population. In most democracies such attempts at demographic engineering are open to judicial review and public scrutiny, though. The debate in India on the Citizenship Amendment Act (CAA), 2019, and the Rohingyas is a case in point.

101. Soon after the government repealed Article 370, Devadas (2019b) suggested that 'Kashmiris have a very strong sense of identity, more than most others. Ego, or a sense of superiority, is a strong part of that identity. *It needs to be nurtured*' (emphasis added). Writing on the first anniversary of the repeal of Article 370, former BBC journalist Altaf Hussain (2020) noted, 'People in Kashmir say that Jammu and Kashmir being the only Muslim-majority state in Hindu majority India should have been treated with the utmost care and positive discrimination. The Indian government has instead chosen to effect demographic change'. A retired bureaucrat referred to 'Kashmiri exceptionalism' forged in the unique context of accession of the princely state (interview, 13 August 2021). On the second anniversary, senior BJP leader Ram Madhav interpreted the Kashmiri expectation differently. He argued, 'There was an insistence on recognising that J&K's merger with India in 1947, although it had other options as a Muslim-majority state, was a concession by its people and, hence, India should perpetually be grateful to them. This had bred a sense of "Kashmiri exceptionalism" in the mainstream political establishment of the state for decades. Article 370 was seen as India's grateful gesture, not for material development, but as a stamp of acknowledgement of that exceptionalism. Successive governments in Delhi chose to ignore this subterranean sentiment and continued to pamper the leadership' (Madhav 2021). See Abdullah (2016) for a Kashmiri perspective on the Valley's exceptionalism.

102. 'The story of Jammu's underrepresentation in the Assembly', as Hari Om (1995b) puts it, 'started in 1951 when the state was kept outside the purview of census operation of that year throughout the country. The immediate consequence of that action was the carving out the constituencies for the Assembly based on the 1941 census. This the government did despite the fact that the complexion of the state population had undergone a sea change owing to the migration of over one lakh Muslims from Kashmir to Pakistan and an influx of over 200000 refugees from the Pakistan occupied Kashmir (PoK), as also the migration of thousands of Hindus and Sikhs from the valley to the Jammu region.'

103. The 1957 and 1967 delimitations did not cover J&K. The 1971 and 2001 delimitations applied only to the parliamentary seats of J&K (Election Commission of India [ECI] 1957: iv, 1963: 141, 1976: 183, 2008a: 165). So it seems Karan Singh's second suggestion regarding delimitation was accepted.

104. The number of seats allocated to the Pakistan-occupied territories remained unchanged at 24 in both 1995 and 2019 when the size of the assembly was expanded.

105. Hari Om (1995b) discusses the delimitations conducted until 1995. While his broad conclusion about the under-representation of Jammu is correct, the use of rounded-off figures in the article magnified the actual loss of Jammu.

106. Drabu (2020e) suggests that the state's 'electoral rolls data has always thrown up lots of inconsistencies. The percentage share of the electorate at the regional level keeps on changing during each election. Sometimes Jammu region registers more electorate than Kashmir region and at others Kashmir region outnumbers Jammu region.... As per the 2002 electoral statistics of the state, Jammu region had two lakh registered electors more than the Kashmir region. But as per the election commission of India data, the Kashmir region had nearly 2.50 lakh more electors than the Jammu region in 2014 [read 2008] assembly elections.' However, in a later piece Drabu (2021b) overlooked the erratic variations and asked, '[E]ven if the voter per constituency is higher in Jammu than in Kashmir, is this situation unique to Jammu Province?' See note 109 of this chapter for erratic changes in Ladakh's electorate.

107. Hari Om (2021c) suggests that 'the number of voters in Jammu province was more as compared with Kashmir *till* 2002' (emphasis added). But this was true only in one year, namely 2002.

108. The size disparity was more pronounced in the case of parliamentary constituencies because Jammu's electorate for the parliamentary elections included the West Pakistan refugees who did not have the right to vote in the assembly elections because they were not state subjects. As per the census, Jammu's share in the state's parliamentary seats was 2.65, 2.70, 2.72, 2.62

and 2.57 as per the 1961, 1971, 1981, 2001 and 2011 censuses, respectively. The 2022 delimitation has for the first time split the five parliamentary seats almost equally between the two regions by adding assembly seats of Punch and Rajouri districts to the erstwhile Anantnag parliamentary constituency.

109. This complaint is no longer valid, though, as the relative population shares of the two districts of Ladakh have reversed in the meantime. A comparison of Figures 4.1a and 4.1c suggests that while Leh's population grew faster than Kargil's between 1961 and 1981, the electorate of the latter grew at twice the rate of the electorate of Leh. On the other hand, between 1981 and 2011, Kargil's population grew at a faster rate than Leh, but its electorate grew at less than half the rate of Leh.

110. We can add that Leh's Nubra constituency has never elected any Muslim candidate either.

111. A political leader had suggested that most reserved ST seats were likely to be allotted in Kashmir since all the SC seats fell in Jammu (interview, 26 July 2021). This would have reduced the number of seats open to non-tribal Kashmiri-speaking population. However, the Delimitation Commission reserved only three seats for STs in Kashmir, including two that anyway elect tribal candidates.

112. Unlike J&K, both Assam and Kerala have had non-Hindu chief ministers, including Muslims. Also, unlike J&K, in other states the chief minister's office is not effectively reserved for one region. This has been a major complaint of the Jammu region and Hindus but is overlooked by the partisans of Kashmir.

113. In his discussion of the relative strengths of the two divisions, Snedden (2017) suggests in passing that 'Jammu may now be more populous than Kashmir'. He does not indicate the source of his information.

114. Several interviewees in Jammu claimed that the overcount of the population of Kashmir and Kashmiri speakers is corroborated by a range of non-census indicators as there are more electricity, water and cooking gas connections, schools, ration cards and voters in Jammu, but surprisingly it has lesser population than Kashmir (interviews, Jammu, 2 June 2016 and 30 June 2018; New Delhi, 1 July 2018; see also Hari Om 2021c).

115. The Chenab-centric discourse is as old as the Kashmir dispute. For instance, the Dixon Plan 'assigned Ladakh to India, the Northern Areas and Pakistan-Occupied Kashmir (POK) to Pakistan, split Jammu between the two [across Chenab], and envisaged a plebiscite in the Kashmir Valley' (Noorani 2002). A similar plan seems to have been 'seriously debated' soon after the 1962 Sino-Indian War (Abdullah 2016: 445). The latest delimitation report, too, is reminiscent of the Dixon Plan insofar as 'this is the first time that a statutory body has de facto recommended bifurcation of an old parliamentary

constituency and combined two culturally distinct Muslim societies to form a new one' (L. Puri 2022; see also Ganai 2022).

116. Ghulam Nabi Azad, too, noted that 'it would be more appropriate to call Akhnoor area as Chenab Valley because ahead of Reasi it is totally hilly region and from Akhnoor it actually appears like Valley where irrigation facilities have also been created with the construction of Ranbir and Partap Canals' (*Daily Excelsior* 2022c; see also Hari Om 2022b).

117. Hari Om (2021a; see also 1995a) argues that the Hindu community, 'which constitutes over 40 per cent of J&K population ... has been contributing to the state exchequer almost 80 per cent revenue every year' but gets 'in return not even crumbs'. Note that Hari Om overestimates the Hindu population share that has never exceeded 32.24 per cent. Even the corrected share of Hindus is unlikely to be close to 40 per cent.

118. 'In the civil secretariat, Jammu's representation was less than 10 per cent.... All 12 corporations of the Jammu and Kashmir government had their headquarters in Srinagar with almost 100 per cent of the employees from the Valley. Most major Central government offices were in Srinagar' (Behera 2002).

119. A BJP MP complained that the state government created 'new districts of Badgam, Pulwama and Kupwara in Kashmir' in 1979 without constituting any commission, while it 'evaded' the recommendations of the Wazir Commission (1983) to create Reasi, Kishtwar and Bahu (Samba) districts in Jammu (GoI 1995b: 19). Chief Minister Ghulam Nabi Azad who created several new districts in the late 2000s admitted that 'though the Wazir Commission had recommended the creation of only four districts – three in Jammu and one in the Kashmir Valley – the changing scenario over the past 30 years made it imperative to reassess the process for creating new districts' (*Outlook* 2006).

120. According to Fayyaz (2017), '1,956 candidates have been selected since 1995 [up to 2014] for the posts of combined competitive services (KAS/KPS), Munsiffs, Assistant Directors in Economics and Statistics (Planning), Range Officers Grade-I and Assistant Conservators of Forest (ACF). Of them, the lion's share of 1291 has gone to Jammu and Kashmir has got less than 50 per cent i.e only 632 posts. Ladakh has bagged 33 posts.' Further, except 1995 'when Rifat Jabeen from the Valley stood on top of the merit list, all the top positions in the subsequent 8 selections have been grabbed by the candidates from Jammu'.

121. Sikhs have also staked a claim to the Tral and Baramula constituencies (Bhakto 2021; J. K. Singh 2021). The All Parties Sikh Coordination Committee (APSCC) decided to stay away from delimitation in the absence of a provision of reservation for religious minorities (Bhakto 2021). The small Christian (community leader, interview, Jammu, 8 May 2022) and Buddhist (community

leader, interview, Paddar, 23 September 2021) communities, too, demanded reservation in the legislature.

122. Three generations of Naqshbandi divines revered by the Gujjars and the Bakarwals have won every election in Kangan.

123. Kashmiri Muslims, too, ask how the share of Muslims in J&K's population decreased between 1961 and 1981, 'a completely reverse trend as compared to other states within India' (Hamid 2011).

124. According to Rahi (2019), there 'are 25 Assembly constituencies where Gujjars constitute 25 percent to 60 percent of franchise', including 'twelve Assembly segments where Gujjar Candidates won elections', whereas in another '14 constituencies, Gujjar/Bakerwal vote can turn the tables' (see also Nazeer 2022). The latter constituencies are mostly in Kashmir. Gujjar interviewees from Baramula had expected that most reserved ST seats would be located in Kashmir. Indeed, there are at least five seats in Kashmir (and Kishtwar) that have substantial tribal populations and could have been reserved. However, the Delimitation Commission reserved three seats in Kashmir, including Gurez and Kangan, that have always elected tribal candidates and are unlikely to elect candidates of the BJP or its potential electoral allies. As Nazeer (2022) puts it, the latest delimitation is a manifestation of the 'tendency to limit tribal politics to Rajouri and Poonch'. The puzzle posed by the BJP's restrained use of the distribution of tribal seats has not received much attention in the media though. For instance, Devadas (2022a) and Swami (2022), who discuss the implications of the delimitation at length, completely ignore what the distribution of reserved ST seats tells us about the political calculus of the BJP.

125. The delimitation report has been criticised on various grounds. First, six out of seven newly created seats that were awarded to Jammu are unequally distributed between the Hindu and Muslim dominated areas. As a result, some of the formerly Muslim majority constituencies have been dismantled, and the average size of the Muslim dominated constituencies seems to have increased (Ashiq 2022a). Second, the Anantnag parliamentary constituency combines unrelated areas. After the redistribution of the assembly seats that added six constituencies to Jammu, it was inevitable that a parliamentary seat would have to be split. There were several possible ways of combining assembly constituencies into a parliamentary constituency spanning Kashmir and Jammu. The Delimitation Commission's choice suggests gerrymandering. L. Puri (2022) points out that 'the two areas are connected by a fair-weather route ... [that] can only be traversed during the 3–4 months of summers.... Otherwise, an elected candidate of the proposed constituency will have to travel 500 km from Rajouri-Poonch to Anantnag district via Jammu plains.' It bears noting that the usually low voting

rates in Anantnag and nearby areas might allow Punch and Rajouri to steal the show in the parliamentary elections. Further, seen from a national security perspective, the choice seems to be driven by New Delhi's compulsions to introduce collective action problems. Not everyone in Jammu is happy though. A Jammu-based Twitter user lamented, 'The enslaved Jammu was demanding separation from Kashmir & state status, saying it alone could end their 75-yr-old slavery and defeat Kashmir Jihad[,] but JK Delimitation Panel de-linked Rajouri-Poonch from Jammu and merged it with Jihad gripped K's [Kashmir's] Anantnag LS constituency.' On the other hand, 'mainstream political parties in Kashmir reject it [the new constituency] calling it an exercise of gerrymandering to benefit BJP and its affiliates. Many see it prelude to Greater Jammu with South Kashmir, for the first [time] becoming part of the Jammu parliamentary constituency' (Ganai 2022; see also *The Tribune* 2022). In any case, there are limits to gerrymandering. Omar Abdullah observed that the BJP conducted 'early delimitation exercise in Jammu and Kashmir to benefit their own party by increasing seats in Jammu but when the final delimitation report came in by default it actually benefits National Conference at various places like Kupwara and Anantnag districts in Kashmir Valley' (S. Hussain 2022). Third, the ST seats have been clustered, denying representation to the tribal community outside the areas where they have a very high concentration of population and have always won elections. Fourth, SCs questioned the use of faulty census data that cost them a seat and the lack of rotation of the reserved seats. Fifth, STs questioned the failure of the commission to reserve a parliamentary seat for the community. Sixth, refugees from Pakistan-occupied Kashmir questioned their under-representation, while West Pakistan Refugees demanded a reserved seat. Seventh, Jammu questioned the logic behind locating almost all the reserved seats for the SCs and the STs in one division (Hari Om 2022a). Last but not least, people are complaining against the arbitrariness of constituency boundaries. The commission addressed some of the criticisms (*Daily Excelsior* 2022b).

126. The conflicts between Shias and Buddhists in Ladakh exemplify the narcissism of the small difference. In Kashmir, the narcissism is coupled with the 'fear of small numbers' linked to the 'anxiety of incompleteness' – that is, '[n]umerical majorities can become predatory and ethnocidal with regard to small numbers precisely when some minorities ... remind these majorities of the small gap which lies between their condition as majorities and the horizon of an unsullied national whole, a pure and untainted national ethnos' (Appadurai 2007: 8).

127. Paharis are a collection of diverse ethnolinguistic and religious groups that came together in the 1970s when the government launched special schemes to support Gujjars and Bakarwals. They have been attacked by both Gujjars and

Dogras for being an opportunistic or artificial group propped up by Kashmiris to divide Jammu.

128. The Shia–Sunni differences in Kashmir play out within certain bounds because of the shared historical and political experience of the two communities and the minority status of Shias (Wani 2010: 185–86; Hussain and Mehdi 2021). Shared ethnolinguistic roots, too, help limit sectarian conflicts.

129. A leader from Kupwara complained about the marginalisation of his district: 'There is a university in central Kashmir, a university in south Kashmir. Yet the authorities chose to locate the central university just 15 kilometres from the existing University of Kashmir. Train service was truncated at Varmul and all the administrative relief systems seem to be scarce in this impoverished district' (Chowdhary 2010a: 17).

130. See note 54 of this chapter for Sheikh Abdullah's explanation of the difference between Kashmir and Jammu in terms of internal cohesion and B. Puri (1983a: 188) for a politically salient example of the acceptance of the hegemony of Sheikh Abdullah (and Kashmir).

131. Madhav (2021) suggests that the reorganisation of the state was New Delhi's way of undermining the Kashmiri hegemony that until recently was nurtured. The introduction of political reservation for STs (to belatedly fulfil constitutional obligations toward marginalised communities) serves that purpose, too. However, Hari Om has, in a series of op-eds, questioned if the reorganisation has reduced Srinagar's dominance and changed New Delhi's attitude.

132. See Englebert et al. (2002) for an elaboration of Clifford Geertz's idea of 'suffocation' and 'dismemberment' in nation states. Shia Kargil laments the dismemberment of the larger Ladakhi Shia realm that includes Baltistan and feels suffocated in a Buddhist-dominated political entity. (An implication of this is that Greater Kashmir cannot simply subsume Kargil as the latter is engaged with its own project of Greater Ladakh.) Almost every community of J&K has similar complaints.

133. Kashmiri Pandits complain that they are seen as Hindus in Kashmir and as Kashmiris in Jammu. Gujjars see Pandits as half-Muslims, while Pakistanis see Kashmiri Muslims as 'Brahmin ke aulaad' (Dulat 2010: 16, 294). Gujjars are seen as partisans of Jammu in Kashmir, Muslims in Jammu and partisans of India along the LoC. Gaddis are seen as Hindus by their Muslim neighbours and as STs by the Hindus of the plains. We can add similar examples from Ladakh. Note also that mainstream Kashmiri politicians such as Omar Abdullah complain that they are seen as nationalists in Kashmir and separatists elsewhere (Vinayak 2020), while Ghulam Nabi Azad is seen as a politician from Jammu in Kashmir and as a Muslim in Jammu (see note 58 of this chapter).

# 5

# The Limits of Law

If you have reasons to suspect that in any area due to any organised movement, the religion is not being truthfully returned, you should record them as actually returned by the respondent....

—Census of India, 2011: Instruction Manual
(Government of India [GoI] n.d.14: 6.52)

## Introduction

We have so far analysed various errors in Jammu and Kashmir's (J&K) census and discussed their larger political context. We showed that some of the errors cannot be accounted for by conventional demographic and non-demographic explanations and are likely to be driven by politically motivated intervention. This chapter examines the legal and administrative contexts of data collection to understand the checks and balances in the process of enumeration. This is mostly ignored in academic as well as governmental discussions on data quality. We will discuss the constraints that circumscribe data collection exercises because of which the punitive provisions of law have failed to prevent the manipulation of the census in J&K and other states.

The chapter first introduces the legal framework determined by the Census Act, 1948, and the Census Rules, 1990, and discusses the administrative machinery that is mobilised to conduct the census. It then explains why erring respondents and enumerators cannot be punished for deliberate errors and argues that it is neither desirable nor practical to provide security to enumerators. We will argue that the punitive measures of the Census Act, 1948, serve as a hollow threat because governments are unwilling to strain their relationship with people and public servants for conducting a once-a-decade exercise. As a result, punitive measures are rarely enforced even though the fines are mild. This is also true of several other common-law countries. The chapter then discusses the implications of the absence of provisions in the Census Act, 1948, and the Census Rules, 1990, for the correction of data. So not only is the Office of the Registrar General of India (ORGI) ill-equipped to deal with mass subversion of the exercise, but it is also unable to correct the data afterwards. The decline in the availability and quality of metadata provided by the ORGI has, however, meant that users do not have adequate background information to assess the quality of census data.

# The Census Act, 1948

In colonial India, a temporary legislation was introduced before every decennial census that lapsed after the completion of the exercise (Maheshwari 1996: 137–39; see also S. Subramanian 1960: 112). Newly independent India introduced a permanent legislation, the Census Act, 1948, to provide a proper legal framework for population censuses. In his introduction to the new law, the first home minister, Sardar Vallabhbhai Patel, noted the difficulties faced in the 1941 census due to communal conflict (Maheshwari 1996: 139–40). The Census Act broadly followed the template of the colonial laws,[1] but it enhanced punishments to 'check the interplay of sectional, religious or communal rivalries' (ibid.: 141). The revised punitive provisions, however, proved to be ineffective. In the very first census conducted after independence, the Hindu–Sikh conflict impaired the language data of Punjab, the Patiala and East Punjab States Union (PEPSU) and Himachal Pradesh, and the disaggregated data for Hindi, Urdu, Punjabi and Hindustani were not published for these areas (GoI 1951: 12–13, 1954d: 2). The punitive provisions of the Census Act have remained ineffective over the years in face of several other instances of manipulation.

The provisions of the Census Act can be classified according to whether they apply to public servants, as defined in Section 5 of the Act, or to others. In the case of public servants, offences are linked to the non-fulfilment of duties, as specified in Section 6. Others, including individuals whose support has been solicited by officials to carry out census and individuals qua respondents, can violate the law if they fail to fulfil the duties assigned to them (Section 7), fail to provide material support for conducting census (Section 7), fail to respond to questions or supply incorrect information (Sections 8 and 10), refuse to provide access to premises (Section 9) and trespass into a census office (Section 11h). The first two situations can arise because the Act empowers public servants to seek the support of others to conduct census. The government is required to compensate the owners of private premises or vehicles. Owners can challenge the quantum of compensation but not deny their personal or material support if asked for. Penalties for offences are specified in Section 11. Confidentiality is guaranteed in Section 15 and is widely advertised (Figure 5.1), which is unlike, say, the Registration of Births and Deaths Act, 1969, that allows scrutiny of records and their use as evidence in courts. The rest of the Census Act lays out the legal and administrative framework within which the punitive provisions can be invoked. The Act is supplemented by the Census Rules, 1990 (amended in 1994, 2009, 2018 and 2022), that fill in administrative details. The Census Act, 1948, was amended in 1950, 1956, 1959, 1963, 1965, 1974, 1976, 1986 and 1994. The last amendment extended the applicability of the Census Act, 1948, to pre-tests, pilot studies, houselisting and housing census and post-enumeration checks (PEC) or post-enumeration surveys (PES), and revised the punitive provisions.

Sections 8 and 11 constitute the core of the Census Act, 1948, from the perspective of checking manipulation during enumeration. Section 8 makes it legally binding to answer the questions asked by enumerators.

> Section 8. (Asking of questions and obligation to answer) (1) A census-officer may ask all such questions of all persons within the limits of the local area for which he is appointed as, by instructions issued in this behalf by the [Union Government] and published in the Official Gazette, he may be directed to ask. (2) Every person of whom any question is asked under sub-section (1) shall be legally bound to answer such question to the best of his knowledge or belief.

> Provided that no person shall be bound to state the name of any female member of his household, and no woman shall be bound to state the name of her husband or deceased husband or of any other person whose name she is forbidden by custom to mention. (Census Act, 1948)

Section 10 ('occupier or manager to fill up schedule') extends Section 8 to institutional households.

The Indian (Census Act, 1948, Section 8), Bangladeshi (Census Order, 1972, Article 7) and Pakistani (General Statistics [Reorganization] Act, 2011, Article 23[4]) census laws make the answering of all but one question compulsory. They offer an exemption only if customs forbid a respondent from mentioning the names of certain relatives. In contrast, the British (Census Act, 1920, amended in 1991 and 2000, Article 8[1A]),[2] American (US Code Title 13, Section 221[c]) and Australian (Census and Statistics Act, 1905, Section 14[3]) censuses provide exemption in the case of religion. The divergence between common-law countries in the West and the Indian subcontinent has its roots in the enumerative practices of the colonial state that readily used macro-ascriptive identity markers such as

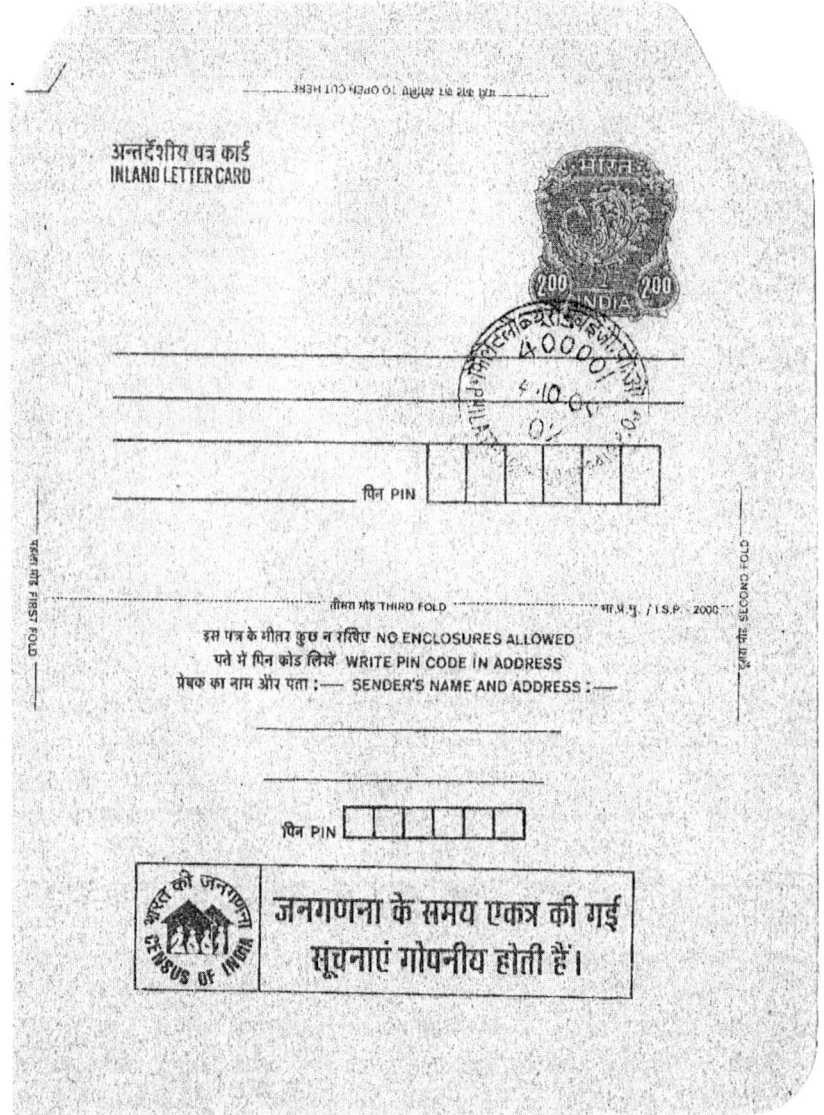

**Figure 5.1** Inland letter card from 2000 advertising the confidentiality guaranteed by the upcoming 2001 census

*Source*: Image by the author.

*Note*: This inland letter card was printed at the India Security Press (ISP) and issued on 4 October 2000 (Kotadia and Gandhi 2002a). It features a monolingual census advertisement (issued only in Hindi): 'जनगणना के समय एकत्र की गई सूचनाएं गोपनीय होती हैं।' (Information collected at the time of census is confidential).

religion to categorise people in colonies even though the state–church separation was observed in white-settler colonies and even in the officially Anglican United Kingdom (Bhagat 2003).

The refusal to answer compulsory questions and the giving of false responses are punishable offences under Section 11 of the Census Act, 1948.

> Section 11. (Penalties) (1d) any person[3] who intentionally gives a false answer to, or refuses to answer to the best of his knowledge or belief, any question asked of him by a census-officer which he is legally bound by section 8 to answer, or ... (1g) any person who, having been required under section 10 to fill up a schedule, knowingly and without sufficient cause fails to comply with the provisions of that section, or makes any false return thereunder, or ... (2) Whoever abets any offence under sub-section (1) shall be punishable with fine which may extend to one thousand rupees.

The Census Act treats *refusing to respond* and *giving incorrect response* as comparable offenses.[4] Moreover, it does not link punishment to the seriousness of the offence. The punishment for refusing to answer one question is the same as that for refusing to answer any question at all. Also, the fine for refusing to respond to a question related to the ownership of bicycle (in the Houselisting and Housing Schedule) is the same as that for refusing to answer crucial questions on migration, fertility and age (in the Household Schedule). The Act also ignores the possibility of collusive manipulation. Section 11(2) mechanically addresses the possibility of collusion by proscribing the abetment of the non-fulfilment of responsibilities, which include refusing to respond, giving incorrect responses and obstructing or interfering with census. But it neither recognises the distinctive (and potentially political) character of collusive manipulation nor acknowledges that the nature and impact of collusion are contingent upon the parties involved.

The instruction manual for enumeration shows a better understanding of collusive manipulation. It instructs enumerators to inform their supervisor in case they come across any organised movement to manipulate response to questions on identity – caste, religion and language. While possibilities of coverage errors (inadvertent omission and double counting by enumerators) (Government of India [GoI] n.d.14: 29) and collusive manipulation leading to content errors (misreporting of identity) (ibid.: 45–46, 55) are acknowledged in the manual, collusive manipulation leading to coverage errors is not discussed. Various census reports acknowledge the possibility of content errors due to the politicisation of numbers (for instance, GoI 2004a: xiv), but even these are silent on the possibility of politically motivated overcounting or over-reporting – that is, coverage errors. The widespread neglect of coverage errors associated with overcounting is examined in the next chapter.

Section 11(1)(b) of the Census Act prescribes punishment for enumerators who deliberately make false entries, but it is not clear if it also applies to cases in which the enumerator knowingly records wrong information supplied by respondents (and does not inform his or her supervisor). As mentioned earlier, training manuals clearly instruct enumerators to record answers 'as actually returned by the respondent' even when they have grounds to suspect that respondents have deliberately offered wrong information. The Act is silent on collusive manipulation by an enumerator and a respondent, but it is in principle covered under Section 11(2).

Last but not least, the possibility of manipulation of data by the government (not merely condoning manipulation at the grassroots by respondents and/ or enumerators) is not covered by the Act. The government can strategically manipulate the categories of enumeration before or after enumeration and/or even alter the collected data. This also highlights another gap in the Act. It does not apply to the complete life cycle of census data (Figure 1.5). For instance, there is no provision in this or any other allied legislation that addresses political interference at the stage of design of questionnaire and choice of methodology, on the one hand, and data processing and the use of data by state (or non-state) actors, on the other.[5]

We have so far discussed the scope of the Census Act and the gaps in its design. It would be instructive to review, if and how, the Act is invoked in courts. Court cases involving the Act can be classified as follows: (*a*) objections to the use of census data for the delimitation of constituencies for elections to various tiers of government; (*b*) objections to the change of names or boundaries of administrative units while census is in progress; (*c*) demands for the regularisation of employees temporarily recruited during census; (*d*) objections to the delegation of authority to enumerate; (*e*) disputes over the status of headcounts carried out during the inter-censal period; and (*f*) demands for the inclusion of more categories of information in the census questionnaire.

Among these only cases under (*a*) question the quality of census statistics (for example, *Chakhesang Public Organisation & Ors. v. The Union of India & Ors.*, WP No. 67 of 2006, Gauhati High Court). However, most cases of the (*a*) type focus on deviations from the guidelines for delimitation and invoke the Act perfunctorily.[6] Three aspects of the cases of the (*a*) type that question alleged lapses during the collection and processing of data and dispute the validity of census statistics are noteworthy. First, such cases question census in instrumental terms – that is, manipulated data are problematic insofar as they affect delimitation. Second, individuals or communities qua manipulators are not the defendants in these cases. Third, like all the other types of cases listed earlier, these cases were filed by non-governmental entities and individuals. In other words, the government has not filed cases under the Act to address instances of manipulation. However, as discussed later in this chapter, there have been instances of the initiation of disciplinary action against public servants through bureaucratic channels.

## Respondents

The design of the census and the associated administrative machinery inadvertently incentivise manipulation by respondents. The extended de facto (synchronous) method of counting that is meant to curb double counting and minimise manipulation engenders mistrust in fractured societies, particularly when the neutrality of the government is suspect. While census has no direct link with the allocation of welfare benefits in most cases, and the marginal benefit to an individual manipulator is often negligible, communities, localities and individuals try to secure future entitlements or protect their existing entitlements by boosting numbers if they fear that the government will be unable, or unwilling, to stop others from manipulating statistics. There is another aspect of the design of the Census Act that has not received much attention. The Act tries to balance diverse interests and gives limited powers to enumerators restricting both their ability to manipulate the answers of respondents and their authority to cross-examine wrong answers. For instance, in the case of the question regarding whether one belongs to an SC or an ST community, enumerators are asked to check if the response agrees with the list provided by the state government but does not allow them to cross-examine the respondent. So when certain non-tribal communities in Maharashtra declared en masse that they belonged to an ST community, enumerators simply noted down their response leading to inexplicable discontinuities in the trends of various parameters.

Further, people try to be consistent across surveys because they are not aware of legal provisions that limit the interlinking of data collected by different government agencies. The popular perception of the government's statistical system as a unified whole is not entirely unreasonable though.[7] First, the same set of the last-mile public servants enumerate residents of an administrative unit in censuses, enrol citizens of the corresponding electoral constituency as voters and identify below-poverty-line (BPL) households in that area. Second, the government cannot overlook major disagreements between data collected by different agencies, even in the absence of legal and formal interconnections. In fact, even where law prohibits linking individual-level data in different databases, there is no sanction against comparisons at the aggregate level. For instance, controlling for natural growth and migration, the electorate and the number of poor households in an area cannot exceed the adult population and the total number of households reported in the census, respectively. Third, government officials themselves do not fully understand the nature of interrelationship of various databases. Bhat (2018: 182) offers the following account of the visit of an enumerator to a household in Kashmir, which suggests that he or she believed that household size reported during the census would decide ration entitlement.

> I visited a household for enumeration. The household was very poor but very humble. There was a lady in the household who had delivered a male boy some two months back but unfortunately, this baby had expired soon after the birth. I just added the expired infant in the household roster, so that an additional member is included in the ration card.

Fourth, the census publicity campaign stresses the developmental and policy implications of the exercise, and the government admits that this affects the quality of data. A census report on Nagaland notes the following in this regard.

> Many equated it [census] with electoral rolls and saw the decadal Census exercise as an opportunity to increase the population in villages and towns to increase the vote bank. Further, loyalty to one's village and community compelled many to record their names in villages despite actually residing elsewhere. These problems were also compounded by the Developmental model followed in the State in which allocation of funds ... is made on the basis of population and households in a village. This naturally led many to try and increase the fund flow into their villages by showing non-existing population and households in the Census records. (GoI 2011h: viii, 2; see also note 35 of Chapter 4)

Fifth, unequal access to the benefits of affirmative action pushes communities to misreport their identity. S. Kulkarni (1991: 207) highlights an instance of 'fraudulent responses [members of non-tribal communities reporting themselves as STs] ... made in the belief that this would entitle the reporting households to the benefits available to the Scheduled Tribes'. Sahoo (2018: 34, 45) draws attention to tribal Christians misreporting themselves as Hindus.

While the popular perception of census may nudge individual respondents to manipulate their responses, we will argue that the government cannot invoke the punitive provisions of the Census Act to check manipulation. Modern censuses assume a 'cooperative relation between a state and its citizens' (Alonso Starr quoted in Prewitt 2010: 239). Invoking punitive provisions will generate a 'backlash against the census that would further depress cooperation' (Prewitt 2003: 15). The recent experience of Australia suggests that even a mild emphasis during census outreach on the punitive provisions can adversely affect public perception of the exercise (Purtill 2017). The controversies around attempts in the United States to include a citizenship question in the 2020 census and the Indian government's attempt to club the register of citizens and the 2021 census further highlight the central role trust plays in government efforts to collect data. There are other reasons, though, as to why the bureaucracy is not keen to invoke punitive provisions. The responsibility of conducting census rests on the regular government staff on the ground. The last thing this staff wants is to destroy the everyday rapport it enjoys with the people for a once-a-decade activity because the bitterness engendered by the implementation of the punitive provisions will linger long after enumeration. Also, 'the threat of punishment' is 'pointless' if 'the administration lacked the means to identify false declarations' (Göderle 2016: 81). To secure conviction it must be established that the information provided is incorrect conditional upon one's 'knowledge or belief'. Similar clauses are found in other laws such as the Registration of Births and Deaths Act, 1969 (Section 23 [1b]) and the Collection of Statistics Act, 2008 (Section 6), and, also, in the census laws of several other common-law countries. An incorrect response to the question on, say, the number of members in a household – that is, coverage errors – can possibly be objectively identified. However, prosecution will be difficult in the case of incorrect response to questions related to identity – that is, content errors.[8] Additionally, in the Indian context, establishing the violation of law will involve a long and tortuous legal process for which the already overstretched administration and voluntary enumerators have no appetite. So, in practice, the punitive provisions of the Census Act are rarely invoked against respondents.

We have so far argued that for various reasons the government cannot check manipulation by respondents. We will argue next that the government cannot even afford to prosecute negligent public servants.

## Bureaucracy

Errors attributed to enumerators can be explained by poor training, moral hazard, social, cultural and political biases, self-interest and acquiescence to administrative fiat or political pressure (Figure 5.2). In most states, enumerators are accused of moral hazard or dereliction of duty insofar as they may not enumerate the

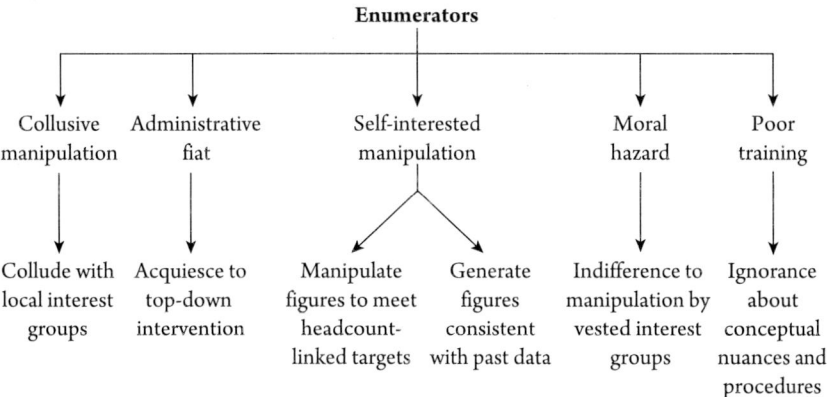

**Figure 5.2**   Errors attributed to enumerators

*Source*: Prepared by the author.

population by visiting every household in their assigned areas.[9] The problem of poor training is often ignored. On the one hand, a great majority of enumerators and other census officials are temporarily requisitioned from other departments, and therefore they neither have prior exposure nor the incentive to invest in training for a once-a-decade activity. On the other hand, the infrastructure for training of enumerators is inadequate. In 2011, the four-level training pyramid connected 90 national trainers to 2.7 million supervisors and enumerators through 725 master trainer facilitators and 54,000 master trainers (GoI 2011b: 10). Few enumerators read the elaborate training manuals, and fewer understand its nuances. There is also hardly any hands-on training for enumerators. As a result, it is commonplace for enumerators to confound de facto and de jure methods of enumeration leading to coverage errors or misunderstand the linkage between caste and tribe, on the one hand, and religion, on the other, leading to content errors. Even the limited training does not help in the case of absentee teachers in remote areas, who subcontract their responsibilities including teaching, census enumeration and voter enrolment to others.[10]

Another problem that has not received much attention relates to the biases of enumerators, which are often aligned with the dominant community. Minorities across India complain that enumerators are generally from outside the community and therefore have little idea of, and still less patience to understand, their distinctive character. This may result in inadvertent or deliberate misclassification (Tikhirs misclassified as members of Yimkhiung tribe in Nagaland, tribes misclassified as non-tribal communities in Karimganj district of Assam) or even undercounting (Kukis and Kacharis in Nagaland). In J&K, tribes such as

Bakarwal, Gujjar, Gaddi and Sippi complain that enumerators often deliberately undercount them (coverage error) or misreport their language and tribal status (content error).

Self-interested manipulation by public servants can take several forms. Officials often try to maintain consistency between different datasets. Concerns in this regard have been expressed with respect to PES in India (GoI 2001a: 225).[11] This is because the government officials who conduct censuses are also responsible for PES. Officials may also manipulate statistics when faced with policy targets related to, for instance, population control (Dandekar 2004: 44–8), agricultural reforms (Jerven 2013: 63, 75–9; Desiere, Staelens and D'Haese 2016), education and health (Sandefur and Glassman 2015), unemployment (Bookman 2013: 57) and economic reforms (Deaton and Kozel 2005: 196).

Further, it can be argued that officials could also manipulate data because a greater population share translates into more funds under their discretion.[12] This possibility needs to be examined in the context of the multiple tiers of bureaucracy. The census involves all four tiers of bureaucracy including senior bureaucratic positions manned by union and provincial civil servants, mid-level bureaucracy recruited within the state and the lowest rung often recruited locally within districts (Figure 5.3).

Government officials are in principle transferrable. So if they increase the population share of the area where they are currently posted, in the future they may find themselves in a jurisdiction whose population share and, by implication, share in government funds is lesser than it should be. We need to distinguish between the higher and lower echelons of the bureaucracy. The bureaucracy at the ORGI is concerned only about timely completion and incident-free enumeration without negative press coverage and political objections. In fact, the author's engagement with officials at the ORGI suggests that they often do not even check the data of the smaller states and communities, which partly explains numerous content errors in their headcount.[13] The Director of Census Operations (DCO) is mostly a union public servant who is not necessarily from the same state. It is noteworthy that until 2011 all DCOs of J&K were Kashmiri and/or Muslim.[14] Senior civil servants posted in districts, who supervise exercises like census and have the power to over-rule lower officials, are the most mobile among government servants. In fact, the complaint we often hear is that these officials seldom enjoy stable tenures and are unable to take their initiatives to the logical conclusion.

The lower rungs of the bureaucracy are on the other hand excessively immobile, being transferred either when they create problems for their seniors or when they are promoted.[15] If supervision is not perfect, it can be argued that lower

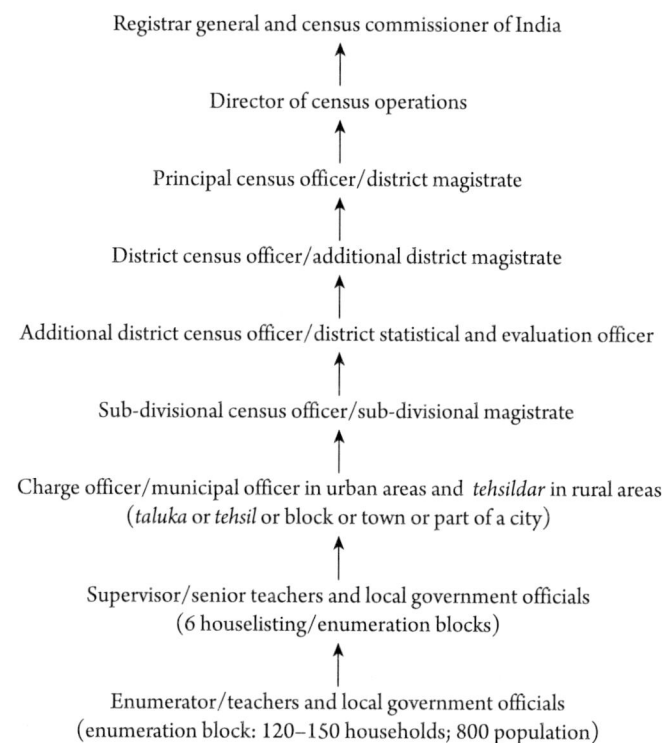

Registrar general and census commissioner of India

↑

Director of census operations

↑

Principal census officer/district magistrate

↑

District census officer/additional district magistrate

↑

Additional district census officer/district statistical and evaluation officer

↑

Sub-divisional census officer/sub-divisional magistrate

↑

Charge officer/municipal officer in urban areas and *tehsildar* in rural areas
(*taluka* or *tehsil* or block or town or part of a city)

↑

Supervisor/senior teachers and local government officials
(6 houselisting/enumeration blocks)

↑

Enumerator/teachers and local government officials
(enumeration block: 120–150 households; 800 population)

**Figure 5.3**   Organisation of census operations

*Source*: Census Rules, 1990; GoI (n.d.14: 3); GoI (n.d.22: 98); GoI (2010a: 2); GoI (2011d: 67).

officials have incentives to inflate headcount as members of local communities and as administrators of welfare schemes and funds. This, however, is not true of teachers who account for the bulk of the enumerators. Teachers are not directly involved in the apportionment or disbursement of welfare benefits.[16] So while the demand for inflating numbers qua officials who administer development funds cannot be ruled out (and this demand could grow as transfer rates drop), fund allocation is in the hands of highly mobile senior officials. Lower officials mostly implement predetermined policies subject to budgetary constraints imposed from above. It could be argued that junior officials can, possibly, keep the people in the dark and inflate population to attract additional development funds. But this cannot be concealed for long as government statistics are published routinely and thoroughly combed where people are dependent upon welfare schemes. So such officials must share gains with community (leaders) and/or senior officials sooner than later or risk being exposed.

Further, officials may voluntarily or otherwise condone manipulation of data by communities. Guilmoto and Rajan (2013: 69) argue that 'the Indian census has become far more democratic and manipulation is often more a political statement now than the outcome of statistical incomprehension'. In other words, people are not passive onlookers, and neither they nor the enumerators are ignorant of the consequences of manipulation. We have seen in Chapters 2–4 that non-state actors can mobilise people qua respondents and also exhort[17] or threaten officials to follow their diktat. In the case of community mobilisation to subvert counting there is, perhaps, very little the enumerator on the ground can do. The last-mile public servant is the most difficult to cross-examine and therefore the easiest to manipulate link in the statistical system, particularly in remote and politically restive areas. A Nigerian federal census officer explained the difficulty in finding the right kind of enumerator under such circumstances:

> If local people were appointed as enumerators, they would presumably assist the local population, as they appeared to have done in the present enumeration just held, to inflate the population artificially to obtain an increased share of amenities. If on the other hand, enumerators were to be appointed from outside, they would meet the difficulties of lack of knowledge of the area and, in many cases, lack of understanding of the language spoken. Furthermore, I imagine that considerable difficulty will be experienced in recruiting people from outside to go among strangers, many of them possibly hostile, with the unpleasant task of exposing a fraud. (Aluko 1965: 391)

P. D. Stracey, who served in the north-east, described the problem of manpower management as 'a set of contradictions incapable of being completely resolved' that affected the whole range of administrative activities.

> Staff were not to be posted in their 'home' districts as far as possible for reasons which, from the old British days, were well known and which in Nagaland were doubly valid because of the tribal clannishness. But how was the message of forest conservancy to be put over if there was no confidence between them and the people, even allowing for the limited utility of the common *lingua franca*, Assamese? So I tried the experiment of putting a man in his own tribal region and even, if necessary, near his village. But the results were generally disastrous! The best results were had with outsiders ... but they were often too timid to make their presence felt. (Stracey 1968: 268–9)

Indeed, experience shows that enumerators and other officials cannot defy entire communities determined to subvert the state and its routine activities.

In remote, politically restive areas bureaucrats from outside can be compelled to ignore manipulation insofar as they are personally vulnerable. But bureaucrats posted in their native districts are additionally vulnerable to manipulation through kinship networks.[18] In fact, governments in most developing countries face such manpower problems in the whole range of their regular operations and general administration rather than just once a decade during censuses.

The census faces manpower constraints at four levels. First, the union government relies upon state governments for door-to-door enumeration. The entire exercise depends on 'honorary labour by [hundreds of thousands of] persons deputed from other public employments' (Conlon 1981: 113). Senior bureaucrats are wary of initiating formal punitive action that will antagonise employees' unions and create problems for the ruling party apart from disturbing a variety of government statistical exercises and developmental and administrative interventions and their evaluation. In the run-up to the 2001 census, the president of the State Employees Conference (SEC) in J&K said that government employees should not be treated as 'sacrificial lambs' (*The Tribune* 2000b). The Registrar General of India and Census Commissioner (RGI&CC) noted that J&K was among the three states where employees' strikes 'just prior to the commencement of [2001] census operations' had 'jeopardized the census calendar' (Banthia 2001a, 2001b: 22; note 34 of Chapter 2). A month before the 2011 census, government employees had threatened to go on strike in J&K because of the non-implementation of the recommendations of the Sixth Pay Commission in the state (*Kashmir Times* 2011h). Second, census officials interviewed in Srinagar, Kohima and New Delhi suggested that they do not have effective administrative control over enumerators, who report to the district administration (see also GoI n.d.13: 32, 1951: 10, 1963a: 5; A. Bose 2000: 1433; *Kashmir Times* 2000d). But we have already noted that the district administration is itself constrained in this matter. Third, a senior census official associated with the 2011 census lamented that the directorate was 'understaffed', while 'those on rolls do not work' (interview, Srinagar, 30 November 2015). Fourth, it seems the government also found it difficult to find a bureaucrat ready to serve as the DCO for the 2001 census in J&K, which meant that the exercise was rudderless until a few months before houselisting (see also note 40 of Chapter 7 for related problems in other states).

## Prosecuting Violations

Given the manpower constraints, the punitive provisions of the Census Act are invoked in the worst-case scenario. On rare occasions, public servants have been discharged on grounds of dereliction of duty or prosecuted for refusing to perform their census duties as happened in the 1961 census of J&K (GoI 1963a: 2, 5). Otherwise, pro forma notices are used to rein in errant public servants. Commenting on enumerators responsible for the first census in Assam, Manipur and Tripura, the superintendent of census operations (SCO) noted that

> [t]he reaction of Supervisors and Enumerators on appointment was one of general reluctance to accept. They could be made to work only on pressure. In several cases notices had to be issued to show cause against prosecution. Issue of the notices was enough in most cases to bring them round. (GoI 1954b: 5)

In addition to the instance from 1961 mentioned earlier, there are hardly any references to punishment in J&K between 1961 and 2011. The DCO for the 1971 census reported that 'in Udhampur, finding that staff in the local Veterinary Unit had been avoiding performance of Census duties, the Charge Superintendent did not hesitate in sending them a formal notice for the start of proceedings under the Census Act' (GoI 1971b: 8). Sometimes even such notices are eschewed, and the senior bureaucracy relies on informal threats. The DCO for the 2001 census in J&K seems to have threatened to transfer those neglecting their responsibilities to Assam (interview, Jammu, 4 December 2019), while the government 'directed heads of departments to withhold salaries of employees who fail to make themselves available for work relating to the census' (Rediff 2000a; *Kashmir Times* 2000a). During the 2001 census, first information reports (FIRs) were filed against a few employees who 'failed to report on duty [in Banihal township, which adjoins Kashmir] despite repeated ultimatums' (*Kashmir Times* 2000h), notices were sent to many employees in Kashmir and a few teachers were reportedly arrested in Tangmarg (*Kashmir Times* 2000d). The notices were not issued by the DCO's office, though, as the state government employees are governed by their respective departments (ibid.). In 2011, 'disciplinary action [was recommended] against some supervisors and enumerators [in Srinagar] for not maintaining coordination with assistant charge officers and charge officers at grass root level and two enumerators have been suspended' (*Kashmir Times* 2011c). Otherwise, the Census Act does not seem to have been invoked even in the case of an enumerator who 'seemed to have *deliberately* recorded in a number of cases a religion different from the one professed by the persons enumerated' (GoI 1963a: 38–39, emphasis added). The errors were corrected, though (ibid.).

Even in the few instances in which action was initiated, there were long delays in implementing disciplinary action. According to the proceedings of the Fifth Conference of the Directors of Census Operations, held as part of the 1991 census, there were 22 disciplinary or vigilance cases, including two from J&K,

pending in 10 directorates (GoI 1992b: 104–05). We do not have the break-up of the cases depending on whether they are disciplinary or vigilance cases. We can roughly identify the rank of the official involved as we have information about the disciplinary authority. The president of India was the disciplinary authority in three cases, the registrar general of India (RGI) was the authority in two cases and the directorates were the authority in the rest. This, possibly, implies that notices were issued to three tiers of the bureaucracy. We have information on the duration of delay in the resolution of 18 out of 22 cases. The mean delay was about one year and four months. In six cases the delay was longer than two years. The small number of cases and the long delays in concluding them suggest that for all practical purposes the existing penal provisions are ineffective. Actually, the senior bureaucracy tasked with completing the census with the help of poorly trained and poorly motivated staff cannot afford to get distracted by tortuous administrative and legal proceedings. To put it differently, senior bureaucrats are primarily concerned about meeting the deadlines than ensuring the quality of data. As the DCO for the 1981 census of Manipur pointed out, the punitive provisions are superfluous once enumeration is in progress:

> It was brought to my notice, during and after the Houselisting operations that due reliance should not be placed upon the competence and sincerity of some of the teachers engaged as enumerators, and that as they were not strictly under the administrative control of the Charge Officers, it was difficult to extract satisfactory census work from them. Punitive measures taken at the eleventh hour were no answer to the problem of completing the enumeration operations at the right time. (GoI n.d.13: 32)

At another level, the Census Act does not hold the political leadership or even the senior bureaucracy accountable. Other laws governing the collection of statistics, too, overlook the possibility that the government can manipulate statistics. Even those that acknowledge that the government could err explicitly state that the punitive provisions do not apply to it.

In short, while there are good reasons to believe that enumerators could be a part of the problem, acting against them is not feasible even if they are found guilty of dereliction of duty. These problems are aggravated in conflict zones such as Kashmir, where both respondents and enumerators are threatened by non-state actors, the administrative machinery has been weakened by years of political unrest and there is a long-standing trust deficit between the government and people. The senior bureaucracy responsible for the census is acutely aware of how fragile the willing participation of people and public servants in such exercises is and seems to be prepared to live with faulty data (in one round) rather than risk a long-term alienation for the sake of a once-a-decade exercise. In this setting, punitive measures cannot be credibly invoked against either respondents or enumerators.[19] We will therefore turn to the security of enumerators next because some of the best-known cases of pervasive manipulation of census have been reported in conflict zones.

## Security of Enumerators

Despite the history of disruption of census in J&K and other politically restive states in the north-east,[20] the government does not formally provide a security cover to enumerators. The difficulties faced in conducting census amidst political unrest notwithstanding, providing security to enumerators is not desirable due to the need to maintain the non-confrontational and non-coercive character of the exercise. Even otherwise providing security is not feasible because enumeration is simultaneously conducted across the country and involves nearly three million government employees.

It is not the case that laws governing surveys do not envisage the provision of security. While the Collection of Statistics (Central) Rules, 1959, did not contain any provision in this regard, the Collection of Statistics Rules, 2011, framed under the Collection of Statistics Act, 2008, stipulate that 'in disturbed areas, the police, the para-military and the armed forces shall provide such assistance as would be required by the concerned statistics officer' (Article 10.3). However, this cannot be used to provide security to census enumerators because the Collection of Statistics Act, 2008, does not cover 'human population census' (Article 32) and, even otherwise, deals with specialised surveys that require small teams of surveyors.

In the absence of the administrative report, it is not clear if enumerators were provided security in Kashmir during the 2001 census.[21] There have been a few occasions in the past, though, when census enumerators were provided with a security cover in conflict zones. But the presence of security forces can alienate both respondents and enumerators. In 1961, 'filled up Houselists' were 'snatched away at gun point' in parts of Manipur because of which 'Houselisting had to be undertaken afresh during the course of enumeration' that in turn was synchronised with 'the intensification of the Manipur Administration's drive against the Naga hostiles' (GoI 1966c: 2, 7, 18). Long after this, people in that area continued to associate census with combing operations (GoI n.d.13: 35). Another example from Manipur is in order. The author spoke to a few census officials sent from New Delhi for the re-verification of the 2011 headcount in three sub-divisions of Manipur's Senapati district. Given the sensitive nature of the area and contentious nature of reverification, the census officials were accompanied by a large contingent of security personnel. More than half a decade later, they could barely conceal the delight about having had the security cover usually meant for senior political figures. It was clear that the presence of security forces had limited their field engagement.

Indeed, the experience of other countries shows that the involvement of security forces in censuses creates more problems than it solves. In the 1973 census of Nigeria, '[t]he census returns were conveyed to Lagos by the military personnel, presumably to frustrate any attempts to falsify the records' (Adepoju 1981: 32). But this did not improve trust in the census, and the differences over the census eventually 'contributed greatly to the fall of [President Yakubu] Gowon's regime in 1975' (ibid.: 29). The involvement of the armed forces in the 2017 Census of Pakistan undermined trust in the exercise (*Dawn* 2017a, 2017c). Security personnel accompanying enumerators of the 2023 census of Pakistan came under attack from insurgents on multiple occasions and, at least, three policemen lost their lives, and many others were injured in these attacks (*Dawn* 2023a, 2023b). In Myanmar, too, the involvement of armed forces in the census compromised the impartiality of the exercise because of their involvement in counter-insurgency operations targeting ethnic minorities.

## A Hollow Threat

In most common-law countries, fines for census violations are barely 1 per cent of the per capita income, and even these low levels of fines are rarely imposed. In India, the magnitude of fine under the Census Act has remained unchanged at INR 1,000 for decades.[22] At present, this amount is less than what a manual labourer would earn in two (urban areas) or three (rural areas) days. However, it must have been more than three years of wages of a manual labourer when the Act was first adopted in 1948.[23] So the opportunity cost of violating the law must have been very high in the 1950s, which is not the case at present. Other laws governing statistics do not prescribe very high fines either.[24]

The fines are not high in other common-law countries such as the US (not more than USD 500, as per the US Code Title 13, Section 221[a–b]), the United Kingdom (not exceeding Level 3 on the standard scale – that is, GBP 1,000, as per Census Act, 1920, Article 8[1]), Australia (10 penalty units or less – that is, AUD 2,100, as per the Census and Statistics Act, 1905, Section 14[1]) and Bangladesh (a prison sentence not exceeding one month and/or a fine not exceeding 'two hundred rupees', as per the Census Order, 1972, Article 13[e]). In Bangladesh, India and the US, and also in several other common-law countries such as Botswana, Fiji, Malaysia and the Bahamas, the fine is currently less than 1 per cent of the per capita gross domestic product (GDP) (Figure 5.4). The fine is more than 1 per cent of the per capita GDP in Singapore. In Australia and the United Kingdom, the fine is about 3 per cent of the per capita GDP. The current levels of fines are indicative of the seriousness with which governments treat violations of census laws. In India, the fine for violating the census law is the same as that for driving a two-wheeler vehicle without registration.

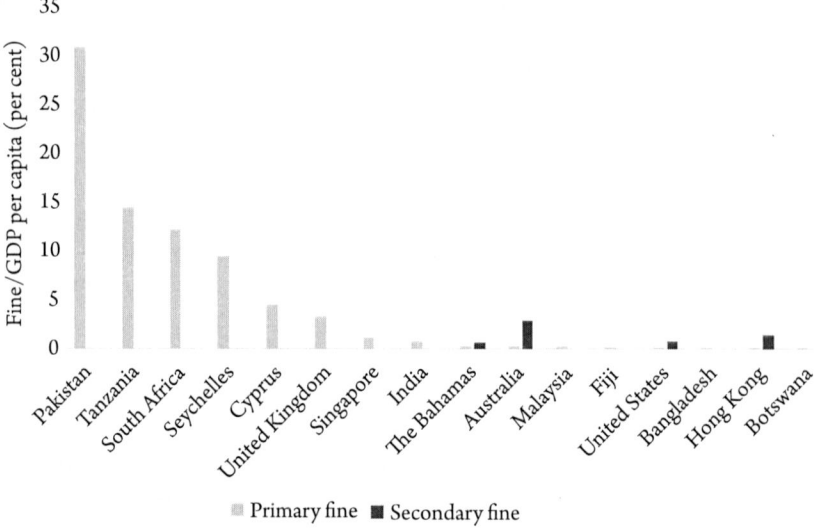

**Figure 5.4**  Census fines relative to the per capita gross domestic product (GDP) of various countries

*Source*: Kumar (2020a: 1148); fines: census and statistics laws downloaded from the webpages of the respective national statistical or legal departments or the United Nations Statistics Division (Laws and Acts on Vital Statistics System, https://unstats.un.org/unsd/vitalstatkb/ KnowledgebaseCategory14.aspx (accessed on 19 January 2019); GDP per capita (current local currency unit, 2017): 'World Bank Open Data', https://data.worldbank.org (accessed on 14 November 2018).

*Note*: The graph shows the ratio of fines (2017) prescribed in the census legislation of various common law countries to their GDP per capita (2017), both expressed in current local currency units. In countries where there is a hierarchy of fines, both the primary (offence: refusing to supply information and, in some countries, also supplying incorrect information) and secondary (offence: in some countries, supplying incorrect information) fines are shown.

The fine is quite high in Pakistan (INR 50,000 and 'may extend to two hundred rupees for each day during which the offence continues' as per the General Statistics (Reorganization) Act, 2011, Article 46) – that is, nearly 30 per cent of the per capita GDP. The relative fines are also very high in South Africa. Pakistan and South Africa are outliers, possibly because their census laws are of recent origin. However, the fines are low in Australia despite periodic revisions.

There is a reason why most governments have not revised fines. The punitive provisions of the census laws are rarely, if ever, used. As noted earlier, there are hardly any recorded cases in India involving the prosecution of individuals qua respondents on grounds of refusing to respond to census questions or giving misleading responses. A few cases of cursory invocation of the punitive provisions can be found in census reports of the earlier decades. A census report for J&K, for instance, noted that

> in majority of the cases the enumerator had not been able to extract information about the respondent's participation in some other economically productive work ... elderly persons of the area and the Government servants by making clear to such respondents the provisions of the Census Act that they were convinced and parted with the information about their participation in secondary/marginal economic activities. (GoI 1985b: 38)

The administrative report of the 1981 census of Manipur highlighted the practical difficulties in invoking the punitive provisions against respondents.

> As the enumerators have been instructed to record faithfully whatever a person answers to the questions put by them there is nothing we can do about these incorrect answers. The law against defaulting enumerators and persons refusing to answer, the census questions, as it stands, appears to be not very effective. Long drawn-out processes of law have been prescribed by the Census Act, for prosecution of Census offences ... provisions of the Census Act require that for the prosecution of a Census offender, sanction has to be sought and obtained first and then a formal complaint has to be lodged before a Court of competent jurisdiction. And once a complaint is lodged, as provided under the Act, the law will take its own course and there is no certainty that the prosecution will result in conviction. No Census Officer who has to do a good deal of work within a limited time is likely to have a sustained interest in prosecuting a Census defaulter. (GoI n.d.13: 39)

Once again, the experience of other common-law countries is in broad agreement with India's experience. In the past seven censuses, the United Kingdom has never prosecuted more than a few hundred individuals for breaking the law (Glaister 2012). For instance, '[i]n 2001, three million people refused to fill in the forms but there were fewer than 40 convictions' (*Deccan Herald* 2011). In Australia, too, not more than a few hundred people have been prosecuted over the past two decades (Purtill 2017). It seems that there has not been any prosecution related to the violation of census laws in the US since 1960 (Reamer 2012). While our discussion is largely focused on English-speaking, common-law countries, the experience of other countries is not quite different. Koch-Weser (2013: 22) notes that in China 'the enforcement of the Statistics Law has been lackluster. In 1994 and 1997, a series of nation-wide inspections uncovered more than 60,000 violations of the Statistics Law. In spite of the huge number of violations, very few officials were punished.' We have already seen how punitive provisions of India's Census Act are ineffective vis-à-vis the bureaucracy.

## Correcting Headcounts

We have so far seen that India's census law is incomplete and inadequate vis-à-vis manipulation by individuals, let alone (politically motivated) collusive manipulation or manipulation through administrative fiat, and its punitive provisions are ineffectual. The problem is compounded by the absence of any provision in the Census Act, 1948, and the Census Rules, 1990, for correcting erroneous data. The training manuals are silent in this regard even when they discuss the possibility of mobilisation to manipulate the composition of headcount. The training manual for the 2011 census offered the following instructions to enumerators in case they came across instances of coordinated misreporting.

> If you have reasons to suspect that in any area due to any organised movement,
>
> 6.52. the religion is;
>
> 6.60. the Scheduled Castes or the Scheduled Tribes are;
>
> 6.76 (i) the mother tongue is
>
> not being truthfully returned, you should record them as actually returned by the respondent and make a report to your Supervisory Officer for verification. (GoI n.d.14: 45–46, 55)

The manual is silent on the administrative procedure for verification and the course of action to be adopted during or after the enumeration if verification confirms collusive manipulation. As far as the enumerator is considered, his or her responsibility is over after reporting the problem. Also, it is not clear how the legal responsibility shall be apportioned among the defendants if collusive manipulation is established.

Further, while the Census Act prescribes punishment in case respondents fail to fulfil their obligations under the law, it does not expect compliance of them after the wrongdoing has been established. In fact, even if a defendant offers to provide correct information in the process of legal proceedings, the Act does not contain any provision for updating the census data. To put it differently, a person who has been charged for providing incorrect responses is not expected to supply the correct information as there is no provision for updating the records after the completion of the revisional round. In the United Kingdom, respondents can offer to comply with the Census Act, 1920, even after filing of a case (Glaister 2012). Other Indian legislation on statistics do not release respondents from the obligation to provide correct information even after conviction – for example, the Collection of Statistics Act, 2008 (Article 15[2]) and the Registration of Births and Deaths Act, 1969 (Article 13[4]). Pakistan's General Statistics (Reorganization) Act, 2011 (Article 46), too, requires compliance and imposes a fine of 'two hundred rupees for each day during which the offence continues'.

Both the Census Act and the Census Rules are silent on how to handle (a) errors during and after enumeration but before the provisional data are published, (b) errors detected between the publication of provisional and final figures and (c) errors detected after the release of final figures. The census adopts ad hoc measures to cope with such situations. A few major categories of responses can be identified.

First, nothing is done in an overwhelming majority of cases. In the absence of administrative reports for the recent censuses, this means that users are not even aware of pulls and pushes that have shaped the headcount. The *minor* and *local* abnormalities that make up the smoothly varying national headcount are not acknowledged, let alone examined and corrected. In Maharashtra, 'a sub-caste of Koshti caste ... reported to the census enumerators in 1981 that they belonged to the Halba/Halbi tribe' after the implementation of the Removal of Area Restrictions (Amendment) Act of 1976 that delinked the Scheduled Areas and the Scheduled Tribes (STs) (Kulkarni 1991: 206) and a legislation that 'cancelled sales of tribal land to non-tribals made after 1957' (Guha 2003: 161–62). The government was aware of the problem well before enumeration and still did not cross-verify obviously erroneous statistics (S. Kulkarni 1991: 206; Guha 2003: 161–62). Not only did the government remain silent in 1981 when the problem first surfaced, it also did not try to rectify the errors in the subsequent censuses. Likewise, the errors in the 2001 headcount of Assam's Karimganj district that were acknowledged by the local administration remained uncorrected in the next census (Agrawal and Kumar 2020a: 309–10). In J&K, census reports did not even discuss the errors we identified in Chapter 3.

Second, senior officers on the ground can reject the data. In 2011, 'census schedules for the three sub-divisions were not signed by the then deputy commissioner of Senapati district [of Manipur] citing abnormal growth rate' (*Sangai Express* 2013). Initially, the ORGI endorsed the decision and released estimated headcounts but later restored the reported figures under political pressure (senior census official, interview, Kohima, 11 December 2018).

Third, records are corrected if the census officials in the state find errors before results are published. In 1961, records were corrected in J&K in a few cases in which 'the religion entered was at variance with the name of the person enumerated' (GoI 1968b: 235). The assumption of a neat mapping between names and religious affiliations is problematic, though. For the most recent census, the DCO of Nagaland (interview, Kohima, 24 June 2013) explained the three-layered checks beginning with correction at the initial stage before the data are sent to the processing centres, followed by manual checks for mismatch in coding after the data are received from the processing centre – say, '[i]f the tribe is Angami, but code is SC [Scheduled Caste], then we can easily correct this. Even if tribe is not mentioned, we can find it out from names. ORGI allows for correction in such cases.' The final round of checks is conducted at the level of the ORGI. Agrawal and Kumar (2020a: 174–79) discuss the numerous content errors in Nagaland's headcount that escaped these multiple layers of checks.

Fourth, once the local administration authenticates and deposits filled-in sheets, the ORGI can only withhold the publication of data under the Census Rules, 1990 (Rule 7). For instance, the disaggregated language data for Punjab, the PEPSU and Himachal Pradesh were not published after the 1951 census due to concerns about the quality of data, among other things. Results for three sub-divisions of the state of Manipur and (only) the general population tables (GPT) for the state of Nagaland were not published after the 2001 census for similar reasons. In each of these cases, the decision was administrative and ad hoc in nature and did not invoke the Census Act or even the Census Rules.

Fifth, the officials in the ORGI pass the blame for errors on to the DCOs, who in turn blame the ORGI. There is no accountability in the system, particularly because the heads of the ORGI and the directorates are associated with the exercise for a limited period on deputation. On the rare occasions when census data kick up a political storm, census medals are withheld (as happened after the 2001 census in Nagaland), and bureaucrats supervising the exercise are shunted out (as happened after the ORGI mistakenly compared the 1991 population of India excluding J&K with the 2001 population that included the state).

Last but not least, in very rare cases, the political leadership helps rebuild trust in the census. The 2011 Census of Nagaland that corrected accumulated errors in the state's headcount is a case in point. Note that there is no guideline in the census legislation for a course correction, and, as a result, it is entirely dependent on the initiative of the political and bureaucratic leaderships.

It is not the case that the census does not estimate errors in its data. It has an internal mechanism in the form of PES to detect both coverage and content errors.[25] The results of this survey are provided only at the level of region – that is, groups of states and union territories (UTs) – which means they cannot be used to correct even state-level data. Moreover, the National Statistical Commission observed that the results may not be reliable since 'the conduct of PEC [known as PES in recent censuses] by the State Government staff raises some doubt on the integrity of the PEC, which should better be entrusted to an independent agency' (GoI 2001a: 225; see also Mitra 1994: 3207 for similar concerns in the 1950s and the 1960s). In other words, it is conducted by the same machinery that conducts enumeration. In any case, the census does not adjust the final figures using the results of the PES, which means the unadjusted estimates continue to hold as the official estimates and feed into other surveys, population projections, policymaking and academic analyses. Few, if any, policymakers and researchers check if their conclusions based on population statistics hold after correcting for the errors estimated by the PES.

Further, there is no mechanism to retrospectively correct errors identified after the PES. By 2005 the Naga civil society had demonstrated that the 2001 census was deeply flawed. Private entities can approach courts to challenge the data, but they cannot initiate proceedings under the Census Act against public servants or private entities as that requires prior sanction of the government.[26] So the court cases were limited to appeals against the use of flawed data in delimitation. The ORGI's response was limited to not publishing the GPT for Nagaland and not awarding census medals while ignoring the demand of the state government and the civil society for a fresh census. The ORGI accepted the demand to verify the headcount in three sub-divisions of Manipur that reported the most anomalous population growth rates, but the verification exercise was not successful (Agrawal and Kumar 2020b). Nagaland's demand for recounting could not be accepted probably because it involved conducting a fresh census in the entire state. As mentioned earlier, in such cases, the government has to wait until the next census to clean up the data. Gradualism helped in Nagaland where the chief minister threw his weight behind the clean-up. In the absence of political support, the wait-and-watch policy failed in Manipur and J&K, where the government benefitted from polarisation around flawed censuses, and led to larger errors in the next census. The experience of other countries also suggests that a cautious approach is desirable. Nigeria's early post-colonial experience shows that the belief that others inflate headcount can prove to be sticky and a hasty recount may not help. A recount was ordered within a year of the 1962 census of Nigeria, which was widely believed to have been rigged in the northern parts of the country. The fresh census failed to undo the distrust, and census figures continued to be disputed, partly because of the suspicion that the regional governments ensured that population figures varied in an acceptable manner vis-à-vis the 1962 census (Ahonsi 1988: 557). The regional distribution of headcount continues to be contentious in Nigeria, and the government has not been able to conduct regular and reliable censuses after 1973.

In the absence of good-quality metadata and a clear, ex-ante framework for corrections, ad hoc ex-post corrections and recounts compound the problem. The experience of Manipur is illustrative in this regard. In both 2001 and 2011, the ORGI released two sets of data – enumerated and estimated (after the census) – for the Mao-Maram, Paomata and Purul sub-divisions of Senapati district that reported abnormally high population growth. After the 2001 census, the civil society and political parties approached the courts seeking re-enumeration of nine sub-divisions, but the ORGI restricted the exercise to three sub-divisions without explaining the rationale behind the choice of this sub-set. The headcount as per the original enumeration, as reported in the provisional population totals (PPT) for 2001, suggested that these sub-divisions grew at the rate of more than 120 per cent between 1991 and 2001. The ORGI excluded these sub-divisions from subsequent publications, and, contrary to the rule that 'Census returns are not permitted to be arbitrarily corrected and that Census figures are published as obtained' (GoI 1966c: 147), it released estimated figures for these sub-divisions. As per the estimated figures, these sub-divisions grew at the uniform rate of 39 per cent during 1991–2001. The ad hoc procedure through which a 39 per cent population growth rate was estimated is neither mentioned in any census publication nor self-evident.

After the 2011 census, the ORGI initially withheld the results of enumeration and reported estimated figures in the PPT and the primary census abstract (PCA). The estimated figures for 2011 arbitrarily cap the decadal growth rate in the three sub-divisions at 23.8 per cent, which equals the growth rate of the rest of the district during 2001–11. However, after initially withholding the enumerated population data due to 'administrative reasons', the ORGI belatedly released the 'finalized' figures in 2014 without any explanation (GoI 2014f).[27] The enumerated figures for 2011 suggest that the population of the three sub-divisions grew at abnormally high rates between 100 and 136 per cent during 2001–11.

The census publications followed varying formats for reporting the population of Manipur. The PPT for India and its supplement released in 2001 provide enumerated figures for both Manipur and Senapati. Later 2001 publications such as the final population totals (FPT) for India report estimated figures. Other census publications such as *The District Census Handbook* for Senapati and the PCA report district-level estimates excluding the three sub-divisions. Still other publications such as the GPT for India, Manipur's administrative atlas, Manipur's census atlas and the *Statistical Handbook of Manipur* report the headcount of Manipur both including and excluding the estimated figures for the three sub-divisions.

The manner of the release of the data of these three sub-divisions highlights two key problems. First, census publications, including those on Senapati district, cryptically refer to administrative and technical difficulties but do not discuss the problems faced during the enumeration and their impact on data quality or explain the procedure followed to arrive at estimated figures. Second, in the absence of clear guidelines for correction, several contradictory reports were released. Both government officials and researchers end up using the wrong data because of the poor quality of metadata that makes it difficult to figure out the relationship between population data released at different points in time.[28] Even the DCO, Manipur, estimated the growth rate during 2001–11 by comparing the 2001 population of Senapati district excluding the three sub-divisions with the 2011 population including them. Recently, in J&K, the National Conference's (NC's) dissent note submitted to the Delimitation Commission (GoI 2022: 26) contained population estimates for 2011 that agree with neither the PPT nor the PCA.

## Concluding Remarks

This chapter introduced the legislation governing census operations and examined the punitive provisions. It argued that the Census Act, 1948, is structurally deficient vis-à-vis politically motivated manipulation by non-state as well as state actors and lacks provisions for correction. The punitive provisions are mild and rarely invoked in most common-law countries because census is conceived of as a non-coercive activity involving the willing participation of respondents. Moreover, the dependence of the union government on the administrative apparatus of the state governments and the voluntary labour of hundreds of thousands of last-mile public servants constrains its disciplinary hand. Further, given the scale of operations and the sensitive nature of the relation between the government and the people, the state cannot even offer a security cover to enumerators or impose penalties on those who violate the Census Act as it does not want to alienate the people and the lowest cadre of employees for completing a once-a-decade exercise, howsoever important it may be. Even otherwise officials do not have the appetite to invoke punitive provisions as the judicial process can be initiated only after prior sanction[29] and is quite time-consuming while the chances of securing conviction are bleak.

A demonstration of the ineffectiveness of the punitive measures of the Census Act and their infrequent use does not, however, question the relevance of the Act. It only shows that like in the case of many other laws around the world, there is a big gap between the law on books and the law in action and that people often rely on informal norms and institutions. Formal laws have a didactic value and serve as outside options that are invoked only when local mechanisms of dispute resolution fail. For instance, during the 2001 census in J&K, the DCO threatened to send erring enumerators to Assam rather than initiate formal proceedings against them.

In light of the persistent limitations of the Census Act, we can argue that any revamp of the legislation and training manuals must consider the context of enumeration that is circumscribed by the conceptual design and operational compulsions of the Census of India that constrains its ability to address the problem of manipulation. The chapter also discussed the absence of a transparent ex-ante framework for the correction of errors detected during enumeration or after the release of data. We pointed out that the ORGI does not even formally integrate the results of the household census and the PES. On the other hand, ad hoc corrections create more problems than they solve. A more appropriate approach to this problem would involve the adoption of clear ex-ante guidelines for correction and the revival of the earlier tradition of detailed descriptive reports that not only supported informed public debate but also ensured that the next census did not begin with a blank slate.

## Notes

1. Following the Government of India Act, 1935, which assigned 'Census' to the 'Federal Legislative List' (No. 16), 'Census' was placed in the Union List (Constitution of India, Schedule VII, Item No. 69).

2. In the United Kingdom, an exemption vis-a-vis disclosing one's religious identity was introduced in 2000, ahead of the 2001 census in which a question on religion was going to be asked for the first time.

3. Unlike in Australia, Bangladesh, India, Pakistan and the United Kingdom, in the United States (US) the legal liability is restricted to respondents 'over eighteen years of age' (US Code Title 13, Section 221[a]).

4. In Bangladesh, Pakistan and the United Kingdom, there is just one category of punishment for respondents who violate the census law. In the US, fines differ depending on whether one 'refuses or willfully neglects ... to answer' (not more than USD 100) (US Code Title 13, Section 221[a]) or 'willfully gives any answer that is false' (not more than USD 500) (US Code Title 13, Section 221[b]). Likewise, in Australia, non-response attracts a fine of one penalty unit (Census and Statistics Act, 1905, Section 14[1]) (that is, AUD 210), whereas false response attracts a fine of 10 penalty units (Census and Statistics Act, 1905, Section 15) (that is, AUD 2,100).

5. Laws governing maps address the misrepresentation or misuse of cartographic data by non-state actors. Under the Criminal Law (Amending) Act, 1990, '[w]hoever publishes a map of India, which is not in conformity with the maps of India as published by the Survey of India, shall be punishable'. Private entities have been prosecuted under these laws (*Hindustan Times* 2015). The government, though, exempts itself from prosecution in such cases. See, for instance, Section 37 of the draft Geospatial Information Regulation Bill, 2016.

6. The use of census statistics for delimitation is contested in most democracies. For instance, in the US, the first country to link delimitation to decennial census, the correction of census statistics used for delimitation and federal redistribution has attracted a lot of litigation (Bradshaw 1996; Hamsher 2005; Prewitt 2003, 2010).

7. The perception of government surveys as interlinked exercises has only strengthened after the introduction of provisions to link access to welfare schemes to digital identity – that is, the Aadhaar number.

8. The possibility of wrong answers due to ignorance cannot be ruled out. Respondents in developing countries often misunderstand questions on employment (see Chapter 7). If respondents are not aware of or do not accept state-sponsored labels for their identity that will affect the data on religion, language, tribe and caste (GoI 1966c: 147, 1981b: 40). Moreover, a shift in self-identification amidst conflict, urbanisation, migration, changes in affirmative action and domicile policies and changes in political economy cannot be ruled out. Some of the content errors discussed in Chapter 3 partly reflect the ethnolinguistic churning within J&K. However, a structural change in identity affects different parts of a community at different times.

Recall the discussion in Chapter 3 on the differences between Kashmir and Jammu with regard to content errors in the data on Gujjars and Bakarwals and the differences between Leh and Zanskar with regard to the shift in linguistic identity. See also the discussion of changes in content errors in the data on Nagaland's Sumi language across censuses (Agrawal and Kumar 2020a: 177–79). It will be difficult to pinpoint right responses amidst such uneven identity shifts across space and time.

9. A 1961 report fleetingly refers to enumerators not visiting all households in parts of J&K (GoI 1968b: 235). For similar complaints in Nagaland, see Agrawal and Kumar (2020a: 161).

10. A civil servant in Manipur observed that teachers posted in remote areas subcontract tasks to proxies (interview, Senapati, 8 October 2019). The author came across a similar problem in remote villages in Nagaland.

11. In Nigeria, 'regional governments' allegedly ensured that population figures reported in 1963 varied in an acceptable manner 'in order not to appear guilty of inflation in the nullified census [of 1962]' (Ahonsi 1988: 557).

12. The planning officer of a district told a census official that the population of a town in his jurisdiction would be around 50,000–60,000 (interview, Srinagar, 30 November 2015), whereas later the 2011 census reported a much smaller population.

13. After the 2001 census, civil society organisations and even the governments of Nagaland and Manipur blamed the ORGI for inadequate supervision and for releasing inaccurate data without proper checks, but none blamed it for manipulating the data per se (Agrawal and Kumar 2020a, 2020b).

14. Superintendents of census operations (SCOs) and DCOs in J&K include M. H. Kamili (1961, Kashmiri, Muslim), J. N. Zutshi (1971, Kashmiri, Hindu), A. H. Khan (1981, Kashmiri, from Banihal in Jammu division, Muslim), A. R. Parray (1991, census not conducted, Kashmiri, Muslim), Feroze Ahmed (2001, Ladakhi, Muslim), Farooq Ahmad Factoo (2011, Kashmiri, Muslim) and G. Prasanna Ramaswamy (2021, census indefinitely delayed, non-Kashmiri, non-Muslim). Khan and Parray were officers of the Kashmir Administrative Service, who were promoted to the Indian Administrative Service (IAS) cadre.

15. The author met a lower-level bureaucrat who was posted in his 'home' area and had been part of the last two censuses in the same sub-division (interview, Senapati, 8 October 2019). This is next to impossible in case of senior officials.

16. In the last two censuses, government teachers facing the threat of closure of their school due to decreasing enrolment may have inflated child population. But this cannot account for any substantial increase in child population because in any given census year only a small fraction of schools is likely to face

this threat. Moreover, teachers are asked to cover areas that are not necessarily coterminous with the area served by their schools.

17. Brass (1974: 77–78) discusses a case from a district of Bihar in which a respondent shaped the enumerator's understanding of how to record the answer to the question on language.

18. Senior bureaucrats are very rarely posted in their home districts, which reduces the possibility of community pressures on them. This is particularly true of the 'backward' districts, where the higher and even mid-level bureaucracy is manned by members of 'advanced' communities from elsewhere. See also note 15 of this chapter.

19. There are hardly any instances of the use of the punitive provisions in either J&K or elsewhere in the country. In other words, the disuse of the punitive provisions of the Census Act cannot be attributed entirely to the disturbed political conditions. It is, however, not the case that all violations of laws governing statistics remain unpunished in J&K. The contrast between census and cartographic statistics is noteworthy insofar as non-governmental entities that misrepresent India's map can be punished. The television channel Al Jazeera had to go off-air for five days in 2015 for repeatedly displaying an incorrect map of J&K (*Hindustan Times* 2015). However, Al Jazeera continues to indulge in cartographic politics by other means, say, by referring to J&K as 'Indian-administered Kashmir' (Al Jazeera 2021).

20. During the 1961 census, two officials were killed by insurgents in the Naga Hills (GoI 1966a: 1–2), and neighbouring Manipur also reported instances of destruction of filled-in forms (GoI 1966c: 2). In 2011, census officials faced difficulties in Longleng (retired census official, interview, Delhi, 15 September 2012) and Tuensang (state government official, interview, Meluri, 2 November 2014) in Nagaland. In J&K, the census office in Srinagar was burnt in the 1990s, and census material was damaged during the 2001 census (Rediff.com 2000a).

21. Swami (2000a) notes the following regarding the security of enumerators during the 2001 census of J&K: 'Few residents of the Kashmir Valley will be surprised when troops in full battle gear come knocking on their doors. But they might be surprised when the soldiers do not demand information on terrorists: and when the nondescript men accompanying them set about filling in a long questionnaire on employment details and household appliances, among other details. Over 10,000 government officials are scheduled to be escorted from door to door through the Valley from late October to conduct the second phase of the 2001 Census survey' (see also *Times of India* 2000a). However, *Kashmir Times* (2000a) points out that 'government employees … refused to

accept security cover'. Later news reports suggested that a section of employees complained that they had been asked to work even as 'the administration failed to provide security' and demanded 'security and insurance cover' (*Kashmir Times* 2000d, 2000i).

22. See the text of the Census Act, 1948, as amended over the years (GoI 2009a).

23. Presently, '[a]ny census-officer or any person lawfully required to give assistance towards the taking of census who refuses to perform any duty imposed upon him by this [Census] Act' is, in case of conviction, liable to be punished 'with imprisonment which may extend to three years' (Section 11). Until 1994, the Act prescribed six months of imprisonment under Section 11. Further, the penalty for contravention of any order regarding requisitioning of premises, vehicles, and so on, for taking census is 'imprisonment for a term which may extend to one year or with fine or both' (Section 7H).

24. The Collection of Statistics Act, 2008, stipulates a fine of INR 1,000 for individuals (INR 5,000 for companies) who refuse to answer or wilfully give a false answer. The Collection of Statistics Act, 1953 (Article 8), did not distinguish between companies and individuals.

25. The ORGI has other mechanisms, too, for checking the quality of data. The Task Force on Quality Assurance and the Special Task Force for Religion and Scheduled Castes and Scheduled Tribes (GoI 2003: v) and the Working Group on Religion (GoI 2004a: 123), for instance, bring together census officials from across the country to validate the results. The effectiveness of these bodies is suspect, though.

26. Nepal's Statistics Act, 2015 (Section 11), explicitly notes that '[t]he Government of Nepal shall be plaintiff in cases under this Act'. In contrast, the Census Act, 1948 (Section 12), of India does not make the union government the sole plaintiff: 'No prosecution under this Act shall be instituted except with the previous sanction of the State Government or of an authority authorized in this behalf by the State Government.' In fact, in India, the union government cannot initiate cases against most of the workforce mobilised to conduct census including enumerators.

27. Discussions with ORGI officials (interview, New Delhi, 21 February 2019) failed to clarify the procedure followed to adjust the growth rate of Manipur's three sub-divisions. One of the most senior officials involved in the 2011 census suggested that the correction in 2001 was driven by 'political diktat' (interview, details withheld). Another senior official, though, maintained that uncorrected data for 2011 were released under political pressure (interview, Kohima, 11 December 2018).

28. See Begum and Miranda (1979: 90) for a related discussion on census in Bangladesh. See also notes 27 and 28 of Chapter 6 for the experience of Bhutan in this regard.

29. Obtaining sanction to initiate action against an employee of the state government is not straightforward as the ORGI or the DCO cannot provide the necessary clearances. As per the Census Act, 1948 (Section 12), the State Government is the only competent authority in this regard. Even senior civil servants qua census officers cannot directly initiate action. The ability of a state government to initiate proceedings and take them to the logical conclusion is circumscribed by the bargaining power of employee unions, among other things. As a result, even though the 1994 amendment to the Census Act introduced a provision for summary trials of errant officials under Section 13, we do not come across prosecutions in later censuses despite irregularities associated with collusive manipulation involving public servants.

# 6

## Growth as Well-Being

India has fully protected the minority Muslims and has given them equal
rights. The rapidly growing population of Muslims in India is a testimony to
the fact that ... minorities are flourishing here.
> —J. P. Nadda, national working president, Bharatiya Janata Party (BJP)
> (Rajya Sabha 2019c: 543)

## Introduction

The quality of census data for Jammu and Kashmir (J&K) is circumscribed by
uncertain weather, armed insurgency, forced migration, regional and communal
polarisation, uneven accounting of armed forces, international conflicts,
bureaucratic moral hazard, ineffective legal and administrative measures to check
manipulation and changes in borders, reference dates, demarcation of snowbound
territory and distribution of mobile population groups. Our ability to understand
the impact of these factors on data quality is limited by the poor quality of metadata.
It is, however, not clear why successive governments have failed to pay attention
to the growing data deficit of J&K. Most recently, the anomalous headcounts of
2001 and 2011 did not prompt a closer scrutiny.[1] We found that the over-
reporting of children (Chapter 3) and an unusually large growth of the slum
population (Chapter 4) in Kashmir account for several intertwined coverage
errors in the reported population of the state. In this chapter, we will show that
in 2011 the number of households was also inflated in Kashmir. These anomalies
inflate the population shares of both Kashmir and Muslims, and the effect is
compounded in the case of the latter by a very sharp rise in the population of
Generic Tribes (Chapter 3). We found that conventional demographic and non-
demographic factors cannot explain these anomalies and argued that in both
2001 and 2011 enumeration was affected by communal propaganda in Kashmir

that fanned fears of demographic marginalisation. Chapters 4 and 5 discussed the political and legal–administrative contexts of the census. The present chapter relates the indifference of New Delhi towards anomalous population statistics of J&K to strong priors that suggest over-enumeration is unlikely, and that too in conflict zones, and to an older tendency to read population growth as a measure of well-being. It then discusses coverage and content errors in census data on Punjab and J&K. It argues that some of the major anomalies in the headcount of J&K are products of deliberate intervention rather than being mere aggregations of random errors at the grassroots.

## The Priors

It is widely believed that over-enumeration is less likely than under-enumeration. Historically, statistics served as 'the eyes and ears of the government' (Rao 1999: 46), and people feared that surveys were means to expand the tax base and enforce conscription.[2] Over the years censuses 'became less concerned with what the people could be obliged to do for the state and more concerned with what the state could do for them' (Coleman 2012: 335), and there was a corresponding 'shift towards willing participation [in government surveys] on the part of the respondents' (Bookman 2013: 51). However, more than a century and a half later, the older suspicion of the government's statistical endeavours as a prelude to a more intrusive presence of the state has survived, especially in the social and geographical margins because of which it is widely believed that under-enumeration is more likely. The protest against the proposed National Register of Citizens (NRC) in India is a case in point.[3] More generally, the difficulty in exhaustively tracking everyone, moral hazard of enumerators and inaccessibility of the margins, too, imply that over-enumeration is less likely. Local customs and religious beliefs can also restrict the reach of the census.[4]

Introducing the first sample verification of headcount in India, the census commissioner, who was aware of over-enumeration in Bengal and Punjab in the 1941 census, argued:

> Omissions were not merely probable, they must have occurred. This is true as much of the Census in India as in all other countries ... as the count has necessarily to be taken over an extended period; and the people move about during the period, one cannot be absolutely sure that the same person was not counted in two different places by two different enumerators, even though very careful precautions had been taken in advance in order to guard against this contingency. Over-enumeration is, therefore, possible though far less likely than under-enumeration. (GoI 1953: 1)

Writing on the centenary of the Census of India, Srivastava (1972: 139) echoed the consensus when he suggested that over-enumeration seems 'to be less frequent than under-enumeration'.[5] Successive Post-Enumeration Surveys (PES) have, indeed, reported positive net omission rates in all regions of the country (GoI 2006c, 2014d). Government debates and academic literature on coverage errors in census have therefore mostly focused on undercount.

Over-enumeration seems even more unlikely in conflict zones such as Kashmir. For instance, demographer Ashish Bose (1991: 31) wrote, 'I do not rule out the possibility of a large undercount in [politically] disturbed districts of Punjab, Jammu & Kashmir and Assam'. While the inaccessibility of physical and social margins and conflict zones and the consequent communication gap and lack of trust suggest that under-enumeration is more likely, these conditions can also lead to over-enumeration. In the 1981 census, in some parts of Manipur people feared that houselisting was a precursor to 'compulsory procurement of paddy' or 'combing operations by the Army', while elsewhere in the state people believed that the exercise was linked to 'permits for distribution of sugar' (GoI n.d.13: 35). Indeed, after the initial reluctance or aversion to enumeration, the country's periphery has over the past few decades witnessed over-enumeration in several states. On the one hand, the government finds it difficult to cross-examine numbers in the periphery due to the weakening of the state's influence with social and physical distance from the administrative centre. On the other hand, excessive dependence on welfare schemes apportioned according to reported headcounts and the measured socio-economic disadvantage pushes communities in the social and geographical margins to over-report headcount. At the national level, peripheral states account for much of the non-coverage in government censuses and surveys, and within these states their respective peripheries account for a large part of the non-coverage (Agrawal and Kumar 2020a: 228–29, 277, 298–99; Kumar 2022c).

The lack of attention to the possibility of overcount is notable given several well-known instances of over-reporting, particularly in the run-up to the communal partition of the country. The communalisation of census that began in the late nineteenth century became entrenched after communal electorates were introduced under the Morley–Minto Reforms (1909). As early as 1911, the first census conducted after the introduction of communal electorates, communities started exaggerating 'numbers to secure more seats' (I. Ahmed 1999: 124). The last colonial census was conducted in 1941, after the delimitation of constituencies under the Government of India Act, 1935. By then 'redrawing the national boundaries ... along with the exchange of populations was ... already up for public discussion' (Dhulipala 2015: 142; see also Prakash 1979).

Two major instances of overcount were reported in 1941 in Punjab and Bengal, among the four provinces and princely states that were later partitioned, where 'there was competition between communities ... with the result that numbers were inflated' (GoI 1953: 1; see also GoI 1954a: 5). Communities were jostling to secure greater political representation and through that a favourable border in mixed population districts. The census commissioner of India found it difficult to 'defeat an excess of zeal' among people, while his counterpart in the princely state of Hyderabad added that the 'whole population was census conscious' and was 'trying to increase their numbers' (Husain 1945: iv). The first census commissioner of independent India characterised the over-reporting in 1941 as 'willful and anti-social' in character (GoI 1953: 1). In Assam, the third state to be partitioned, the Muslim League government was 'accused of scheming to have the 1941 Census falsely show a higher Muslim population' through manipulation of categories of enumeration (Phanjoubam 2016: 179–180; see also Maheshwari 1996: 133–34).[6] J&K, which was partitioned *after* 1947, did not suffer the subversion of counting in 1941.

While introducing the bill for a permanent census legislation, Sardar Vallabhbhai Patel, the home minister, referred to the poor quality of the 1941 census data used to partition provinces. During the debate on the bill, M. Ananthasayanam Ayyangar and Pattabhi Sitaramayya elaborated how headcount was inflated in Bengal (Maheshwari 1996: 139–41). As a result, the Census Act, 1948, contained stricter penalties than its colonial counterpart to insulate statistics from communal rivalries (ibid.). The new law was, however, followed by serious content errors in the language data of Punjab, the Patiala and East Punjab States Union (PEPSU) and Himachal Pradesh in the 1951 census (GoI 1951: 13), even if the coverage errors of the 1941 census were corrected in 1951 (GoI 1954a: 5).[7] Sample verification of the 1951 census could not be conducted in Punjab, the PEPSU and Himachal Pradesh (GoI 1953: 3). The punitive provisions of the Census Act were made stricter in 1994, but once again this was followed by politically motivated over-reporting in 2001 in Nagaland and the hills of Manipur. Unequal manipulation by communities triggered contests that forced the deferment of delimitation in Manipur and Nagaland which was being conducted after a long gap of three decades (Kumar 2022b) and resulted in the excess allocation of funds to Nagaland under the Fourteenth Finance Commission (Agrawal and Kumar 2020a). In short, there have been several major instances of over-enumeration, including in J&K, which have affected policymaking; yet governments maintain that over-enumeration is unlikely.

# A Measure of Well-Being

The fixation of post-1945 governments on population control is another factor that deflected attention from both the possibility of over-enumeration and the older tendency of governments to treat population growth as a measure of good governance, peace and progress.[8] Kenneth Prewitt draws attention to the latter in the context of early post-colonial United States (US):

> The nation's leaders were influenced by the French political philosopher Rousseau, who wrote that the most certain sign of the prosperity of a government is 'the number and increase of population'. Political thinking in the young America held that a growing population 'was both a product of and a tribute to the blessings which America enjoyed – blessings not only of Nature, but of government, economy, and society'. A flourishing population was a comforting sign that the new form of Republican government was working.... If the nation's leaders were disappointed when the 1790 Census count was lower than expected, it was a disappointment soon set aside. Rapid population growth quickly became America's demographic story. (Prewitt 2003: 6)

The information sought in the early censuses of the US further confirms that its founding fathers were not worried about population growth driven by higher fertility. The 1790 census 'did not seek information on the ages of women, and only sought the numbers of men above and below sixteen years of age, that is men above and below military age. Such limited data would have been quite inadequate to calculate nuptiality, fertility, or other predictors of natural increase' (Guha 2003: 156). The late eighteenth-century Malthusian obsession with population growth is absent here.[9] Population growth was also used in the US as a yardstick of the welfare and security of minorities. John C. Calhoun, who served as the vice president between 1825 and 1832, suggested that '[a]bolition, far from improving the condition of "the African", worsened it: where "the ancient relations" between the races had been retained, the condition of the slaves had improved "in every respect – in *number*, comfort, intelligence and morals"' (Rose 1991: 685, emphasis added).[10]

The use of population growth as a yardstick of good governance in the benign Rousseauian sense has not yet gone out of fashion. Commenting on the success of liberal internationalism, Root (2013: 66) approvingly notes, 'As victors of the Second World War and then the cold war, Western democracies entered the twenty-first century poised to champion the policies that between 1960 and 2001 were so successful that the world population doubled and economic activity increased sixfold.'

Similar arguments are not unheard of in contemporary India. Wajahat Habibullah, a retired Indian Administrative Service (IAS) officer of the J&K cadre and India's first chief information commissioner, pointed out that the share of Muslims in J&K's population 'has steadily increased' since 1981 and that districts of 'the fast-growing Kashmir Division' reported the 'highest population growth rate' as per the 2011 census (Habibullah 2020: 26–27). Habibullah attributed the fast population growth to 'the turnaround in health and life expectancy amongst Kashmiris' due to '[t]he investment of Sheikh Abdullah's government in public health and education in the '70s and early '80s' (ibid.).[11]

On several occasions, S. C. Jamir, one of the founders of the state of Nagaland and five-time chief minister, drew attention to the fact that 'Nagaland has the highest growth rate of population in [the] country' (Jamir n.d.1: 21, also 62). He was aware of the implications of the high population growth for an economy dependent upon shifting agriculture (Jamir 1996: 10–11) and for a revenue-deficit state unable to absorb the educated unemployed (Jamir 1999: 69–70). However, he often sounded as if high population growth rate was a sign that the Nagas were doing well within the Union of India. In his address on the 50th Republic Day celebrations, Jamir pointed out that

> development has not been confined only to infrastructural growth. We have made progress in our productivity and the overall quality of life. If Nagaland today has the highest growth rate of population is it not because we are more healthy than our forefathers and is it not because we live under a more amenable situation that supports life instead of destruction under harsh and primitive conditions? (Jamir n.d.2: 10; see also Jamir 1998: 130)

A few years later, Jamir's inaugural speech at the third Hornbill Festival (2002) made the connection between population growth and well-being more explicit and located it in the middle of a variety of other growth rates.

> The present generation who had not seen the condition of our land prior to the Statehood may find it difficult to appreciate these changes but if you compare even some of the development indicators the contrast is obvious. *In 1963 the people of Nagaland comprised a small population of 3,69,200 persons. Today we have grown more than fivefold to almost two million people.* We had hardly 169 kilometers of surfaced road and some other minor roads constructed mainly for administrative and security purposes. Today we have over 9,860 kilometers of road length reaching all the towns and villages of the State.... In 1963 we had only 735 kilowatts per hour installed capacity of electricity generation. Today …

we have an installed capacity to generate over 100 megawatts of power. This is nearly 140-fold increase since 1963. (Jamir 2003: 208, emphasis added)[12]

While population growth as a measure of good governance and security survived into the twentieth century, applications were increasingly limited to illiberal contexts.[13] In colonial Rwanda, Belgian authorities collected statistics as 'part of their obligations as Trustees of the area under the League of Nations and later the United Nations' (Uvin 2002: 149). The statistical endeavour 'fulfilled the political function of demonstrating to the League of Nations and then to the UN the good effects of Belgian trusteeship and mandate: increasing population figures *demonstrated* that the natives were not starving and were thus doing fine' (ibid.: 150, emphasis added).[14] In colonial India, similar anxieties seem to have dictated the 'practice of eliminating the effects of catastrophes while estimating the "normal" level of mortality', but this 'seriously jeopardized the reliability of the actuarial mortality estimates' (Maharatna and Sinha 2011: 8). More recently, supporters of the Communist Party of China have presented Xinjiang's 'population growth' as the 'best' answer to the 'Western smear campaign on Uygurs'. They argue:

> [T]he overall population in Xinjiang continued to grow steadily from 2010 to 2018. During this period, the population of permanent residents increased by 13.99 percent, among which the Uygur population increased by 25.04 percent, and the Han population 2.0 percent. Clearly, the growth rate of the Uygur population is nearly twice that of the overall residents and is way higher than that of the Han population. The Uygur population has increased by more than 2.5 million people in merely eight years. What kind of "genocide" is this?... [I]t is ridiculous to turn a blind eye to the facts and use population issues as a new lever to smear China. (Yan and Wuqing 2020)

Population growth of a community as a yardstick of its well-being has also been applied to minorities in post-colonial India as suggested in the epigraph to this chapter. Addressing an international audience in Washington, DC, R. N. Ravi, the then chairman of the Joint Intelligence Committee,[15] argued that '[t]he usual notions of minorities and their alienation are not valid in the Indian context'[16] and defended India's treatment of minorities using statistics on population growth. He said:

> Going by the usual definition of minority, their population is over 260 million i.e., over 21 per cent of India's population. There are over 180 million Muslims in India.... The *higher population growth* of smaller communities – in the last

60 years they have grown double in terms of population share – is *a credible indicator* of their ease and sense of stake in a happy co-existence with the rest. (*Outlook* 2015, emphasis added)

More recently, commenting on the condition of Muslims in independent India, a senior police official highlighted population growth as a measure of progress and security:

Today, they [Muslims] are *far more numerous than ever*. There are a much higher number of mosques, madrasas and maulvis, and a much greater display of religious symbolism in dress and appearance than ever. There has actually been an intensification of Islamic religiosity in the public sphere. As for the worldly matters, today's Muslim is better fed, better clothed, better educated, and most well-off in history. What an ordinary Muslim has to eat and wear today was unimaginable for her ancestors at any time in the past. At the same time, a stable secular and democratic polity, free from turbulence and internecine conflicts of the past, has made life more secure for everyone, including Muslims. (Hoda 2020, emphasis added)

Ravi and Hoda, however, represent a more recent innovation to a more-than-a-century-old demographic propaganda that uses the same headcounts to arrive at a very different conclusion. The Hindu nationalists have all along maintained that the large and fast-growing Muslim minority is a threat because of its separatist tendencies. They add that the Muslim population share has grown in India even as the share of non-Muslims has sharply decreased in both Pakistan and Bangladesh. This communal anxiety has roots in the colonial period. The colonial censuses and administrative practices not only solidified hard religious boundaries[17] but also provided avenues for concretely imagining religious aggregates. British officials beginning with Hunter (2002 [1876])[18] routinely offered communal interpretations of statistics that served as cues for propagandists. The census commissioner of Bengal for 1891 calculated the time it would take Hindus to disappear from Bengal due to the higher growth rate of Muslims (Jones 1981: 91). Commenting on the variations in Bengal's population, the superintendent of census operations (SCO) for 1911 noted that 'in different parts of the province this [natural growth] largely depends on strength of Mussalmans, who, as is well known, are more prolific than Hindus' ('O Malley 1913: 63). The SCO of Assam for the 1931 census highlighted that 'the invasion of a vast horde of land-hungry Bengali immigrants, mostly Muslims' and concluded that it was 'sad but by no means improbable that in another thirty years Sibsagar district will be the only part of Assam in which the Assamese will find himself at home' (Hutton 1933: 65–66). Such partisan and alarmist interpretations of census data were commonplace in colonial census reports and occassionally even after independence (see, for instance, GoI 1964f: 248–60). The colonial administration used census to make India governable that required, among other things, understanding and documenting the diversity of the country and representing it in mutually irreconcilable terms. British census officials were 'aware about the role of census statistics on religion flaring up communal divisions in the country' (Bhagat 2001: 4353) but argued that 'the census cannot, however, hide its head in the sand like the proverbial ostrich but must record as accurately as possible facts as they exist and there is no question of the existence of communal differences which are reflected at present in political constituencies' (J. H. Hutton quoted in ibid.). Unfortunately, the colonial officials did not merely 'record' facts; they routinely included statistically suspect conclusions in official reports that fanned communalism. The self-defence of colonial administrators is particularly questionable because, as noted earlier in Chapter 5, the early-twentieth-century British censuses did not cover religion.

The demographic propaganda of Hindu nationalists assumed a stable form after the 1901 census in U. N Mukherji's pamphlet, *Hindus: A Dying Race* (1909), which highlighted the declining population share of Hindus (Datta 1999; Bhagat 2001).[19] For Mukherji the census was a kind of 'Disaster Clock' that measured the distance of Hindus from extinction.[20] *Saviour of the Dying Race* (1924), inspired by Mukherji's pamphlet and written by Swami Shraddhananda of the Arya Samaj, was the other major pamphlet (Datta 1999; Bhagat 2001). It is noteworthy that such pamphlets appeared in Bengal and Punjab, the two provinces where Muslims constituted a majority. These two provinces also witnessed pervasive competitive over-reporting in the last colonial census conducted in 1941 and were among the four provinces that were partitioned. Not coincidentally, all these states have also been at the heart of the Hindu nationalist propaganda in the recent decades. The colonial-era pamphleteering did not stop with the partition of India. In 1979, more than three decades after independence, the Akhil Bharat Hindu Mahasabha published *They Count Their Gains, We Calculate Our Losses* to forewarn the Hindus in the run-up to the 1981 census (Prakash 1979). More recently, a Chennai-based think tank Centre for Policy Studies published *Religious Demography of India* (Joshi, Srinivas and Bajaj 2003), which represents the most extensive statement in this tradition.[21] The book was first published in 2003 with a foreword by the then minister of home affairs, L. K. Advani, whose point of departure was a quote by Augustus Comte: 'Demography is destiny' (ibid.: xv).[22] Advani also released the book (Dasgupta 2003) that was later updated to include data from the 2001 census. After it lost the 2004 general elections and the next government released the 2001 census data on religion, the Bharatiya Janata Party (BJP) published a collection of articles on the changing religious demography, including a piece by Joshi, Srinivas and Bajaj (BJP n.d.). In short, the treatment of the higher growth rate of Muslims as a proof of the well-being of the community and as evidence of the threat posed by the community are interlinked. A section of Hindu nationalists uses growth statistics for specific locations, or 'mini-Kashmirs', as a measure of the threat to 'indigenous' religions (see note 97 of Chapter 4), even as another uses the growing Muslim population share at the national level as evidence of the well-being of minorities in India (Rajya Sabha 2019c: 543; NDTV 2023).

## State Interventions

We have so far discussed examples of ex-post *interpretations* of larger populations as testimonies of success of government policies, which is different from ex-ante *interference* to alter the population through incentives that affect fertility and migration or push people to align closely with the state's preferred identity and ex-post *manipulation* of numbers. We turn next to these latter types of interventions.

There was no tribal majority province in newly independent India. This presented the Constituent Assembly with a dilemma as the principle of 'one person, one vote' would have meant complete marginalisation of sparsely populated tribal areas within legislatures dominated by densely populated non-tribal areas. The criterion of uniform constituency size was relaxed to ensure the representation of tribes (Lok Sabha 1999: vol. 4, 659–74). Moreover, the entry of outsiders and land transactions involving outsiders were regulated through colonial-era laws (Lok Sabha 1999: vol. 9, 967–1084). These provisions continued even after tribal areas in the periphery were granted statehood and have, in fact, become even more stringent in states such as Nagaland.[23] Until recently, J&K, too, was also allowed to restrict the rights of non-state subjects such as the West Pakistan refugees.

The union government has tried to maintain demographic balance even in non-peripheral regions. It has been accused of indifference in the face of under-enumeration of migrants or misclassification of their mother tongue, which inflates the population share of the titular majorities of the respective states. Moreover, to preserve the interstate distribution of population and electoral power, the government encouraged faster fertility decline in high-fertility states by dividing 'national contraceptive acceptance targets ... on the basis of a formula based on population size, social and economic situation, and previous family planning performance' (Gwatkin 1979: 40). It also froze the interstate delimitation of parliamentary constituencies for a period of 25 years in 1976, which was extended for another 25 years after the 2001 census, to protect the interests of states that have been relatively successful in controlling fertility. So until 2031 the interstate distribution of parliamentary seats will be based on the 1971 census (until recently the federal redistribution formulae were also linked to the 1971 census). Further, in six peripheral states, including J&K, even the intrastate delimitation was deferred to until after the first census taken after 2026 – that is, the 2031 census – to preserve the intrastate distribution of political power.

On the other hand, the government has tried to boost the fertility of select micro-minorities. The National Population Policy 2000 suggests that '[m]any tribal communities are dwindling in numbers' and 'may need information and counseling in respect of infertility' instead of 'fertility regulation' (GoI 2000e: 25). Socio-economically marginal groups are not the exclusive targets of the patronising attitude of the state. The Parsis, or Zoroastrians, have also attracted the attention of the state qua preservationist.[24] The denial of contraception to some of the micro-minorities has been criticised though.[25]

Otherwise, demographic interventions of the state in the twentieth century have mostly involved the manipulation of *reported* numbers.[26] Sunni-majority Saudi Arabia did not release census data for a long time as it feared that 'publishing an exact count (showing their own population to be smaller than many supposed) might encourage enemies to invade the country or promote subversion [in the Shia-dominated oil-rich Eastern province]' (Alonso and Starr 1985: 96). Turkmenistan did not release the results of the 2012 census that seems to have shown the population to be lesser than claimed by the government (Goble 2015). In fact, even the US, which is presently the third most populous country, was initially 'worried that a small population would tempt ... enemies to military action' (Prewitt 2003: 6). Bhutan, among the smallest countries, seems to have overstated its population before joining the United Nations (UN) (Chandrasekharan 2013).[27]

Even these isolated examples of boosting population, concealing smaller numbers or overstating numbers, represent a receding tendency. The post-Second World War period has witnessed manipulation of statistics to attract investment and development aid or deflect public criticism over, say, the falling growth rate of national income, rising unemployment and inflation or the impact of reforms (Bookman 2013: 57; Deaton and Kozel 2005; Chandrasekhar and Ghosh 2011, 2013; Desiere, Staelens and D'Haese 2016; Agrawal and Kumar 2020a: 5). Instances of manipulation of the overall population of the country are often limited to understatement, rather than overstatement, as in Bangladesh (Ahsan 2011), because '[a]rithmetically speaking, it is a battle over the size of a denominator – many indicators of economic development are expressed as a proportion of the total population. Politically, a small population is a nice thing to have' (*The Economist* 2011).[28]

China seems to have overestimated the number of births averted through family planning policies (Merli and Raftery 2000; Goodkind 2011). India, too, overestimated the success of its birth control policies (Dandekar 2004: 44–48). During the Emergency (1975–77), '[p]opulation control and family planning' was added to List III (Concurrent List) of the Seventh Schedule through the Constitution (Forty Second) Amendment Act, 1976, and the government encouraged forced sterilisations across India. The politicisation of fertility, with the ruling party and the prime minister's son taking direct interest in sterilisation, translated into unrealistic targets.

> The official 1976–77 sterilization targets allocated among the states in this manner totaled approximately 4.3 million acceptors. Then, one after another, the states began to declare they could do better. All but three of India's major states (the exceptions being Assam, Kerala, and Jammu and Kashmir) raised their targets. These self-proclaimed targets totaled over 8.6 million, twice the central ministry's original figure, with individual states and territories raising their targets by as much as three to six times. (Gwatkin 1979: 40–44)

To conclude, coverage errors attributed to governments are also associated with undercounting as post-colonial governments in the developing world have been anxious to showcase their success in reining in high population growth rates. So this too supports the consensus referred to earlier that undercounting is more likely than overcounting.

Other than cases of undercounting noted in the preceding discussion, instances of manipulation of census statistics by governments have mostly involved altering the composition of the reported population that manifests as content errors. Political transitions have, in particular, been associated with severe content errors due to the abrupt re-categorisation of population through administrative fiat.[29] The temptation to enhance numbers of the dominant group through re-categorisation depends on the context though. There are three possibilities. First, government intervention could be aimed at making the dominant community appear less threatening – for example, Muslims in the Ottoman Empire and Serbs in Yugoslavia (Bookman 2013: 57–58). In India, too, the census offers far more data on sects among Hindus compared to other religious communities. Second, intervention could boost the (relative) size of the dominant group. Apartheid-era South Africa was concerned about 'the increasingly unfavourable black–white population ratio' and manipulated both the census and internal political geography (Lipton 1972: 257).[30] Third, in the early twentieth century, there was 'a strong suspicion' that Tutsis were deliberately undercounted in Tutsi-dominated Rwanda 'to keep the privileged class as small as possible' (Uvin 2002: 155).

## Assuaging Anxious Majorities?

Independent India discouraged assertions of religious identity and recognised linguistic assertions as legitimate. This pushed religious minorities such as Sikhs and Muslims to repackage their claims in linguistic terms (Brass 1974: 46), while retaining the religious flavour by foregrounding the script of their respective scriptures.[31] As a result, not only did the major communal conflicts re-emerge in a linguistic avatar, but their relation with census also survived the political transition. The disaggregated data for the main languages of Punjab, the PEPSU and Himachal Pradesh were not released after the 1951 census to contain communal tensions (GoI 1954d: 2), and it was 'recommended that the mother tongue question not be canvassed' in specific areas in the next census (Mitra 1994: 3209). While the suggestion was not implemented, people continued to align their reported linguistic affiliation with their religious identity deepening the communal polarisation that reflected as anomalies in the census data on language.[32] The content error in language data resolved after a majority returned their mother tongues correctly in 1991. However, in the meantime, the government seems to have intervened to mitigate the Sikh anxiety about their separate (religious) identity and ability to politically dominate Punjab, the only state of the country where they constitute a majority. Between 1971 and 1991, Punjab's Sikh population grew faster than the state despite a lower fertility than that of Hindus. The high growth rate of Sikhs in Punjab reported between 1981 and 1991 cannot be explained by the possible migration of Sikhs from other parts of the country after the anti-Sikh riots of 1984 because the Sikh population outside the state reported a higher growth rate. In 2001, the Sikh population growth rate dropped sharply below the state's growth rate. The growth rate of Sikhs was twice that of Hindus as per the 1991 census, but the 2001 census suggested that Hindus grew twice as fast as Sikhs (Table 6.1).

Gill (2007: 248) suggests that the government possibly manipulated data to appease Sikhs. He argues:

**Table 6.1**   Population share and growth of Sikhs in Punjab and India

| Share and growth rate | Community | 1971 | 1981 | 1991 | 2001 | 2011 |
|---|---|---|---|---|---|---|
| | Sikhs in Punjab | 0.6022 | 0.6075 | 0.6295 | 0.5991 | 0.5769 |
| | Hindus in Punjab | 0.3754 | 0.3693 | 0.3446 | 0.3694 | 0.3849 |
| Population share | Sikhs in India* | 0.0194 | 0.0196 | 0.0199 | 0.0191 | 0.0176 |
| | Sikhs of Punjab in Sikhs of India* | 0.7952 | 0.7879 | 0.7860 | 0.7686 | 0.7778 |
| | Sikhs in Punjab | | 0.2499 | 0.2518 | 0.1429 | 0.0968 |
| Decadal growth rate | Hindus in Punjab | | 0.2188 | 0.1273 | 0.2874 | 0.1867 |
| | Punjab | | 0.2389 | 0.2081 | 0.2010 | 0.1389 |
| | Sikhs in India* | | 0.2615 | 0.2548 | 0.1689 | 0.0838 |

*Source*: Table REL-0300 of Census of India, 2011; GoI (2004a) for the earlier censuses.
*Note*: (*a*) 1951 and 1961 have been omitted as Punjab included Chandigarh and Haryana. (*b*) The 1971 figures for India exclude Sikkim. (*c*) *India excludes Assam and Jammu and Kashmir (J&K) in all the years as census could not be conducted in these states in 1981 and 1991, respectively. The two account for less than 1.25 per cent of the Sikh population of India. (*d*) The 2001 figures for India exclude Mao Maram, Paomata and Purul sub-divisions of Senapati district of Manipur.

The tactics adopted in terms of under-reporting of data on new religious persuasions as well as their characteristic categorisation in terms of various sects in Punjab is indicative of the role of political factors. In spite of the fact that the numbers of adherents of these sects, particularly the Radhaswamis, have been continuously on the rise in recent decades, these are grossly underplayed in census returns so that the Sikhs do not become adequately aware of the erosion of their population base. In contradistinction, the population figures for the Sikhs seem to be somewhat inflated in censuses from 1961 to 1991.[33] Similar processes could also not be ruled out completely in areas having notable concentration of tribal and other minority people. (Ibid.)

Gill adds that 'the census gives figures for the Hindu Radhaswamis only while it is silent concerning Radhaswamis from the Sikh community despite the fact that spread of this sect has been greater among the Sikhs' (ibid. 2007: 247).[34] After the insurgency subsided in the early 1990s, the next census revealed a sharp decrease in the growth rate of Sikh population. Gill suggests that the 'only plausible explanation of this could be that the figures relating the Sikh population got somewhat inflated for political reasons' (ibid.). He argues:

> It is helpful to the ruling class politically as it does not disturb the Sikh community whose base is systematically being eroded in Punjab through the spread of such sects.... Interestingly, the Sikh political leadership itself also seems to find non-recording of this type of data in its own interest since it helps avoid not only the resentment among the Sikh masses in this connection, but it also allows the obtaining Sikh political leadership to stay on the favourable side of India. (Ibid.: 248)[35]

A similar concern has been expressed in the context of J&K, where Hindus complain that New Delhi has condoned the inflation of the Kashmiri Muslim population. Kashmir's experience is different from that of Punjab, where the government allegedly condoned the reclassification of the population of heterodox sects that resulted in content errors. However, content errors resulting from government intervention cannot explain the change in the relative population shares of the main religions and regions of J&K reported in the census because, as discussed in Chapter 3, there are hardly any heterodox or unclassified religious groups that may have been inadvertently or deliberately misclassified. We showed that coverage errors explain the sharp increase in the Muslim population share. Just as governments in developing countries may be inclined to under-report headcount to claim success in population control, communities competing for the public pie overstate their headcount with an eye on the distribution of development funds, delimitation of electoral constituencies and demarcation of administrative units. Communities also stress the need to maintain the relative population shares when their share is declining relative to some historical datum (the dual majority of Kashmiris and Muslims in post-colonial J&K) or power-sharing arrangement (seat share of Kashmir in the state assembly). Government response in such situations depends on the local context and is not necessarily circumscribed by the universal temptation to reduce the denominator.

The 1981 census is the most reliable census conducted over the past few decades. It was conducted by the popularly elected government of Sheikh Abdullah. The state was relatively politically stable at that time, and census had not yet come under sustained communal attack. Enumeration was conducted between 20 April and 5 May (Table 2.3), which means the Darbar and Bakarwals would have moved back to Srinagar by the time the census was conducted. So it is highly unlikely that Kashmir's population was undercounted in 1981. However, the 2001 census revealed a sharp increase in the population share of Kashmir, which is counterintuitive.

First, the Kashmir division's population share increased despite the exodus of Hindus and, to a lesser extent, Sikhs in the 1990s. In fact, Muslims, too, migrated in substantial numbers. A few thousand political migrants – that is, Muslims who had to leave Kashmir due to their political or ideological affiliation[36] – and a much larger number of non-political Muslim migrants[37] will also have to be accounted for to estimate the magnitude of migration in the early 1990s. Observers unanimously agree that Jammu district emerged as the single biggest magnet for economic migrants within the state (Drabu 2020f, Jamwal 2020a). The large-scale migration out of Kashmir to Jammu and other states reflects in growing net out-migration from the state (Table 3.6) and increasing Kashmiri-speaking and Kashmir-born population in Jammu (Table 4.5). The armed conflict also reduced the migration of, for instance, Bakarwals to Kashmir (Javaid Rahi, interview, 13 August 2021).

Second, it has been argued that the heavy deployment of central armed forces explains the increase in Kashmir's population share reported in the 2001 census. Since non-Muslims are not under-represented in the armed forces, their deployment should have increased the population share of non-Muslims in the state, and it should not have tilted the balance in favour of Kashmir because there was a heavy deployment in Jammu as well. Moreover, as discussed in Chapter 2, armed forces do not seem to have been properly counted in Kashmir during the 2001 census, and therefore their deployment cannot explain the increase in Kashmir's population share in 2001.

Third, Kashmir's population share increased even though the 2001 census was conducted during winter when the Darbar and nomadic communities had moved to Jammu. In 1971, enumeration was conducted between 10 March and 31 March. In 1981, enumeration was conducted between 20 April and 5 May (Table 2.3). Fourth, all but one census official interviewed by this author argued that political unrest in 2000–01 should have resulted in undercounting in Kashmir. In fact, the provisional population totals (PPT) for the 2001 census carried a detailed note in this regard. It noted that

> while people in Jammu Province barring a few militancy infested pockets, by and large, welcomed census-taking after a gap of twenty years their counterpart in Kashmir Valley were sceptical about the whole exercise. But surprisingly while the entire population of the State had wholeheartedly cooperated in the first phase i.e. at the time of Houselisting which was conducted in May 16–June 5, they got up to negative postures in the second phase. More belligerency was in evidence in Kashmir Province than in areas under the Jammu Province. (Ibid. 2001b: 25)

It added that

> the people of the State were found divided on the importance of the data that would emerge as a result of the field operations. There was enthusiasm and vigour in Jammu city and other parts of the Jammu province, whereas complete indifference was evident in major parts of Kashmir Valley towards census-taking in the State. It is no secret that people in the Valley are sceptical to any work undertaken by the Government agencies. Neither the Central nor the State Government has found acceptance among a large section of population. They won't go by the substance of the exercise but see who is the initiator or the sponsor of the activity ... when militant outfits in the company of some politicians spoke against the population enumeration on one ground or the other, the print media also supported this cause. (GoI 2001b: 25–26)

The indifference or even hostility towards census among sections of Kashmir, coupled with the disturbed political conditions that severely restricted the free movement of enumerators should have resulted in undercounting and a decrease rather than an increase in Kashmir's population share.

Fifth, double counting cannot explain the increase in Kashmir's population share. A senior census official in the directorate of census operations, Srinagar, pointed out that settlers in urban areas get counted in their native villages, too, due to benefits linked to residence in backward areas. He suggested that a large fraction of '2–3 lakh employees posted outside their native places' are likely to be enumerated in both their place of posting and native place (census official, interview, Srinagar, 30 November 2015).[38] There is no reason why there should be more double counting in Kashmir, where unlike the Jammu division most districts are well connected to the divisional headquarter in Srinagar and certain location-specific categories of affirmative action have limited applicability (see, for instance, Maqbool 2023). It is also not clear why there should be more double counting in 2011 compared to 2001.[39]

We have shown that births, deaths and other conventional demographic and non-demographic factors cannot account for coverage errors driven by politically motivated manipulation. We will further argue that unlike in Nagaland the coverage errors in J&K did not result merely from the bottom-up aggregation of erroneous responses at the grassroots. There are several clues in this regard. First, the population shares of north, central and south Kashmir have remained unchanged between 2001 and 2011 despite migration within Kashmir to Srinagar and the out-migration of the entire Hindu community and, to a lesser extent, Sikh and Muslim communities of the Valley to Jammu and other parts of the country. On the other hand, despite steady in-migration, both Jammu district and the Mountainous region of which it is a part have seen a decline in their population share within both the state and the Jammu division. Also, the rate of urbanisation of Jammu district stagnated, and its slum population sharply contracted even as Kashmir's urbanisation and slum populations increased. Like the sharp decline in the child sex ratio (CSR) and the increase in the child population share, the higher rates of growth of urban and slum population were also pervasive across Kashmir.

Second, as discussed in Chapter 4, as per the 2001 census, the population share of Kashmir grew to just about the level that would support 46 seats assigned to it in the delimitation of assembly constituencies carried out six years ago in 1995.

Third, Bhat (2018: 182) identifies 'three narratives from the Enumerators' that he suggests 'are representatives of what most other enumerators did during the Census'. We will quote him at length here as he was one of the first to flag this problem soon after the provisional data for the 2011 census were released.

1. I was *told by my supervisor* that enumerated population in the assigned village is less as compared to the expected population. I was asked to add 20 more household members in the village. So I revisited some of the households where either there was a pregnant lady in the last trimester or where there was a death in the recent past. All third trimester pregnant ladies were shown to have delivered a live birth. If the lady already had a daughter, this one was shown to be a son and if it was the first pregnancy, it was generally recorded as a son.

2. I visited a household for enumeration. The household was very poor but very humble. There was a lady in the household who had delivered a male boy some two months back but unfortunately, this baby had expired soon after the birth. *I just added the expired infant* in the household roster, so that an additional member is included in the ration card.

3. I was *asked to add few members* and I revisited a household to include some more members. During, the revisit I found a household who had very recently lost the breadwinner in an accident. The expired person was enumerated as a living person to meet the target. (Ibid., emphasis added)

Third-trimester pregnancies and recent deaths can at best account for about 50,000 additional population that is less than a tenth of our conservative estimates of overcount in Kashmir. The importance of Bhat's account lies elsewhere.

It is interesting that the non-existent children who were enumerated were in many cases yet to be delivered third-trimester pregnancies. Likewise, the enumeration of the very recent deaths is also noteworthy. This is an important clue regarding the involvement of those who understood the implications of intervention at one stage of the census for its detection at subsequent stages. Counting third-trimester pregnancies in February and very recent deaths meant that the PES, which was going to be carried out from April through July – that is, a month after the enumeration – would most likely fail to detect the inflation of headcount.

So it was not just poor respondents trying to get an extra ration or a job card. The large and pervasive coverage errors across all the districts of Kashmir and significant anomalies in a variety of parameters in the 2011 census could not have been possible unless the political and/or bureaucratic leadership intervened or at least condoned politically motivated manipulation. Bhat (2018: 182) argues that enumerators served as the link between the Kashmiri elite discourse on the imminent adverse demographic change directed by New Delhi and the everyday concerns of people at the grassroots. He suggests that

> the general public was interested in over-enumerating the household members in order to get more subsidized commodities and they were fully supported by the Census staff in this business by inventing novel ways to add nonexistent souls. This helped them to keep happy both the separatists and the household heads. (Ibid.)

Fourth, in 2011, compared to the rest of the country, J&K reported a much larger increase in the number of households between the houselisting and household phases of census with only Goa and Lakshadweep being close to it. In 2001, the increase in the number of households in J&K between the two phases was comparable to the rest of the country. In both 2001 and 2011, Kashmir reported a larger increase in the number of households between the two phases of census than Jammu. Srinagar stood out within Kashmir in this regard (Figure 6.1). The sharp rise in the number of households in Kashmir in the household phase compared to the houselisting phase suggests a narrow window during which the manipulation of headcount took place – namely, the household phase of census.[40] Following Bhat (2018: 182), we can add that figures would have been manipulated towards the latter half of the household phase and during the revisional round.[41] In the absence of metadata on census in snowbound areas we cannot conclusively establish the aforementioned claim by testing the hypothesis that there were systematic differences in data quality between (a) snowbound areas and the rest of the areas, (b) communities counted in Kashmir's snowbound areas and the rest of Kashmir,

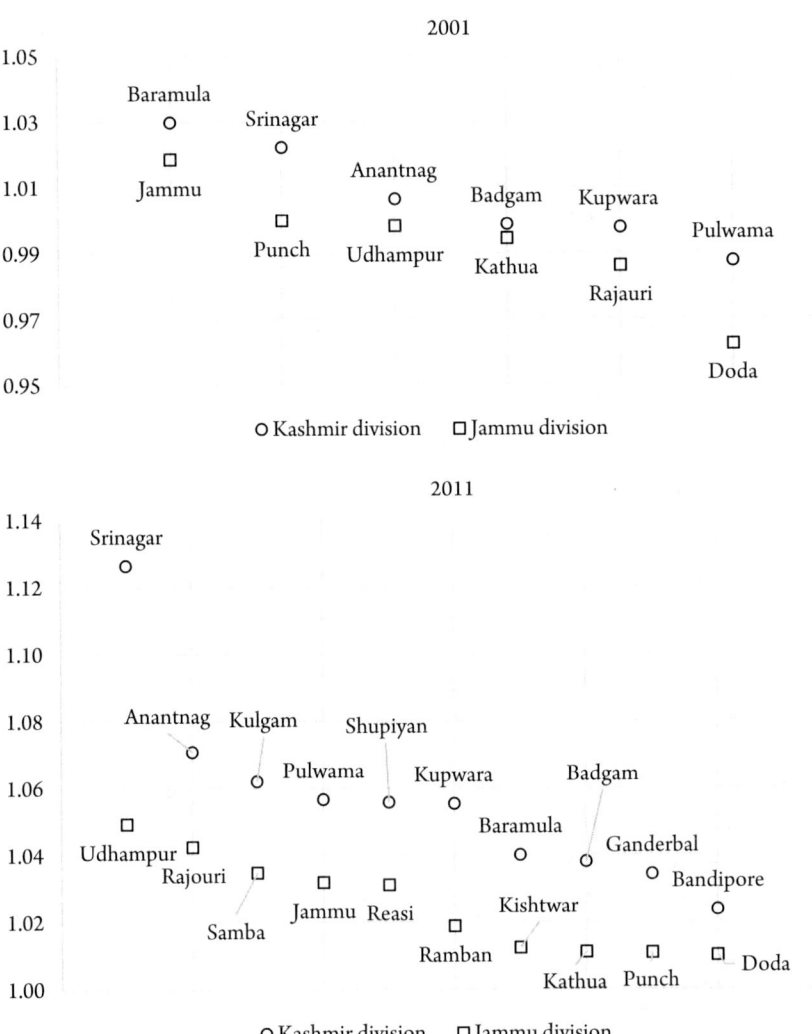

**Figure 6.1** Ratio of households in the household phase to houselisting and housing phase, 2001 and 2011

*Source*: Prepared by the author using table H - 3A appendix of Census of India, 2001; table HH-3 of Census of India, 2011; primary census abstracts (PCA) of 2001 and 2011.

*Note*: The figures for both phases exclude institutional and houseless households.

(*c*) communities counted in Jammu's snowbound areas and the rest of Jammu and (*d*) communities counted in snowbound areas of both Kashmir and Jammu and the rest of the population.

As noted earlier, health department officials had right from the beginning questioned the reliability of the 2011 census, particularly the data on the CSR. The then chief minister ordered an enquiry into the mismatch between the data on the CSR from the health department and the 2011 census. The findings were not made public. As argued earlier, once the data began to be used for regional comparisons, the initial concern about the skewed sex ratio disappeared.[42] In fact, neither the then registrar general of India (RGI) nor the then home secretary recall having received any formal complaint from the state regarding irregularities in the census or errors in the census data. In Nagaland and Manipur, the state governments had repeatedly and formally questioned the reliability of the 2001 census in communication with the union government as well as in submissions to courts.

The discussion so far suggests that the inflation of headcount in J&K was driven by a combination of mundane pressures linked to access to welfare schemes that apply across states, and the appeals to both enumerators and respondents to help sustain the numerical dominance of Kashmir within J&K. The chief minister was aware of this (see, for instance, Jameel 2010b), and it is unlikely that the union government was not.[43] Several interviewees in Jammu complained that New Delhi was primarily concerned about the completion of census and allowed Kashmir to dictate the outcome. It can be argued that New Delhi is obsessed with sustaining routines such as conducting censuses and elections to nourish its self-image as a modern state and failed to nudge Srinagar to build consensus within the state on census qua public good. Instead, it allowed the politically restive Valley to shape the collective self-portrait generated by the census as long as the exercise was allowed to complete.

## Concluding Remarks

This chapter discussed the widely held belief that under-enumeration is more likely than over-enumeration because of which cases of overcounting in the social and geographical peripheries are ignored. This strong prior is complemented by an older tendency of governments to treat growing numbers as a sign of well-being, security and progress. The chapter argued that amidst the post-colonial obsession with smaller denominators, this tendency survives vis-à-vis the social and geographical peripheries. We then saw how the Indian government, political leaders and bureaucrats have viewed the population growth of minorities and provinces dominated by such groups. We also discussed cases of coverage and content errors influenced by the government's attitude towards the population of minorities. We asked if the large anomalies in Kashmir's census that cannot be accounted for by conventional demographic and non-demographic factors can be explained by politically motivated manipulation condoned, if not aided, by the government.

Having examined the context of enumeration and both demographic and non-demographic factors that can explain flawed data, we can now try to assign the responsibility for various types of errors to different actors or census processes (Figure 6.2). First, there are errors such as the misclassification of 'Gujari' speakers as speakers of a dialect of Khandeshi and 'Ponchi' speakers as speakers of a dialect of Gujarati and the emergence and disappearance of Halam speakers. It is unlikely that actors within the state are responsible for these errors that are not politically salient. Instead, these errors reflect decaying checks and balances within the Office of the Registrar General of India (ORGI) even as chances of misclassification have grown as a lot of data processing now happens through computers. We have seen similar content errors in Nagaland and several other states (Agrawal and Kumar 2020a: 174–79, 301–03, 309–10). Second, the sharp changes in the headcount of speakers of languages of Ladakh were driven by communal mobilisation among both Shias and Buddhists but did not involve the inflation of headcount. Religious figures, civil society leaders and government employees such as schoolteachers working independently of the state effected abrupt and large changes in the population shares of closely related languages. These local actors influenced both respondents and enumerators of their respective communities. It is also noteworthy that at least in Leh and Zanskar the mobilisation was a product of a longer-term debate on protecting the Ladakhi Buddhist identity. In the case of Kargil and Sanku, the census data suggest that the process of communal mobilisation would have begun after the 1981 census. Third, in 2011, the over-enumeration of the child population, particularly males, across Kashmir was driven by communal appeals of influential public figures that shaped the perspective of public servants and lay-respondents. In fact, the polarising debate on demography that began in 2000 never fully subsided and was periodically revived by one or the other controversy in Kashmir. In 2001, the inflation of headcount in Kashmir manifested as increase in the household size and did not involve lay-respondents and must have been driven entirely by the bureaucracy and/or the political leadership. In 2011, the inflation took three forms: inflation of family size through the addition of fictitious children (Chapter 3), over-reporting of slum population (Chapter 4) and over-reporting of the number of households (Chapter 6). A combination of non-communal, mundane incentives to over-report household size such as headcount-linked entitlements to subsidised ration and demographic anxiety[44] generated these errors. The pervasive inflation of headcount across Kashmir suggests that it was

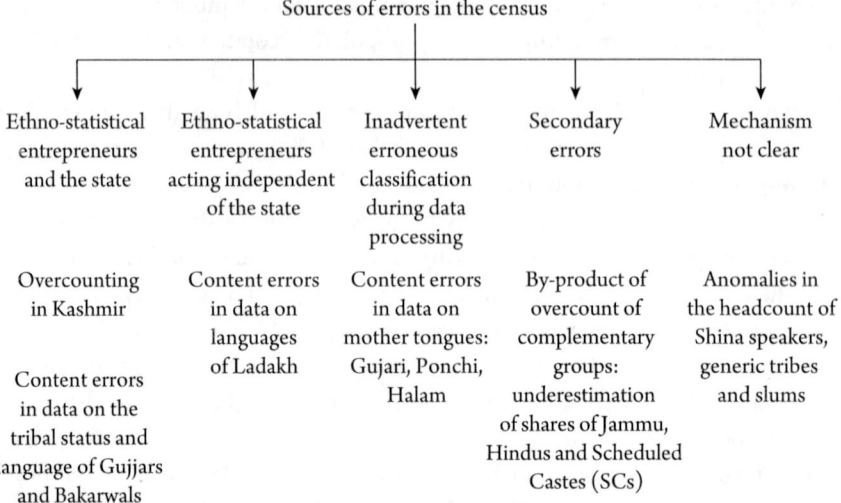

**Figure 6.2**   Sources of errors in the census of Jammu and Kashmir (J&K)

*Source*: Prepared by the author.

condoned, if not aided, by the senior bureaucracy and the political leadership. The over-enumeration of Kashmir's population remained unexamined because the received wisdom suggests that over-enumeration is unlikely in conflict zones. The misreporting of Gujjars and Bakarwals as non-tribals in Kashmir, too, involved government servants and other actors at the local level who helped shape census outcomes in light of the long-standing political preferences of the majority non-tribal Kashmiri-speaking community. Fourth, the falling population shares of the Jammu region, Mountainous region, Hindus, and (Hindu and Sikh) Scheduled Castes (SCs) and Hindu Scheduled Tribes (STs) are by-products of overcounting of complementary regions and communities. However, it can possibly be argued that Jammu district's population was undercounted because despite steady in-migration its urbanisation stalled, slum population contracted and its population share within the state and the Jammu division decreased. Lastly, both the rationale and mechanism of inflation of the headcount of Generic Tribes and the contraction of Shina speakers are not clear. Overstatement of the population of Generic Tribes added to the Muslim population, but the overcount was almost evenly distributed across divisions. The contraction of Shina and the eighteen-fold increase in the Dardi (Kashmiri)-speaking population of Gurez adds hardly anything to Kashmiri language's headcount even if it expands its reported geographical footprint and helps achieve a higher degree of homogeneity within the Kashmir division.

# Notes

1. J&K is, though, not the only state whose data deficit has failed to attract the attention of the government, media and academia. Other states in the ethno-geographic periphery such as Nagaland and Manipur have suffered a similar neglect (Agrawal and Kumar 2020a, 2020b; Kumar 2022c). Also, recall the discussion in Chapter 1 about the general neglect of data quality in policy and academic discourses.

2. Jerven (2013: 73) points out that underestimation of the population was commonplace in colonial Sub-Saharan Africa. In 1921 and 1931, the Census of India was affected by non-cooperation movements, which resulted in under-enumeration, particularly in the western provinces (GoI 1954a: 4–5; Maheshwari 1996: 116). Sub-nationalist conflicts in the post-colonial period have also been associated with boycott calls against censuses as in Nagaland (1961) and Kashmir (1991, 2001).

3. While the government is treating the potential impact of the mistrust generated by the NRC on data collection from a law-and-order perspective (*Times of India* 2022), Pronab Sen, former chief statistician of India, fears that the trust deficit could hamper data collection over the next decade (Magazine and Sasi 2020). It has, in fact, already affected the 78th round of the NSS.

4. The British censuses were obsessed with fertility and, by implication, women (Guha 2003: 156) and were to that extent not welcome in societies where customs required reticence on such matters. Indeed, under-enumeration of women has been a persistent problem in India. To address this problem, the census legislations of the successor states of British India exempt respondents from mentioning the names of female relatives. In Mizoram, certain 'cults and sects of Christians' were reluctant to participate in the 2011 census because '[g]etting identified with UID numbers would mean getting identified by the beast (the devil)' (*Assam Tribune* 2010). More recently, in Nagaland the church had to intervene when certain groups claimed prophecies against Aadhaar and digital identity (*Morung Express* 2021a). The Naga Students' Federation, though, offered a more nuanced critique of Aadhaar. It argued that Aadhar was a 'big threat to Naga customary law and identity' and opposed 'such drastic nationalising policy upon the Nagas pending the final Indo-Naga solution'. It added that the 'free enjoyment of social benefits within the fabric of collective social identity is sacred to the Nagas and any denial thereof in the name of identity numerisation is inimical to the social and religious practice of the Nagas' (*Morung Express* 2017).

5. Popular debates, too, stress under-enumeration. Recently, a politician belonging to the ruling Bharatiya Janata Party (BJP) 'claimed that contrary to official figures, India's actual population was around 150 crore [1.5 billion],

considering the Rohingya refugees and illegal Bangladeshis living in India, as
well as "unaccounted Indians" (*Morung Express* 2020).

6. The Muslim League's manoeuvring in Assam would have resulted in content,
   and not coverage, errors. Even in other provinces, unlike the Indian National
   Congress (INC) that did not take any clear stand on census, the Muslim
   League campaigned to ensure that Muslims reported Urdu as their mother
   tongue and did not report their caste or sect in response to the question on
   religion (Maheshwari 1996: 113–14, 116–36).

7. The census commissioner noted that 'in few places [where] ... people got
   excited about the language question, they succeeded only in spoiling the
   language returns; [but] it did not occur to them to inflate numbers' (GoI 1953:
   1; see also GoI 1951: 13). It is not clear if there were systematic differences
   between the princely states and the British-ruled territory of Punjab with
   regard to content and coverage errors.

8. There are several proverbs and blessings in modern Indian languages such as
   Hindi that link well-being to the growth of family.

9. Thomas Malthus published *An Essay on Population* in 1798, which nudged
   the British government to conduct decennial censuses (Bhagat 2003: 686).
   Malthus is referred to at length in the report of the 1941 census of J&K. Even
   though the higher population density implied divergence between the attitudes
   of the United States (US) and Western Europe toward population growth, not
   everyone in Europe shared the Malthusian pessimism. Marquis de Condorcet, a
   contemporary of Malthus, 'expected that fertility rates would come down with
   "progress of reason" so that greater security, more education ... would restrain
   population growth' (Sen 2001: 9). However, it took a long time for this line of
   argument to be accepted as it went against the conventional understanding of
   growth as a measure of prosperity and progress. For instance, Sri Aurobindo,
   an Indian spiritual leader, found it puzzling that 'it is only countries backward
   in development and education which keep up the old [high] rate of increase' of
   population leading to a counter-intuitive situation where '[t]he unfit tend to
   multiply, the fit to be limited in propagation'. This was 'an abnormal state of
   things which indicates something *wrong* in modern civilisation. But, whatever
   the *malady* is, it is not peculiar to Hindus or to India, but a worldwide *disease*'
   (Sri Aurobindo 1909, emphasis added).

10. Those opposed to the abolition of slavery also highlighted data, which showed
    that 'the rate of insanity amongst free blacks was 11 times higher than that of
    slaves' and hence 'blacks were congenitally unfit for freedom' (Rose 1991: 685).

11. Wajahat Habibullah's claim is problematic for, at least, three reasons. First, he
    does not mention the source of information on life expectancy at the level of
    regions of J&K. Second, it is not clear why the policies of the Sheikh Abdullah

government selectively benefitted Kashmir and Muslims more than other communities and regions. Abdullah's land reforms, for instance, benefitted the weaker sections across the state. Third, in recent decades improvement in life expectancy has been associated with decreasing total fertility rate (TFR).

12. Nagaland reported its highest growth rate in the 2001 census conducted during Jamir's tenure, but his interpretation of numbers did not affect the headcount. There is no evidence that Jamir or his administration tried to inflate the state's headcount. In discussions with his former protégés-turned-foes and many others there was not even the slightest hint that his government might have interfered (former finance minister, interview, Dimapur, 11 April 2013; president, Naga People's Front, interview, Kohima, 15 November 2013; member of parliament and former chief minister of Nagaland, interview conducted by Ankush Agrawal, New Delhi, 27 May 2015). Over-reporting was driven by socio-economic and political competition at the level of villages and tribes (Agrawal and Kumar 2020a). Further, at that time, most formulae governing federal-level distribution of resources used the 1971 population. The Eleventh Finance Commission (2000–05), which used the 1971 population data, submitted its report on 7 July 2000 – that is, before the houselisting phase of the 2001 census was completed. So there was no immediate incentive for Jamir to manipulate the headcount. Moreover, the next assembly election was due in 2003, and Jamir was not sure of winning due to the growing opposition from civil society organisations. It bears emphasising that Jamir's own tribe lost the numbers games, while his party lost the next election (see also note 36 of Chapter 4).

13. Population growth, or growth of population share, as a measure of the 'vitality of the community' (Jones 1981: 90) remains popular among those obsessed with ethnic or religious demography of societies. While the position of partisans of Hindus and Muslims is discussed in Chapter 4, and also in this chapter, sections of Christians have also been worried about their population share in India.

14. Morland (2018: 59) attributes colonial Sri Lanka's 'early population growth' in the nineteenth century 'to the relatively advanced socioeconomic status'.

15. R. N. Ravi was also the government's interlocutor for the Naga peace process in 2015 and was later appointed as the governor of Nagaland and then Tamil Nadu. Note that, to begin with, relatively junior figures in the establishment such as Ravi started using growth statistics to defend the current BJP government against domestic and international criticism of its treatment of minorities. This was later picked up by senior figures. We have already noted a 2019 statement of the working president of the ruling party (Rajya Sabha 2019c: 543). Most recently, the union minister of finance, Nirmala Sitharaman,

told an audience at the Peterson Institute for International Economics, Washington, DC, that 'India has the second-largest Muslim population in the world, and that population is only growing in numbers. If there is a perception, or if there's in reality, their lives are difficult or made difficult with the support of the state ... will the Muslim population be growing than what it was in 1947?' and added that '[e]very minority has been dwindling in its number ... decimated in Pakistan' (NDTV 2023).

16. Around this time, a Jammu-based lawyer filed a petition challenging 'the unfairness and discrimination of the State towards the communities in the state of Jammu and Kashmir which are eligible to be notified as minorities' (*India Today* 2017b). Several such petitions were filed in different courts. The Supreme Court dismissed one such petition filed by a lawyer and BJP leader because '[l]anguages are restricted state wise. Religions don't have state borders. We have to take a pan India approach' – that is, Muslims should be treated as a minority even in Muslim-majority J&K and Lakshadweep (*Business Standard* 2019b). This decision does not seem to have settled the question. Another petition in the Supreme Court sought revocation of the law that allows 'Muslim, Christian, Sikh, Buddhist and Parsee communities' to 'enjoy benefits meant for minorities even in those states [in the north-east and the north-west] where they are in majority' (*The Print* 2021). The petitioner argued that 'the Hindus ... are a minority in several north-eastern states besides Punjab and Jammu and Kashmir. However, the Hindu community is deprived of the benefits that are available to the minority communities in these states' (ibid.). In response to this petition, the union government informed the Supreme Court that 'State governments can declare any religious or linguistic community, including Hindus, as a minority within the said state' (NDTV 2022c), which effectively implies that there can be state-specific religious minorities. But two months later the union government informed the court that 'the power is vested with the central government to notify minorities' (NDTV 2022d). Most recently, the Supreme Court suggested that religious and linguistic minorities should be identified at the level of states, which means Hindus and Hindi speakers are entitled to enjoy constitutional protections qua minority in, say, Lakshadweep (Rajagopal 2022).

17. See comparisons of pre-colonial and early colonial censuses (Guha 2003; Peabody 2001).

18. In a highly influential apologia, *The Indian Musalmans*, in response to a question of Viceroy Lord Mayo ('Are the Indian Musalmans bound by their religion to rebel against the Queen?'), Hunter (1876 [2002]) used statistics to prove that the colonial administration was unfair to Muslims, which was the cause of their unrest. Hunter completed his book on 23 June 1871 – that is, before the results of the first countrywide statistics were published.

19. The genesis of the Muslim position on census can be traced back to the late nineteenth century. Soon after the 1881 census, Syed Ahmed Khan, a highly influential nineteenth-century reformer, began talking in terms of the majority–minority binary ('[a]lthough the number of Mohammedans is less than that of the Hindus ... yet they must not be thought insignificant or weak') and suggested that democracy was not desirable for Muslims of India (Guha 2010: 73). The Muslim League formed in 1906 presented the government with a threefold demand: Muslim representation across 'the entire system of governmental patronage' commensurate with their population share, reduction of the size of Hindu population by classifying the 'uncivilized portions of the community' separately and additional representation for Muslims in light of their 'political importance and value of the contribution ... to the defence of the Empire' (Jones 1981: 89). Later, the League asked Muslims to declare Urdu as their mother tongue and avoid mentioning their caste or sect in response to census questions (Maheshwari 1996: 113–14, 116–36).

20. One of the earliest visualisations of census as a disaster clock in the Indian context is found in an 1891 census report on Bengal (Jones 1981: 91). See Joshi, Srinivas and Bajaj (2003: figure 2.1) for a twenty-first-century iteration of census qua disaster clock for Hindus. Visualisation of data on population as a time bomb has become commonplace. See A. Gupta (2017) for J&K and the *Morung Express* (2018b) for Nagaland.

21. In 2022, J. K. Bajaj, one of the authors of *Religious Demography of India*, was appointed as the chairman of the Indian Council of Social Science Research (ICSSR) and conferred the Padma Shri award for his contribution to literature and education.

22. The ORGI is part of the Union Ministry of Home Affairs. After the 2001 census, the ORGI released a detailed report on religion for the first time (GoI 2004a), which triggered further enquiries, including by the Sachar Commission.

23. Tripura seems to be one of the major instances of the failure of the policy of allowing minorities to dominate their traditional territory. But indigenous tribes accounted for barely 53 per cent of Tripura's population even as early as 1901 (Government of Tripura 2021). This slender majority was irreversibly eroded by the massive influx of Bengalis from East Pakistan after the partition of British India.

24. The following summarises the government's approach to Parsis: 'The Parsi population deserves an exceptional but definite mention and place in this volume due to their very small numbers not only in India but also in the world ... a clear visible but extremely unfortunate decline of a rich civilization of Zoroastrians and its people. It is apparent from 2001 census results that urgent and drastic interventions are required by all concerned including possibly by

the government and definitely the Parsi community leaders to ensure survival of Parsi population in India. Fertility improvement innovative initiatives rather than fertility control measures adopted by the community so far are possibly the need of the hour before it reaches a point of no return' (GoI 2004a: xxv). Despite criticism from within the community (Chari 2014), the government remains committed to the project that has helped boost the community's fertility (Shelar 2019; A. Pandit 2021). Such measures have a limited efficacy, though, and, in fact, deflect attention from the need for deeper reforms such as allowing women married to non-Parsis to retain their membership of the community (K. Srivastava 2020).

25. 'Rather than granting them their autonomy and rights to address historical injustices, this perception has led to disastrous State Government interventions in the name of their "preservation". One such scheme has been the State policy disallowing members of PVTGs [Particularly Vulnerable Tribal Groups] from availing of sterilisation schemes in government hospitals. Tribes ... have been denied permanent methods of contraception in an attempt by the State to encourage population growth in the face of their apparently dwindling numbers' (GoI 2014b: 61).

26. The state's partisan approach to demography may also reflect in how numbers are presented – for example, 'in the volume entitled Report of the 1911 Census, the discussion on the Hindus covered 11 pages, on the Muslims four pages, on the Jains one page, and on the Sikhs a half page only. On the other hand, the Christians with a population of 3,876,203 only in 1911 were allocated nine and a half page' (Gill 2007: 240).

27. There have been inexplicable changes in Bhutan's reported population. The government first admitted that a 'significant discrepancy exists between the population totals reported by the United Nations and those provided by the Royal Government of Bhutan. The estimates of the United Nations place Bhutan's population at 1.9 million in 1997. The UN's estimate is a projection based on RGB's [Royal Government of Bhutan's] estimate of 1.035 million in 1969 [that is, two years before the grant of UN membership].... The World Bank, in its computation of per capita GNP [gross national product] for Bhutan, takes into account the lower population total provided by RGB, whereas for other calculations (such as enrollment rates), the UN continues to use the UN estimate of 1.9 million for 1997' (RGB 2000: 37–38). However, later it explained away the discrepancy by suggesting that 'population figures for those early years [based on censuses in 1969 and 1980] were large because of the large influx of migrants from neighboring countries, particularly labor migrants at the instance of construction coinciding with the large-scale development work on infrastructure expansion. With the development

emphasis changing to human resource development the large [number of] migrants moved out and in 1996 the total population was found to be around 600,000' (RGB 2012: 1).

28. Policymakers are not necessarily obsessed with the denominator of indicators. This is evident from, for instance, the fact that the release of the Indian census data on identity has to be approved by the Cabinet Committee for Political Affairs (GoI 1993: 20, 2011i: 111–12), whose perspective is unlikely to be limited to boosting per capita figures. Governments are also concerned about other things such as the optimal size of workforce and the pool of potential recruits for the armed forces. As a result, even countries with a declining population may interfere with statistics amidst growing pronatalist rhetoric if government policies to boost fertility are ineffectual. Most recently, China delayed the release of the results of the 2020 census amidst concerns about the contraction of population and the lacklustre success of the relaxation of the one-child policy (Kawate 2021; Yu 2021). Also, concerns about population growth need not be anchored to a developmental vision. Apartheid South Africa was alarmed at the higher growth rate of Africans because that rendered segregation increasingly difficult and eroded the raison d'être of the regime (Lipton 1972). Bhutan's government excluded alleged immigrants to increase the population share of the dominant community. The 2005 census in Bhutan reported only the natural growth rate of population in the year preceding the reference date but not the population growth rate (RGB 2006: 17) because contrived corrections had rendered censuses incomparable.

29. Political transitions have also been associated with coverage errors. Punjab and Bengal in British India witnessed overcounting in the last colonial census (GoI 1954a: 5), whereas Sub-Saharan Africa struggled with 'overestimation in postcolonial censuses' (Jerven 2013: 73). There were sharp variations in India's tribal population around decolonization driven by political considerations and administrative decisions rather than demographic changes (Raza and Ahmad 1990: 6). More recently, Nagaland and Manipur witnessed over-reporting of headcount in the run-up to the 2002 delimitation (Agrawal and Kumar 2020a, 2020b).

30. Apartheid South Africa carved out homelands, among other things, to reduce its Black population. See Morland (2018: 39–40) for other examples of changing borders to shed undesirable population. Assamese-speaking Hindus seem to have abetted Bengali-speaking Muslim-majority Sylhet's secession to establish their demographic dominance in post-colonial Assam (Phanjoubam 2016).

31. The partisans of autonomy, or independence, in Kashmir champion Kashmiri ethno-regional identity tethered to a Perso-Arabic script for the Kashmiri language and Urdu as the official language and the lingua franca. The exodus of Pandits rendered the linguistic and religious dimensions of the identity of Kashmir coterminous.

32. '[A] large majority of the Hindu population of Punjab, along with a section of the rural Sikh scheduled castes [SCs], had been strongly mobilised to incorrectly record Hindi instead of Punjabi as their mother tongue' (Gill 2007: 244; see also GoI 1954d).

33. Horowitz (2000: 178), first published in the early 1980s, noted, 'Fearful as they are of extinction, the Sikhs have had in recent decades the highest rate of population increase of any group in the [*sic*] Punjab, a rate one-third higher than the all-India rate.' Gill's account suggests that the high growth rate reflects state intervention.

34. The difference in how sects clubbed under religions are reported is, possibly, an indicator of the government's inclination to represent the Hindu population as fragmented and thus less threatening to the minorities. While the proportion of sects in the reported population of Hindus is larger than other religions, the absolute figures are quite small.

35. In 2021, senior INC leader Ambika Soni turned down an offer to become Punjab's chief minister. She made her 'stand clear that for Punjab there has to be a Sikh face' (NDTV 2021). More recently, Sunil Jakhar alleged that he was not elevated to the position of Punjab's chief minister because the INC wanted a Sikh to lead the government (M. Malik 2022). This suggests that at least a section of the INC believes that it is important to ensure that Sikhs do not feel cornered in the only Sikh-majority province. In 2002, for similar reasons, the INC accepted Mufti Mohammad Sayeed's demand to lead the INC–People's Democratic Party (PDP) alliance in J&K (see also note 58 of Chapter 4) even though the then BJP-led union government allegedly preferred 'a Congress government over a PDP government' (Dulat 2010: 269–70).

36. A Kashmiri Muslim interviewee had to leave his village in the early 1990s and could return only in the mid-2010s. There is a Jammu and Kashmir Political Migrant Front that represents 'political migrants staying in various security zones' (*The Tribune* 2015b). In 2010, there were about 2,000 Muslim political migrant families from Kashmir in Jammu (*Kashmir Life* 2010). 'A big chunk of Muslim migrants were activists of nationalist parties, relatives of political leaders and employees. They migrated following the kidnapping of Prof. Mushir-ul-Haq, Vice-Chancellor of Kashmir University, and the spurt of killing of political personalities' (Hari Om 1998: 109, see also 108).

37. Evans (2002: 25–26) suggests that Muslims migrated out of Kashmir in substantial numbers. Peer (2009: 58) suggests, 'Parents saw getting their children out of Kashmir as the solution. The rich were sending their children to Europe and North America; the middle and the lower middle class chose all sorts of colleges and universities in Indian cities and towns from Bangalore to Balia.'

38. Double counting accounted for a sizeable part of the overcount of Nagaland's headcount in 2001 (Agrawal and Kumar 2020a: 202–03, 228–29, 242).

39. The director of census operations (DCO) for the 2001 census ruled out double counting because 'people were scared of being counted even once' and added that the census schedule is so long that one would wave away enumerators by saying 'we have already answered' (interview, Jammu, 4 December 2019).

40. Even, in 2001, the results may have been manipulated during the household phase. In the middle of the 2001 census, Swami (2000a) reported that '[s]ources at the Registrar-General's office told Frontline that a "very preliminary" study had shown a rise in the population from 59 lakhs in 1981 to a little over twice that number in line with projections made by experts' (see also *Kashmir Times* 2000d). (The actual population as per the 2001 census was 10.14 million rather than 11.8 million.) Another news report filed a week later adds to our suspicion. In face of repeated disruptions and extensions of the deadline, the government toyed with the idea of extrapolating from the houselisting phase completed earlier that year: 'Insiders said the government is thinking of taking the preliminary data, collected in the first phase, into consideration and calculate the total population. Some officers are opposing this plan because it will not bring the desired results' (*Kashmir Times* 2000i). Note that the ORGI has to, in any case, extrapolate population for logistical purposes including requisitioning manpower and ordering stationery. The number of forms for the household round was, for instance, estimated in 1971 by adding three per cent to houselisting headcount to account for the growth of population and adding another ten per cent for reserve (GoI 1971b: 18). The requirement for the houselisting phase was in turn determined by population projections based on the last census.

41. Three interviewees alleged that various officials posted in Kashmir told them that they had seen overwriting on 2011 census forms there (interview, Jammu, 2 June 2016; interview, Jammu, 30 June 2018; interview, New Delhi, 1 July 2018), which in their opinion was evidence of manipulation. A Kashmir-based researcher suggested that errors entered the data when the enumerators shared their schedules with their supervisors who instructed them to go back and add numbers by inflating the household size (interview, 21 July 2021). It is not

clear if this may have necessitated overwriting as page totals are in principle entered after the end of the revisional round unless enumerators complete the forms including totals during the first visit.

42. A Kashmiri researcher who had followed the 2011 census closely told this author that both the bureaucracy and the media were not open to questions regarding the quality of census data (interview, 21 July 2021).

43. A retired civil servant observed, 'Every element of governance in Kashmir, particularly since the theft of the holy relic from Hazratbal in 1963, has been supervised by the IB [Intelligence Bureau]. Every CM [chief minister] has acted under the "advice" of the Jt [Joint] Director IB headquartered in Srinagar … in light of what I have said, it [IB being aware of manipulation] is a possibility' (interview, 12 January 2021). Another retired civil servant added, 'The Delhi Durbar has no idea about the ground realities in Kashmir. The only reality they consider is the one put forward by the IB' (interview, 10 December 2015). A former intelligence official, though, ruled out the direct interference of the IB in census operations due to the shortage of manpower because of which it cannot depute officers to the directorates of census operations (interview, 16 August 2021). Indeed, in 2013, the government informed the parliament that about 30 per cent of the positions in the IB were vacant (GoI 2013b). Whether the IB was aware of the subversion of the census could not be ascertained. A senior official of the military intelligence suggested that they, too, did not follow the process to be able to comment upon it (interview, Udhampur, December 2019). However, the armed forces seem to be interested in the demography around their camps (*Greater Kashmir* 2018) and in tracking the movement of people in border districts (see Agrawal and Kumar 2020a: 184 for Nagaland). Recently, after a protracted encounter with militants in Punch, the army seems to have proposed to allot 'new house numbers to individual hearths' and record 'the names and other details of those who dwell there, including the mobile phone numbers of all family members' (M. S. Pandit 2021).

44. A former official of the Ministry of Home Affairs associated with J&K around 2011 suggested that 'separatists … could have interfered … by bringing pressure on the local personnel' (interview, 24 April 2023), while a retired officer of the J&K cadre added that 'the insurgency and Muslim fundamentalism could quite possibly have affected' the outcome 'because of the biases affecting the collectors and compilers of data' (interview, 23 February 2016).

# Part IV
# Reforms

# 7

# Reinventing the Census

[I]t was not administratively possible to achieve the desired result by using any kind of force... against the wishes of the people without whose active cooperation nothing was possible.

—Feroze Ahmed, director of census operations (DCO), Jammu and Kashmir (J&K) (Government of India [GoI] 2001b: xi)

[W]e must take immediate steps for creating 'Census consciousness' among the people.

—J. N. Zutshi, DCO, J&K (GoI 1971b: 3)

While referring to the need of adequate publicity, I feel compelled to sound a note of caution. It must not be overdone.

—J. N. Zutshi, DCO, J&K (GoI 1971b: 8)

We cannot spoil the future of a whole generation [by starting demographic competition] for [winning] just one parliamentary seat.

—Former member of legislative assembly (interview, Leh, 19 September 2019)

## Introduction

We examined coverage and content errors in the census data for Jammu and Kashmir (J&K) and discussed their administrative, legal and political contexts at different levels of aggregation. We found that the over-reporting of children (particularly male children), a large increase in the slum population and a large increase in the number of households between the houselisting and household phases of census in Kashmir explain several interconnected anomalies in the 2011 census of J&K, including the drop in the child sex ratio (CSR), the rise in the

share of child population, the rise in the population share of Kashmir within the
state and the drop in the corresponding population shares of the Jammu division
and groups concentrated almost entirely in Jammu such as the Scheduled Castes
(SCs). Further, errors in data on non-scheduled languages, dialects of scheduled
languages and tribes were also examined and attributed to political mobilisation,
unintentional misclassification and a large increase in the population of Generic
Tribes. Even conservative estimates of politically motivated over-reporting of
the headcount, which do not account for (*a*) the over-reporting in 2001 that
manifested in the abnormal increase in mean household size of Kashmir and
(*b*) the over-reporting of the population aged 10–14 years in 2011, suggest that
the 2011 census overestimated Kashmir's population by about 10 per cent.

We argued that the inability of the government to conduct reliable censuses in
Kashmir – reflected in the cancellation of the exercise on two occasions, ad hoc
changes to the reference date in other censuses and the contested nature of the
data – can be explained from three different perspectives. First, statistics are widely
seen as insignia of modernity with their symbolism often being more important
than the content. Despite concerns about the difficulty in conducting census
amidst political unrest, the government rushed to resume censuses (and elections)
in the late 1990s to repair its self-image as a modern state and signal the return of
normalcy after years of turmoil. The collection of data was therefore an end rather
than a means to something else. Unsurprisingly, the quality of data did not receive
much attention. Since the resumption of censuses and elections qua routines
of the state was equated with the normalisation of New Delhi's sovereignty in
Kashmir and the restoration of India's nationhood and its international image,
partisans of independence opposed censuses and elections in the 1990s and the
early 2000s.[1] In 2011, however, they tried to use the census to improve Kashmir's
bargaining power within the state. This reorientation in Kashmir vis-à-vis the
census is in agreement with the experience of similar groups in the north-east
that in recent decades began to interfere in government surveys in favour of their
preferred communities. The government does not seem to mind such partisan
interference in conflict zones because it enables the authorities to stage 'event[s]
of normalcy', as Mufti (2023) would put it, without any opposition.

Second, the supply and quality of government statistics are compromised
in divided societies with dysfunctional democratic institutions and stagnant
economies because of collective action problems and resource constraints that
undermine the provision of public goods. J&K's data deficit is therefore conjoined
with its democracy deficit (dysfunction of participative institutions that are
supposed to build consensus over public goods including government statistics)
and development deficit (lack of internal resources for financing public goods).

Third, census generates a collective self-portrait of the society. Disputes over the portrayal are inevitable where, as Sumantra Bose (2007: 191) puts it, 'the "self" is deeply divided, even fractured'. The competing discourses of self-determination and national integration reduce the 'self' and 'nation' in J&K to statistically determinable entities whose political affiliations and commitments are automatically fixed by the census. Winning censuses is thus seen by the supporters of Kashmir's independence as a step towards sustaining the existing distribution of political power and, eventually, winning the plebiscite. Jammu's interest in data is limited to undercutting the dominance of Kashmir. New Delhi, on the other hand, is keen on sustaining its image as a modern democracy and is therefore primarily interested in normalising the routines of statehood in conflict zones. In fact, as discussed in Chapter 6, we can add that the union government was perhaps not averse to a collective portrait that enhanced the standing of Kashmir and Muslims to de-emphasise the perceived threat of demographic change.

After the provisional results of the 2011 census were released, the opinion in Srinagar was divided over the acceptance of the anomalous census figures. While the increase in the population share of Kashmir (and Muslims) brought relief, the dramatic decline in the CSR across the Valley was seen as bad for Kashmir's image and was, in fact, seen as a matter of immense shame. The Omar Abdullah-led state government focused its attention exclusively on the parameter that affected the *image* of Kashmir – namely, the poor CSR. The state government initially questioned the census data on the CSR and even decided to carry out its own investigations, but the findings were not made public. The abnormal increase in the child population share that was conjoined with the abnormal decrease in the CSR was ignored in the debate in Kashmir. Selective reading is what 'an internationally reputed think tank' based in Chennai, too, did in its 'scientific study on the religious demography of the country' (Chellappan 2018). The think tank 'found that the root cause behind the peaking of secessionism in the northern State is due to the drastic change in the religious demography' (ibid.). It focused on the sharp rise in the child population share that *proved* that Muslims had higher fertility rates, which it claimed was part of a conspiracy against Hindus. Several news reports approvingly covered the study, but none of them noted that the think tank completely ignored the sharp drop in the CSR and that the 2011 child population statistics were being retrospectively used to explain the armed insurgency that began more than two decades ago.[2]

A senior health official posted in Kashmir in 2011 told the author that large-scale foeticide tipping the CSR across Kashmir was quite unlikely, given the absence of a strong bias against girl child, and an increase in child population share was also unlikely due to the improvement in correlates of fertility (interview, 12 August 2021). Most leaders in Kashmir though focused on foeticide as an immoral, un-Islamic act. Omar Abdullah, the chief minister, vowed to punish those behind foeticide. He added that 'the punishment of Almighty for those indulging in this immoral and sinful activity is heavy and eternal' (*Greater Kashmir* 2011b). Abdullah was aware of the communal mobilisation in the run-up to the census in Kashmir (see, for instance, Jameel 2010), but he overlooked the possibility that manipulation may have skewed the sex ratio even as his health department disputed the census data. The Jammu and Kashmir Liberation Front (JKLF) chairman, Yasin Malik, referred to the drop in the CSR as 'highly shameful'. He said:

> With the advent of Islam, this immoral practice was done away with at every place in the world where people embraced Islam. According to the Quran and traditions of Islam, foeticide is a grave unpardonable sin equivalent to murder. We cannot claim to be Muslims while indulging in heinous crime of foeticide. Murder of a female child was in vogue during the period of ignorance in Arabs but it was strictly forbidden by Islam. If we restart this immoral practice, we have no right to call ourselves as civilized people. (*Greater Kashmir* 2011c)

On another occasion, Malik lamented that 'the truth (female foeticide) has exposed the hollowness in our society. Our heads hang in shame' (Ishfaq-ul-Hassan 2011). In one of his customary Friday sermons, the chairman of a faction of the All Parties Hurriyat Conference, Mirwaiz Umar Farooq, condemned *dukhtar kushi* (female foeticide). He said, 'It is a matter of shame that Kashmiri Muslims are aborting their girl children' (*Asian Age* 2011). He also 'called upon all imams and scholars to launch a vigorous awareness campaign against female feoticide [*sic*]' (Ishfaq-ul-Hassan 2011). As noted in the earlier chapters, Yasin Malik and Mirwaiz Umar Farooq had contributed to the communalisation of the public debate around the 2011 census.

Several letters to the editor published in May 2011 in the *Greater Kashmir* also framed the decline in the CSR in terms of non-adherence to Islam. A letter writer suggested that '[i]f we have to live as per the principles of Islam, then we must put an end to the practice of illegal abortions'. Another offered a more expansive statement of the Islamic perspective, pointing out that '[k]illing of daughters and giving preference to sons is just going against the very teachings of Islam. As Muslims we must know the worth of women' and curiously added that 'the most precious thing in the world is a virtuous woman'. Yet another lamented that '[i]t is really sad to see women killing their daughters in their wombs' because 'our religion has never taught us to do so'.

Kashmir, though, is not the only place where leaders and people reframed data deficit in moral and religious terms and transformed an institutional failing into a source of collective social or national shame. Data deficit has been a source of shame in places as diverse as China (Ghosh 2020: 58–59, 112), Nigeria (Aluko 1965: 371) and Nagaland in north-east India (Agrawal and Kumar 2020a: 286–87). In Christian-majority Nagaland, the flawed 2001 census was followed by moral self-doubt rather than a demand for institutional reforms.

This concluding chapter uses the insights from the earlier chapters to suggest a multi-pronged approach to address some of the key problems facing censuses in twenty-first-century India. The chapter first discusses how statistics have become an integral part of debates between communities in conflict and locates J&K's data deficit in relation to its democracy and development deficits. It then discusses the impact of data quality on policymaking using examples related to federal redistribution and delimitation of constituencies. Given the limited efficacy of legalistic and technocratic interventions vis-à-vis improving data quality, it suggests statistical reforms and a non-legalistic approach to address the politicisation of census by locating J&K's experience in comparison with similar problems in other states. It uses a game-theoretic framework to integrate the experience of J&K, Manipur and Nagaland, which, respectively, stand for the whole range of responses of state governments to the politicisation of data – namely, ignoring the problem, blaming one community and resolving through public consultation.

## Your Facts, My Facts

Deep divisions over *facts* about the past and the present are a major obstacle to reconciliation in J&K. Assorted statistics, often of dubious provenance, are used as context-free add-ons in debates whose conclusions have been arrived at even before looking at evidence that merely serves the purpose of providing a veneer of objectivity. Since we have already examined population statistics and delimitation at length, a few other examples related to conflict, territory and place names are discussed here.

While New Delhi uses the successful completion of elections to claim normalcy, pro-independence groups highlight low voting rates. When the elected leaders lean too far towards autonomy, New Delhi reminds them that they do not represent the people as voting rates were low. Likewise, when New Delhi claims that the insurgency is confined to a very small part of the overall area of the state, pro-independence groups remind us that this area accounts for the majority of the state's population.

The government finds itself in a bind over the estimate of the strength of insurgency. A large estimate would discredit the counterinsurgency strategy, but a small one would compel the government to prematurely wind up the massive counterinsurgency apparatus. Khurram Parvez of the Jammu and Kashmir Coalition of Civil Society claimed that 'half' of the Indian army is based in the state and asked if the government needed '700,000 soldiers to fight 150 militants' (A. Ashraf 2016). Pakistani sources claim that about a million troops are posted in the state (Philip 2019). Shukla (2018) argues that '[t]o say that half the army is deployed in J&K is a pure delusion' (see note 45 of Chapter 2 and note 26 of Chapter 3).

The most contentious statistics, though, relate to riots in Jammu in the late 1940s, the exodus of Pandits in the early 1990s and terrorism and insurgency since 1989. When Pandits highlight their exodus in the 1990s, Kashmiri Muslims draw attention to their own out-migration to the plains of western Punjab in the *late nineteenth* century under the Hindu Dogra rulers.[3] While the Hindus of Jammu remind Kashmir of the exodus of Pandits, Kashmiri Muslims draw attention to violence during partition that targeted the (non-Kashmiri-speaking) Muslims of Jammu and caused massive displacement. The Hindus of Jammu question the reliability of statistics of partition violence, present the violence as a reaction to what happened in Pakistan-occupied territories and western Punjab and assail the fleeting, politically motivated concern of Kashmiri Muslims for the Muslims of Jammu. Indeed, as Shekhar Gupta (2022) puts it, a 'great tragedy' has been reduced 'to argument over scorecard of killings' (see also Jamwal 2013: 27–28).[4] Two things are noteworthy about these numbers games. First, the discourse is driven by 'the selective historical memories of each community – of another's violence', with the 'current or historic victimhood … [serving as] one's claim to virtue', which together entail the 'need to show the other who was superior through retaliation' (Devadas 2019a: 17, 27). Second, each community uses statistical–cartographic scaffoldings to support its case. However, to *enliven* the case they also equate themselves with a globally well-known example of an oppressed community. Kashmiri Muslims flesh out their demographic fears by equating Kashmir with Palestine (see note 24 of Chapter 4), while Pandits refer to themselves as victims of Apartheid (Panun Kashmir n.d.2: 13). Further analysis of the statistics of the Kashmir conflict is omitted for want of space, but we will discuss a recent controversy over riots in Jammu in the late 1940s.

Writing after the reorganisation of the state, Thapar (2019) pointed out that 'we hear a lot about what happened to Kashmiri Pandits in 1990…. But not a word is spoken about what happened to the Muslims in Jammu five decades earlier.' He suggested that 'Jammu just after Independence in 1947 … was a

Muslim-majority city. But literally, in weeks, communal riots, mass killings and forced migration turned it into a Hindu-majority one.' He added that researchers suggest that 'between 70,000 and 237,000 Muslims were killed … a further 500,000 displaced'.[5] Hari Om (2019b) pointed out that Thapar's figures were grossly inaccurate because as per the 1941 census the population of Jammu city was barely 50,370 (including about 15,000 Muslims).[6] He asked, 'How could about 35,000 Hindus in Jammu City kill 70,000 to 237,000 Muslims in Jammu City and expel 500,000 other Muslims from Jammu City?'[7] Thapar seems to have confounded Jammu city and Jammu province.[8] In any case, as discussed earlier, mechanical comparisons of population shares covering the decade of partition are problematic because the census was not conducted in 1951, a few districts lost territory due to the tribal invasion, there was a large-scale cross-border transfer of population and those left behind flocked to areas where their respective communities were dominant.[9] When R. R. Sharma, the former vice-chancellor of the University of Jammu, repeated Thapar's claims, Hari Om (2020) once again highlighted the flawed statistics and criticised the 'vilification campaign against Jammu'.[10]

Our next example is related to cartographic statistics. Panun Kashmir suggests that Pandits are demanding *merely* 3.87 per cent (8,600 square kilometres) of the entire territory of J&K for a homeland for Pandits (Panun Kashmir n.d.3). The partisans of Kashmiri Muslims sidestep the reason behind the demand for a separate homeland or separate colonies for Pandits and instead link the demand to Kashmir's bête noire, demographic change:

> The geographical area of the proposed homeland identified in a map circulated by Panun Kashmir … covers about 10,600 sq. km[11] of the Kashmir Valley's total area of about 15,500 sq. km, which means 68.38 per cent of the Valley is claimed for a mere 4 per cent population (as per 1981 census). In plain language, Panun Kashmir does not only want over 68 per cent of the total area of Kashmir but also throw millions of Muslims out of it to claim a separate and exclusive Hindu Homeland. The claimed area is the most fertile region of Kashmir and is its rice bowl. Further, it includes almost all the water sources of Kashmir … as well as its main glaciers … nearly all tourist destinations…. Since the homeland is sought to be governed by the Constitution of India with no Article 370, it would be open for Hindus from different parts of India to acquire property and settle there. (K. B. Ahmad 2017: 284–85)

The debate over a homeland for the Pandits in Kashmir brings to light the largely ignored sub-national cartographic dimension of the Kashmir conflict. The visualisations of 'Greater Kashmir' cut across Ladakh and Jammu, challenging the status quo maintained by New Delhi. Panun Kashmir's map of homeland, on the other hand, splits Kashmir into two and challenges *both* New Delhi and Srinagar. There is a key difference between 'Greater Kashmir' and 'Panun Kashmir' though. 'Greater Kashmir' is visualised as an aggregation of territories populated by communities that nominally share a religion, while the visualisations of 'Panun Kashmir' are reminiscent of ethno-cartographic endeavours in the hill states of the north-east.

The two sides also spar over place names, with Pandits claiming the state government has encouraged the renaming of places with Hindu-sounding names. Anantnag, also referred to as Islamabad(-Anantnag) is, perhaps, the foremost site of this nominal conflict.[12] K. B. Ahmad (2017: 255) suggests that the 'name of not a single place or a Hindu symbol in Kashmir has been "changed into [the] Islamic lexicon"'. He adds that the allegations in this regard rest on just one example – namely, 'Islamabad, a South Kashmir town which they claim was originally named Anantnag', but there was 'no historical evidence supporting the assumption'. However, at least since 2011, the district administration's website carries the following note that is found in *all* editions of *The District Census Handbook* (1961–2011).

> The district as well as its headquarter town are also called Islamabad. Regarding this second name no mention is to be found in the old chronicles of Kashmir. It is however, said that the name of Islamabad was assigned to the town by one Islam Khan who was the Governor of Kashmir during the Mughal rule in 1663 A.D., but the change in its nomenclature proved temporary and during the reign of Gulab Singh the town as well as district again *resumed* their old name, Anantnag, but still the name Islamabad is Popular among common masses, though officially the name Anantnag is used. (Government of Jammu and Kashmir [GoJK] 2021a, emphasis added)

The 'alien' name of Anantnag has attracted a lot of commentary in Kashmir because signboards of shops and buses plying between Srinagar and 'Islamabad' attracted the ire of the armed forces in the early 1990s (Deva 2015; Peer 2009: 47).[13] The Kulgam district administration's website is another site of nominal contests. Earlier it noted the following on the etymology of 'Kulgam':

> Tazkira Sadat-i-Simanania, compiled by Swaleh Reshi, a reputed scholar and poet of 13th century, contains the name of place as 'Shampora' which was later on renamed as Kulgam by Syed Hussain Simnani (RA) when beholden with the myriad number of Canals and streams flowing through the village.

The word Kulgam denotes 'Kul' meaning thereby the 'Whole' and the 'Gam'
in Arabic to 'teach righteousness'. (GoJK 2018)

At least since March 2019, Kulgam's website also carries the following note that
replaces the contrived Arabic etymology with the widely accepted Kashmiri
etymology that is also referred to in the first *District Census Handbook* of Kulgam.

Tazkira Sadat-i-Simanania, compiled by 13th-century scholar and poet
Swaleh Reshi, gives the name of place as 'Shampora'. Syed Hussain Simnani
later renamed it 'Kulgam' (kul for 'clan' and gam for 'village' in Sanskrit).
(GoJK 2021b)

Such nominal battles over facts are routinely fought across the state.[14] The
Tourism Department 'issued advertisements in newspapers to celebrate Kashmir
Fort festival from May 23 to 25 [2016] at Hari Parbat in Srinagar. In the
advertisement, they also wrote Koh-e-Maran in place of Hari Parbat.' This 'deep
rooted conspiracy' was unanimously condemned by the Pandits (*Financial Express*
2016). As discussed in Chapter 1, Kashmiri Muslims, too, complain against the
marginalisation of their heritage in government propaganda. Most recently, the
rechristening of a number of schools, colleges and roads by the administration
to commemorate those who sacrificed their lives in wars with Pakistan or
fighting the secessionist insurgency (*Greater Kashmir* 2021) revived the nominal
competition.[15] The National Conference's (NC's) dissent note submitted to the
Delimitation Commission questions the obliteration of constituencies with
names of historical significance (Government of India [GoI] 2022).

The sub-national disputes on the ownership of territory and its appellation
have international counterparts involving India, Pakistan and China that map
the erstwhile princely state to consolidate the territory under their control and
reiterate claims on the territory that they believe justly belongs to them or is
strategically important. Pakistan calls the Indian part of Kashmir as 'Indian-
held Kashmir', while the Pakistani part is 'Pakistan-occupied Kashmir' for
Indians. Cross-border infiltration encouraged by the Pakistani army and border
incursions of the Chinese army are attempts to challenge the cartographic
status quo, which are complemented by irredentist counter-mapping driven by
strategic calculus and ideological commitments rather than surveys.[16] Kashmiris
from the Indian side who 'illegally' cross the Line of Control (LoC), too,
challenge the division of territory between India and Pakistan. The international
cartographic politics is not confined to the formal diplomatic arena. When
Pakistan's national security advisor refused to forward a request for medical visa

to the Indian High Commission, the then minister of external affairs, Sushma Swaraj, tweeted that 'POK [Pakistan-occupied Kashmir] is an integral part of India. Pakistan has illegally occupied it. We are giving him [the applicant from PoK] visa. No letter required' (NDTV 2017). On another occasion, when a Kashmiri student sought help through social media, Swaraj replied, 'If you are from J&K state, we will definitely help you. But your profile says you are from "Indian occupied Kashmir." There is no place like that.' The student changed his Twitter location to 'Jammu and Kashmir' (*Hindustan Times* 2018).

It is not the case that maps and statistics always serve as weapons in the hands of the state and vested interests. Urla (1993: 818) points out that '[i]n the hands of the socially or politically disenfranchised, numbers may also be a language of social contestation, a way that ethnic groups, women, and minorities can make themselves visible, articulate their "differences" from the dominant society, and make claims upon the state and its services' (see also Appadurai 2012: 639; Goderle 2016: 87; Taylor 2016: 13). She adds that 'at stake in minority concerns with statistics are not only competing claims to resources but also competing claims to truth' (Urla 1993: 819). Activists based in Kashmir have used statistics to question New Delhi's claims of normalcy by highlighting otherwise ignored statistics on civilian detentions, injuries, deaths and disappearances. They turn to scale-jumping and ally with national and international media and inter- and non-governmental human rights organisations to counter New Delhi's attempt to encase Kashmir as a law-and-order problem within a neat hierarchy of scales bound by international borders. Pandit activists employ similar strategies vis-à-vis Srinagar to question Kashmiri Muslim portrayals of the Valley as a land of communal harmony.

## A Triple Deficit

States such as Nagaland suffered insurgency since the inception and began with hardly any infrastructure or revenue from tourism and export of primary products. In contrast, J&K started off with a much larger infrastructure base and pool of trained workforce and well-developed tourism, handicraft and horticulture sectors. However, the state government failed to raise sufficient resources. Commenting on the finances of the state in 1969–70, Bhattacharjee (2016: 142) pointed out that 'about two thirds of its total receipts then came from gross central transfers including loans'. A decade later – that is, before the onset of the insurgency – B. Puri (1981: 845) pointed out that J&K received 'the highest per capita Central aid and loans in the country. The percentage contribution of Central devolutions to the state's total revenues is also the highest.'

Puri added that the state 'gets about 78 per cent of its total receipts financed by the Centre, where the Centre directly finances many expensive projects, where annual visitors equal almost 40 per cent of its population and army expenditure inflates local incomes and, above all, where a uniquely charismatic leadership is in power should have by now taken its economy to the take-off stage' (ibid.: 846). The condition worsened after the 1980s. The ratio of the state's own tax revenue to gross central devolution and transfers was 0.14 in 2001 compared to 0.2 for special-category states and 1.28 for other states (Agrawal and Kumar 2020a: 317). For the more recent period, it was reported that 'Jammu and Kashmir has received 10 per cent of all Central grants given to states over the 2000–16 period, despite having only one per cent of the country's population' (Raghavan 2016). Decades of preferential treatment has, however, failed to reduce J&K's dependence on federal funds. Bhattacharjee (2018: 25) argues that this is true for most of the erstwhile special-category states because 'the benefits flow in perpetuity without any accountability or performance monitoring of the states. This has made them overwhelmingly dependent on central funding.'[17] The large federal transfers reflect in poverty reduction though. 'Grants under the special category status' helped J&K 'to expand the network of government activities and public services. Through direct and indirect employment, it has been able to bring down poverty level, like in other special category states, to less than the national average' (Bhattacharjee 2016: 140).

Indeed, despite the persistent and large shortfall in revenue, the state has maintained a workforce that is much larger than that of other states. Former finance minister Haseeb Drabu pointed out that 'with 4.5 lakh people already in government service (Bihar, with nearly 10 times the population, has only 4 lakh employees), one out of every seven in the working population is a government servant. Add the pensioners, and every third person will be on the payrolls of the government!' (Drabu 2020a). To these we can add those engaged as contractors, suppliers and casual labourers by various departments and the armed forces. Preferential federal redistribution, rather than the strength of the local economy, allowed the state government to support a bureaucracy twice as large as the average Indian state despite a much smaller tax base (Agrawal and Kumar 2020a: 317).[18]

While the tax base of the governments in peripheral states such as J&K remains small, the lack of opportunities outside the public sector forces them to continuously expand public employment that constrains their ability to build infrastructure and deepens their dependence on federal transfers. Yet, despite massive overstaffing, the growth of public employment cannot keep up with the burgeoning ranks of unemployed college graduates. In J&K, the problem of unemployment of the educated youth has persisted over a long period. Land redistribution and the spread of education under Sheikh Abdullah and Bakshi Ghulam Mohammad destabilised traditional hierarchies and generated huge aspirations in Kashmir. Later, 'Green Revolution ... multiplied Kashmir's paddy production and a White Revolution ... spurred dairying. But young Kashmiris ... were not interested in agriculture. Waving degrees, they demanded jobs they considered respectable' (Devadas 2019a: 197–98). Even by the 1980s, '[t]here were no jobs to be had ... except through bribery' (ibid.: 198). When Farooq Abdullah returned as the chief minister in the late 1990s, his government created more jobs 'even though the government was already overstaffed twice over' (ibid.: 341). Two decades later, as one of the last finance ministers of the state, Drabu (2020a) highlighted the impossibility of satisfying the demand for government jobs: 'At an average rate of retirement of 3 per cent, and much higher rate of growth of youth entering the job market, someone would do well to estimate the job market dynamics!'[19] In short, the state's own revenue cannot finance even a fraction of the public employment and public goods. The large and persistent gap between public spending and revenue and the local economy's inability to generate sufficient jobs is a measure of the development deficit.

With the private sector crowded out of the economy by a bloated public sector and stifled by insurgency and infrastructural bottlenecks, capturing the public pie by securing a favourable delimitation of electoral constituencies, the creation of smaller administrative units and access to affirmative action and welfare benefits appears to be an attractive 'economic' activity. Communities turn to ethno-statistical entrepreneurship to fix numbers and boundaries used for the distribution of public resources that is based on the reported population and measured socio-economic disadvantage. This further undermines the public goods system. On the other hand, the under-supply and poor quality of government statistics – that is, data deficit – hamper development planning. Poor development, in turn, reduces the fiscal legroom to implement statistical reforms.

J&K has also been a long-standing site of democracy deficit that is reflected in, among other things, the inability of the government to hold elections on schedule (in the early 1990s and after 2018), the violation of human rights by the agencies of the state, curbs on news media, the appointment of former intelligence, army, police and Ministry of Home Affairs (MHA) officials as governors, the deployment of armed forces and central paramilitary forces in civilian areas and the retention of contentious laws such as the Armed Forces (Jammu and Kashmir) Special Powers Act (AFSPA), 1990, despite popular opposition. The development deficit of J&K compounds its democracy deficit rooted in domestic (and international) challenges to India's sovereignty in Kashmir and a heavy-handed counterinsurgency. It not only deepens economic disparities that undermine democratic politics but also reduces the opportunity cost of disruptive protests. The democracy deficit in turn feeds back into development deficit because policies made by institutions that do not enjoy the trust of people are deficient in both inputs from all sections as well as legitimacy.

More importantly, democracy deficit also contributes to data deficit as chronic public unrest and insurgency disrupt surveys and undermine the consensus on the production and use of statistics. Moreover, those who lose share in political power and government spending due to poor-quality data lose trust in the impartiality of the state that in turn deepens the democracy deficit. Ethno-statistical entrepreneurship spawned by the *triple* deficits of *data*, *democracy* and *development* politicises numbers and reduces politics to a numbers game (Figure 7.1). Ethno-statistical entrepreneurs potentially include politicians, bureaucrats, civil society and community leaders, journalists and even researchers, who seek to streamline identity, manipulate statistics and redraw administrative and electoral borders to improve their (community's) access to political power and material well-being.

Ethno-statistical entrepreneurship (including counter-mapping) spiked across India's periphery after the sharp decline in violence in the late 1990s and the early 2000s. The 1991 census was completed after the end of the Cold War but just before the liberalisation of the economy. During the Cold War, nationalist insurgencies involved the rejection of international borders as well as national censuses and elections. Later censuses – 2001 and 2011 in Nagaland and Manipur and 2011 in Kashmir – were conducted in an atmosphere largely free of direct threats to census operations. But that did not mean indifference to census. Instead of fighting to change international borders and rejecting national censuses, non-state actors began to manipulate intra-national borders and numbers to improve their (community's) bargaining power within the international borders. Census and maps emerged as new avenues for older conflicts aided by the fact that the 2001 census was going to be followed by a long-awaited delimitation of electoral constituencies. In J&K, for instance, census became 'just a pretext for a religious war, led by the competing elites' (Swami 2000).[20] However, even in 2001 the Hizbul Mujahideen's threat to census was couched in *statistical* (opposition to de facto method of counting) and *developmental* (fear of obstruction of 'avenues of development for Muslims' if they lose majority) terms.

**Figure 7.1**   Data, development and democracy deficits

*Source*: Prepared by the author.

*Note*: See Agrawal and Kumar (2020a: fig. 1.2) for an earlier iteration of this figure.

## Policy Implications

To tackle the massive unrest triggered by the Amarnath controversy, the union government launched a scheme called 'Udaan' to train up to 100,000 young Kashmiris following the recommendations of the Expert Group to Formulate a Jobs Plan for the State of Jammu and Kashmir, headed by C. Rangarajan (2010–11). Devadas (2018: 207) pointed out that the expert group failed to resolve the discrepancies it noted between the figures of the employment bureaus in the state and of the National Sample Survey Office regarding the number of unemployed in the state[21].... [a proper investigation would have revealed] that several of those who described themselves as 'unemployed' in fact had jobs in the private sector, or ran small businesses, or had substantial incomes from horticulture. (Ibid.: 208)

He argued that 'it was clear that the [Rangarajan] committee ... had misunderstood the *meaning* of unemployment in the context of Jammu and Kashmir' (ibid.: 206, emphasis added). In Kashmir, a job 'quite often meant a secure and pensionable government job, one that turned one into a part of the network of power and influence' and 'was about a designation, so that one could feel important'. Moreover, 'security and the pension of a government "job" were much in demand ... also among those looking for a "suitable" groom' (ibid.: 208).

The Prime Minister's High Level Committee for Preparation of Report on Social, Economic and Educational Status of the Muslim Community of India, chaired by Justice Rajinder Sachar, relied upon the 43rd to 61st rounds of the National Sample Survey (NSS) and the 2001 census without accounting for the variations in the quality of the data over time (GoI 2006b). Agrawal and Kumar (2020a) show that the results of the NSS are not reliable for J&K because of erratic changes in the coverage. We can add that the National Sample Survey Office (NSSO) does not account for mobile population groups and snowbound areas that are inaccessible for almost half of the year during two sub-rounds. Together these areas and communities accounted for about 10 per cent of the undivided state's population and spanned all three divisions. The failure of the NSSO to make adequate arrangements for a sizeable segment of the population is partly explained by the poor quality of metadata published by the census. The NSSO is expected to update its sampling frame after every census, but the Office of the Registrar General of India (ORGI) releases metadata in bits and pieces without following any pre-declared calendar. So when the NSSO changes the sampling frame after a census, it does not have access to the necessary metadata. This has in the past resulted in errors in the sampling frame, multipliers and population projections used by the NSSO (Agrawal and Kumar 2020a: 160, 261, 290; Kumar 2021a). In earlier chapters, we have shown that the recent censuses of J&K are unreliable, which means that even in the absence of the aforesaid problems the sampling frame would be flawed.

The next set of examples is related to census-based redistribution. After independence, key political and economic exercises such as interstate delimitation and federal redistribution tracked the latest census results.

When the government introduced severe measures to control population growth in the mid-1970s, various federal redistribution formulae were hastily uncoupled from the census and 1971 was fixed as the reference year for population in federal redistribution formulae.[22] This was followed until the Thirteenth Finance Commission. In 2015, the Fourteenth Finance Commission divided the weight of population criterion between the 1971 (17.5 per cent) and 2011 (10 per cent) populations. The Fifteenth Finance Commission switched to the 2011 census. The change in the population baseline from 1971 to 2011 favours the more populous but economically stagnant northern states. However, the commission added a new performance-based criterion that measures the progress made 'in moving towards the replacement rate of population growth'. The weight of population criterion is now divided between the 2011 population (15 per cent) and performance-based criterion (12.5 per cent). The use of the 2011 population baseline reduces the resource share of the less populous but economically vibrant southern states, but they gain because of criteria that reward good governance and success in controlling fertility. Moreover, the 1971 census is still used to estimate demographic performance that is the reciprocal of total fertility rate (TFR) scaled by the 1971 population.

It is widely believed that the use of government statistics eliminates 'discretion in the allocations' (GoI 2013a: iv, 9, 26–27). This overlooks distortions due to errors in government statistics and the choice of one among multiple sources of government statistics. The Fifteenth Finance Commission could have used the TFR from the census, the National Family Health Survey (NFHS) or the Sample Registration System (SRS). The TFR of J&K estimated from the 2011 census is 3.0, the highest among the large states and the second highest among all the states (GoI 2020b: 6.53). The SRS estimate was 1.9 (Table 3.7). The commission itself notes that the state's TFR was 2.4 as per NFHS-3 and 2.0 as per NFHS-4 (see annex GI.3 of the report), and yet it accepted a much higher estimate based on the flawed census. J&K gained in federal redistribution due to over-reporting of the population but lost due to the overestimation of the TFR.

The Delimitation Commission constituted after the reorganisation of J&K used the 2011 census data. As per the uncorrected population, Kashmir should get 51 seats, with the remaining 39 seats going to Jammu. However, Jammu would get at least 41 seats even in the case of conservative corrections. The commission awarded 47 seats to Kashmir and 43 to Jammu. The procedure followed by the commission is not clear because it used a combination of the uncorrected 2011 census population and other factors such as terrain that can be assessed only when a detailed report is released.

# Minimalist Reforms

In most developing countries, policymaking is hamstrung due to the unavailability of reliable and timely information as well a shortage of resources. Despite major advances in the early decades after independence in building a robust framework for population censuses, rolling out large-scale sample surveys and creating a federal statistical system, the whole range of India's government statistics began to falter by the 1970s. Two examples of the growing discontent are in order: one from the field of economic planning and the other from social justice. B. S. Minhas resigned from the Planning Commission in 1973 due to differences over 'distorted or juggled' data that were used 'to present an unrealistically bright picture of India's prospects' in the run-up to the Fifth Five-Year Plan (1974–79) (Weinraub 1973). Later, in that decade, B. K. Roy Burman dissociated himself from the Second Backward Classes Commission that failed to take all available facts into account (Roy Burman 1992: viii, 1998: 3178). The degeneration of the census seems to have started in the mid-1970s when federal policies were uncoupled from the latest headcount. Both timeliness and accuracy cease to be priorities when the data are not going to be used in high-stakes exercises.

A series of developments since then including the growing politicisation and deinstitutionalisation of policy making, polarisation of the polity and the society and economic reforms have aggravated the data deficit. The liberalisation of the economy, in particular, had a far-reaching impact on government statistics.[23] First, structural adjustment restricted the funding of politically unimportant statistical agencies affecting both recruitment and overhaul of their systems in the 1990s.[24] This, for instance, affected the NSS, until recently the most important source of household-level socio-economic information.[25] Similar manpower constraints in other wings of the government affected the supply and quality of administrative statistics across the board. These constraints were particularly severe at the level of state governments. The union government belatedly addressed this through grants to states through the Thirteenth Finance Commission that called for building the statistical capacity of the district administration (GoI 2009b: 37, 224). Later the union government extended support to states under a scheme called Support for Statistical Strengthening (*Morung Express* 2018a).[26] The aforesaid budget constraints do not seem to have impacted the 2001 census though it was conducted a decade after the introduction of liberalisation. By then the government finances had considerably improved. The budget allocation for census increased from INR three billion to INR twelve billion in the decade to 2001 (Banthia 2001a).

Second, at a more fundamental level, economic liberalisation also altered the structure of the underlying political economy by withdrawing the government from several sectors, reducing regulatory controls and reporting requirements, allowing greater integration with international markets and supply chains and unleashing rapid technological change. These developments rendered the statistical systems designed in the 1950s out of sync with the changed circumstances. So the colonial statistical system that was overhauled after independence to provide information for economic planning in a federal democracy had to be further reformed after economic liberalisation. The government addressed the problem belatedly by constituting the National Statistical Commission (NSC) in 2000 to suggest comprehensive reforms. The NSC submitted its report in 2001, but the recommendations were implemented in a piecemeal fashion over the next decade.[27] In the meantime, the already weak statistical institutions turned dysfunctional due to manpower constraints and the growing politicisation of statistics. The experience of the NSSO exemplifies this. Commenting on the contested nature of the design of the NSS in the 1990s, Deaton and Kozel (2005: 190, 196) point out that 'the political right had an interest in showing low poverty, and the political left in showing high poverty, and this undoubtedly intensified the debate on survey design and led to the unfortunate compromise design that temporarily undermined the poverty monitoring system'. A decade later, the 66th round of the NSS (2009–10) 'showed that employment generation fell significantly short of the target of the 11th Five Year Plan' (Agrawal and Kumar 2019). The result of this round came under attack from '[s]ome highly placed officials' who 'decided that the data must be wrong, and castigated the NSSO for its faulty investigative methods' (Chandrasekhar and Ghosh 2011), and 'the NSSO was encouraged to break from tradition and generate one more, large sample survey of employment relating to 2011–12' (Chandrasekhar and Ghosh 2013). 'Ahead of the 2019 polls, the Narendra Modi government delayed the release of the results of the 2017 Periodic Labour Force Survey that contested its employment claims' (Agrawal and Kumar 2019).

Third, at yet another level, the withdrawal of the state added urgency to capture public resources in the economically stagnant ethno-geographic periphery that was not equipped to harness the fruits of liberalisation. These also happened to be the regions where nationalist insurgencies degenerated into sub-nationalist conflicts over redistribution after the end of the Cold War. The union government partly resolved this problem through an ad hoc solution – namely delinking federal redistribution and the 2001 census – but this merely shifted the locus of distributional conflicts from the federal level to the provincial level and, in fact, intensified intrastate conflicts.

The quality of census data has also been affected by political interference in addition to the poor training of enumerators, the weakening of the earlier system of checks and balances on the ground, the growing use of technology in data processing with insufficient quality checks and grassroots manipulation, often with the complicity of politicians, officials and a variety of other ethno-statistical entrepreneurs. Further, the growing delays in the release of data compounded the problems posed by the dwindling supply of metadata. The release of census data on religion was delayed twice during the last two decades, first by the Bharatiya Janata Party [BJP] in 2004 and then the Indian National Congress (INC) in 2013 (Kumar 2015b). Several other census tables for 2011 were also delayed and issued without metadata (Agrawal and Kumar 2020c; Kumar 2021a).

The resultant data deficit has undermined both policymaking and routine administration and eroded the government's capacity to deal with sudden developments such as the Covid-19 pandemic. Attempts to predict the magnitude of reverse migration triggered by the pandemic were constrained by the belated release of the 2011 census data on migration in instalments, while the government admitted that it had failed to adequately monitor reverse migration. The government was additionally constrained by the unavailability of data on consumption and unemployment for the period immediately before the pandemic as the NSSO has not been able to provide reliable estimates after 2011–12. So, presently, the government is planning with a decade-old data. This is problematic because of the growing migration, digitalisation and globalisation of the economy, and, most recently, the triple shocks of demonetisation, haphazard transition to the Goods and Services Tax (GST) regime and the pandemic that have disrupted the pre-2011 patterns and trends of economic activities.

The growing data deficit has led to calls for greater engagement with non-governmental data and newer forms of data generated in the digital space that have already begun to overshadow conventional sources of information in some fields. The conventional census, whose roots go back to the late eighteenth century, seems outdated. On the one hand, the census is already yielding ground to other means of accounting population in parts of Europe where governments now rely on administrative statistics to estimate their populations. On the other, both our personal and public lives are increasingly dominated by newer forms of data. An elaboration of the continued significance of the census in the twenty-first century is therefore in order before we discuss possible reforms. Enacted two years before India became a republic, the Census Act, 1948, is among the very few legislations deeply embedded in the Constitution, which makes human population census a sine qua non of power- and resource-sharing at all levels. Until the 1970s, the redistribution of constituencies used data from the latest census, but the Constitution (Forty-Second Amendment) Act, 1976, suspended decennial delimitation to protect the interests of states that were relatively successful in meeting family-planning targets. Attempts to conduct delimitation ahead of schedule in 1990 and 1996 were unsuccessful (Sivaramakrishnan 1997: 3281). While the reference date for the intrastate delimitation of assembly constituencies was eventually changed to 2001 in 2003, interstate delimitation of parliamentary seats remains linked to the 1971 census. In six states, even intrastate delimitation proved to be contentious and had to be deferred. In five other states, the intrastate delimitation was based on the 2001 census with amendments to accommodate the demands of indigenous communities (Agrawal and Kumar 2020a: 306; Kumar 2022b). A similar change in the reference date for population statistics used in the Finance Commission's formula for federal redistribution generated a lot of controversy. The Fourteenth Finance Commission was the first to partially shift away from the 1971 benchmark. While the Fifteenth Finance Commission switched completely to the 2011 census for the estimate of population, the 1971 population is still used in the case of performance-based criterion linked to fertility. Several states protested these revisions in federal redistribution formulae. So even attempts to update the decades-old reference dates for census-based distributions have proven to be contentious. The barriers to completely discarding the familiar and widely accepted conventional censuses as the basis for distributing power within the union seem to be insurmountable in the foreseeable future. The protests surrounding the Citizenship Amendment Act (CAA) and the National Register of Citizens (NRC) have further demonstrated the continued political salience of census in our public life, particularly for communities in the social and geographical margins.

While the census will remain politically unassailable, statistical agencies cannot ignore the newer types of data. Private corporations and information companies are generating a lot of data, and researchers, data journalists and amateur fact-checkers are using the same to question government sources. Government agencies are themselves generating digital data, and critics highlight the inconsistency between the conventional and the newer sources of government data. For instance, questions were raised when Aadhaar enrolment was found to be less than census population in Kashmir (census official, interview, Srinagar, 30 November 2015; note 46 of Chapter 3) and Nagaland (Agrawal and Kumar 2020a: 183). More importantly, the government can itself not accept a serious inconsistency between data generated by different departments, even if these datasets are not formally interconnected. While individual-level data are rarely compared, the aggregate statistics of different databases are subjected to consistency checks. For instance, the Election Commission of India (ECI) compares the electorate and the adult population as per the census (ECI 2017). So statistical agencies will have to eventually integrate conventional and digital databases, which is challenging as they have not yet been able to integrate most of conventional databases and at times even different parts of the same survey – for instance, the state and national samples of the NSSO. But this is not the key problem facing sources of government statistics such as the census.

The lack of clarity about the purpose of census and the deteriorating trust in the exercise that has been reduced to an arena of political and communal contests in several states are the key challenges. The reforms have so far been limited to the introduction of digital technologies. Ashish Bose (2000: 1433) suggested that '[m]odern technology applied to data which are less than reliable will not guarantee a good census.' Indeed, in the absence of deeper reforms, the growing use of new technologies has been associated with growing delays in the release of data (Agrawal and Kumar 2020c), declining availability and quality of metadata (Kumar 2021a) and persistent and serious coverage and content errors (Chapter 3; see also Agrawal and Kumar 2020a: ch. 4).[28] It is also noteworthy that the coverage of newer databases such as Aadhaar is deficient in areas where conventional databases are also deficient (ibid.: 297). We will outline a multi-pronged reform below, focusing on some of the long-standing concerns about census schedules, the role of bureaucracy and the trust gap with people.

## Schedules

There were 14 questions in the household census schedule in 1951, which grew to 29 by 2011 (GoI n.d.7). Between 1961 and 2011, the share of government spending in national income grew from 5.55 to 11.08 per cent. The correlation between the expansion of census schedule and the growth of government spending's share in national income, which has been observed in other countries as well (Coleman 2012: 335), is driven by the fact that governments need more data both to raise taxes as well as rationalise higher spending. T. N. Srinivasan (2003: 306) throws light on the institutional context of demands to include more questions in survey schedules. He reminisced that the NSSO's governing council

> was often asked either to add more questions to already long questionnaires, or to collect new data. However, those making such demands did not think of the cost of adding questions, including in terms of the quality of the response. Nor, for that matter, did they say what question they would take out in order to maintain the length of the questionnaire.

This applies to the growth of census schedules, too. Indeed, reforms should begin with the rationalisation of the length of census schedules, which will require revisiting the rationale for census (Kumar 2020d).

Census schedules should be restricted to information required to fulfil constitutional, administrative and statistical obligations (Figure 7.2). The constitutional obligations of the census include providing data for distributing public offices (and conducting elections) and resources among administrative units,[29] implementing affirmative action policies[30] and formulating policies for linguistic minorities.[31] Policymakers and administrators require data on socio-economic parameters disaggregated across the three tiers of the federal polity for planning public infrastructure and services. The statistical purpose is limited to providing inputs for the subsequent stages of the same census as well as future censuses (note 40 of Chapter 6), estimation of vital statistics (GoI 1978a: 2) sampling frame for surveys (Agrawal and Kumar 2020a: 252–65) and population projection (ibid.: 128–29). The census schedules should be trimmed to limit the scope of questions to fulfil the aforementioned requirements.

Cutting down the length of unwieldy schedules has several advantages. First, it will help improve data quality by allowing focused training of enumerators and reducing their workload on ground. Second, the tradition of preparing detailed administrative reports has decayed. Such reports are not available for the last two censuses, which happen to be the most controversial in J&K. For the last census, even the complete report on general population totals (GPT) has not been issued. These reports used to share information on difficulties faced while conducting census, changes in the classification of rural and urban areas,

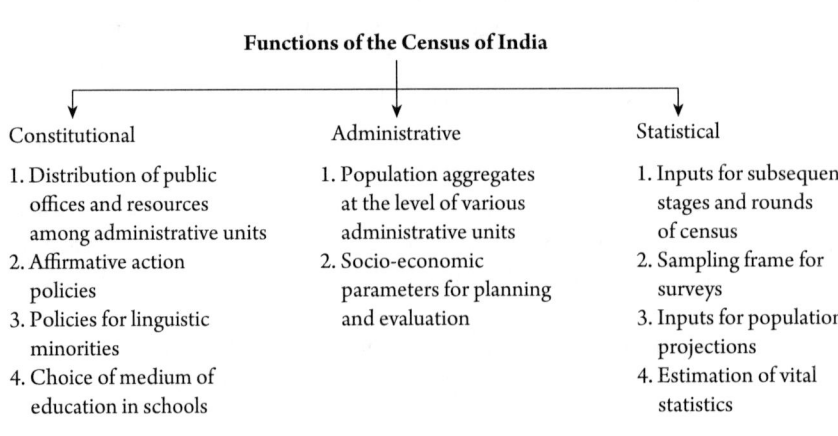

**Functions of the Census of India**

| Constitutional | Administrative | Statistical |
|---|---|---|
| 1. Distribution of public offices and resources among administrative units | 1. Population aggregates at the level of various administrative units | 1. Inputs for subsequent stages and rounds of census |
| 2. Affirmative action policies | 2. Socio-economic parameters for planning and evaluation | 2. Sampling frame for surveys |
| 3. Policies for linguistic minorities | | 3. Inputs for population projections |
| 4. Choice of medium of education in schools | | 4. Estimation of vital statistics |

**Figure 7.2**    Functions of the Census of India

*Source*: Prepared by the author.

inter-censal territorial realignments, changes in reference dates, accounting of mobile populations, political disturbances during census and interstate territorial disputes, among other things. In the absence of metadata, government officials themselves end up misinterpreting data or using wrong data.[32] The reduction in the volume of data to be processed will allow senior officials to pay greater attention to data quality and revive the earlier tradition of detailed descriptive reports that were key sources of metadata. The availability of administrative reports will also ensure that the next census does not begin with a blank slate.[33] The reduction of the workload should also make room for more disaggregation and cross-tabulation of non-standard and miscellaneous categories and a more careful scrutiny of outliers in the data.[34] Third, it will cut down processing time and help reduce the growing delays in the release of data. Fourth, shorter schedules will seem less invasive and assure respondents uncomfortable with sharing too many details. With lesser information shared, it will also be difficult for data miners armed with latest technologies to disaggregate and map the data on to localities jeopardising the confidentiality of respondents. So this will contribute to the protection of the confidentiality guaranteed in the Census Act, 1948 (Article 11[1][b], 15) and enhance trust in the exercise.[35] Last but not least, reduced workload will hopefully help revive the tradition of rich ethnographic studies. The earlier censuses were followed by a rich array of studies (Mitra 1994) that not only supplemented the metadata but also offered inputs for potential improvements. Similar studies are needed to improve the accuracy of data on, say, migration. The patterns and trends of the migration of the Bakarwals have changed over the past five decades. There is a need for ethnographic studies to understand these afresh and redesign census procedures accordingly.

While the benefits of shorter schedules are evident, the choice of questions to be dropped can be contentious. Therefore, as a first step, we will identify only those questions that have no relation to the core obligations identified earlier (Figure 7.2). Nearly half of the Houselisting and Housing Schedule is devoted to questions on household amenities and assets. This information is not needed to fulfil any of the obligations listed earlier and can be more appropriately collected through sample surveys. Thus, these questions can be dropped. Likewise, a quarter of the questions in the Household Schedule probe characteristics of workers, which can be dropped except for the question on occupation. It is unfair to expect inadequately trained enumerators to assess unemployment when even full-time, trained surveyors, armed with more nuanced questionnaires, struggle in this regard. For similar reasons, the issue of disability is perhaps better studied using specialised surveys.[36]

## Accountability

Since human population census is the bedrock of government statistics, the government will have to address the lack of accountability at the heart of the exercise. But let us begin with an issue that has not received much attention in the recent times but has been raised intermittently since the colonial period. The First Schedule to the Allocation of Business Rules, 1961, assigns the task of taking census to the Ministry of Home Affairs (GoI 2019c: 91). This is problematic as this ministry is also responsible for internal security and counterinsurgency. Even more problematic is the fact that the ORGI needs the approval of the Cabinet Committee on Political Affairs (CCPA) before releasing data on 'very sensitive' data on religion, language, caste and tribe (GoI 1993: 20, 2011i: 112). It would help to separate the ORGI from the Union Ministry of Home Affairs and place it under the Ministry of Statistics and Programme Implementation (MOSPI), which is responsible for national sample surveys and national accounts.[37] The proposed integration will, however, require deep reforms as the MOSPI has itself suffered a loss of credibility.

However, more than finding the right ministry for the ORGI, there is a need to establish accountability within the system. The bureaucracy lacks an incentive to invest in the exercise unless the political leadership makes census a priority as happened in Nagaland in 2011. Senior officials in districts who supervise the exercise and the junior staff who enumerate are not the primary users of the data; nor is census their primary responsibility. Even the directors of census operations (DCOs) are appointed for a limited period on deputation often without any prior exposure. So, as a DCO put it, 'my feeling was that "I joined the Census Department as a naive and conducted the census as an expert"' (GoI 2001b: xii).

Discussions with senior census officials across the country suggest four responses to questions about data quality. First, the incumbent DCO refuses to comment on the questions one raises as they pertain to the period before he or she assumed office. (It bears noting that a large part of the data collected during one's tenure is published after one's deputation as the DCO or the registrar general of India (RGI) ends.) The former DCO expresses an inability to comment as he or she is no longer in that office and redirects one to the current incumbent. Second, the DCOs pass the blame to the district administration and the RGI, who, in turn, blames the DCOs and the district administration. Third, when confronted with errors, senior officials suggest that they did not receive any (formal) complaint. Fourth, incumbents explain the errors by referring to the poor quality of the past census data.

The deployment of manpower is a key area that needs attention. As long as India follows the canvasser method,[38] the last-mile public servants, who are conscripted to carry out enumeration, are indispensable. Most other databases, except electoral rolls, involve relatively small teams of trained surveyors. However, census requires the simultaneous mobilisation of a very large workforce across the country, with preliminary preparations beginning nearly four years before the reference date. Commenting on the manpower crisis facing the census, Ashish Bose (2000: 1433) pointed out that the opportunity cost of public servants' time has grown tremendously and asked, 'Is it possible to conduct a good census ... through an army of unmotivated, disinterested and even hostile school teachers and petty bureaucrats?' (see also A. Bose 1991: 31).

We have already discussed the trimming of schedules to reduce the enumerators' cognitive burden and workload. Several interviewees in Jammu proposed the induction of non-native enumerators. This is entirely impractical. Mobilising such a large workforce from elsewhere for the census in Kashmir when the Jammu region and other states are also in the process of conducting the census is not feasible. Even otherwise, language barriers and a lack of familiarity with the geography will prove to be insurmountable barriers. Moreover, the extensive deployment of 'outsiders' in a conflict zone will require elaborate security arrangements that will further undermine trust in the exercise. It will also undermine the standing of local officials in their other routine engagements with the people. In the end there is no escape from the local administration. The quality of the census *cannot* exceed that of the general administration. Reforms should focus on simplifying the process of enumeration. For instance, given the improvement in both the presence of administration in snowbound areas and means of communication, the practice of separately enumerating snowbound areas can be discontinued (see also notes 62 and 73 of Chapter 2). Note that this was followed in the 2001 census of J&K on an ad hoc basis (Table 2.3) and should be formally implemented in snowbound areas across states. Further, enumerators may be assigned duties in neighbouring blocks, or *tehsils* – that is, outside the administrative unit they are currently posted in and preferably not in the administrative unit they live in or belong to.[39]

The recruitment of senior bureaucracy, too, needs to be revisited. It has been proposed to upgrade 'the post of the census commissioner to the additional secretary', but, as Banthia (2004: 3862) rightly points out, the current arrangement is more appropriate as it allows much needed stability at the helm. Unlike additional secretaries who have a tenure of two years, a joint secretary serves in his or her post for five years. However, it would perhaps help to appoint DCOs and the census commissioners from within the statistical bureaucracy rather than depend upon civil servants on deputation. This will also help avoid delays in the appointment of DCOs due to the unwillingness of state governments to depute officers, who are anyway not willing to join this unglamorous office.[40] In J&K, the government will have to additionally address the possible differences between the two regions. A journalist recalled that 'one of the census officials [in Jammu] told me that they do not trust the figures received from Srinagar' (interview, Jammu, 2 June 2016). On the other hand, Kashmiri enumerators do not trust their counterparts in Jammu (Bhat 2018: 181).

## Rebuilding Trust

The reforms discussed so far cannot address the growing trust deficit amidst the communalisation and politicisation of census. There is a need for a renewed outreach to engage various stakeholders, including respondents, for which the ORGI has over the decades developed a broad template (Figure 7.3). As we will see later in this section, the intensity and sincerity of the outreach varies across censuses (for example, Nagaland in 2001 and 2011) and across states in a census (for example, Nagaland, on the one hand, and Manipur and J&K, on the other, in 2011).

Further, since the performance of a statistical agency with respect to timeliness affects users' perception about other dimensions of data quality (United Nations Economic Commission for Europe 2018: 10), the problem of delays needs urgent attention. Political interference is most evident in delays in the release of data before elections and delays in turn fuel partisan politics.[41] The delays in the release of religion, language and migration data have increased steadily since 1981 (Agrawal and Kumar 2020c). The 2001 and 2011 religion data were belatedly released due to parliamentary elections, possibly to avoid controversies (Kumar 2015b). However, instead of checking the politicisation of the data, the delay fuelled speculation and made room for communal propaganda. The 2001 post-enumeration survey (PES) report was also delayed 'to avoid needless political controversies' while delimitation was in progress (A. Bose 2008: 16).[42]

Provisional targets for the publication of main reports used to be set in advance in some of the earlier censuses (see, for instance, Visaria 1977: 206). The timeliness of census data has deteriorated since then, with most data for the 2011 census being released after enormous and unjustified delays – that too as Microsoft Excel files without any mention of the official date of release, let alone metadata.[43] Using the example of the manner in which the 2011 data on

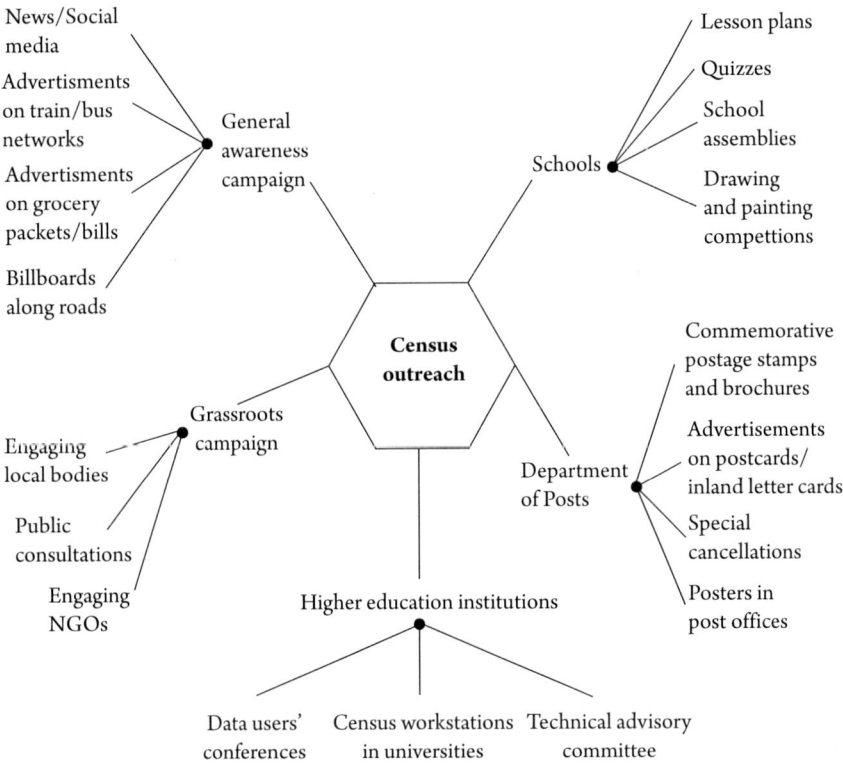

**Figure 7.3**  Outreach activities of the Office of the Registrar General of India (ORGI)

*Source*: Prepared by the author.

religion were released, we can say that '[o]nce there is a delay in releasing data it is impossible to find a politically non-controversial opportunity for release' (Kumar 2015b). Precommitment to a calendar of release is therefore necessary to eliminate political interference and curb the growing delays in the release of data. The Census Rules should be amended to specify, among other things, cut-off dates for the release of data.[44] While tracking delays in the release of data is important, it is equally important to check if the data are being used by the government. India has been conducting censuses regularly but after 1971 it has not been able to use the data for, say, the delimitation of constituencies and federal redistribution.[45]

While improving the timeliness of census publications will enhance trust in the data, there is no substitute to a well-designed outreach program. Stakeholders need to be made aware that they do not stand to gain much from manipulation. In fact, manipulation only deepens the mistrust between competing groups that exacerbates the pre-existing conflicts, even as it upends planning and administrative processes.[46] In J&K, the inflation of the headcount compounded the dissatisfaction of Jammu (Hindus) without adding much to Kashmir's seat share. The experience of other states, such as Nagaland and Manipur, and countries, such as Nigeria and Pakistan, corroborates this.

We will probe Nagaland's experience to understand how to rebuild trust. In 2001, Nagaland's hill tribes inflated their headcounts to avoid the loss of constituencies to Dimapur located in the plains allegedly dominated by 'outsiders'. Despite massive over-reporting, they barely managed to limit Dimapur's gain to four seats – that is, the seats it would have gained as per the 1991 census. The state government formally rejected the 2001 census in a letter to the ORGI. Such a rejection has no impact on the status of the published data though. A 2005 household survey raised doubts about the reliability of census figures (Hazarika 2018: 292–93), and the chief minister admitted that the state's population was inflated (Hazarika 2005). The administration, though, did not wake up until 2009, when a sample survey conducted in six districts found fewer people compared to 2001 (*Nagaland Post* 2009). There was also a growing realisation within the government that flawed statistics hampered planning (DCO, Nagaland, interview, Kohima, 19 September 2012). This is when the state government stepped in. In 2009, two years before the census, the state government convened a consultative meeting including civil society, church, student and tribal organisations, village elders and political parties (Government of Nagaland 2009). The government did not blame any community or region for the mess and impressed upon all that a reliable census is necessary for both development planning and political and social harmony. At no point did the government try to facilitate dialogues between the competing communities. The participants unanimously agreed that earlier censuses were flawed and resolved to educate their respective constituencies. The consultative meeting built a consensus against communally motivated interference in the census. The census authorities on their part engaged the civil society and people. Further, as far as possible the administration avoided appointing enumerators in their own localities, and enumerators approached the relevant signatory to the 2009 resolution if they faced resistance from a community.

Almost all communities refrained from manipulation even though the 2011 census extended Dimapur's potential gain from delimitation from four to six seats in a 60-member assembly. The correction was so severe that Nagaland became the first state to report a 'contraction' of population after 1951. The population of a few tribes and districts contracted by more than half, and the subsequent electoral roll revision pushed several constituencies back to the 1990s in terms of the size of the electorate. Nagaland's demographic somersault – decades of unusually high population growth followed by a sudden 'contraction' – eliminated, to a large extent, the anomalous growth in the *reported* population until 2001.

The switch did not happen due to any change in the punitive provisions of the Census Act, 1948 (last amended in 1994) or any political or socio-economic change. In discussions with census officials and political leaders, the author did not come across any hint that the use of punitive measures was contemplated. The government mechanically invoked the threat of punishment. It also announced that enumeration would be cross-checked using church records and would be followed by the collection of biometric data, but neither was part of the official calendar of the census. Three phrases capture what helped in the 2011 census of Nagaland: public consultations, greater transparency and improved vigilance. Nagaland's experience suggests that it helps to educate and engage lay-respondents and users right from the early stages of the lifecycle of a census. The outreach for the 2011 census, which is among the better censuses conducted in the state, began two years before the household phase.

The importance of a transparent consultative process becomes evident when we compare Nagaland with Manipur and J&K, which faced similar problems. Several Naga-dominated hill sub-divisions of Manipur reported very high population growth in 2001. Political parties opposed to delimitation based on the flawed census approached the union government and courts. Unlike its Nagaland counterpart, the Manipur government held only one community responsible for the problem and tried to correct the problem through coercive administrative mechanisms – namely, a selective recount. In the end, Manipur failed to resolve the problem, and over-reporting persisted in the 2011 census. J&K, where the problems in the 2001 census remained unacknowledged, faced a bigger problem in 2011, with widespread over-reporting across Kashmir.

Kumar (2020a, 2023a) uses elementary games to understand the manipulation of headcount in a multi-ethnic setting. The games have multiple equilibria and communities can move from one equilibrium to another without the threat of legal sanction. He therefore argues against imposing the existing penalties, let alone introducing newer penalties, and suggests that confidence-building measures are key to addressing the trust deficit that drives competitive manipulation. Following Kumar (2020a, 2023a), we will represent the interaction between two communities during enumeration using a using a one-shot simultaneous-move game of complete information in which each community has two pure strategies – *over-report* and *not over-report*. The choice of a simultaneous-move game is justified as the Indian census follows the extended de facto (synchronous) method of enumeration. Complete information follows from the fact that the communities have known each other for a long time. Also, this is a one-shot game as census is a decennial exercise, and not every round is linked to the redistribution of power and resources.

A community is worse off in this game if it does not over-report, while others do so because it loses electoral seats and share in development funds linked to population. Further, the chances of a fresh census are slim when other communities have over-reported as the majority may prefer the new status quo. Nevertheless, over-reporting is costly because officials must be bribed or coerced to manipulate records. Moreover, in addition to the risk of detection and punishment by authorities, there is a risk of conflict with communities denied a fair share of electoral seats and development funds because of the flawed statistics. This is reflected in several objections to delimitation of electoral constituencies filed in courts. When all communities over-report, each incurs the cost of manipulation (to maintain its population share) and bears the risk of conflict with rival communities as well as of detection by authorities. In addition, development planning is vitiated. The following conditions govern the ordering of outcomes.

1. A community is better off over-reporting if others over-report – that is, (over-report, over-report) is preferred to (not over-report, over-report).

2. The outcome represented by (over-report, over-report) is not better than (not over-report, not over-report) because the status quo persists under the former as the state is compelled to use old statistics, and the effort to manipulate may go waste.

3. 1 and 2 imply that over-reporting by both communities (over-report, over-report) can neither be the best outcome nor the worst outcome.

4. (Not over-report, not over-report) cannot be the worst outcome because under the status quo communities do not incur the cost of manipulation. Otherwise, communities would perversely prefer simultaneous manipulation even though this entails costs without gains.

5. Over-reporting when others do not over-report (over-report, not over-report) is preferred to (over-report, over-report).

Given these conditions, depending on the relative ordering of (not over-report, not over-report) and (over-report, not over-report), only two orderings of the outcomes are possible, which relate to different types of players.

1.  Type I: (not over-report, not over-report) ≻ (over-report, not over-report)
    ≻ (over-report, over-report) ≻ (not over-report, over-report)
2.  Type II: (over-report, not over-report) ≻ (not over-report, not over-report)
    ≻ (over-report, over-report) ≻ (not over-report, over-report)

These two orderings can support three games depending on how the communities rank the outcomes (Figure 7.4). In the first two games, the players are paired with their own types (I–I and II–II). The last game pairs players of different types (I–II or II–I).


Game I (Type I–Type I): There are two pure-strategy Nash equilibria in this game that resembles games of coordination. Under the first, neither community over-reports, whereas under the second both over-report. The former Pareto dominates the latter. This game has a mixed-strategy equilibrium as well that is Pareto-dominated by (not over-report, not over-report). The mixed-strategy equilibrium can be interpreted in three ways: (*a*) each community randomises across censuses and/or regions, (*b*) each community comprises two types of individuals with different propensities to manipulate and (*c*) at least one community has a small private information.


Game II (Type II–Type II): This game resembles the Prisoner's Dilemma and has just one pure-strategy equilibrium in which both communities over-report (over-report, over-report). This outcome is, though, Pareto-dominated by (not over-report, not over-report). The equilibria of Game I–II (Type I–Type II or Type II–Type I) resemble that of Game II.

The widespread manipulation of the 2001 census of Nagaland, when almost all communities over-reported their numbers, relates to the (over-report, over-report) equilibrium of Game I. The outcome of the 2011 census of Nagaland relates to the (not over-report, not over-report) equilibrium of Game I because almost all communities refrained from engaging in manipulation.[47] As discussed earlier, this equilibrium switching happened due to the state government's awareness campaign rather than any punitive action. So, Game I rather than Game II, which involves punishment to achieve Pareto improvement, better captures the interaction between communities during the census.[48]

The mixed-strategy equilibrium of Game 1 relates to (*a*) the 2001 and 2011 censuses of Manipur, where only a few communities in the hill districts over-reported their populations, (*b*) the 1991 census of Nagaland, where only a few

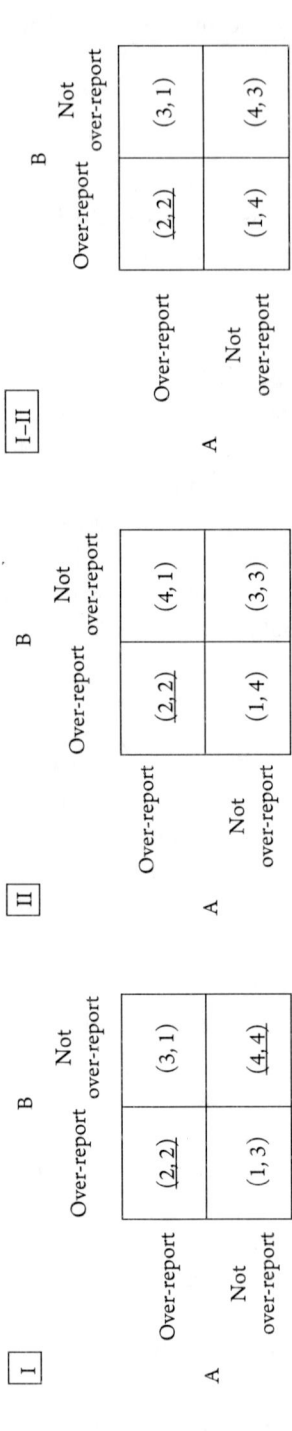

**I**

| A \ B | Over-report | Not over-report |
|---|---|---|
| Over-report | (<u>2, 2</u>) | (3, 1) |
| Not over-report | (1, 3) | (<u>4, 4</u>) |

**II**

| A \ B | Over-report | Not over-report |
|---|---|---|
| Over-report | (<u>2, 2</u>) | (4, 1) |
| Not over-report | (1, 4) | (3, 3) |

**I–II**

| A \ B | Over-report | Not over-report |
|---|---|---|
| Over-report | (<u>2, 2</u>) | (3, 1) |
| Not over-report | (1, 4) | (4, 3) |

**Figure 7.4** Manipulation games

*Source:* Kumar (2020a).

*Note:* (*a*) A and B denote communities. (*b*) $4 \succ 3 \succ 2 \succ 1$. (*c*) I, II and I–II identify different games based on the type of players paired. (*d*) The pure-strategy equilibria are underlined.

communities over-reported their populations, and (*c*) the 2001 and 2011 censuses of J&K, where over-reporting of population was mostly confined to Kashmir.

# Concluding Remarks

The United States introduced decennial censuses to facilitate the delimitation of electoral constituencies, which effectively amounts to apportioning federal resources across jurisdictions and communities in proportion to their population. Later federal redistribution and the 'enforcement of non-discriminatory elections' were formally linked to the census (Bhagat 2003: 686). Eventually the census came to determine the redistribution of a substantial proportion of public resources. Alonso and Starr (1985: 94) point out:

> In 1940 ... only about nine percent of the federal budget [of the United States (US)] was distributed as grants to states and localities. But by the early 1970s, that percentage had nearly doubled, and today such funds ... make up about one-fifth of state and local revenues. Where these billions get spent is largely determined by congressionally mandated formulas using statistics.

Alonso and Starr add that 'the U.S. government ... spends $1.4 billion a year collecting statistics' (ibid.: 93). A decade later, T. Porter (1995: 28) pointed out that the United States spent 'about 6 percent of the gross national product' on measurement.[49] Cross-country evidence from developed countries suggests that census schedules grew with the share of government spending in national income (Coleman 2012: 335). The Planning Commission and the Finance Commission in India linked a very large proportion of federal funding to government statistics. With so much tied to numbers, government spending on statistics has grown, and, at the same time, communities and governments find it difficult to stoically face reported changes in numbers.

The growing spending on data has not meant better understanding and utilisation, though. Roy Burman (1992, 1998) highlights the disuse of the data on caste by government commissions constituted to formulate affirmative action policies. T. N. Srinivasan (2003: 306) draws attention to the insufficient use of macroeconomic data. Agrawal and Kumar (2020a: 42–51) discuss at length the case of Nagaland, where even the most basic statistics are not used or are widely misunderstood in the academia, media and government. We discussed the poor understanding of the area and population statistics for J&K.

Likewise, the growing spending on data does not automatically translate into an improvement in their quality, at least in developing countries like India.[50]

In particular, the results of the census in India are shaped by a combination of factors, including inadequate training for enumerators, bureaucratic moral hazard, political interference, weak non-governmental capacity to assess data quality and a lack of interest in data quality in the academia. Another major factor that vitiates statistics relates to competition over the public pie, which includes various elected bodies, public sector employment and public institutions of higher education that are distributed according to reported population shares and measured socio-economic disadvantage.

This book presented the story of the census in J&K where bureaucratic compromises, weather and political uncertainties, changing itineraries of 'nomadic' tribes, internal displacement of communities, disproportionately large public sector, ignorance of statistical, legal and administrative foundations of the census among respondents and enumerators, competing methodological pressures on the design of the census, partisan interference, political turf wars, international conflict, communal mobilisation, omissions, misrepresentations and misinterpretations in government publications, and partisan or poorly informed news reports and academic analyses shape numbers that are used by extremists in Kashmir and Jammu to assemble threatening statistical caricatures of the 'other' that can serve as easy targets of manufactured hate. Stricter laws and harsher penalties and technological improvisations cannot address the problem of data deficit that is enmeshed in democracy and development deficits. The revisions to the Census Act have so far been piecemeal and incremental in nature and have mainly focused on legal and administrative problems. For instance, in 1994, the Act was extended to cover pre-tests, pilot studies, houselisting and housing census and post-enumeration checks (PEC) and PES. However, the Census Rules were amended a decade and a half later in 2009 to clarify that the relevant schedules 'may not be notified in the Official Gazette'. The Act and Rules need to be amended to lay out clear guidelines to facilitate ex-post correction, avoid political interference and tackle politically motivated collusive manipulation.[51] While legalistic solutions have limited efficacy vis-à-vis manipulation, amendments that plug in the gaps identified in Chapter 5 will have a didactic value insofar as they will make problems transparent and (bureaucratically) legible and will, hopefully, serve as focal points for an informed public debate on the problems.[52]

In the end, there is no alternative to improving the accountability of the administration,[53] building trust in the impartiality of public institutions and facilitating dialogue between communities. Trust in key public institutions also requires bolstering the autonomy of the Comptroller and Auditor General (CAG) of India and the judiciary, the independence of the academia and media, the freedom of speech and the right to information (RTI). These are broadly constitutive of a modern democracy. The CAG, the the Central Information Commission (CIC) and media are particularly important as they help cross-examine administrative statistics. However, the integrity of these institutions has deteriorated over the past decade, which contributes to democracy deficit. Both the availability and quality of the CAG reports have, for instance, declined over the last decade (Upadhyaya 2020).

We discussed a minimalist agenda for reforms in this chapter that called for rationalising the length of census schedules, a pre-commitment to a calendar for releasing data, enhancing bureaucratic accountability and redoubling efforts to engage respondents. Additional reforms will require careful deliberation as the Census Act, 1948, is deeply intertwined with India as a modern nation state. Even seemingly technical reforms can have a much wider constitutional political impact. Consider, for instance, the choice of the extended de facto synchronous (hereafter de facto) method of counting. In several states including J&K, objections to the methodology of census are essentially related to the *de facto* method of counting (Agrawal and Kumar 2020a: 238, 323; *The Tribune* 2000b; GoI 2001b: 29; *Kashmir Times* 2011b). To be more precise, the accounting of migrants is the bone of contention. The *de jure* method would be more acceptable to the complainants in such cases, but it cannot be introduced in the census. The *de jure* method will pose enormous conceptual and operational challenges. Moreover, the *de jure* method will perhaps undercut the foundation of the secular nationalist project and the non-denominational state by indirectly foregrounding *identity* rather than the *current place of residence* as the primary variable of interest to the government. Concerns about the *de facto* method will have to be addressed through awareness campaigns. On the other hand, certain proposals for change will require a re-evaluation of long-standing assumptions and practices. The treatment of snowbound areas in the census reflects an outdated understanding of space. Likewise, the current practice of enumeration of the 'nomadic' (and boat) population reflects the normative bias of the census in favour of settled populations (see note 56 of Chapter 2) and awkwardly applies the de facto method of counting to mobile populations. The counting of mobile populations and mapping them on to administrative units should be governed by considerations related to the provision of public goods. The solution arrived at for nomadic populations will have much wider relevance for the accounting of seasonal migration between, say, eastern and southern states of India. A proposal for change in the accounting of such categories of population would require a broad consensus among both the migrant and host communities and their respective state governments.

To conclude, reinventing the census for the twenty-first century requires an understanding of data in relation to development and democracy, as artifacts dynamically shaped by their socio-economic and political contexts. It requires an investment in building consensus over government statistics as public goods. In the absence of such a consensus, legalistic interventions and technocratic innovations will secure marginal improvements in data quality without being able to rescue data from partisan politics.

## Notes

1. The importance of censuses and elections can be gauged from the fact that the Right to Education Act, 2009, bars deployment of teacher for 'any non-educational purposes' except 'the decennial population census, disaster relief duties or duties relating to elections' (Article 27).

2. Another example of flawed analysis is in order. Ashraf and Karthik (2020) present a misleading analysis of the threat of demographic change by treating all migrants including intrastate migrants as interstate migrants, which resulted in a highly exaggerated estimate of outsiders eligible for domicile under revised rules. Jamwal (2022: 392) extrapolated their claim to suggest that the number of migrants 'could be doubled within a few years attracted by the right to get a domicile certificate'. Aiyar (2023) went a step ahead and claimed that 'there is even fear that a Bihari might emerge as the leader of Jammu as a result of the registration of some 25 lakh outside labour as voters, made possible by the precipitate withdrawal of article 35A safeguards'. At least, in the first two years there has been no serious interest even among de facto state subjects settled elsewhere in the country to secure domicile under the relaxed guidelines (*Indian Express* 2021b, 2021c). See Chapter 3 for a discussion of the data on migration.

3. Kashmiri Muslims often refer to the 1891 census that reported Kashmir-born Muslims in Punjab (see, for instance, K. B. Ahmad 2017: 80–81). Abdullah (2016: 45–46, 339) offers a brief but graphic account of the plight of Kashmiri Muslim migrants in Punjab.

4. In most conflict zones, the scale of atrocities is deeply politicised. An unfortunate incident of rape of women in Konan and Poshpora in 1991 turned into a ground for statistical battles. Wajahat Habibullah, who carried out the preliminary investigation of the incident, noted that he 'thought perhaps the entire village had decided to say they were raped so that the victims do not have to live alone with this blot' (Jaleel 2013). B. G. Verghese, too,

made a similar point (*Kashmir Times* 2000e). Three decades later, the partisans of the government and its opponents remain stuck to their preferred numbers – no rape versus mass rape. In Bangladesh, 'the orthodox nationalistic discourse about the 1971 war' maintains that three million people lost their lives in the war of independence, and attempts to question the figure are readily labelled as 'anti-liberation' mindset (Bergman 2014). 'Three million deaths' has acquired an iconic status and is intertwined with the self-imagination of Bangladesh.

5. According to Syed Ali Shah Geelani, 'in late 1947, some five lakh Muslims were killed by Dogra forces and Hindu chauvinists in Jammu' (Sikand 2010d). Abdullah (2016) offers detailed remarks on the communal conflict in Jammu but does not provide estimates of the loss of lives or displacement.

6. As per the 1941 census, Jammu's urban population excluding the cantonment was 50,379.

7. In 1941, Jammu town's population included 15,920 Muslims – 15,806 Muslim (Others) and 114 Muslim (Shias) – compared to 30,564 Hindus (Government of Jammu and Kashmir [GoJK] 1943b: 10–11, 1943c: 102, 341) – that is, Jammu was a Hindu-majority city. The population share of the Muslims in Jammu district (Jammu city) dropped from 39.59 (31.60) to 10.03 (17.24) per cent between 1941 and 1961 (ibid.; GoI 1966d: 15, 42).

8. Commenting on *The Kashmir Files*, a controversial movie based on the exodus of Pandits, the Kerala unit of the Indian National Congress (INC) tweeted, 'Even in the 1948 communal riots after partition, over 1,00,000 Kashmiri Muslims were killed in Jammu, but there were no retaliatory killings of Pandits' (*India Today* 2022). This is unlikely to be true. According to the 1941 census, there were only 4,976 Kashmiri speakers, including Hindus and Muslims in Jammu and Kathua districts (GoJK 1943c: 306–07).

9. 'The tribal raids of 1947 resulted in the mass movement of population from areas on the other side of the Cease-Fire Line as also from one district to another within the territory administered by the state. There has been a tendency among different communities to concentrate in the districts where they already claimed a predominant share in the population' (GoI 1968b: 242).

10. The focus on statistics in the debate on partition-era violence in Jammu diverts attention away from its social and political contexts. While Ved Bhasin blames the princely state that leaned on Hindu extremist organisations for instigating and executing violence in Jammu, he adds that Sheikh Abdullah 'didn't intervene or could not … perhaps his feeling was that the Muslims in Jammu were not his supporters' (S. Khan 2009). For the same reason Sheikh Abdullah seems to have 'supported ceasefire [when the Indian army was advancing]' because the people on the other side had 'never accepted him as the leader' (ibid.).

Abdullah's autobiography hints at an early rupture with Jammu (Abdullah 2016: 124–25).

11. Panun Kashmir's map of homeland mentions 8,600 square kilometres – that is, 55.48 per cent of Kashmir (Panun Kashmir n.d.3).

12. Sheikh Abdullah's autobiography, too, refers to Anantnag as both 'Islamabad' (Abdullah 2016: 89, 91, 93, 106, 438–39) and 'Anantnag' (ibid.: 107, 301, 451–52). It is not clear if the original Urdu text uniformly uses one or the other appellation and if that uniformity in the original text was imposed by the scribe or the copy editor.

13. When the author started one of his trips to Anantnag, the gentleman who offered travel support informed his family that he was going to 'Islamabad-Anantnag'. A news report referred to Anantnag as Islamabad (Anantnag) (M. Ali 2011a). Such hyphenated and parenthetical usages reflect the contested nature of space in conflict zones (see Agrawal and Kumar 2020a: 92, 102, for Nagaland).

14. In his autobiography, Sheikh Abdullah noted that Suleiman Hillock was renamed as Shankaracharya Hill by the Dogra rulers (Abdullah 2016: 30).

15. Nominal contests over space are commonplace in conflict zones. See Agrawal and Kumar (2020a: 92, 101–03, 120) for Nagaland.

16. The tentative border dividing the territories administered by India and Pakistan was called the Ceasefire Line before being rechristened as the Line of Control (LoC) in the Simla Agreement of 1972. The border between J&K and Pakistani Punjab is known as the international boundary or working boundary.

17. 'During the post-1953 era, the GoI took a conscious decision to allow and indeed encourage Kashmiris to steal from the government since it was felt that this would lead to greater loyalty to India! Strangely enough, Kashmiri Muslims rationalise corruption in government by saying that stealing from the Indian occupying force is not haraam!' (retired Indian Administrative Service [IAS] officer, interview, 10 December 2015; also political scientist, interview, Jammu, 8 May 2022). In fact, Sheikh Abdullah's autobiography abounds in references to this problem. Such corruption is widespread in the north-east, too (Chasie 2000).

18. At present, the sanctioned strength of the IAS cadre for the union territory [UT] of J&K is 137 compared to 652 for Uttar Pradesh, which is about 20 times more populous (*Economic Times* 2022a). A decade ago, the sanctioned strength for Uttar Pradesh was 592 compared to 137 for J&K (GoI 2011k).

19. Agrawal and Kumar (2020a: 314) discuss a similar crisis facing Nagaland, where about 125,000 persons are employed by the state government out of a population of more than 2 million.

20. In Northern Ireland, where 'the census data has often been used to advance sectarian arguments', 'demography had become even more politicised, a case perhaps of "war by other means"' (Anderson et al. n.d.). In Palestine, census is 'as important as the intifada.... It is a civil intifada' (Greenberg 1997).

21. The Rangarajan Committee suggested that it would 'be useful to view the NSS number of unemployed as the baseline number for strategizing on the number of jobs that need to be created. On the other hand the J&K numbers on the "job seekers", could be used as the aspirational ceiling number' (GoI 2011g: 13).

22. It is not clear why intrastate delimitation was also frozen in the mid-1970s. This not only distorted resource allocation within states but also entrenched resistance to any future delimitation as it would involve drastic changes, while a whole generation had grown up unaccustomed to periodic revisions. In 2002, the government repeated the mistake and once again froze intra-state delimitation along with interstate delimitation for an extended period. Note that in the 2002 delimitation only assembly constituencies were delimited.

23. The twelfth decennial population census, including the revisional round, concluded on 5 March 1991 – that is, more than four months before the formal inauguration of economic reforms. So the first post-liberalisation census was conducted in 2001.

24. Shetty (2012: 43) argued that the 'the loss of pride in statistics as a profession, particularly in public administration and public institutions' has compounded the shortage of staff due to resource crunch as the government finds 'it difficult to attract sufficient numbers of candidates and fill staff positions created after the regular attrition'. We can add that there has been a decline of teaching of statistics in the country, even as the better qualified candidates now have more lucrative employment options in the private sector. Together these two drastically limit the pool of candidates available to the statistical agencies of the government.

25. For manpower constraints faced by the NSSO, see Bhalla (2014) and Chandrasekhar and Ghosh (2011). Shetty (2012: 43) suggested that 'sample sizes have been drastically reduced because of the inadequacy of the organisational strength at the NSSO or at the CSO [Central Statistical Office] level. As a result, both sample and non-samples errors are feared to be high.'

26. See also Agrawal and Kumar (2020a: 34) for the role of international bodies and aid organisations in nudging developing countries, including India, to reform their statistical systems.

27. For critiques of the NSC, see Srinivasan (2003) and Vidwans (2002a, 2002b).

28. The Socio Economic and Caste Census (SECC) was perhaps the first large-scale Indian survey that used computer-assisted personal interviewing, but the data were affected by content errors and were not released. In neighbouring Pakistan, the 2023 census was conducted using tablet computers, which were supposed to be the answer to all that ails the exercise in the country since 1971. But the turn to technology did not help as the older problems related to inadequate remuneration for enumerators, logistical constraints, poor design of schedules, inter-provincial mistrust and lack of security in insurgency-hit areas remained unattended. Further, the enumerators struggled with faulty digital equipment and software glitches, amidst concerns about the confidentiality of data.

29. The census is referred to in the following articles of the Constitution that govern the distribution of seats: Articles 55 (Manner of Election of the President), 81 (Composition of the House of the People) and 170 (Composition of the Legislative Assemblies). Later amendments added references to the census in Articles 82 (Readjustment after each census), 243 (The Panchayats) and 243P (Municipal bodies).

30. Articles 330 (Reservation of seats for Scheduled Castes and Scheduled Tribes in the House of the People) and 332 (Reservation of seats for Scheduled Castes and Scheduled Tribes in the Legislative Assemblies of the States), too, refer to the census.

31. Articles 350 (Language to be used in representations for redress of grievances), 350A (Facilities for instruction in mother-tongue at primary stage) and 350B (Special Officer for linguistic minorities) do not refer to the census but mandate policies that require data on language.

32. We have already discussed how the ORGI miscalculated the growth rate of Muslims in India during 1991–2001. See Jamwal (2020b) on how the government misinterpreted census data on J&K's languages. For examples from Nagaland and Manipur, see Agrawal and Kumar (2020a, 2020e).

33. During a visit to the library of a Directorate of Census Operations, this author learnt that its collection did not include earlier administrative reports. The neglect of administrative reports is despite the acknowledgement of the importance of these reports for 'planning a future census' and 'in bringing out several sidelights of the operations' and exhortation to DCOs to 'express your views freely and offer critical comments' (GoI 2010c). A three-volume administrative report was planned for the 2011 census at the level of the RGI and the DCOs (ibid.). With great difficulty the author was able to locate only the second volume of the administrative report prepared by the RGI for the 2011 census.

34. Fostering a better understanding of the data requires greater attention to non-standard (say, institutional population, slum population, and second- and third-language speakers) and miscellaneous (say, 'Other Languages' and 'Generic Tribes') categories. Disaggregation and cross-tabulation of these categories will resolve puzzles related to, say, the abrupt shifts in language data of Ladakh and Gurez; the stable share of Sikhs in Kashmir's population; ballooning of unclassifiable migrants; and the drop in Ladakh's sex ratio and the rise in its Hindu and Sikh population during 2001–11. Census reports should also include a separate chapter on outliers, such as J&K's child population statistics in 2011 or growth rate of Nagaland's population in 2001.

35. The switch to a digital census requires a comprehensive review of respondents' confidentiality as well as safeguards against the misuse of unit-level data by governmental and non-governmental agencies. In 2022, the government notified databases and computer resources related to the census and the National Population Register (NPR) as critical information infrastructure under the Information Technology Act, 2000 (amended in 2008) (*The Print* 2022a). Critical information infrastructure refers to 'Computer Resource, the incapacitation or destruction of which, shall have debilitating impact on National Security, Economy, Public Health or Safety'. The notification partly addresses the misuse by non-governmental agencies. A month later, in response to questions in the Lok Sabha, the government clarified that 'no cyber attack or data hacking incident has been observed in the Data Centre or the Disaster Recovery Sites of the Office of Registrar General and Census Commissioner of the country' (*The Print* 2022b).

36. The question of the scope of census was discussed even in the run-up to the first post-colonial census. P. C. Mahalanobis suggested that 'while a complete headcount for the minimum demographic details, such as age, sex, marital status, relationship to the head of the household and composition of the latter would still be necessary for a variety of administrative purposes including delimitation of electoral constituencies, details of social, economic and cultural characteristics of the population for development policy and planning purposes had best be collected on a small, statistically sound and representative sample basis' (Mitra 1994: 3207). R. A. Gopalaswami, the then census commissioner, did not agree because 'in a population of multi-lingual, multilevel and multi-ethnic communities with unevenly distributed conglomerates at diverse levels of development, it would not be possible to draw satisfactory multi-stage, multi-level statistical samples – random or systematic – that would satisfy the requirements of micro- and macro-planning' (ibid.).

37. The Committee on Legislative Measures in Statistical Matters noted that the ORGI 'opposed … amendments on the Census Act, 1948, and the Registration of Births and Deaths Act, 1969' to bring census under the jurisdiction of the NSC and include census in the list of core statistics – that is, 'statistics of national importance' (GoI 2011i: 32–33). The ORGI argued that the proposed amendments violated the established practice that 'only the Ministry which is administering the Act is empowered to propose amendments to the Acts through a Bill which is presented in the Parliament' (ibid.: 111). It added that census was 'a large-scale administrative time bound exercise and cannot be treated merely as statistical one', that it involved the collection of 'sensitive data' and that the amendments proposed by the committee would undermine the orderly conduct of census, which provides critical inputs to key functions of the government, by diluting 'statutory powers' of the ORGI (ibid.: 111–12).

38. As per Rule 6D of the Census (Amendment) Rules, 2022, in the next census 'a person may fill-up [*sic*], complete and submit the census schedule through self-enumeration'. In fact, even before this amendment the Indian census combined the canvasser and householder methods of enumeration, but the latter was restricted to the institutional population.

39. A 2009 consultative meeting convened by the Nagaland government to discuss the faulty 2001 Census resolved that in the following decennial census '[n]o enumerator will be allowed to do enumeration where … his own tribe are in majority' and that the participants would extend 'active support and cooperation to census operations' (Government of Nagaland 2009). While the resolution was in no way binding upon the ORGI, it allowed more room for the DCO in matters related to the assignment of enumerators and dealing with opposition on the ground. The then DCO recalled several instances where community volunteers honoured the resolution. Note that deputing officials outside their areas may not necessarily help. In 1962, the Nigerian government seems to have tried to prevent ethnic collusion by exchanging enumerators across regions (Ahonsi 1988: 555, 557). In the 1991 census, '[n]o commissioner was put in charge of areas where he or she came from' (Okolo 1999: 323). However, these censuses failed to earn the trust of people and proved to be deeply divisive.

40. The scale of the problem of the non-availability of directors in states is evident from the following detailed note in the 2001 provisional population totals (PPT): 'Assam saw three Directors during a span of two months just before the population enumeration was to commence. After the sudden voluntary retirement of the regular Director in December 2000, the Director of Census Operations, Sikkim was given the additional charge of Assam until a suitable replacement was found at the end of January 2001. Unfortunately, the State

Government of Meghalaya could not provide the services of a full-fledged Director. In March 2000, the then Assam Director was given the additional charge of Meghalaya and on his retirement, Director of Census Operations, Arunachal Pradesh held the additional charge of Meghalaya. In Nagaland, Andaman and Nicobar Islands, Lakshadweep and Pondicherry the Directors of Census Operations were available only in an ex-officio capacity. In a few states the Directors joined very late posing immense problems of coordination. The Director of Census Operations, Chandigarh, Union territory, oversaw the Houselisting Operations in the Maharashtra Directorate as the Director of Census Operations, Maharashtra could assume charge only in May, 2000. It is extremely crucial to have all the Directors of Census Operations in position at least two years prior to the actual commencement of the census' (GoI 2001d: 4).

41. A former census commissioner of Pakistan noted that the 'delay and postponement led to [the] politicisation of the 1991 census' in Pakistan (A. H. Khan 1998: 481).

42. In neighbouring Pakistan, post-enumeration checks (PEC) were not conducted after the 2017 census (Wazir and Goujon 2019) due to political divisions. See also note 6 of Chapter 5 for the US experience regarding the correction of census estimates.

43. 'The census report of 1931 was easily the most comprehensive of all such reports complete with maps, statistics, ethnographic accounts and provinces' reports together with a three-volume all-India report' (K. S. Singh 1996: 141). A wide range of qualitative reports were also released after the first two post-colonial censuses (Mitra 1961, 1994; Mohanty and Momin 1996). The general population tables of the 1961 census were by far the most detailed (GoI 1964a). Thereafter, the qualitative output steadily declined even though several summative reports were released around the census centenary in 1971. For the longer-term reduction in the descriptive component of census reports, see T. Porter (1995: 36-37) for France and Sundar (2000: 116) for colonial India.

44. See Agrawal and Kumar (2020c) for desirable and feasible timetables for the release of data. Also, note that the proposed National Policy on Official Statistics called for the production and publication of core statistics 'as per pre-announced calendar, free from Government influence ... along with critical analysis regarding the quality of data and implication of the use of data in policy making and administration' (GoI 2018c: Sections 5.1.7 [1], [7]). Core statistics do not include census though.

45. In the United States, the seats of the House of Representatives could not be reapportioned after the 1920 census because the Conservatives alleged that the population of rural areas was underestimated (Prewitt 2003: 16).

46. As Basu (1997: 10) points out, 'the use of demographic arguments in favour of communalism [exclusive communitarian projects or goals] can lead to exactly the kind of political and social instability that demographic events themselves can'.

47. Kumar (2020a) models the correction of population in the 2011 census of Nagaland using a *correction* game that slightly modifies the manipulation game used here. While the modified game is more appropriate and richer, its key predictions are same as the game used here. Also note that the simple model presented here can accommodate the case of unequal over-reporting by communities.

48. The superfluous role of punitive measures can be confirmed with the help of a fuller model in which communities inflate headcounts to increase their share of seats in the legislature and development funds and are mindful of the costs of manipulation including fines (Kumar 2023a). Also, following Basu (2015), we can argue that the Nagaland government's intervention in 2009 did not create a new equilibrium. Rather it helped the society shift from one equilibrium (widespread over-reporting) to another, pre-existing equilibrium (no over-reporting) by creating a new focal point.

49. It is difficult to arrive at similar estimates for developing countries due to poor accounting and largely informal character of the economy. In the recent years, this difficulty has been compounded by the fact that increasingly large volumes of data are being generated by processes and institutions outside the control or supervision of governments.

50. The net omission rate detected through the PES increased from 18.0 and 17.6 in 1981 and 1991, respectively, to 23.31 in 2001 and 22.98 in 2011 (GoI 2014d: 2). The ORGI cryptically mentions that this variation could be a product of 'the manner in which PES was conducted' – that is, the quality of the survey.

51. The most recent example of piecemeal reforms relates to the indefinitely delayed 2021 census. The government has introduced digital devices for data collection and made provision for self-enumeration. However, privacy concerns and questions about the quality of self-enumerated data remain unaddressed.

52. Note that the reforms suggested here are different from occasional suggestions of both census officials and academics to make the law stricter and invoke it more often. For instance, the census commissioner for 1981 noted the following: 'The 1981 census experience clearly indicates the urgent need for a hard look at the Census Act. Unlike the palmy days when the count was on one day or when the Collectors could order all dogs to be chained and lamps to be lit and placed in windows to help enumerators, in today's world we have found that *constant vigilance was necessary at every stage merely to ensure that*

*enumerators were available and carrying out the operations as required.* No doubt, under the statute functionaries can be punished but the stipulations in the law are not firm enough with regard to this matter' (GoI 1981d: 7, emphasis added). He suggested, 'If the census has to be carried out successfully next time it is *clear* that the law has to be considerably stiffened' (ibid., emphasis added). As discussed in Chapter 5, the government is severely constrained by political and other practical considerations in its relationship with grassroots public servants who serve as enumerators. In any case, the census commissioner's wish was partly fulfilled in 1994 when the Census Act was amended to tighten the punitive provisions, but this was followed by some of the most egregious instances of manipulation. Commenting on the inflation of the headcount of the Halba or Halbi tribe in Maharashtra due to the fraudulent responses of non-tribals who wanted to claim tribal status, Kulkarni (1991: 207) attributed the problem to, among other things, the government's unwillingness to invoke the punitive provisions against respondents. In Chapter 5, we explained why most democracies do not take recourse to punishment. Moreover, it is not clear how the mere awareness of the laws can deter misreporting when the government lacks both the ability as well as resolve to prosecute en masse violations of law. Despite the evidence of ineffectiveness of the punitive measures in checking non-compliance with the Census Act, the government still believes that it can use the threat of punishment to address problems arising out of lack of trust (see note 3 of Chapter 6).

53. A corollary to the fact that the quality of enumeration is a function of administrative capacity and trust in public institutions, among other things, is that we need studies of the quality of census data that are sensitive to the spatial and temporal variations in the quality of general administration and the relation between the government and people. More broadly, such studies will have to account for the fact that the quality of statistical institutions is shaped by both material and political conditions (see note 20 of Chapter 1 and note 64 of Chapter 2).

# Appendix

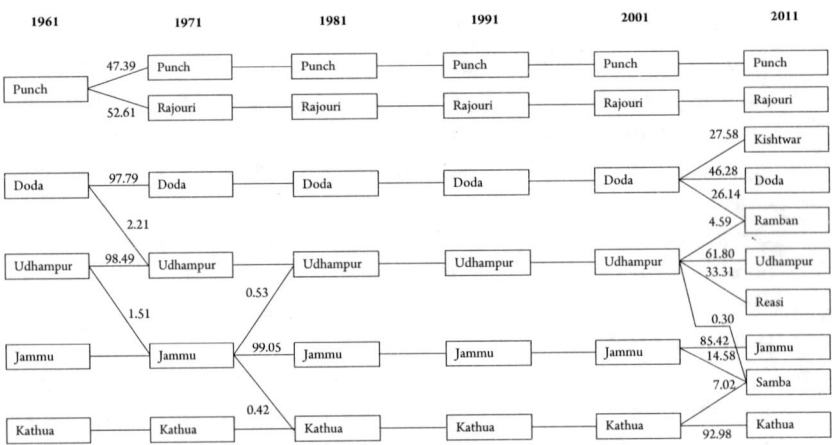

**Figure A.1**  Evolution of districts of Jammu

*Source*: Prepared by the author using Kumar and Somanathan (2009), general population tables (GPT) for 1961–2001, Table A-2 of Census of India, 2011, and district census handbooks for 2011.

*Note*: (*a*) The figures for population transfer between 2001 and 2011 are based on the information about the creation of new districts and further details available in district census handbooks as the GPT report for the 2011 census of J&K has not yet been released. (*b*) The transfer of population from Jammu to Kathua and Udhampur between 1971 and 1981 due to the realignment of the international border around Akhnoor does not seem to have involved any transfer of territory. Jammu's area decreased even as the area of Kathua did not increase. A comparison of GPT reports of the 1971 and 1981 censuses suggests that as per the estimates provided by the surveyor general, the area of Udhampur increased by 1 square kilometre (GoI 1972a: 25, 1983a: 68). However, appendix 1 to table A1 of 1981 (GoI 1983a: 83) suggests that the area of Udhampur district increased by 2.08 square kilometres due to the transfer of one part village from Samba *tehsil* of Jammu district to Ramnagar *tehsil* of Udhampur district. Further, a comparison of table A2 of the 1971 and 1981 censuses suggests a larger increase in population of Udhampur (GoI 1972a: 70, 1983a: 104) than indicated by the appendix to table A2 of 1981 (GoI 1983a: 105), which is possibly explained by the fact that the population of certain villages of Akhnoor *tehsil* was adjusted in Udhampur after 1971. See also note 5 of Chapter 2 and Table 2.1. (*c*) Samba district was carved out of Jammu and Kathua. A village was also transferred from Udhampur to Samba. Table A-2 of the 2011 census suggests that the population of this village was 2,215, but the primary census abstract (PCA) for 2001 suggests that the population of this village was 2,660. We have followed Table A-2. (*d*) Nakarkote and Titri villages were transferred from Pakistan-occupied territories of J&K to Punch after the 1971 war – that is, after the 1971 census (GoI 1983a: 85). (*e*) Numbers adjacent to lines denote the share of the population (as per the preceding census) of the parent district that was transferred to a new or an existing district before the next census.

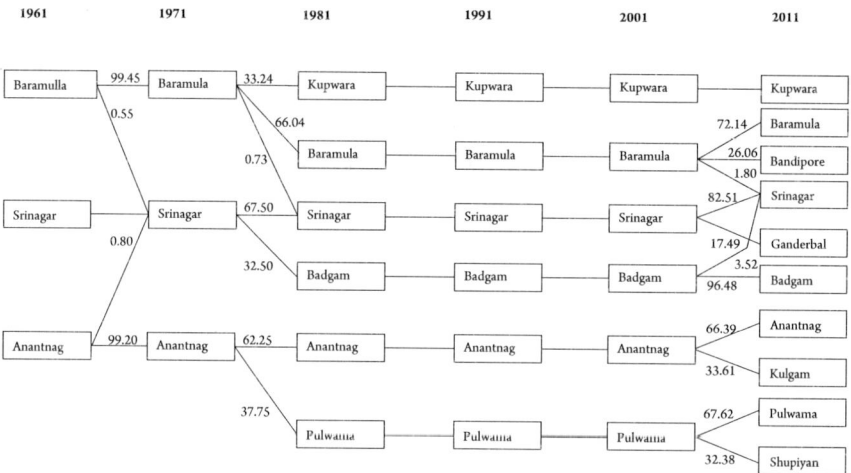

**Figure A.2**  Evolution of districts of Kashmir

*Source*: Prepared by the author using Kumar and Somanathan (2009), general population tables (GPT) for 1961–2001, Table A-2 of Census of India, 2011, and district census handbooks for 2011.

*Note*: (*a*) The figures for population transfer between 2001 and 2011 are based on the information about the creation of new districts and further details available in district census handbooks as the GPT report for the 2011 census of J&K has not yet been released. (*b*) Gasla village was transferred from Pakistan-occupied territories of J&K to Kupwara after the 1971 war – that is, after the 1971 census (GoI 1983a: 85). (*c*) Numbers adjacent to lines denote the share of the population (as per the preceding census) of the parent district that was transferred to a new or an existing district before the next census.

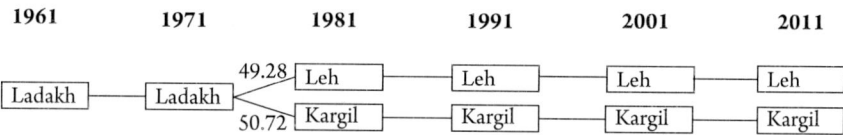

**Figure A.3**  Evolution of districts of Ladakh

*Source*: Prepared by the author using Kumar and Somanathan (2009), general population tables (GPT) for 1961–2001, Table A-2 of Census of India, 2011, and district census handbooks for 2011.

*Note*: (*a*) Certain villages were transferred from Pakistan-occupied territories of J&K to Ladakh after the 1971 war – that is, after the 1971 census. These include Turtok, Chulungkha, Taksi and Thanga villages added to Leh, and Bodagam, Hundermon, Hundermon Brok and Buzber villages added to Kargil (GoI 1983a: 85). See also note 6 of Chapter 2. (*b*) Numbers adjacent to lines denote the share of the population (as per the preceding census) of the parent district that was transferred to a new or an existing district before the next census.

# Bibliography

## Select Interviews

Ahmed Ali Fayyaz, independent journalist, 28 July 2021 (phone).

Ajay Chrungoo, Panun Kashmir, Jammu, 2 June 2016, 12, 14, 21 July 2021 (phone) and 19 April 2023 (phone).

Altaf Hussain, veteran journalist, Srinagar, 9 and 11 July 2021 (phone), 29 May 2022.

Anuradha Bhasin Jamwal, *Kashmir Times*, Jammu, 14 September 2021 (phone), 10 May 2022.

Arun Joshi, *The Tribune*, Jammu, 2 June 2016.

Asgar Ali Karbalaie, former member of legislative assembly (Kargil), Kargil, 20 September 2019, 22 November 2019.

Basheer Ahmed Bhat, Population Research Centre, Srinagar, 30 May 2022.

C. Chandramouli, former registrar general of India, 6 August 2021 (phone).

Census official, directorate of census operations, Srinagar, 30 November and 1 December 2015.

Census officials, Office of the Registrar General of India (ORGI), New Delhi, 21 February 2019.

Christian community leader, Jammu, 8 May 2022.

Community leader (Barwala), Jammu, 7 May 2022.

Community leader (Mahasha), Jammu, 10 May 2022.

Community leader (Megh), Jammu, 10 May 2022.

Community leader (Ravidas), Jammu, 8 May 2022.

Dipankar Sengupta, economist, University of Jammu, 9 May 2022.

Director, military intelligence, Udhampur, 2019.

*Drussu*, Pulwama, 30 May 2016.

---

Italics identify census villages (spellings as per the PCA) where the author interacted with the staff of primary health centres and/or *anganwadi*s.

Employees union leader, Baramula, 19 and 20 July, 19 September 2021 (phone).

*Faqir Gujri*, Srinagar, 23 and 25 May 2016.

Farooq Ahmad Factoo, former director of census operations (Jammu and Kashmir, 2011), Srinagar, 30 May 2022.

Feroze Ahmed, former director of census operations (Jammu and Kashmir, 2001), Jammu, 4 December 2019, 22 October 2021 (phone).

Feroze Khan, chief executive councilor, Ladakh Autonomous Hill Development Council (Kargil), Kargil, 20 September 2019.

*Filler*, Kishtwar, 27 June 2018.

Former intelligence official, 21 March 2021, 16 and 26 August 2021 (phone).

Former intelligence official, 13 August 2021 (phone).

Former legislator, Leh, 20 November 2019.

Former union home secretary, New Delhi, 11 September 2019.

Former union home secretary, 27 June 2023 (phone).

Ghulam Sayedin, ex-councillor (Chuchot), Leh, 20 November 2019.

*Hardo Pandove*, Shopian, 30 May 2016.

Hari Om, Jammu, 30 June 2018, 13 September 2019, 26 September 2019, 21 and 25 September 2020.

Harinder Baweja, *Hindustan Times*, 15 September 2021 (phone).

Hekali Zhimomi, director of census operations (2011 census), Nagaland, Kohima, 19 September 2012, 24 and 25 June 2013.

Independent journalist, 20 August 2021 (phone).

Indian Administrative Service (IAS) officer, 10 February 2021 (phone).

Jammu-based lawyer, New Delhi, 1 July 2018.

Javaid Rahi, secretary, Tribal Research and Cultural Foundation, Jammu, 4 December 2019, 13 August, 19 September 2021 (phone), 7 May 2022.

*Kilshi Pain*, Gurez, 27 May 2016.

Labha Ram Gandhi, West Pakistan Refugees Action Committee, Jammu, 3 June 2016.

Lama Chosphel Zotpa, New Delhi, 2 December 2019.

Lawyer, Jammu, 13 July 2021 (phone).

*Mamoosa*, Pattan, Baramula, 29 May 2016.

Manipur state government official, Senapati (Manipur), 8 October 2019.

Masood Butt, Tibetan Public School, Srinagar, 30 May 2022.

Member, All India Confederation of SC/ST/OBC Organisations, Anantnag, 5 August 2021 (phone).

Metongmeron Ao, director of census operations (2001 census), Nagaland, Kohima, 9 April 2013.

Municipal commissioner, Municipal Corporation, Srinagar, 17 October 2022.

Nagaland state government official, Meluri, 2 November 2014.

Nasir Shabani, lecturer, Kargil, 23 November 2019.

Nawang Tundup, general secretary, Himalayan Buddhist Cultural Society, Paddar, 23 September 2021 (phone).

Noor Ahmed Baba, political scientist, Srinagar, 28 May 2022, 16 October 2022.

P. Stobdan, former ambassador, Leh, 18 September 2019.

P. T. Kunzang, Ladakh Buddhist Association (LBA), Leh, 19 September 2019.

Phunsok Tashi, executive councillor for tourism and Zanskar affairs, Kargil, 20 September 2019.

Praveen Swami, Network18, 12 July 2021, New Delhi (phone).

*Qumroo*, Khansahib, Badgam, 24 May 2016.

R. K. Kalsotra, All India Confederation of SC/ST/OBC Organisations, Jammu, 14 September 2019, 4 December 2019, 7 May 2022.

Rajiv Chunni, S. O. S. International: An Organization for PoK Displaced Persons, Jammu, 2 June 2016.

*Rajpur Kamila*, Nowshera, 30 June 2018.

Rekha Chowdhary, political scientist, Jammu, 8 May 2022.

Researcher, Srinagar, 21 July 2021, 23 October 2021 (phone).

Retired census official, Delhi, 15 September 2012.

Retired census official, Directorate of Census Operations (J&K), 28 July and 30 July 2021 (phone).

Retired Indian Administrative Service (IAS) officer, 10 December 2015, 3 and 4 January 2016, 23 February 2016, 19, 21 and 22 June 2016, 13 August 2021, 3 September 2021 and 6 May 2023 (email).

Retired Indian Administrative Service (IAS) officer, 10 December 2015, 3 and 4 January 2016, 23 February 2016, 13 August 2021, 3 September 2021 (email).

Retired Indian Administrative Service (IAS) officer, Chennai, 22 April 2023, 24 April 2023 (email) and 28 April 2023 (phone).

Nawang Rigzin Jora, former member of legislative assembly (Leh), Leh, 19 September 2019; Jammu, 3 December 2019.

Retired state government official, Punch, 21 September 2021 (phone).

*Sarak*, Doda, 26 June 2018.

Satinder Sahani, retired Indian Administrative Service (IAS) officer, 13 August 2021 (phone).

Senior census official, Kohima, 11 December 2018.

Senior census official, New Delhi, 7 April 2012.

Senior health official, Jammu and Kashmir, 12 August 2021 (phone).

Shah Md Chaudhary, lawyer, Jammu, 13 September 2019.

State government official, Ind, Ramban, 28 June 2018, 30 June 2018 (phone).

*Sherepora*, Kulgam, 31 May 2016.

*Sumbal Bala*, Kangan, Ganderbal, 28 May 2016.

Sumit Hakhoo, *The Tribune*, Jammu, 2 June 2016.

Sunni community leader, Leh, 21 September 2019.

Taj Mohiuddin, former member of legislative assembly, Uri, 26 July 2021 (phone).

Tribal Leader, Bhaderwah, 5 December 2019, 7 September 2021 (phone).

Tsering Samphel, Former member of legislative assembly, Leh, 21 September 2019.

T. Stanzin, Kargil, 22 November 2019.

Wajahat Habibullah, retired Indian Administrative Service (IAS) officer, 12 January 2021 (email).

Zafar Ali Khatana, former vice chairman, Advisory Board for Gujjars and Bakerwals, 24 September 2021 (phone).

Zulfiqar Majid, *Deccan Herald*, Srinagar, 23 May 2016.

## Government Sources

Banthia, J. K. 2001a. 'Mobilising Support for India's Census: Constraints and Challenges'. Paper presented at UNFPA/PARIS21 International Expert Group Meeting on Mechanisms for Ensuring Continuity of 10-Year Population Censuses: Strategies for Reducing Census Costs, Pretoria, 26–29 November.

———. 2001b. 'Mobilising Support for India's Census: Constraints and Challenges'. Presentation at UNFPA/PARIS21 International Expert Group Meeting on Mechanisms for Ensuring Continuity of 10-Year Population Censuses: Strategies for Reducing Census Costs, Pretoria, 26–29 November.

———. 2004. 'Census Goldmine'. *Economic and Political Weekly* 39(35): 3862.

DD News. 2020. 'Mausam'. Doordarshan, 11 May.

Election Commission India (ECI). 1957. 'Delimitation of Parliamentary and Assembly Constituencies Order, 1956 (Corrected up to 31 March 1957)'. New Delhi: Election Commission India.

———. 1963. 'Delimitation Commission Order No. 1'. *Gazette of India*, no. 50, New Delhi, 21 March.

———. 1976. 'The Delimitation of Parliamentary and Assembly Constituencies Order, 1976'. New Delhi: Election Commission India.

———. 2008a. 'Delimitation of Parliamentary and Assembly Constituencies Order, 2008'. New Delhi: Election Commission India.

———. 2008b. *Changing Face of Electoral India: Delimitation 2008*, vols. 1–2. New Delhi: Election Commission India.

———. 2017. *Electoral Statistics Pocket Book 2017*. New Delhi.

European Commission. 2015. 'Standard Eurobarometer 83: Europeans and Economic Statistic'. Directorate-General for Communication. https://europa.eu/eurobarometer/surveys/detail/2099. Accessed on 6 January 2023.

Government of India (GoI). n.d.1. 'Tashkent Declaration January 10, 1966'. New Delhi: Ministry of External Affairs.

———. n.d.2. 'Simla Agreement July 2, 1972'. New Delhi: Ministry of External Affairs.

———. n.d.3. 'The Relevance and Ramifications of Census'. New Delhi: Office of the Registrar General and Census Commissioner of India.

———. n.d.4. 'Objects, Essential Features and Utility of Census'. New Delhi: Office of the Registrar General and Census Commissioner of India.

———. n.d.5. 'India at a Glance'. New Delhi: Office of the Registrar General and Census Commissioner of India.

———. n.d.6. 'Jammu & Kashmir Data Highlights: The Scheduled Tribes. Census of India 2001'. New Delhi: Office of the Registrar General and Census Commissioner of India.

———. n.d.7. 'Census of India 2001, General Population Tables: Jammu & Kashmir (Table A-1 to A-4)'. Srinagar: Directorate of Census Operations, Jammu and Kashmir.

———. n.d.8. 'Census 1971, Series 8, Jammu and Kashmir, Digest of Population Statistics'. Srinagar: Directorate of Census Operations, Jammu and Kashmir.

———. n.d.9. *Urban Frame Survey*. New Delhi: National Sample Survey Office.

———. n.d.10. 'Census 2011: Interesting and Fun Facts'. Kohima: Directorate of Census Operations, Nagaland.

———. n.d.11. 'Census of India 1961 Volume III, Assam Part II-A, General Population Tables'. Shillong: Superintendent of Census Operations, Assam.

———. n.d.12. 'Brief History of Census'. New Delhi: Office of the Registrar General and Census Commissioner of India.

———. n.d.13. 'Census of India 1981 Series 13, Part 1A, Manipur Administration Report: Enumeration'. Imphal: Director of Census Operations, Manipur.

———. n.d.14. 'Census of India 2011: Instruction Manual for Updating of Abridged Houselist and Filling up of the Household Schedule'. New Delhi: Office of the Registrar General and Census Commissioner of India.

———. n.d.15. 'Census of India 1971 Series 8-Jammu & Kashmir, Part II-C(i) & Part V-A, Distribution of Population by Religion and Scheduled Castes, Jammu & Kashmir'. Srinagar: Directorate of Census Operations, Jammu and Kashmir.

———. n.d.16. 'Compendium of India's Fertility and Mortality Indicators, 1971–2013'. New Delhi: Office of the Registrar General and Census Commissioner of India.

———. n.d.17. 'Primary Census Abstract Data Tables (India & States/UTs – District Level), 2011'. New Delhi: Office of the Registrar General and Census Commissioner of India.

————. n.d.18. 'Census of India 2011: Houselisting and Housing Census Schedule'. New Delhi: Office of the Registrar General and Census Commissioner of India.

————. n.d.19. 'Census of India 2011: Household Census Schedule'. New Delhi: Office of the Registrar General and Census Commissioner of India.

————. n.d.20. 'Census of India 2021: Houselisting and Housing Census Schedule'. New Delhi: Office of the Registrar General and Census Commissioner of India.

————. n.d.21. 'Census of India 2021: Household Schedule Pre-Test'. New Delhi: Office of the Registrar General and Census Commissioner of India.

————. n.d.22. 'Census of India 1981, Series 31, Mizoram, Part I-A & B, Administration Report (Enumeration and Tabulation)'. Directorate of Census Operations, Mizoram.

————. 1951. 'Census of India. 1951, Volume VIII, Punjab Sub Zone (Punjab, PEPSU, Bilaspur, Delhi, and Himachal Pradesh) Administration Report Part I: Enumeration'. Simla: Superintendent of Census Operations, Punjab Zone.

————. 1952. 'General Report No. 1 on the First round, October 1950 – March 1951'. New Delhi: Department of Economic Affairs, Ministry of Finance.

————. 1953. 'Census of India 1951, Paper No. 1 of 1953, Sample Verification of the 1951 Census Count'. New Delhi: Office of the Registrar General and Census Commissioner of India.

————. 1954a. 'Census of India 1951, Paper No. 6, Estimation of Birth and Death Rates in India during 1941/50–1951 Census'. New Delhi: Office of the Registrar General and Census Commissioner of India.

————. 1954b. 'Census of India, 1951 Volume XII Assam, Manipur and Tripura, Administrative Report'. Shillong: Superintendent of Census Operations, Assam, Manipur and Tripura.

————. 1954c. 'Census of India, 1951 Volume XII Assam, Manipur and Tripura Part I-A Report'. Shillong: Superintendent of Census Operations, Assam, Manipur and Tripura.

————. 1954d. 'Census of India, Paper No. 1, Languages 1951 Census'. New Delhi: Registrar General and Census Commissioner of India.

————. 1957. 'Census of India, 1951 Volume I, India, Part I-A – Report'. New Delhi: Registrar General and Census Commissioner of India.

————. 1961. '1961 Census: Sorting and Compilation Instructions'. New Delhi: Office of the Registrar General and Census Commissioner of India.

————. 1963a. 'Census of India 1961, Volume VI – Part VIII – A, Jammu and Kashmir, Administration Report (Enumeration)'. Srinagar: Superintendent of Census Operations, Jammu and Kashmir.

————. 1963b. 'Census of India 1961, Volume III – Part VIII-A, Enumeration, Administration Report Assam'. Shillong: Superintendent of Census Operations, Assam.

————. 1963c. 'Census of India 1961, Volume XVI, West Bengal & Sikkim Part VIII, Administration Report VIII-A Enumeration'. Calcutta: Superintendent of Census Operations, West Bengal and Sikkim.

————. 1963d. 'Census of India 1961, Volume XV, Uttar Pradesh Part VIII A, Administration Report on Enumeration'. Lucknow: Superintendent of Census Operations, Uttar Pradesh.

————. 1964a. 'Census of India 1961, Volume I, India, Part II-A (i): General Population Tables'. New Delhi: Office of the Registrar General and Census Commissioner of India.

————. 1964b. 'Census of India 1961, Volume VI, Jammu and Kashmir Part II-A, General Population Tables'. Srinagar: Superintendent of Census Operations, Jammu and Kashmir.

————. 1964c. 'Census of India 1961, Volume VII, Kerala Part II-A, General Population Tables'. Trivandrum: Superintendent of Census Operations, Kerala and the Union Territory of Laccadive, Minicoy and Amindivi Islands.

————. 1964d. 'Census of India 1961 Volume XVI, West Bengal & Sikkim Part II-A General Population Tables'. Calcutta: Superintendent of Census Operations, West Bengal and Sikkim.

————. 1964e. 'Census of India 1961 Volume XV, Uttar Pradesh Part II-A, General Population Tables'. Lucknow: Superintendent of Census Operations, Uttar Pradesh.

————. 1964f. 'Census of India 1961, Volume III, Assam, Part I-A, General Report'. Shillong: Superintendent of Census Operations, Assam.

————. 1965. 'Census of India 1961 Volume VI Jammu and Kashmir Part II-C Cultural and Migration Tables'. Srinagar: Superintendent of Census Operations, Jammu and Kashmir.

————. 1966a. 'Census of India 1961, Volume XXIII, Nagaland, Part II-A, General Population Tables'. Kohima: Superintendent of Census Operations, Nagaland.

————. 1966b. 'Census of India 1961, Volume XV, Uttar Pradesh, Part I-A (i), General Report on the Census'. Lucknow: Superintendent of Census Operations, Uttar Pradesh.

————. 1966c. 'Census of India 1961, Volume XXII, Manipur, Part I-A, General Report Including Subsidiary Tables'. Imphal: Superintendent of Census Operations, Manipur.

————. 1966d. 'Census of India 1961, Jammu & Kashmir, District Census Handbook 7, Jammu District'. Srinagar: Superintendent of Census Operations, Jammu and Kashmir.

————. 1966e. *The Handbook of India*. New Delhi: Department of Tourism, Ministry of Transport and Communication.

———. 1968a. 'Census of India 1961, Volume VI, Jammu and Kashmir Part I-A (i), General Report'. Srinagar: Superintendent of Census Operations, Jammu and Kashmir.

———. 1968b. 'Census of India 1961, Volume VI, Jammu and Kashmir Part I-A (ii), General Report'. Srinagar: Superintendent of Census Operations, Jammu and Kashmir.

———. 1969. 'Census of India 1961, Volume XI, Mysore, Part VIII-B, Administration Report (Tabulation)'. Mysore: Superintendent of Census Operations.

———. 1970. 'Census of India 1961, Village Survey Monograph of Kharboo, Vol-VI Part VI No. 13'. Srinagar: Directorate of Census Operations, Jammu and Kashmir.

———. 1971a. 'Census of India 1971 Series 8-Jammu & Kashmir Paper 1 of 1971, Population Totals (Provisional)'. Srinagar: Directorate of Census Operations, Jammu and Kashmir.

———. 1971b. 'Census of India 1971. Series 8-Jammu & Kashmir, Part VIII-A, Administration Report on Enumeration'. Srinagar: Directorate of Census Operations, Jammu and Kashmir.

———. 1971c. 'Census Centenary'. New Delhi: Indian Posts and Telegraphs Department, Government of India.

———. 1971d. 'India Counts Itself: Operation Census is a Way to a Scientific Analysis'. Films Division, Ministry of Information and Broadcasting.

———. 1972a. 'Census of India 1971, Series 8, Jammu & Kashmir Part II-A General Population Tables'. Srinagar: Directorate of Census Operations, Jammu and Kashmir.

———. 1972b. 'Census of India 1971, Series 22, West Bengal Part VIII-A, Administrative Report-Enumeration'. Calcutta: Directorate of Census Operations, West Bengal.

———. 1972c. 'Census of India 1971, Series 8, Jammu & Kashmir Part VIII-A, Administration Report on Enumeration Jammu & Kashmir'. Srinagar: Directorate of Census Operations, Jammu and Kashmir.

———. 1974a. 'A Portrait of Population. Jammu and Kashmir, Census of India 1971'. Srinagar: Directorate of Census Operations, Jammu and Kashmir.

———. 1974b. 'Census 1971 Parts X-A & B Town & Village Directory, Village & Townwise Primary Census Abstract, Series 8, Jammu & Kashmir, District Census Handbook, Rajauri District'. Srinagar: Directorate of Census Operations, Jammu and Kashmir.

———. 1975. 'Census of India 1971, Series 1, India, Part II-A (i), General Population Tables'. New Delhi: Office of the Registrar General and Census Commissioner of India.

———. 1976a. 'Census of India 1971, Series 1, India, Part II-C (i), Social and Cultural Tables'. New Delhi: Office of the Registrar General and Census Commissioner of India.

———. 1976b. 'Census of India 1971, Series 8, Jammu and Kashmir Part I-A, General Report'. Srinagar: Superintendent of Census Operations, Jammu and Kashmir.

———. 1977. 'Census of India 1971, Series 22, West Bengal Part I-A, General Report'. Calcutta: Superintendent of Census Operations, West Bengal.

———. 1978a. 'A Portrait of Population (Manipur)'. Imphal: Directorate of Census Operations, Manipur.

———. 1978b. 'Census of India 1971, Series 8, Jammu and Kashmir Part I-B, General Report'. Srinagar: Superintendent of Census Operations, Jammu and Kashmir.

———. 1981a. 'Census of India 1981, Jammu & Kashmir, Paper I of 1981, Provisional Population Totals'. Srinagar: Directorate of Census Operations, Jammu and Kashmir.

———. 1981b. 'Census of India 1981, Jammu & Kashmir, Paper I of 1981, Supplement, Provisional Population Totals'. Srinagar: Directorate of Census Operations, Jammu and Kashmir.

———. 1982a. 'Census of India 1981, Series 17, Punjab, Part II, General Population Tables'. Chandigarh: Directorate of Census Operations, Punjab.

———. 1982b. 'Census of India 1981, Series 15, Nagaland, Part I (A & B), Administration Report on Enumeration and Tabulation'. Kohima: Directorate of Census Operations, Nagaland.

———. 1983a. 'Census of India 1981, Series – 8, Jammu & Kashmir, Part II – A, General Population Tables'. Srinagar: Directorate of Census Operations, Jammu and Kashmir.

———. 1983b. 'Census of India 1981, Series – 5, Gujarat, Part II – A, General Population Tables'. Gandhinagar: Directorate of Census Operations, Gujarat.

———. 1985a. 'Census of India 1981, Series – 8, Jammu and Kashmir, Paper I of 1985, Household Population by Religion of Head of Household [up to *tehsil* and town level]'. Srinagar: Directorate of Census Operations, Jammu and Kashmir.

———. 1985b. 'Census of India 1981, Series – 8, Jammu and Kashmir, Paper I-A, Administrative Report – Enumeration'. Srinagar: Directorate of Census Operations, Jammu and Kashmir.

———. 1985c. 'Census of India 1981, India, Part II-A (i), General Population Tables'. New Delhi: Office of the Registrar General and Census Commissioner of India.

———. 1987. 'Census of India 1981 Series 8 Jammu and Kashmir Paper 1 of 1987 Households and Household Population by Language Mainly Spoken in the Household'. Srinagar: Directorate of Census Operations, Jammu and Kashmir.

———. 1989a. 'Census of India 1981, A Portrait of Population. Jammu and Kashmir'. Srinagar: Directorate of Census Operations, Jammu and Kashmir.

———. 1989b. 'Fertility in India: An Analysis of 1981 Census Data'. Occasional Paper No. 13 of 1988, Office of the Registrar General and Census Commissioner of India, New Delhi.

———. 1989c. 'Proceedings of the First Conference of Directors of Census Operations of 1991 Census (6–9 November 1989), New Delhi'. New Delhi: Office of the Registrar General and Census Commissioner of India.

———. 1990a. 'Census of India 1981, Series-I India Part IV B(i), Population by Language/Mother-Tongue (Table C-7)'. New Delhi: Office of the Registrar General and Census Commissioner of India.

———. 1990b. 'Proceedings of the Second Conference of Directors of Census Operations of 1991 Census (5–8 February 1990), New Delhi'. New Delhi: Office of the Registrar General and Census Commissioner of India.

———. 1990c. 'Proceedings of the Third Conference of Directors of Census Operations of 1991 Census (24–26 October 1990), Panaji'. New Delhi: Office of the Registrar General and Census Commissioner of India.

———. 1991a. 'Census of India 1991, Series – 1, Paper 1 of 1991, Provisional Population Totals'. New Delhi: Office of the Registrar General and Census Commissioner of India.

———. 1991b. 'Proceedings of the Fourth Conference of Directors of Census Operations of 1991 Census (8–10 May 1991), New Delhi'. New Delhi: Office of the Registrar General and Census Commissioner of India.

———. 1992a. Census of India 1991 Series - 1 India Paper 1 of 1992 Vol. I. Final Population Totals. New Delhi: Office of the Registrar General and Census Commissioner of India.

———. 1992b. 'Fifth Conference of Directors of Census Operations of 1991 Census (13–15 May 1992), Shimla: Agenda Notes'. New Delhi: Office of the Registrar General and Census Commissioner of India.

———. 1993. 'Proceedings of the Sixth Conference of Directors of Census Operations of 1991 Census (22–24 April 1993)'. New Delhi: Office of the Registrar General and Census Commissioner of India.

———. 1994. 'Census of India 1991, India, Part II-A (I), General Population Tables'. New Delhi: Office of the Registrar General and Census Commissioner of India.

———. 1995a. 'Census of India 1991, District Census Handbook, Ahmadnagar'. Mumbai: Directorate of Census Operations, Maharashtra.

———. 1995b. 'Committee on Petitions (Tenth Lok Sabha) Twenty Fourth Report, C.B.-I No. 194 Vol. XXIV'. New Delhi: Lok Sabha Secretariat.

————. 1995c. 'Census of India 1991, Series-1 India, Paper 1 of 1995, Religion'. New Delhi: Office of the Registrar General and Census Commissioner of India.

————. 1996. 'Census of India 1991, Population Projections for India and States 1996-2016, Report of the Technical Group on Population Projections Constituted by the Planning Commission'. New Delhi: Office of the Registrar General and Census Commissioner of India.

————. 1997. 'Census of India 1991: A Portrait of Population'. Panaji: Directorate of Census Operations, Goa.

————. 1999. 'Census of India 2001, First Conference of the Directors of Census Operations (23–25 September 1999), New Delhi: Agenda Notes'. New Delhi: Office of the Registrar General and Census Commissioner of India.

————. 2000a. 'Home Minister Stresses the Importance of Census for Planning'. Press Information Bureau, 14 November.

————. 2000b. 'Census of India 2001, Second Conference of the Directors of Census Operations (3–5 February 2000), New Delhi: Agenda Notes'. New Delhi: Office of the Registrar General and Census Commissioner of India.

————. 2000c. 'Census of India 2001. Proceedings of the Second Conference of the Directors of Census Operations (3–5 February 2000), New Delhi'. New Delhi: Office of the Registrar General and Census Commissioner of India.

————. 2000d. 'Census of India 2001. Proceedings of the Third Conference of the Directors of Census Operations (26–27 May 2000), New Delhi'. New Delhi: Office of the Registrar General and Census Commissioner of India.

————. 2000e. *National Population Policy 2000*. New Delhi: National Health Mission.

————. 2001a. *National Statistical Commission*, vols. 1–2. New Delhi: Ministry of Statistics and Programme Implementation.

————. 2001b. 'Census of India 2001, Series 2 Jammu & Kashmir Paper – 1 of 2001, Provisional Population Totals'. Srinagar: Director of Census Operations, Jammu and Kashmir.

————. 2001c. 'Census of India 2001'. New Delhi: Department of Posts.

————. 2001d. 'Census of India 2001, Series 1, India, Provisional Population Totals, Paper – 1 of 2001'. New Delhi: Office of the Registrar General and Census Commissioner of India.

————. 2003. 'Census of India 2001 Series 2: Jammu &. Kashmir, Final Population Totals (State, District, Tehsil and Town)'. Srinagar: Directorate of Census Operations , Jammu and Kashmir.

————. 2004a. 'Census of India 2001, The First Report on Religion Data'. New Delhi: Office of the Registrar General and Census Commissioner of India.

———. 2004b. 'Census of India 2001, Series-14, Nagaland. District Census Handbook Part – A & B Phek District, Village & Town Directory, Village and Townwise Primary Census Abstract'. Kohima: Directorate of Census Operations, Nagaland.

———. 2005a. 'General Population Tables: India, States and Union Territories, (Tables A-1 to A-3), Part I'. New Delhi: Office of the Registrar General and Census Commissioner of India.

———. 2005b. 'Series – 1, Census of India 2001, Slum Population (640 Cities and Towns Reporting Slums), Vol. I'. New Delhi: Office of the Registrar General and Census Commissioner of India.

———. 2006a. *Population Projections for India and States 2001–2026: Report of the Technical Group on Population Projections Constituted by the National Commission on Population*. New Delhi: Ministry of Health and Family Welfare.

———. 2006b. *Social, Economic and Educational Status of the Muslim Community of India: A Report*. New Delhi: Prime Minister's High-Level Committee, Cabinet Secretariat.

———. 2006c. *Census of India 2001: Report on Post Enumeration Survey*. New Delhi: Office of the Registrar General and Census Commissioner of India.

———. 2007. 'Census of India 2001 Paper 1 of 2007 Language India, States and Union Territories (Table C-16)'. New Delhi: Office of the Registrar General and Census Commissioner of India.

———. 2009a. *Manual on Vital Statistics*. New Delhi: Central Statistical Organisation.

———. 2009b. *Thirteenth Finance Commission 2010–2015*, vol. 1: *Report*; vol. 2: *Annexes*. New Delhi: Finance Commission, Government of India.

———. 2010a. 'Circular No.: 9 (Preparation of District Census Plan) CEN/32/ KAR/2010 (Date: 09.03.10)'. Bengaluru: Office of the Director of Census Operations.

———. 2010b. 'Census of India 2011 - Circular No. 8 (Formation and identification of Slum Enumeration Blocks for Slum Demography)'. January 20. New Delhi: Office of the Registrar General and Census Commissioner of India.

———. 2010c. 'Census of India 2011 - Circular No. 25 (Drafting of the Administrative Report for Census 2011)'. November 29. New Delhi: Office of the Registrar General and Census Commissioner of India.

———. 2011a. 'Census of India 2011, Provisional Population Totals, Paper 1 of 2011, Madhya Pradesh, Series 24'. Bhopal: Directorate of Census Operations, Madhya Pradesh.

———. 2011b. 'Census of India 2011, Provisional Population Totals, Paper 1 of 2011, Series 1 India'. New Delhi: Office of the Registrar General and Census Commissioner of India.

———. 2011c. 'Census of India 2011, Administrative Atlas of India'. New Delhi: Office of the Registrar General and Census Commissioner of India.

———. 2011d. 'Census of India 2011, Provisional Population Totals, Paper 1 of 2011, Jammu & Kashmir, Series 2'. Srinagar: Directorate of Census Operations, Jammu and Kashmir.

———. 2011e. 'Census of India 2011'. New Delhi: Department of Posts.

———. 2011f. 'Department of Jammu & Kashmir Affairs'. Ministry of Home Affairs, 2 May. https://web.archive.org/web/20110508161640/http://mha.nic. in/uniquepage.asp?Id_Pk=306. Accessed on 17 April 2021.

———. 2011g. *Report of the Expert Group to Formulate a Jobs Plan for the State of Jammu and Kashmir.* New Delhi: Prime Minister's Office. https://archivepmo. nic.in/drmanmohansingh/getdoc.php?id=G9XT9077.pdf. Accessed on 5 January 2023.

———. 2011h. 'Census of India 2011, Provisional Population Totals, Paper 1 of 2011, Series 14 Nagaland'. Kohima: Directorate of Census Operations, Nagaland.

———. 2011i. *Report of the Committee on Legislative Measures in Statistical Matters as Adopted by the National Statistical Commission.* New Delhi: National Statistical Commission Secretariat.

———. 2011j. 'Sanctioned Strength of IAS Officers'. Ministry of Personnel, Public Grievances and Pensions, 30 November.

———. 2012. 'National Data Sharing and Accessibility Policy'. *Gazette of India*, 17 March, no. 11. New Delhi: Department of Science and Technology.

———. 2013a. *Report of the Committee for Evolving a Composite Development Index of States.* New Delhi: Ministry of Finance.

———. 2013b. 'Lok Sabha Unstarred Question No. 2379'. New Delhi: Ministry of Home Affairs.

———. 2014a. *Mapping the Adverse Child Sex Ratio in India: Census 2011.* New Delhi: Office of the Registrar General and Census Commissioner of India; New Delhi: United Nations Population Fund, India.

———. 2014b. *Report of the High-Level Committee on Socio-Economic, Health and Educational Status of Tribal Communities of India.* New Delhi: Ministry of Tribal Affairs.

———. 2014c. 'Vital Statistics of India Based on the Civil Registration System 2011'. New Delhi: Office of the Registrar General and Census Commissioner of India.

———. 2014d. 'Census of India 2011, Report on Post Enumeration Survey'. New Delhi: Office of the Registrar General and Census Commissioner of India.

———. 2014e. 'Census of India 2011, Jammu & Kashmir, Series-02, Part XII-B, District Census Handbook Kupwara (Village and Town Wise Primary Census Abstract [PCA])'. Srinagar: Directorate of Census Operations, Jammu and Kashmir.

———. 2014f. 'Order No. 9/25/2013-CD (Cen), dated 07-01-2014'. New Delhi: Office of the Registrar General and Census Commissioner of India.

———. 2014g. 'Census of India 2011, Nagaland Series-14, Part XII-B District Census Handbook Phek Village and Town Wise Primary Census Abstract (PCA)'. Kohima: Directorate of Census Operations, Nagaland.

———. 2015a. 'Sex Ratio (0–6 Years) for India and States/Union Territories Based on Census 2001 and Goal for Eleventh Plan Period 2011–12'. 9 July. https://data.gov.in/catalog/sex-ratio-0-6-years-india-and-states-union-territories-based-census-2001-and-goal-eleventh. Accessed on 13 October 2015.

———. 2015b. *The Constitution of India (as on 9th November, 2015)*. New Delhi: Legislative Department, Ministry of Law and Justice.

———. 2015c. *Report of the Fourteenth Finance Commission*. New Delhi: Fourteenth Finance Commission.

———. 2015d. 'Over 1.20 Crore Bogus Rations Cards Deleted during the Last Three Years'. 8 December. New Delhi: Food and Public Distribution, Ministry of Consumer Affairs.

———. 2018a. *Children in India 2018: A Statistical Appraisal*. New Delhi: Ministry of Statistics and Programme Implementation.

———. 2018b. 'Census of India 2011 Paper 1 of 2018 Language India, States and Union Territories (Table C-16)'. New Delhi: Office of the Registrar General and Census Commissioner of India.

———. 2018c. 'National Policy on Official Statistics (Draft)'. New Delhi: Ministry of Statistics and Programme Implementation.

———. 2019a. 'Jammu and Kashmir Reorganisation (Removal of Difficulties) Second Order, 2019'. *Gazette of India*, no. 3585, 2 November.

———. 2019b. 'Maps of Newly Formed Union Territories of Jammu Kashmir and Ladakh, with the Map of India'. Ministry of Home Affairs. https://www.mha.gov.in/whatsnew/maps-of-newly-formed-union-territories-of-jammu-kashmir-and-ladakh-map-of-india-0. Accessed on 23 December 2022.

———. 2019c. 'Government of India (Allocation of Business) Rules, 1961 (as amended up to Amendment Series no. 348, dated 5 February 2019)'. New Delhi: Cabinet Secretariat.

———. 2020a. *The Constitution of India (as on 9th December, 2020)*. New Delhi: Legislative Department, Ministry of Law and Justice.

———. 2020b. *Finance Commission in Covid Times: Report for 2021–26*, vol. 1: *Main Report*. New Delhi: Fifteenth Finance Commission.

———. 2020c. *Finance Commission in Covid Times: Report for 2021–26*, vol. 4: *The States*. New Delhi: Fifteenth Finance Commission.

———. 2022. 'Notification, No. 282/J&K/2022 (Vol.II) Dated: 14th March, 2022'. *Jammu and Kashmir Official Gazette*, vol. 134, nos. 50–51, 14 March.

————. 2023. *A Treatise on Indian Censuses since 1981*. New Delhi: Office of the Registrar General and Census Commissioner of India.

Government of Jammu and Kashmir (GoJK). n.d.1. 'Ethnographic Profile of Tribes in Jammu and Kashmir'. Tribal Affairs Department. https://tribalaffairs.jk.gov. in/ethonographic.pdf. Accessed on 6 August 2021.

————. n.d.2. *Digest of Statistics 2013–14*. Srinagar: Directorate of Economics and Statistics, Planning and Development Department.

————. 1912. 'Census of India 1911, Volume XX Kashmir, Part I Report'. Lucknow: Newul Kishore Press.

————. 1923. 'Census of India 1921, Volume XXII Kashmir, Part II Tables'. Lahore: Mufid-i-'Am Press.

————. 1933a. 'Census of India 1931, Volume XXIV Jammu & Kashmir State Part I-Report'. Jammu: Ranbir Government Press, Government of Jammu and Kashmir.

————. 1933b. 'Census of India, 1931 Volume XXIV Jammu & Kashmir State Part II-Tables'. Jammu: Ranbir Government Press, Government of Jammu and Kashmir.

————. 1933c. 'Census of India, 1931 Volume XXIV Jammu & Kashmir State Part III-Administrative'. Jammu: Ranbir Government Press, Government of Jammu and Kashmir.

————. 1943a. 'Census of India, 1941 Volume XXII, Jammu and Kashmir, Part IV, The Administration Report'. Jammu: Ranbir Government Press, Government of Jammu and Kashmir.

————. 1943b. 'Census of India, 1941 Volume XXII, Jammu and Kashmir, Part III, Village Tables and Housing Statistics'. Jammu: Ranbir Government Press, Government of Jammu and Kashmir.

————. 1943c. 'Census of India, 1941 Volume XXII, Jammu and Kashmir, Parts I & II, Essay and Tables'. Jammu: Ranbir Government Press, Government of Jammu and Kashmir.

————. 1968. *Report of the Jammu & Kashmir Commission of Inquiry*. Jammu: Government Press.

————. 1993. *Digest of Statistics 1991–92*. Srinagar: Directorate of Economics and Statistics, Planning and Development Department.

————. 2001. *Ladakh: The Land of Celebrations*. Srinagar: Jammu and Kashmir Tourism, Government of Jammu and Kashmir.

————. 2003. *The Constitution of Jammu and Kashmir*. https://jkdat.nic.in/pdf/ Rules-Costitutionof-J&K.pdf. Accessed on 5 January 2023.

————. 2007. *Manual of Jammu and Kashmir Election Law*, 8th edition. Jammu: Compilation Cell, Department of Law, Government of Jammu and Kashmir.

————. 2018. 'District Profile'. Kulgam District, 5 January. https://kulgam.nic.in/ history. Accessed on 5 January 2018.

————. 2020. 'About J&K'. InvestJK, Department of Industries and Commerce, Jammu and Kashmir. https://www.investjk.in/welcome/aboutjk. Accessed on 26 August 2021.

————. 2021a. 'About District: History'. Anantnag district, 10 August. https://anantnag.nic.in/history. Accessed on 26 August 2021.

————. 2021b. 'About District: History'. Kulgam District, 10 August. https://kulgam.nic.in/history. Accessed on 26 August 2021.

————. 2021c. 'List of KAS Officers as on 9/19/2021'. General Administration Department. https://jkgad.nic.in/leftMenu/KAS_Officers.aspx. Accessed on 19 September 2021.

————. 2021d. 'Civil List of Jammu Kashmir Police Service (JKPS) as on 19.03.2021'. Home Department. jkhome.nic.in/pdf/Civil%20list%20of%20JKPS%20Revised%2019.03.2021.pdf. Accessed on 19 September 2021.

Government of the Union Territory of Jammu and Kashmir. 2021. 'J&K at a Glance'. 7 February. https://www.jk.gov.in/jammukashmir. Accessed on 17 April 2021.

Government of Nagaland. 2009. 'Resolutions Adopted in the Consultative Meeting on Census 2011 Held at the Zonal Council Hall, Kohima, on 30th September, 2009' (courtesy: Director of Census Operations, Nagaland).

Government of Nigeria. n.d. 'History of Population Censuses in Nigeria'. Abuja: National Population Commission. https://www.nationalpopulation.gov.ng/about-us/history-of-population-censuses-in-nigeria. Accessed on 8 March 2021.

Government of Pakistan. n.d. 'Area and Population of Administrative Units'. Pakistan Bureau of Statistics. https://www.pbs.gov.pk/sites/default/files/population_census/Administrative%20Units.pdf. Accessed on 6 March 2021.

Government of Tripura. 2021. 'Demographic Features'. Tripura State Portal. https://tripura.gov.in/demographics. Accessed on 13 March 2022.

Grierson, G. A. 1907. *Linguistic Survey of India*, vol. 9, part 3. Calcutta: Office of the Superintendent of Government Printing.

————. 1909. *Linguistic Survey of India*, vol. 3, part 1. Calcutta: Office of the Superintendent of Government Printing.

————. 1919. *Linguistic Survey of India*, vol. 8, part 2. Calcutta: Office of the Superintendent of Government Printing.

Husain, M. 1945. *Census of India 1941*, vol. 21, part 1: *Report*. Hyderabad-Deccan: Government Central Press.

Hutton, J. H. 1933. *Census of India, 1931*, vol. 1: *India*, part 1: *Report*. Delhi: Manager of Publications.

International Institute for Population Sciences (IIPS). 1995. *National Family Health Survey (MCH and Planning), India, 1992–93*. Mumbai: International Institute for Population Sciences.

International Institute for Population Sciences (IIPS) and ICF International. 2017a. *National Family Health Survey (NFHS-4), India, 2015–16: Jammu and Kashmir*.

Mumbai: International Institute for Population Sciences and ICF International.

———. 2017b. *National Family Health Survey (NFHS-4), India, 2015–16: India.* Mumbai: International Institute for Population Sciences and ICF International.

International Institute for Population Sciences (IIPS) and Macro International. 2009. *National Family Health Survey (NFHS-3), India, 2005–06: Jammu and Kashmir.* Mumbai: International Institute for Population Sciences and Macro International.

Jamir, S. C. n.d.1. 'Chief Minister S.C. Jamir Speeches Feb '93 to August '94'. Kohima: Directorate of Information and Public Relations, Government of Nagaland.

———. n.d.2. 'Speeches of Mr. S.C. Jamir Chief Minister Nagaland 1999'. Kohima: Directorate of Information and Public Relations, Government of Nagaland.

———. 1996. 'Speeches of Mr. S.C. Jamir Chief Minister Nagaland 1994-1996'. Kohima: Directorate of Information and Public Relations, Government of Nagaland.

———. 1998. 'Speeches of Mr. S.C. Jamir Chief Minister Nagaland January to December 1997'. Kohima: Directorate of Information and Public Relations, Government of Nagaland.

———. 1999. 'Speeches of Mr. S.C. Jamir Chief Minister Nagaland 1998'. Kohima: Directorate of Information and Public Relations, Government of Nagaland.

———. 2003. 'Speeches of Mr. S. C. Jamir Chief Minister Nagaland 2002–03'. Kohima: Directorate of Information and Public Relations, Government of Nagaland.

Kumar, Radha, M. M. Ansari and Dileep Padgaonkar. n.d. *Group of Interlocutors for J&K: A New Compact with the People of Jammu and Kashmir.* https://www.mha.gov.in/sites/default/files/J%26K-InterlocatorsRpt-0512.pdf. Accessed on 6 January 2023.

League of Nations. 1936. 'Convention on Rights and Duties of States Adopted by the Seventh International Conference of American States. Signed at Montevideo, December 26th, 1933'. *Treaty Series, Treaties and International Engagements Registered with the Secretariat of the League of Nations* 165(3801–24): 19–42.

Lok Sabha. 1986a. *Lok Sabha Debates (English Version), Fifth Session (Eighth Lok Sabha), Eighth Series,* vol. 13, no. 5. New Delhi: Lok Sabha Secretariat.

———. 1986b. *Lok Sabha Debates (English Version), Seventh Session (Eighth Lok Sabha), Eighth Series,* vol. 22, no. 17. New Delhi: Lok Sabha Secretariat.

———. 1991. *Lok Sabha Debates, First Session (Tenth Lok Sabha),* vol. 2, no. 17. New Delhi: Lok Sabha Secretariat.

———. 1999. *Constituent Assembly Debates, Official Report,* vols. 1–9. New Delhi: Lok Sabha Secretariat.

————. 2000. *Lok Sabha Debates (English Version), Fourth Session (Thirteenth Lok Sabha), Thirteenth Series*, vol. 8, no. 3. New Delhi: Lok Sabha Secretariat.

————. 2019. *Seventeenth Series, First Session, 2019/1941 (Saka) No. 37, Tuesday, 6 August 2019 / Shravana 15, 1941 (Saka)*, vol. 4. New Delhi: Lok Sabha Secretariat.

Ministry of Defence (MoD). 2011. *The India Pakistan War of 1965: A History*. New Delhi and Dehradun: History Division, Ministry of Defence, Government of India; Natraj Publishers.

————. 2014. The India Pakistan War of 1971: A History. History Division, Ministry of Defence, Government of India: New Delhi and Natraj Publishers: Dehra Dun. 'O Malley, L. L. S. 1913. *Census of India, 1911*, vol. 5: *Bengal, Bihar and Orissa and Sikkim*, part 1: *Report*. Calcutta: Bengal Secretariat Book Depot.

Porter, A. E. 1933. 'Census of India, 1931. Volume V. Bengal & Sikkim Part I Report'. Calcutta: Central Publication Branch.

Rajya Sabha. 2019a. 'Rajya Sabha Official Report (Floor Version), Vol. 249, No. 33, Monday 5 August 2019 14 Shravana, 1941 (Saka)'. New Delhi: Rajya Sabha Secretariat.

————. 2019b. 'Rajya Sabha Official Report (English Version), Vol. 249, No. 33, Monday 5 August, 2019 14 Shravana, 1941 (Saka)'. New Delhi: Rajya Sabha Secretariat.

————. 2019c. 'Rajya Sabha Synopsis of Debate, Wednesday, 11 December 2019'. New Delhi: Rajya Sabha Secretariat.

Rahi, Javaid. 2010. 'Re-Schedule Census Dates: JK Gujjars to Registrar General of India Jammu'. Press note on memorandum submitted to the Office of the Registrar General of India (ORGI), New Delhi, 7 February.

Royal Government of Bhutan (RGB). 2000. *Bhutan National Human Development Report 2000*. New Delhi: Planning Commission Secretariat. http://hdr.undp.org/en/reports/national/asiathepacific/bhutan/bhutan_2000_en.pdf. Accessed on 25 October 2013.

————. 2006. *Results of Population & Housing Census of Bhutan 2005*. Thimpu: Office of the Census Commissioner Royal Government of Bhutan. https://www.nsb.gov.bt/download/5058. Accessed on 7 January 2023.

————. 2012. *Statistical Yearbook of Bhutan 2012*. Thimphu: National Statistics Bureau. http://www.nsb.gov.bt/publication/files/pub10pp3748yo.pdf. Accessed on 25 October 2013.

Survey of India (SoI). 2017. Political Map of India, Seventh Edition (1:4,000,000). New Delhi: Survey of India.

————. 2019a. Political Map of India, Eighth Edition (1:4,000,000). New Delhi: Survey of India.

————. 2019b. Political Map of India, Ninth Edition (1:4,000,000). New Delhi: Survey of India.

———. 2020. *Political Map of India, Tenth Edition* (1:4,000,000). New Delhi: Survey of India.

Survey of Pakistan (SoP). 2020. *Political Map of Pakistan, Fifth Edition* (1:3,000,000). Rawalpindi: Survey of Pakistan.

Srivastava, S. C. 1972. *Indian Census in Perspective.* Census Centenary Monograph No 1. New Delhi: Office of the Registrar General and Census Commissioner of India.

Sample Registration System (SRS). 2002. *SRS Bulletin: Sample Registration System*, vol. 36, no. 2. New Delhi: Office of the Registrar General and Census Commissioner of India.

———. 2007. *SRS Bulletin: Sample Registration System*, vol. 42, no. 1. New Delhi: Office of the Registrar General and Census Commissioner of India.

———. 2014. *Sample Registration System: Statistical Report 2013* (report no. 1 of 2014). New Delhi: Office of the Registrar General and Census Commissioner of India.

———. 2017. *Sample Registration System: Statistical Report 2016.* New Delhi: Office of the Registrar General and Census Commissioner of India.

———. 2019. *SRS Bulletin: Sample Registration System*, vol. 51, no. 1. New Delhi: Office of the Registrar General and Census Commissioner of India.

United Nations (UN) Economic Commission for Europe. 2018. *Value of Official Statistics: Recommendations on Promoting, Measuring and Communicating the Value of Official Statistics.* Geneva: Task Force on the Value of Official Statistics, United Nations Economic Commission for Europe.

United Nations (UN) Secretariat. 2010. *Post Enumeration Surveys: Operational Guidelines, Technical Report.* New York: World Population and Housing Census Programme, Department of Economics and Social Affairs.

Zonal Educational Officer. 2019. 'List of Employees Belonging to ST Category Working in Zone Chandanwari'. Chandanwari: Zonal Educational Officer (Courtesy: Md Iqbal).

## Secondary Sources

Abdullah, Sheikh Mohammad. 2016. *The Blazing Chinar: Autobiography*, 2nd edition. Srinagar: Gulshan Books Kashmir.

Abramson, David. 2002. 'Identity Counts: The Soviet Legacy and the Census in Uzbekistan'. In *Census and Identity: The Politics of Race, Ethnicity, and Language in National Censuses*, edited by David I. Kertzer and Dominique Arel, 176–201. Cambridge, UK: Cambridge University Press.

Adepoju, Aderanti. 1981. 'Military Rule and Population Issues in Nigeria'. *African Affairs* 80(318): 29–47.

Agarwala, Tora. 2021. 'Online "Census" for Assamese Muslims Launched by Indigenous Associations'. *Indian Express*, 16 April.

Agrawal, Ankush, and Vikas Kumar. 2013. 'Nagaland's Demographic Somersault'. *Economic and Political Weekly* 48(39): 69–74.

———. 2014. 'Infirmities in NSSO Data for Nagaland'. *Economic and Political Weekly* 49(12): 20–22.

———. 2017a. 'Cartographic Conflicts within a Union: Finding Land for Nagaland in India'. *Political Geography* 61: 123–147.

———. 2017b. 'NSSO Surveys along India's Periphery: Data Quality and Implications'. Working Paper No. 9, Azim Premji University, Bengaluru.

———. 2019. 'Politics Should Not Meddle with Our Official Statistics'. *Mint*, 20 November, 15.

———. 2020a. *Numbers in India's Periphery: The Political Economy of Government Statistics*. Cambridge and New York: Cambridge University Press.

———. 2020b. 'Manipur's Population Conundrum'. *Economic and Political Weekly* 55(48): 48–56.

———. 2020c. 'Delays in the Release of India's Census Data'. *Statistical Journal of the International Association for Official Statistics* 36(1): 217–230.

———. 2020d. 'One Foot in the City, One in the Village: India's Urban Poor and Their Rural Bonds'. South Asia@LSE, London School of Economics, 26 May.

———. 2020e. 'Anomalies in Manipur's Census, 1991–2011'. Working Paper No. 16, Azim Premji University, Bengaluru.

Ahmad, Khalid Bashir. 2017. *Kashmir: Exposing the Myth Behind the Narrative*. New Delhi: SAGE Publishing.

Ahmad, Mudasir. 2018. 'Two Women Who Crossed Over from PoK Win Panchayat Posts in Kashmir'. *The Wire*, 14 November.

Ahmad, Mukhtar. 2000a. 'Hizb Commander Announces Conditional Ceasefire in J&K'. Rediff, 24 July.

———. 2000b. 'Hizb "Bans" J&K Census'. Rediff, 3 September.

———. 2000c. 'JK May Extend Deadline for Second Phase of Census'. Rediff, 11 October.

Ahmed, Ishtiaq. 1999. 'The 1947 Partition of Punjab: Arguments Put Forth before the Punjab Boundary Commission by the Parties Involved'. In *Region and Partition: Bengal, Punjab and the Partition of the Subcontinent*, edited by I. Talbot and G. Singh, 116–7. Oxford and New York: Oxford University Press.

———. 2020. 'Jinnah: His Successes, Failures and Role in History'. New Delhi: Penguin Random House India.

Ahonsi, Babatunde A. 1988. 'Deliberate Falsification and Census Data in Nigeria'. *African Affairs* 87(349): 553–62.

Ahsan, Mohammad Badrul. 2011. 'A Census without Consensus'. *Daily Star*, 22 July.

Aiyar, Mani Shankar. 2023. 'All Is Not Well in Jammu and Kashmir'. *The Wire*, 9 February.

Alam, Jawaid. 2006. *Jammu and Kashmir 1949-64: Select Correspondence between Jawaharlal Nehru and Karan Singh*. Gurgaon: Penguin Books.

Ali Ahmed, A. H. n.d. 'The Fifth Population Census in Sudan: A Census with a Full Coverage and a High Accuracy'. New York: Statistics, Department of Economic and Social Affairs, United Nations.

Ali, Jehangir. 2022. '"Eroding Culture": Ladakh Admin Criticised for Removing Urdu Requirement from Revenue Dept Jobs'. *The Wire*, 13 January.

Ali, Muddasir. 2011a. 'Jammu Leads in Literacy'. *Greater Kashmir*, 8 April, 1, 10.

———. 2011b. 'JK under Radar for Female Feticide'. *Greater Kashmir*, 24 April, 1, 10.

———. 2011c. 'Govt "Challenges" Census Figures'. *Greater Kashmir*, 24 May, 1, 10.

Al Jazeera. 2010. 'India Moves to Improve Kashmir Life'. 13 July.

———. 2021. 'Indian Air Force Base in Kashmir Hit by Explosions'. 27 June.

Alonso, William, and Paul Starr. 1985. 'A Nation of Numbers Watchers'. *Wilson Quarterly* 9(3): 92–96.

———. 1987. *The Politics of Numbers*. New York: Russell Sage Foundation.

Aluko, S. A. 1965. 'How Many Nigerians? An Analysis of Nigeria's Census Problems, 1901–63'. *Journal of Modern African Studies* 3(3): 371–92.

Ambale, Suhas. 2020. 'Insecurities about Demographic Change Are Not a Problem Just in Kashmir Valley'. *Indian Express*, 31 October.

Anderson, Benedict. 2006. *Imagined Communities: Reflections on the Origin and Spread of Nationalism*. London and New York: Verso.

Anderson, James, Ian Shuttleworth, Chris Lloyd and Owen McEldowney. n.d. 'Political Demography: The Northern Ireland Census, Discourse and Territoriality'. https://sp.ukdataservice.ac.uk/doc/5362/mrdoc/pdf/q5362uguide.pdf. Accessed on 20 October 2021.

Anderson, Siwan, and Debraj Ray. 2012. 'The Age Distribution of Missing Women in India'. *Economic and Political Weekly* 47(47–48): 87–95.

Appadurai, Arjun. 2007. *Fear of Small Numbers: An Essay on the Geography of Anger*. Calcutta: Seagull Books.

———. 2012. 'Why Enumeration Counts'. *Environment & Urbanization* 24(2): 639–41.

Ashiq, Peerzada. 2010. 'Geelani Census Salvo: Write Indian-Held against Nationality Column'. *Hindustan Times*, 10 April.

———. 2021. 'Ending the Shift between Jammu and Kashmir'. *The Hindu*, 10 July.

———. 2022a. 'Delimitation Draft: Average Population Size of Kashmir Constituencies Go Up against Jammu Seats'. *The Hindu*, 10 February.

———. 2022b. 'Intercepted Messages Signal Grim Times for J&K'. *The Hindu*, 15 May.

Ashraf, Ajaz. 2016. '"Do You Need 700,000 Soldiers to Fight 150 Militants?": Kashmiri Rights Activist Khurram Parvez'. *Scroll.in*, 21 July.

Ashraf, Ajaz, and Vignesh Karthik K. R. 2020. 'Why J&K's Demography Will Change Beyond Belief'. *NewsClick*, 31 May.

Ashraf, M. 2011. 'Kashmir's Dichotomy: Hindu Past and Muslim Present'. *Kashmir Times*, 21 January, 6.

*Asian Age*. 2011. 'Mirwaiz Wife Gives Birth to 2nd Daughter'. 7 July, 5.

*Assam Tribune*. 2010. 'Census Hits Roadblock in Mizoram – Superstition!' 19 June.

Attane, I., and Y. Courbage. 2000. 'Transitional Stages and Identity Boundaries: The Case of Ethnic Minorities in China'. *Population and Environment* 21: 257–58.

Aung, San Yamin. 2018. 'Still No Date for Release of Census Findings on Ethnic Populations'. *Irrawaddy*, 21 February.

Attane, I., and Y. Courbage. 2000. 'Transitional Stages and Identity Boundaries: The Case of Ethnic Minorities in China'. *Population and Environment* 21: 257–58.

Badhana, Mohd Iqbal. 2021. 'Memorandum with Regard to Provision of Reserved Seats for ST in Distt Baramulla for Assembly Seats'. All India Confederation of SC/ST/OBC Organisations (Unit: Baramulla), AIC/DB/21/14, 7 July 2021.

Bakula, Kushok. 1953. 'Statement to the Press on Maulana Masudi's Reference to Ladakh in His Speech in the Indian House of the People on February 17, 1953'. 5 March.

Barrier, N. G. 1981. *The Census in British India: New Perspectives*. New Delhi: Manohar Publications.

Bashir, Abid. 2011. 'Census 2011: Lesser Girls in More Babies'. *Rising Kashmir*, 7 April.

Basu, Alaka Malwade. 1997. 'The "Politicization" of Fertility to Achieve Non-Demographic Objectives'. *Population Studies* 51(1): 5–18.

———. 2009. 'Over-Demonizing the International Population Movement'. *Population Studies* 63(2): 187–93.

Basu, K. 2015. 'The Republic of Beliefs: A New Approach to "Law and Economics."' World Bank Policy Research Working Paper No. 7259, World Bank, Washington, DC.

Baweja, Harinder. 1994. 'Pakistan's Failure in Geneva Dampens Mood in Kashmir, Demoralises Militants'. *India Today*, 31 March.

———. 2000a. 'J&K CM Farooq Abdullah Tries to Play Autonomy Card, Strengthen His Hold on Valley'. *India Today*, 10 July.

———. 2000b. 'J&K Autonomy Report: CM Farooq Abdullah Loses Chance to Upstage Hurriyat Conference'. *India Today*, 17 July.

Bradshaw, S. T. 1996. 'Death, Taxes, and Census Litigation: Do the Equal Protection and Apportionment Clauses Guarantee a Constitutional Right to Census Accuracy?'. *George Washington Law Review* 64(2): 379–413.

British Broadcasting Corporation (BBC). 2001. 'Leading Kashmir Militant Killed'. 25 July.

———. 2011. 'The "Killing Fields" of Kashmir'. 23 May.

———. 2019. 'Kashmir: Why India and Pakistan Fight Over It'. 8 August.

Begum, Sharifa, and Armindo Miranda. 1979. 'The Defectiveness of the 1974 Population Census of Bangladesh'. *Bangladesh Development Studies* 7(3): 79–106.

Behera, Navnita Chadha. 2002. 'A Signal from Jammu'. *Frontline*, 8 November.

———. 2006. *Demystifying Kashmir*. Washington, DC: Brookings Institution.

Beigh, Javed. 2021. 'Pakistan Betrays Kashmiris by Refusing Use of Airspace for Srinagar–Sharjah Flight'. News18, 4 November.

Bergman, David. 2014. 'Questioning an Iconic Number'. *The Hindu*, 24 April.

Bhagat, R. B. 2001. 'Census and the Construction of Communalism in India'. *Economic and Political Weekly* 36(46–47): 4352–356.

———. 2003. 'Role of Census in Racial and Ethnic Construction: US, British and Indian Censuses'. *Economic and Political Weekly* 38(8): 686–91.

———. 2006. 'Census and Caste Enumeration: British Legacy and Contemporary Practice in India'. *Genus* 62(2): 119–34.

Bhakto, Anando. 2021. 'The All Party Sikh Coordination Committee Asks Members of the Sikh Community to Abstain from Jammu and Kashmir's Delimitation Exercise to Redraw the Constituencies in the Union Territory'. *Frontline*, 24 February.

Bhalla, Sheila. 2014. 'Behind the Post-1991 "Challenge" to the Functional Efficiency of Established Statistical Institutions'. *Economic and Political Weekly* 49(7): 43–50.

Bharatiya Janata Party (BJP). n.d. 'Ideological Prejudice Manifests in the Public Life'. https://www.bjp.org/images/publications/ce.pdf. Accessed on 6 January 2023.

Bhat, Bashir Ahmad. 2011. 'Where Have They Gone? Searching Missing Girls in Jammu and Kashmir'. *Greater Kashmir*, 12 May.

———. 2014. 'Reality behind Declining Child Sex Ratio in Kashmir India'. Poster Session 6: Population Aging; Gender, Race and Ethnicity, Population Association of America 2014 Annual Meeting, Boston, 2 May. https://paa2014. populationassociation.org/papers/140336. Accessed on 7 January 2023.

———. 2016. 'Trends and Patterns of Antenatal and Delivery Care in Jammu and Kashmir: An Analysis of Health Management and Information System'. In *Maternal and Child Health in India: A Compendium of Studies Conducted by the Population Research Centres*, edited by C. R. K. Nair and K. S. James, 123–40. New Delhi: Ministry of Health and Family Welfare Government of India.

————. 2018. 'Reality Behind Declining Child Sex Ratio in Jammu and Kashmir'. In *Emerging Issues in Maternal, Neonatal and Child Health in India: A Compendium of Studies Conducted by the Population Research Centres 2016–17*, 169–86. New Delhi: Ministry of Health and Family Welfare, Government of India.

Bhatia, Mohita. 2020. *Rethinking Conflict at the Margins: Dalits and Borderland Hindus in Jammu and Kashmir*. New Delhi: Cambridge University Press.

Bhattacharjee, G. 2016. *Special Category States of India*. New Delhi: Oxford University Press.

————. 2018. 'Is the Special Category Status Really Dead?' *Economic and Political Weekly* 53(20): 25–29.

Bhattacharya, Sabyasachi. 2003. *Vande Mataram: The Biography of a Song*. New Delhi: Penguin Books.

Biemer, Paul P. 2010. 'Total Survey Error Design, Implementation, and Evaluation'. *Public Opinion Quarterly* 74(5): 817–848.

Bookman, M. Z. 2013. *The Demographic Struggle for Power: The Political Economy of Demographic Engineering in the Modern World*. Oxon: Routledge.

Boräng, F., A. Cornell, M. Grimes and C. Schuster. 2017. 'Cooking the Books: Bureaucratic Politicization and Policy Knowledge'. *Governance* 31(1): 7–26.

Bose, Ashish. 1991. *Population of India: 1991 Census Results and Methodology*. Delhi: B.R. Publishing Corporation.

————. 2000. 'Beyond a Headcount: Preparing for Census of 2001'. *Economic and Political Weekly* 35(17): 1433–34.

————. 2003. 'Census in Snowbound Areas'. *Economic and Political Weekly* 38(28): 2932–34.

————. 2004. 'Census Goldmine: Dissemination of 2001 Data'. *Economic and Political Weekly* 39(32): 3595–97.

————. 2005. 'Beyond Hindu–Muslim Growth Rates: Understanding Socio-Economic Reality'. *Economic and Political Weekly* 40(5): 370–74.

————. 2008. 'Accuracy of the 2001 Census: Highlights of Post-Enumeration Survey'. *Economic and Political Weekly* 43(22): 14–16.

Bose, Sumantra. 2003. *Kashmir: Roots of Conflict, Paths to Peace*. Cambridge, MA: Harvard University Press.

————. 2007. *Contested Lands: Israel-Palestine, Kashmir, Bosnia, Cyprus, and Sri Lanka*. Cambridge (MA) and London: Harvard University Press.

Brass, Paul R. 1974. *Language, Religion and Politics in North India*. New York: Cambridge University Press.

*Business Standard*. 2015. 'JK Govt to Provide Ration to People as per 2011 Census'. 23 April.

————. 2019a. 'Mayawati Hails Revocation of Special Status to Jammu and Kashmir'. 6 August.

———. 2019b. 'Religion Has No State Border: SC Junks Plea on Minority Status for Hindus'. 17 December.

Chakraborty, Tanika, and Sukkoo Kim. 2010. 'Kinship Institutions and Sex Ratios in India'. *Demography* 47: 4.

Chamie, J. 1994. 'Demography: Population Databases in Development Analysis'. *Journal of Development Economics* 44: 131–46.

Chandra, K. 2009. 'Caste in Our Social Imagination'. *Seminar* 601 (September).

Chandrasekhar, C. P., and Jayati Ghosh. 2011. 'Latest Employment Trends from the NSSO'. *Hindu Business Line*, 12 July.

———. 2013. 'The Employment Bottleneck'. *Hindu BusinessLine*, 8 July.

Chandrasekharan, S. 2013. 'Bhutan: Discrepancy in Population Figures, Need for Caution?' Update no. 102, South Asia Analysis Group, New Delhi.

Chari, Mridula. 2014. 'We Are Not Pandas: Irate Parsis Criticise Ads Urging Them to Procreate'. *Scroll.in*, 11 November.

Chasie, Charles. 2000. *The Naga Imbroglio (A Personal Perspective)*, 2nd edition. Kohima: Standard Printers and Publishers.

Chaube, S. K. 2012. *Hill Politics in Northeast India*, 3rd edition. Hyderabad: Orient Blackswan.

Chaudhury, Pradipta. 2005. 'Does Caste Indicate Deprivation?' *Seminar* (549).

———. 2012. 'Political Economy of Caste in Northern India, 1901-1931'. Working Paper, Centre for Economic Studies and Planning, Jawaharlal Nehru University, New Delhi.

Chellappan, Kumar. 2018. 'Kashmir Unrest Fallout of Changes in Religious Demography: Experts'. *The Pioneer*, 18 June.

Chhibber, Maneesh. 2020. 'Jammu and Kashmir's Biannual Darbar Move Is Bleeding India, Must Stop Now'. *The Print*. 10 June.

Chowdhary, Rekha. 2000. 'Autonomy Demand Kashmir at Crossroads'. *Economic and Political Weekly* 35(30): 2599–603.

———. 2010a. 'Identity Politics and Regional Polarisation in J&K'. *Economic and Political Weekly* 45(19): 15–20.

———. 2010b (ed.). *Identity Politics in Jammu and Kashmir*. New Delhi: Vitasta.

———. 2016. *Jammu and Kashmir: Politics of Identity and Separatism*. Oxon: Routledge.

———. 2019. *Jammu and Kashmir: 1990 and Beyond: Competitive Politics in the Shadow of Separatism*. New Delhi: SAGE Publishing.

Choudhary, Zafar. 2007. 'Across the LoC: Gujjars Face an Identity Crisis'. Article No. 2214, Institute of Peace and Conflict Studies, New Delhi.

———. 2008. 'Being Muslim in Jammu'. *Economic and Political Weekly* 43(34): 11–14.

———. 2011. 'Understanding the Gujjar–Pahri Faultline in J&K: A Gujjar Perspective'. Special Report No. 106, Institute of Peace and Conflict Studies Conflict Alert, New Delhi.

Centre for Monitoring Indian Economy (CMIE). 2020. 'Consumer Pyramids Household Survey Sample Survival & Response Rate'. 11 February. New Delhi: Centre for Monitoring Indian Economy Pvt Ltd.

Centre for Policy Studies (CPS). 2016. 'Religion Data of Census 2011: XV Jammu and Kashmir Jammu and Kashmir: Hindus No More Have a Place in the Valley'. Centre for Policy Studies, Chennai, 29 February.

———. 2017. 'Religion Data of Census 2011: A-2 Jammu and Kashmir – Is There a Demographic Component of the Proxy War in the Kashmir Valley?'. 30 July.

Cohn, Bernard S. 1987. 'The Census, Social Structure and Objectification in South Asia'. In *An Anthropologist among the Historians and Other Essays*, edited by Bernard S. Cohn, 224–54. New Delhi: Oxford University Press.

Coleman, David. 2012. 'The Twilight of the Census'. *Population and Development Review* 38(S1): 334–51.

Conlon, Frank F. 1981. 'The Census of India as a Source for the Historical Study of Religion and Caste'. In *The Census in British India: New Perspectives*, edited by N. G. Barrier, 103–17. New Delhi: Manohar Publications.

Curzon, George. 1908 (1907). *The Romanes Lecture, Frontiers*. Oxford: Clarendon Press.

Dabla, Bashir Ahmed. 2012. 'Sociological Dimensions and Implications of the Kashmir Problem'. In *The Parchment of Kashmir History, Society, and Polity*, edited by Nyla Ali Khan, 179–210. New York: Palgrave Macmillan.

*Daily Excelsior*. 2021. 'Several deputations call on LG'. 18 January.

———. 2022a. 'LBA Youth Wing Demands Anti-Conversion Law'. 13 January.

———. 2022b. 'Poonch Assembly Seat Is Now Open, Rajouri Reserved; Suchetgarh Restored but Goes to SCs'. 26 February.

———. 2022c. 'Everyone Affected with Pak Sponsored Terrorism in J&K: Azad'. 21 March.

Dandekar, V. M. 2004. 'Forty Years after Independence'. In *Indian Economy: Problems and Prospects*, edited by Bimal Jalan, 33–83. New Delhi: Penguin Books.

Dar, Umer Maqbool. 2011. 'APDP Begin Headcount of Missing'. *Greater Kashmir*, 11 March.

Das, Shaswati. 2019. 'With J&K Bifurcation, India Gets Its First Buddhist Dominated Union Territory'. *Mint*, 7 August.

Das Gupta, Jyoti Bhusan. 1968. *Jammu and Kashmir*. The Hague: Martinus Nijhoff Publishers.

Dasgupta, Swapan. 2003. 'Book Review: "Religious Demography of India" by A. P. Joshi, M. D. Srinivas and J. K. Bajaj'. *India Today*, 19 May.

Datta, Pradip Kumar. 1999. *Carving blocs: Communal Ideology in Early Twentieth-Century Bengal*. New Delhi: Oxford University Press.

*Dawn*. 2017a. 'Criticism of Army Role in Census Rejected'. 17 April.

———. 2017b. 'Muslim, Christian, Hindu, Ahmadi or "Other": The Census, a Source of Fear and Hope for Minorities'. 22 May.

———. 2017c. 'UN Body Assails Census Data Sharing with Nadra, Army'. 24 September.

———. 2021a. 'CCI Decides to Start New Census by Year-End'. *Dawn*, 13 April.

———. 2021b. 'Census Controversy'. 1 June.

———. 2023a. 'Policeman Martyred in Attack on Census Team in DI Khan'. 8 March.

———. 2023b. 'Two Policemen Martyred in Attacks on Census Teams in KP'. 14 March.

Deaton, A., and V. Kozel. 2005. 'Data and Dogma: The Great Indian Poverty Debate'. *World Bank Research Observer* 20(2): 177–99.

Debroy, Bibek. 2020. 'A Country of the Young? Not Really'. *The Week*, 25 January.

*Deccan Herald*. 2011. 'Fine of 1,000 Pounds if Britons Fail to Answer Census Queries'. 22 February.

Deshpande, Satish. 2021. 'Who's Afraid of a Caste Census?' *Indian Express*, 14 August.

Deshpande, Satish, and Mary E. John. 2010. 'The Politics of Not Counting Caste'. *Economic and Political Weekly* 45(25): 39–42.

Desiere, S., L. Staelens and M. D'Haese. 2016. 'When the Data Source Writes the Conclusion: Evaluating Agricultural Policies'. *Journal of Development Studies* 52(9): 1372–87.

Desrosières, Alain. 2013. 'The History of Statistics as a Genre: Styles of Writing and Social Uses'. *Bulletin de Me´thodologie Sociologique* 119: 8–23.

Deva, Waheed-u Zaman. 2015. 'Islamabad, Not Anantnag'. *Greater Kashmir*, 14 March.

Devadas, David. 1988. 'Buddhists Flare Up at Missionaries and Administration in Ladakh'. *India Today*, 15 July.

———. 2007. *In Search of a Future: The Story of Kashmir*. New Delhi: Penguin Books.

———. 2018. *The Generation of Rage in Kashmir*. New Delhi: Oxford University Press.

———. 2019a. *The Story of Kashmir: Geopolitics, Politics, Society, Culture and Changing Aspirations* (Kindle edition). Self-published.

———. 2019b. 'J&K and Ladakh: Layered, Nuanced Governance?' *Economic Times*, 23 August.

———. 2022a. 'J&K Delimitation Proposals: Is BJP Trying to "Engineer" a Win?' *The Quint*, 9 February.

———. 2022b. 'J&K's Kashmiri Pandit Crisis: Why the Govt Is Clueless about What to Do'. *The Quint*, 4 June.

Dhulipala, Venkat. 2015. *Creating a New Medina: State Power, Islam, and the Quest for Pakistan in Late Colonial North India*. New Delhi: Cambridge University Press.

*Daily News and Analysis (DNA)*. 2016. 'Agenda for Alliance: Full text of the agreement between PDP and BJP'. 23 May.

Donthi, Praveen. 2019. '"In J&K's Reorganisation, Kargil Is the Biggest Loser": Asgar Ali Karbalai'. *The Caravan*, 30 October.

Dorje, Chhewang. 2019. 'Why District Status for Zanskar Sub-Division?' *Daily Excelsior*, 30 March.

Drabu, Haseeb. 2019. 'Was Special Status a Development Dampener in J&K?' *Mint*, 8 August.

———. 2020a. 'Domicile Domiciliated'. *Greater Kashmir*, 9 April.

———. 2020b. 'Delimitation: The Context'. *Greater Kashmir*, 2 July.

———. 2020c. 'Delimitation II: Of Hindus and Muslims'. *Greater Kashmir*, 6 July.

———. 2020d. 'Delimitation III: The Area Angle'. *Greater Kashmir*, 9 July.

———. 2020e. 'Delimitation in J&K: A Suggested Approach'. *Greater Kashmir*, 19 July.

———. 2020f. 'Decoding Darbar Move'. *Greater Kashmir*, 23 April.

———. 2021a. 'J&K Delimitation: Go by the Population Rule'. *Hindustan Times*, 1 March.

———. 2021b. 'Area Not a Right Criteria'. *Daily Excelsior*, 10 July.

Drèze, Jean, and Anmol Somanchi. 2021. '"Bias It Is": CMIE Chief's Defence of CPHS Survey Elicits Fresh Critical Response from Jean Drèze, Anmol Somanchi'. *Economic Times*, 6 July.

Dulat, A. S. 2010. *Kashmir: The Vajpayee Years*. Noida: HarperCollins Publishers.

Dutta, Amrita Nayak. 2020. 'PoK Weather Forecasts Will Now Be a Regular Feature on DD, AIR News Bulletins'. *The Print*, 8 May.

Ebenstein, Avraham. 2010. 'The "Missing Girls" of China and the Unintended Consequences of the One Child Policy'. *Journal of Human Resources* 45(1): 87–115.

*Economic Times*. 2022. 'Centre Fast-Tracks Process for IAS Cadre Rules Change, Sets Up Committee for Direct Recruit IAS Officers'. 24 January.

*Economic and Political Weekly*. 2021. 'State and Its Anxiety of Caste Census'. *Economic and Political Weekly* 56(33): 8.

*Encyclopedia Britannica*. 2018. 'Census'. 12 July.

Englebert, Pierre, Stacy Tarango and Matthew Carter. 2002. 'Dismemberment and Suffocation: A Contribution to the Debate on African Boundaries'. *Comparative Political Studies* 35: 1093–118.

Evans, Alexander. 2002. 'A Departure from History: Kashmiri Pandits, 1990-2001'. *Contemporary South Asia* 11(1): 19–37.

*Express Tribune*. 2010. 'Mirwaiz Urges India to Withdraw Troops'. 10 May.

Fareed, Rifat. 2022. 'India's Modi Promises Investments on Kashmir Visit'. Al Jazeera, 24 April.

Fawehinmi, Feyi. 2018. 'The Story of How Nigeria's Census Figures Became Weaponized'. *Quartz Africa*, 6 March.

Fayyaz, Ahmed Ali. 2013. 'Census Indicates Alarming Level of Foeticide in Jammu & Kashmir'. *The Hindu*, 12 June.

———. 2017. 'Jammu Beating Valley in All PSC Selections since 1995'. *State Times*, 23 January.

———. 2021. 'When Sheikh Abdullah Was in Jail, Syed Ali Shah Geelani Was Eyeing Power with Congress'. News18, 6 September.

*Financial Express*. 2016. 'Hari Parbat Turns into "Koh-e-Maran," Kashmiri Pandits Fume'. 21 May.

Ganai, Naseer. 2014. 'Farooq Abdullah and Syed Ali Shah Geelani Swap Their Roles as Dove and Hawk in J&K Politics'. *India Today*, 22 June.

———. 2021. 'BJP's Jammu Conundrum: Traders Protest against "Outsiders" Running Away with Business Like East India Company'. *Outlook*, 11 October.

———. 2022. 'Delimitation Commission Report: Is It Greater Jammu or Greater Kashmir?'. *Outlook*, 13 February.

Geertz, Clifford. 1973. *The Interpretation of Cultures: Selected Essays*. New York: Basic Books.

Ghosh, Arunabh. 2020. *Making It Count: Statistics and Statecraft in the Early People's Republic of China*. Princeton, NJ: Princeton University Press.

Gill, Mehar Singh. 2007. 'Politics of Population Census Data in India'. *Economic and Political Weekly* 42(3): 241–49.

Glaister, D. 2012. '120 People Convicted for Not Filling in Census Form'. *The Guardian*, 27 January.

Goble, P. 2015. 'Unpublished Census Provides Rare and Unvarnished Look at Turkmenistan'. *Eurasia Daily Monitor* 12(26). https://jamestown.org/program/unpublished-census-provides-rare-and-unvarnished-look-at-turkmenistan. Accessed on 23 November 2022.

Göderle, Wolfgang. 2016. 'Administration, Science, and the State: The 1869 Population Census in Austria-Hungary'. *Austrian History Yearbook* 47: 61–88.

Goodkind, Daniel. 2011. 'Child Underreporting, Fertility, and Sex Ratio Imbalance in China'. *Demography* 48(1): 291–316.

*Greater Kashmir*. 2011a. 'Parliament Delegation Expresses Concern over Female Feticide'. 26 April, 1.

———. 2011b. 'Decline in Girl Child Ratio Damaging Phenomenon: Omar'. 17 May, 3.

———. 2011c. 'Malik for Sustained Campaign against Female Foeticide'. 17 May, 3.

———. 2017. 'Why Is it Critical to Kashmir?'. 12 August.

———. 2018. 'Army Seeks Info on Populations in Vicinity of Its Camps'. 27 February.

———. 2019. 'Post August 5, Kashmiris Fear Demographic Change: CCG'. 12 December.

———. 2021. '76 J&K Schools, Colleges, Roads to Be Named after "Martyrs, Eminent Personalities"'. 30 October.

Greenberg, Joel. 1997. 'Palestinian Census Ignites Controversy over Jerusalem'. *New York Times*, 11 December.

Guha, Ramachandra (ed.). 2010. *Makers of Modern India*. New Delhi: Penguin Books/Viking Press.

Guha, Sumit. 2003. 'The Politics of Identity and Enumeration in India c. 1600–1990'. *Comparative Studies in Society and History* 45(1): 148–67.

Gul, Khalid. 2008. '"NC, PDP Serve Delhi's Interests"'. *Greater Kashmir*, 14 June.

Guilmoto, Christophe Z. 2009. 'The Sex Ratio Transition in Asia'. *Population and Development Review* 35(3): 519–49.

Guilmoto, Christophe Z., and S. Irudaya Rajan. 2002. 'District-Level Estimates of Fertility from India's 2001 Census'. *Economic and Political Weekly* 37(7): 665–72.

———. 2013. 'Fertility at the District Level in India: Lessons from the 2011 Census'. *Economic and Political Weekly* 48(23): 59–70.

Gupta, Anil. 2017. 'Demographic Changes Make Jammu a "Ticking Time Bomb"'. *Daily Excelsior*, 12 January.

Gupta, Madhvi, and Pushkar. 2010. 'Ethnic Diversity and the Demand for Public Goods: Interpreting the Evidence from Delhi'. *Economic and Political Weekly* 45(43): 64–72.

Gupta, Radhika. 2023. *Freedom in Captivity: Negotiations of Belonging Along Kashmir's Frontier*. New Delhi: Cambridge University Press.

Gupta, Shekhar. 2011. 'This Death in Pakistan'. *Indian Express*, 8 January.

———. 2022. 'Looking beyond "Kashmir Files," Catharsis & Closure Need Justice, for All Cases of Mass Injustice'. *The Print*, 19 March.

Gwatkin, Davidson R. 1979. 'Political Will and Family Planning: The Implications of India's Emergency Experience'. *Population and Development Review* 5(1): 29–59.

Habibullah, Wajahat. 2020. 'Kashmir: The Tragedy'. *Social Scientist* 48(7–8): 17–28.

Hamid, Peerzada Arshad. 2011. 'Counting Kashmiris'. Himal, April. https://old.himalmag.com/component/content/article/4355-counting-kashmiris.html. Accessed on 7 February 2023.

Hamsher, D. 2005. 'Counted out Twice – Power, Representation and the Usual Residence Rule in the Enumeration of Prisoners: A State-Based Approach to Correcting Flawed Census Data'. *Journal of Criminal Law and Criminology* 96(1): 299–328.

Hari Om. 1990. 'State of Jammu: Alienated from the Valley'. *The Statesman*, 20 June.

———. 1995a. 'Jammu, the ISI Target'. *The Hindu*, 25 February.

———. 1995b. 'Jammu Cries for Justice'. *The Hindu*, 18 May.

———. 1998. *Beyond the Kashmir Valley*. New Delhi: Har Anand Publications.

———. 2013. 'India's Forgotten Nationals'. *The Pioneer*, 22 July.

———. 2017a. 'Tunnel on Mughal Road will Bring Only Disaster'. *State Times*, 18 March.

———. 2017b. 'Where Is Article 370 Now That Rohingyas and Bangladeshis Are Being Settled in Jammu and Ladakh?' *Swarajya*, 1 September.

———. 2019a. 'Darbar Move: KN Pandita Stabs Jammu from the Back-I'. *State Times*, 7 November.

———. 2019b. 'Why Is Karan Thapar 100 pc Wrong in Calling Hindus of Jammu Murderers of Muslims?' 24 September.

———. 2020. 'Vilification Campaign against Jammu'. *State Times*, 5 December.

———. 2021c. 'Moves Afoot to Disintegrate & Subjugate Jammu'. *State Times*, 2 June.

———. 2021d. 'Just One Report in Turkish Media & Section on J&K Temples Deleted from Official Website'. *State Times*, 4 June.

———. 2021e. 'But Where Is "Chenab Valley"?' *State Times*, 10 July.

———. 2021f. 'Pitting Jammu against Jammu to Defeat Jammu'. *State Times*, 14 July.

———. 2022a. 'How Delimitation Report Further Dis-Empowers Already Neglected Jammu'. *State Times*, 20 March.

———. 2022b. 'Azad Calls the Kashmiri Bluff; Says It's Wrong to Term Mountainous Ramban-Doda-Kishtwar as Chenab Valley'. *State Times*, 8 April.

———. 2022c. 'The Jammu Files'. *State Times*, 13 April.

Hartwig, Jochen. 2006. 'On Spurious Differences in Growth Performance and on the Misuse of National Accounts Data for Governance Purposes'. *Review of International Political Economy* 13(4): 535–58.

Hazarika, Sanjoy. 2005. 'Rio Tosses Interim Solution Idea'. *The Statesman*, 24 December.

———. 2018. *Strangers No More: New Narratives from India's Northeast*. New Delhi: Aleph Book Company.

———. 2020. 'Development, Data and Democracy: On Rebuilding the Credibility of India's Statistical System'. Panel Discussion, Bangalore International Centre, 9 November. https://bangaloreinternationalcentre.org/event/data-development-and-democracy. Accessed on 6 January 2023.

Heine, Klaus, and Erich Oltmanns. 2016. 'Towards a Political Economy of Statistics'. *Statistical Journal of the IAOS* 32: 201–09.

Hill, Matthew, and Nishank Motwani. 2017. 'Language, Identity and (In)Security in India–Pakistan Relations: The Case of Kashmir'. *South Asia: Journal of South Asian Studies* 40(1): 123–45.

Himalayan Buddhist Cultural Society, Paddar (HBCS). 2021. 'Memorandum to Micro Minority Buddhist of J&K'. 8 July (Courtesy: Dr Nawang Tundup).

*Hindustan Times*. 2000. 'Census to Begin in Jammu and Kashmir amidst Tight Security'. 10 September.

———. 2015. 'Govt Takes Al Jazeera Off Air over Kashmir Map Row'. 22 April.

———. 2018. 'Student with "Indian-Occupied Kashmir" in Twitter Bio Seeks Sushma Swaraj's Help; No Place Like That, She Says'. 10 May.

———. 2022. 'Home Secretary Ajay Bhalla Reviews J&K Security Situation Following Terror Threat to Kashmiri Pandits'. 6 December.

Hoda, Najmul. 2020. 'Indian Muslims Must Rewrite Their Victim Mindset to Be Indispensable in India's Rise'. 17 August.

Horowitz, Donald L. 2000. *Ethnic Groups in Conflict*, 2nd edition. Berkeley, CA: University of California Press.

Hunter, W. W. 2002 [1876]. *The Indian Musalmans (Introduction by Bimal Prasad)*. New Delhi: Roopa & Co.

Huntington, Samuel P. 1996. *The Clash of Civilizations and the Remaking of World Order*. New York: Simon & Schuster.

Hussain, Altaf. 2020. 'How Does Kashmir See the Revocation of Article 370?' *Kashmir Life*, 5 August.

Hussain, Ashiq. 2016. 'Valley Heats Up over Domicile, Separatists to Hit the Streets'. *Hindustan Times*, 23 December.

Hussain, Shahla. 2021. *Kashmir in the Aftermath of Partition*. New Delhi: Cambridge University Press.

Hussain, Syed Khalid Hussain. 2022. '"Everything Can't Be India's Responsibility…": Omar Abdullah Defends S Jaishankar's Remark Bashing Pakistan'. Zee News, 15 December.

Hussain, Syed Jaleel, and Syed Eesar Mehdi. 2021. 'Insecurity, Identity and Resistance: Contours of Shia Political Discourse in Kashmir'. *Economic and Political Weekly* 56(3): 51–55.

*India Today*. 2017a. 'Yogi Adityanath Compares Kairana "Exodus" to Kashmir: Will Election Commission Take Notice'. 13 February.

———. 2017b. 'Muslims a Minority in Jammu and Kashmir? Supreme Court Asks Centre, State to "Sit Together" and Decide'. 27 March.

———. 2022. 'Rahul Ji, Your Grandmother Felt Differently, Says Vivek Agnihotri on Kashmir Files Row'. 14 March.

*Indian Express*. 2011. 'J&K Literacy Rate Soars to 68%'. 8 April.

———. 2012. 'Aadhar Helps Weed Out Fake Ration Cards in Andhra'. 18 December.

———. 2017. 'Jammu Trader Organisation Threatens to Kill Illegal Bangladeshi, Myanmar Immigrants unless Deported'. 8 April.

———. 2019. 'Rupture in History, Stitching a Future'. 6 August.

———. 2020. 'J&K Realty'. 29 October.

———. 2021a. 'J&K Launches Survey of Nomadic Population'. 30 May.

———. 2021b. 'Only 2 Persons Bought 2 Properties in J-K Post Article 370 Abrogation: Govt to Parliament'. 10 August.

———. 2021c. 'Few Takers for J&K Domicile Offer, Govt Extends Deadline, Sends Teams to Camps to Register'. 14 September.

———. 2022. 'J&K to Host G-20 Meetings in 2023'. 24 June.

Iqbal, Fida. 2011. 'Growing in Numbers Only'. *Greater Kashmir*, 28 April, 7.

Iqbal, Sehar. 2021. *A Strategic Myth: 'Underdevelopment' in Jammu and Kashmir*. New Delhi: Tulika Publishers.

Ishfaq-ul-Hassan. 2011. 'Girl Child Brings Omar Abdullah Separatists Together'. *Daily News and Analysis*, 18 May.

Islam, M. J. 2018. 'Anantnag or Islamabad? What Is the Actual Name of This South Kashmir District?' *Greater Kashmir*, 15 February.

Jalal, Ayesha. 2008. *Partisans of Allah: Jihad in South Asia*. Cambridge (MA) and London: Harvard University Press.

Jaleel, Muzamil. 2013. 'Habibullah Breaks Silence: Govt Deleted Key Portions of My Report on J&K Mass Rape Case'. *Indian Express*, 7 July.

Jameel, Yusuf. 2010a. 'J&K Sceptical about Census, Senses "Conspiracy"'. 18 May.

———. 2010b. 'Muslims Worry over Tally's Implications'. *The National*, 22 May.

Jamwal, Anuradha Bhasin. 2013. 'A Moon of Many Shades'. *Economic and Political Weekly* 48(17): 26–29.

———. 2020a. 'Jammu, the Pawn on the Kashmir Chessboard'. *The Hindu*, 2 June.

———. 2020b. 'Are There 26 Lakh Hindi Speakers and Only 20,000 Urdu Speakers in J&K?'. *Kashmir Times*, 17 October.

———. 2022. *A Dismantled State: The Untold Story of Kashmir After Article 370*. Gurugram: HarperCollins Publishers.

Janus, Thorsten. 2013. 'The Political Economy of Fertility'. *Public Choice* 155(3–4): 493–505.

Jayachandran, Seema. 2017. 'Fertility Decline and Missing Women'. *American Economic Journal: Applied Economics* 9(1): 118–39.

Jerven, Morten. 2013. *Poor Numbers: How We Are Misled by African Development Statistics and What to Do about It*. Ithaca, NY: Cornell University Press.

Jha, R., and A. Sharma. 2003. 'Spatial Distribution of Rural Poverty: Last Three Quinquennial Rounds of NSS'. *Economic and Political Weekly* 38(47): 4985–93.

John, Mary E. 2011. 'Census 2011: Governing Populations and the Girl Child'. *Economic and Political Weekly* 46(16): 10–12.

Jones, Kenneth W. 1981. 'Religious Identity and the Indian Census'. In *The Census in British India: New Perspectives*, edited by N. G. Barrier, 73–101. New Delhi: Manohar Publications.

Joshi, Arun. 2020. 'Jammu shaping narrative for J&K'. *The Tribune*, 6 January.

Joshi, A. P., M. D. Srinivas and J. K. Bajaj. 2003. *Religious Demography of India*. Chennai: Centre for Policy Studies.

Joshi, Binoo. 2001. 'Census in J&K Completed Successfully After 20 Years'. *The Pioneer*, 16 January.

Janagoshthiya Samannay Parishad, Assam (JSPA). 2021. 'Janagoshthiya Samannay Parishad Census'. Guwahati: Janagoshthiya Samannay Parishad, Assam. https://jspacensus.com/index.php. Accessed on 19 April 2021.

Kalsotra, R. K. 2021. 'Memorandum of Demands Presented to Delimitation Commission on Behalf of Confederation, J&K'. All India Confederation of SC/ST/OBC Organisations (Jammu), AICSCT/18, 8 July 2021.

———. 2022. 'Amended Demands Presented to the Commission'. All India Confederation of SC/ST/OBC Organisations (Jammu), AICSCT/48, 4 April 2022.

Kansakar, Vidya Bir Singh. 1977. *Population Censuses of Nepal and the Problems of Data Analysis*. Kirtipur, Nepal: Centre for Economic Development and Administration, Tribhuvan University.

Karmakar, Rahul. 2010. 'Census Weapon for Battle in NE'. *Hindustan Times*, 22 April.

———. 2018. 'Mizos Urged to Have More Children'. *The Hindu*, 16 June.

*Kashmir Life*. 2010. 'The Great Number Game'. 2 May.

*Kashmir Times*. 1990. 'BJP, RSS Suggest Balancing of Population in Kashmir'. 20 April.

———. 2000a. 'Census Operations Resumed as Govt. Withholds Salary of Errant Employees'. 10 November, 1, 8.

———. 2000b. 'Census Important for Planning: Advani'. 15 November, 1.

———. 2000c. 'Panun Kashmir (M) Submits Memorandum to Census Commissioner of India'. 14 October.

———. 2000d. 'Census Ghost Again Haunts Employees'. 23 October, 1, 7.

———. 2000e. 'Tough Time for Vergese over Konanposhpora'. 2 October, 1, 6.

———. 2000f. '2nd Phase of Census Operation Concludes'. 5 October, 6.

———. 2000g. 'Census Work in Kathua Completed'. 7 October, 3.

———. 2000h. 'FIR against Employee on Census Issue'. 4 October, 6.

———. 2000i. 'Census Operations Grounded'. 2 November, 1, 8.

———. 2000j. 'Census Authorities Urged for Nomad's Enumeration'. 8 November, 8.

———. 2000k. 'Readers' Forum: Census – A Mockery [A Letter to the Editor by Haji Zamir Khan]'. 24 November, 4.

———. 2000l. 'Gujjars for Representation in Talks'. 12 December.

———. 2011a. 'Gujjars for Special Census of Nomadic Tribes in J&K'. 31 January, 6.

———. 2011b. 'APSCC Demands Special Cell in Census'. 25 February, 3.

———. 2011c. 'Enumerators Suspended for Not Conducting Census'. 24 February, 5.

———. 2011d. 'Census 2011 Concludes in J&K'. 1 March, 1, 11.

———. 2011e. 'BJP, JSM Stage Walkout: "More Voters, Lesser Panchayats in Jammu"'. 15 March, 11.

———. 2011f. 'How Many Killed in Summer Unrest: One Question, Two Answers'. 24 March.

———. 2011g. 'ST Status for Paharis, Kolis, Gaddis Taken Up with Centre: Sakina'. 26 March, 5.

———. 2011h. 'Employees Threaten Strike'. 6 January, 3.

———. 2011i. 'Valley-Based Pandits Demand Facilities at par with Migrants'. 7 March, 3.

———. 2020. 'J&K's Demographic Change in Virtual World; MHA Website Reduces Valley's Population Over 10 Times'. 18 January.

*Kashmir Walla*. 2018. 'We Request People to Vote and Defeat India: Qazi Yasir'. 7 October.

Kasturi, Kannan. 2015. 'Comparing Census and NSS Data on Employment and Unemployment'. *Economic and Political Weekly* 50(22): 16–19.

Kaur, Ravinder, and Charumita Vasudev. 2019. 'Son Preference and Daughter Aversion in Two Villages of Jammu'. *Economic and Political Weekly* 54(13): 13–16.

Kaur, Ravinderjit. 1996. *Political Awakening in Kashmir*. New Delhi: APH Publishing Corporation.

———. 2010. 'Religion and Identity Politics of Sikhs of Kashmir'. In *Identity Politics in Jammu and Kashmir*, edited by Rekha Chowdhary, 229–43. New Delhi: Vitasta.

Kawate, Iori. 2021. 'China Census Called into Question over 14m "Mystery" Children'. *NikkeiAsia*, 13 May.

Kazmi, Syed Muhamad Abbas. 1996. 'The Balti Language'. In *Jammu, Kashmir & Ladakh: Linguistic Predicament*, edited by P. N. Pushp and K. Warikoo. New Delhi: Himalayan Research and Cultural Foundation and HarAnand Publications. http://koshur.org/Linguistic/7.html. Accessed on 6 January 2023.

Kearney, Robert. 1985. 'Ethnic Conflict and the Tamil Separatist Movement in Sri Lanka'. *Asian Survey* 25(9): 898–917.

Kelkar, Vijay, and Ajay Shah. 2019. *In Service of the Republic: The Art and Science of Economic Policy*. Penguin Random House India. Kindle edition.

Kertzer, David I., and Dominique Arel (eds.). 2002a. *Census and Identity: The Politics of Race, Ethnicity, and Language in National Censuses*. Cambridge, UK: Cambridge University Press.

————. 2002b. 'Censuses, Identity Formation, and the Struggle for Political Power'. In *Census and Identity: The Politics of Race, Ethnicity, and Language in National Censuses*, edited by David I. Kertzer and Dominique Arel, 1–42. Cambridge, UK: Cambridge University Press.

Keynes, J. M. 1927. 'The British Balance of Trade, 1925-27'. *Economic Journal* 37(148): 551–65.

Khan, Akhtar Hassan. 1998. '1998 Census: The Results and Implications'. *Pakistan Development Review* 37(4): 481–89.

Khan, Ali. 2011. 'Kashmir's "Missing Girls"'. *Siasat Daily*, 20 May.

Khan, Saman Ghani. 2018. 'Mosaic Nation: What Made the Census Flawed and Controversial'. *Herald*, 1 July.

Khan, Shahnawaz. 2009. 'Riots Changed J&K Politics [first part of a two-part interview with Ved Bhasin]'. *Kashmir Life*, 3 October.

Khan, Syed Ahmad. 2010 (1882). 'Politics and Discord'. In *Makers of Modern India*, edited by Ramachandra Guha, 59–68. New Delhi: Viking Books.

Koch-Weser, I. N. *The Reliability of China's Economic Data: An Analysis of National Output*. U.S.–China Economic and Security Review Commission Staff Research Project, 2013. https://www.uscc.gov/sites/default/files/Research/TheReliabilityofChina%27sEconomicData.pdf. Accessed on 10 January 2023.

Kotadia, Surendra, and Arun Gandhi. 2002a. *Indian Inland Letter Cards with Advertisements & Slogans*, 3rd edition. Mumbai: Surendra Kotadia.

————. 2002b. *Indian Inland Post Cards with Advertisements & Slogans*, 3rd edition. Mumbai: Surendra Kotadia.

Krätke, Florian, and Bruce Byiers. 2014. 'The Political Economy of Official Statistics: Implications for the Data Revolution in sub-Saharan Africa'. Discussion Paper No. 5, PARIS21 Partnership in Statistics for Development in the 21st Century, Paris.

Krishna, Sankaran. 1994. 'Cartographic Anxiety: Mapping the Body Politic in India'. *Alternatives: Global, Local, Political* 19(4): 507–21.

Kulkarni, Sharad. 1991. 'Distortion of Census Data on Scheduled Tribes'. *Economic and Political Weekly* 26(5): 205–08.

Kulkarni, Sumati. 2004. 'Inputs and Process: An Inside View'. *Economic and Political Weekly* 39(7): 652–58.

Kumar, Narender. 2023. 'Rajouri Terror Attack Is a Warning!' Rediff.com, 4 January.

Kumar, Sanjay, and K. M. Sathyanarayana. 2012. 'District-Level Estimates of Fertility and Implied Sex Ratio at Birth in India'. *Economic and Political Weekly* 47(33): 66–72.

Kumar, Vikas. 2015a. 'Use of Statistics in Politics Not New in India'. *Deccan Herald*, 5 March, 11.

————. 2015b. 'The Dumbing Down of Data'. *The Hoot*, 16 September.

————. 2017a. 'Demonetisation in Numbers: How Statistics Were Used'. *The Hoot*, 12 November.

————. 2017b. 'How the Pew Report on Modi was Covered'. *The Hoot*, 19 November.

————. 2017c. 'Covering NCRB Data: How Newspapers Fared.' *The Hoot*, 6 December.

————. 2020a. 'Census Laws and the Quality of Census Data: The Limits of Punitive Legislation'. *Statistical Journal of the International Association for Official Statistics* 36(4): 1143–60.

————. 2020b. 'Are We Well-Informed about the Country's Periphery?' *The Wire*, 26 October.

————. 2020c. 'Why the Delimitation Exercise in Jammu and Kashmir Calls for Caution'. *The Wire*, 25 September.

————. 2020d. 'Census 2021 Will Be Delayed. It Gives Modi Govt Time to Bring Long-Pending Reforms'. *The Print*, 13 August.

————. 2021a. 'Usability of India's Government Statistics'. *Statistical Journal of the International Association for Official Statistics* 37(1): 429–45.

————. 2021b. 'India Cannot Forge Bonds with Southeast Asia Ignoring Issues of Its Northeast Region'. *The Wire*, 11 February.

————. 2022a. 'Poor Data Hamstrings Gender Equity Reporting in India'. East Asia Forum, 29 March.

————. 2022b. 'Delimitation'. In *The Routledge Companion to Northeast India*, edited by Jelle J. P. Wouters and Tanka B. Subba, 115–20. Oxon and New York: Routledge.

————. 2022c. 'Government Statistics'. In *The Routledge Companion to Northeast India,* edited by Jelle J. P. Wouters and Tanka B. Subba, 200–09. Oxon and New York: Routledge.

————. 2023a. 'When Census Is an Election: A Game-Theoretic Analysis of Over-Reporting of Headcount'. In *Power and Responsibility: Interdisciplinary Perspectives for the 21st Century in Honor of Manfred J. Holler,* edited by Martin Leroch and Florian Rupp, 373–393. Switzerland: Springer International Publishing.

————. 2023b. *Waiting for a Christmas Gift: Essays on Politics, Elections and Media in Nagaland.* Dimapur: Heritage Publishing House.

————. 2023c. . 'The Census of India and the Postal System in Post-Colonial India: An Overview'. *Signet (Quarterly Journal of The Philatelic Congress of India)* 45(1) (Special edition, February): 48–50.

————. 2023d. 'Understanding NFHS-5 Data on Sex Ratio'. Manuscript.

————. n.d. 'The Public Outreach of the Census of India and Data Quality'. Manuscript.

Lamb, Alistair. 1991. *Kashmir: A Disputed Legacy 1846-1990*. Hertfordshire: Roxford Books.

Lipton, M. 1972. 'The South African Census and the Bantustan Policy'. *World Today* 28(6): 257–71.

Madhav, Ram. 2021. 'Abrogation of Article 370 Was the End of "Kashmiri Exceptionalism"'. *Indian Express*, 6 August.

Magazine, Aanchal, and Anil Sasi. 2020. 'Red Flags by Data Panel Chief Pronab Sen: "Mistrust, Attacks May Contaminate Economic Data for Next 10 Years"'. 13 February.

Mahalanobis, P. C. 1965. 'Statistics as a Key Technology'. *American Statistician* 19(2): 43–46.

Maharatna, A., and A. Sinha. 2011. 'Long-Term Demographic Trends in North-East India and Their Wider Significance, 1901–2001'. Occasional Paper No. 26, Institute of Development Studies, Kolkata.

Maheshwari, S. R. 1996. *The Census Administration under the Raj and After*. New Delhi: Concept Publishing Company.

Maini, K. D. 2011. 'Understanding the Gujjar–Pahri Faultline in J&K: A Pahri Perspective'. Special Report No. 106, Institute of Peace and Conflict Studies, New Delhi.

Majid, Zulfiqar. 2008. 'SASB "Unites" Mirwaiz, Geelani'. *Greater Kashmir*, 14 June.

———. 2009. 'Former CM G M Shah Is No More: State Holiday Announced'. *Greater Kashmir*, 6 January.

———. 2022. 'Kashmiri Pandit Farmer Shot Dead by Militants in Kashmir'. *Deccan Herald*, 16 August.

Malik, Irfan Amin. 2020. 'As Fewer Locals Get Key Admin Posts, Kashmiri Muslims Fear Planned Demographic Change'. *The Wire*, 8 October.

Malik, Monika. 2022. 'I Was Popular Pick for CM Post, but Cong Rejected It over My Religion: Jakhar'. *The Pioneer*, 3 February.

Manhotra, Dinesh. 2019. 'Minor Typo Deprives Kolis of ST Benefits for 28 Yrs'. *The Tribune*, 28 January.

Mari Bhat, P. N. 2002a. 'On the Trail of "Missing" Indian Females I: Search for Clues'. *Economic and Political Weekly* 37(51): 5105–18.

———. 2002b. 'On the Trail of "Missing" Indian Females II: Illusion and Reality'. *Economic and Political Weekly* 37(52): 5244–63.

Maqbool, Umer. 2023. 'Are J&K's Reservation Laws Being Changed to Further Marginalise Kashmir?'. *The Wire*, 7 March.

Mari Bhat, P. N., and A. J. Francis Zavier. 2007. 'Factors Influencing the Use of Prenatal Diagnostic Techniques and the Sex Ratio at Birth in India'. *Economic and Political Weekly* 42(24): 2292–2303.

Masood, Bashaarat. 2010. 'Separatists Support Census, Say Muslims Losing Majority Status'. *Indian Express*, 5 April.

McMillan, Alistair. 2000. 'Delimitation, Democracy, and End of Constitutional Freeze'. *Economic and Political Weekly* 35(15): 1271–76.

Merli, M. Giovanna, and Adrian E. Raftery. 2000. 'Are Births Underreported in Rural China? Manipulation of Statistical Records in Response to China's Population Policies'. *Demography* 37(1): 109–26.

Meyer, B. D., W. K. C. Mok and J. X. Sullivan. 2015. 'Household Surveys in Crisis'. *Journal of Economic Perspectives* 29(4): 199–226.

*Mint*. 2009. 'The Sex Ratio Numbers Game'. 18 August.

———. 2021. '168 Rohingyas Living Illegally in Jammu Sent to Jail'. 7 March.

Mir, Hilal. 2021. 'Gov't Website Declares 97% Muslim Kashmir "Predominantly Hindu"'. Anadolu Agency, 11 May.

Mishra, Ashutosh. 2019. 'Leh Celebrates New Union Territory Status'. *India Today*, 9 August.

Mitra, Asok. 1961. 'Bye-Products of the Census of 1961'. *Economic Weekly* 13(27–29): 1061–79.

———. 1994. 'Census 1961: New Pathways'. *Economic and Political Weekly* 29(51–2): 3207–21.

Mohanty, S. K., Günther Fink, Rajesh K. Chauhan and David Canning. 2016. 'Distal Determinants of Fertility Decline: Evidence from 640 Indian Districts'. *Demographic Research* 34: 373–406.

Mohanty, S. P. 1996. 'Significance of Census Data in the Context of Contemporary Indian Society'. In *Census as Social Document*, edited by S. P. Mohanty and A. R. Momin, 162–67. Jaipur: Rawat Publications.

Mohanty, S. P., and A. R. Momin. 1996 (eds.). *Census as Social Document*. Jaipur: Rawat Publications.

Morgenstern, Oskar. 1973. *On the Accuracy of Economic Observations*. Princeton, NJ: Princeton University Press.

Morland, Paul. 2018. *Demographic Engineering: Population Strategies in Ethnic Conflict*. Oxon: Routledge.

*Morung Express*. 2017. 'Imposition of Aadhaar a Threat to Naga Customary Law and Identity'. 1 November.

———. 2018a. 'Mizoram Receives Rs 15.18 Crore for Strengthening State Statistical System'. 18 July.

———. 2018b. 'Nagas Are Living in a Human Time Bomb'. 25 August.

———. 2020. 'Bring Ordinance to Control Population: BJP Leader to PM'. 11 July.

———. 2021a. 'CBCC Advises Citizens not to be Swayed by "False Prophecies"'. 17 January.

———. 2021b. '"Statistical Data Should be Submitted on Regular Basis"'. 28 October.

Mudliar, Preeti. 2020. 'Broken Data: Repairs in the Production of Biometric Bodies'. In *Lives of Data: Essays on Computational Cultures from India*, edited by Sandeep Mertia, 89–98. Amsterdam: Institute of Network Cultures.

Mufti, Mehbooba. 2023. 'SAARC, Not G20, Is What Will Make India a "Vishwaguru"'. *The Wire*, 14 May.

*Nagaland Post*. 2009. 'Nagaland's Inflated Census'. 21 September.

Nahata, Pallavi. 2021. 'CMIE Survey Limitations May Bias Unemployment Data, Says Pronab Sen'. *Bloomberg Quint*, 29 June.

Nair, Harsh V. 2017. 'What about Us? Kashmiri Pandits Prod Supreme Court over Its Order to Review "84 Riots Cases"'. *India Today*, 21 August.

National Conference. 1950. *Naya Kashmir*, translated by Ved Bhasin. Jammu: Maqtab Printing Press.

Nazeer, Zubair. 2022. 'Setting Right the Focal Point of J&K Tribal Politics'. *The Hindu*, 15 February.

NDTV. 2011. 'This Summer, Kashmir to Fight Murderers of Another Kind'. 19 May.

———. 2017. 'Need No Pak Letter for Patient From PoK, It Is India's: Foreign Minister Sushma Swaraj'. 18 July.

———. 2021. '"Punjab Chief Minister Must Be Sikh Leader": Ambika Soni Rejects Offer'. 19 September.

———. 2022a. '"Beware! UP May Become Kashmir, Bengal": Yogi Adityanath Ahead of Vote'. 10 February.

———. 2022b. 'Muslims Make Up 35% Of Assam, Not Minority: Chief Minister Himanta Sarma'. 16 March.

———. 2022c. 'States Can Declare a Community Minority, Centre Tells Supreme Court'. 28 March.

———. 2022d. 'Supreme Court's "Different Stands" Remark on "Minorities" Tag for Hindus'. 10 May.

———. 2023. 'Nirmala Sitharaman's Reply To Query On "Violence Against Muslims" In India'. 11 April.

Nehru, Jawaharlal. 1994 [1946]. *The Discovery of India*. New Delhi: Oxford University Press.

*New Indian Express*. 2010. 'J&K: Geelani Sees Plot to Change Muslim Majority'. 13 May.

Noorani, A. G. 2000. 'Questions about the Kashmir Ceasefire'. *Economic and Political Weekly* 35(45): 3949–58.

———. 2002. 'The Dixon Plan'. *Frontline*, 25 October.

———. 2006. 'Facing the Truth'. *Frontline*, 6 October.

———. 2008. 'Why Jammu Erupts'. *Frontline*, 26 September.

———. 2011. *Article 370: A Constitutional History of Jammu and Kashmir*. New Delhi: Oxford University Press.

———. 2015. 'BJP Raj in Kashmir?' *Greater Kashmir*, 14 March.

———. 2019. 'Jammu's Crisis'. 5 January.

Oberoi, S. S. 2004. 'Ethnic Separatism and Insurgency in Kashmir'. In *Religious Radicalism and Security in South Asia*, edited by Satu P. Limaye, Mohan Malik and Robert G. Wirsing, 171–91. Hawaii: Asia-Pacific Center for Security Studies Honolulu.

Oddie, G. A. 1981. 'Christians in the Census: Tanjore and Trichinapoly Districts, 1871–1901'. In *The Census in British India: New Perspectives*, edited by N. G. Barrier, 119–49. New Delhi: Manohar Publications.

Okolo, Abraham. 1999. 'The Nigerian Census: Problems and Prospects'. *American Statistician* 53(4): 321–25.

Organisation for Economic Cooperation and Development (OECD). 2007. 'Glossary of Statistical Terms'. https://stats.oecd.org/glossary/index.htm. Accessed on 20 February 2021.

*Outlook*. 2006. 'Eight New Districts in Kashmir'. 6 July.

———. 2010. 'Gujjars Appeal to PM For Special Package'. 6 June.

———. 2015. 'Modi Govt Says Notions of Minorities "Not Valid" in Indian Context'. 23 February.

Pais, Jesim, and Vikas Rawal. 2021. 'CMIE's Consumer Pyramids Household Surveys: An Assessment'. *India Forum*, 10 August.

Palmer-Jones, R., and K. Sen (2003). 'What Has Luck Got to Do with It? A Regional Analysis of Poverty and Agricultural Growth in Rural India'. *Journal of Development Studies* 40(1): 1–31.

Panandiker, V. A. Pai, and P. K. Umashankar. 1994. 'Fertility Control and Politics in India'. *Population and Development Review* 20(Supplement): 89–104.

Pandit, Ambika. 2021. 'As Jiyo Parsi Scheme Delivers, Community Sees Record Births'. *Times of India*, 14 June.

Pandit, M. Saleem. 2021. 'Poonch Infiltrators Believed to Have Trekked Down to Shopian'. *Times of India*, 4 November.

Pandita, Rahul. 2017. *Our Moon Has Blood Clots: The Exodus of the Kashmiri Pandits*. Penguin Random House India. Kindle edition.

Panun Kashmir. n.d.1. 'A Homeland for Kashmiri Pandits'. https://panunkashmir. org/facts.html. Accessed on 13 September 2021.

———. n.d.2. 'Kashmiri Panditon ke liye Kashmir Ghati mein Homeland Kyon?'. https://archive.org/details/HomelandKyonPanunKashmir. Accessed on 6 January 2023.

————. n.d.3. Panun Kashmir [Map] (courtesy: Ajay Chrungoo).

Pargal, Sanjeev. 2021. 'Delimitation Comm Seeks Help of SoI for Mapping Different Areas'. *Daily Excelsior*, 18 September.

Peabody, Norbert. 2001. 'Cents, Sense, Census: Human Inventories in Late Pre-Colonial and Early Colonial India'. *Comparative Studies in Society and History* 43(4): 819–50.

Peer, Basharat. 2009. *Curfewed Night*. New Delhi: Penguin Random House India. Digital edition.

Perwez, Shahid, Robin Jeffery and Patricia Jeffery. 2012. 'Declining Child Sex Ratio and Sex Selection in India: A Demographic Epiphany?' *Economic and Political Weekly* 47: 33, 73–77.

Pfeffermann, Danny. 2015. 'Methodological Issues and Challenges in the Production of Official Statistics'. *Journal of Survey Statistics and Methodology* 3: 425–83.

Phanjoubam, Pradip. 2016. *The Northeast Question: Conflicts and Frontiers*. Oxon: Routledge.

Philip, Snehesh Alex. 2019. 'What Imran Khan Says Is 9 Lakh Soldiers in Kashmir Is Actually 3.43 Lakh Only'. *The Print*, 12 November.

Philipsen, Dirk. 2015. *The Little Big Number: How GDP Came to Rule the World and What to Do about It*. Princeton, NJ: Princeton University Press.

Porter, Theodore M. 1995. *Trust in Numbers: The Pursuit of Objectivity in Science and Public Life*. Princeton, NJ: Princeton University Press.

Prabhakara, M. S. 2012. *Looking Back into the Future: Identity & Insurgency in Northeast India*. New Delhi: Routledge.

Prakash, Indra. 1979. *They Count Their Gains, We Calculate Our Losses*. New Delhi: Akhil Bharat Hindu Mahasabha.

Presser, Stanley. 2021. 'Data Collection vs Data Construction: How Methodology and Substance are Inextricably Interwoven'. NCAER Data Innovation Centre Methodology Seminars, National Council of Applied Economic Research, New Delhi, 18 March.

Preston, S. H., P. Heuveline and M. Guillot. 2001. *Demography: Measuring and Modeling Population Processes*. Oxford: Blackwell Publishing.

Prewitt, K. 2003. *Politics and Science in Census Taking*. New York: Russell Sage Foundation; Washington, DC: Population Reference Bureau.

————. 2010. 'The U.S. Decennial Census: Politics and Political Science'. *Annual Review of Political Science* 13: 237–54.

Puri, Balraj. 1974. 'Schizophrenia in Jammu?' *Economic and Political Weekly* 9(6–8): 185–87.

————. 1981. 'Jammu and Kashmir: What Is Wrong with Kashmir's Finances'. *Economic and Political Weekly* 16(19): 845–46.

———. 1983a. 'The Era of Sheikh Mohammad Abdullah I'. *Economic and Political Weekly* 18(6): 186–90.

———. 1983b. 'The Era of Sheikh Mohammad Abdullah II'. *Economic and Political Weekly* 18(7): 230–33.

———. 1999. 'Jammu and Kashmir: How Government Scuttled Regional Autonomy'. *Economic and Political Weekly* 34(23): 1400–01.

———. 2004. 'Jammu and Kashmir: Permanent Resident Bill Questionable Legal, Moral and Political Basis'. *Economic and Political Weekly* 39(14–15): 1456–57.

———. 2008. 'Jammu and Kashmir: The Issue of Regional Autonomy'. *Economic and Political Weekly* 43(34): 8–11.

Puri, Luv. 2022. 'Delinking 1950 Report from J&K Delimitation'. *The Tribune*, 23 February.

Purtill, J. 2017. 'What Fine? No One Yet Prosecuted over 2016 Census "Boycott"'. Australian Broadcasting Corporation, 2 March.

Radhakrishnan, Vignesh. 2019. 'Is Jammu and Kashmir Underdeveloped as Stated by Amit Shah?' *The Hindu*, 7 August.

Raghavan, T. C. A. Sharad. 2016. 'J&K Gets 10% of Central Funds with Only 1% of Population'. *The Hindu*, 24 July.

Rahi, Javaid. 2019. 'Delimitation, Reservation for STs Rekindle New Hopes for Gujjars, Bakerwals'. *Daily Excelsior*, 28 August.

Rai, Mridu. 2018. 'The Indian Constituent Assembly and the Making of Hindus and Muslims in Jammu and Kashmir'. *Asian Affairs* 49(2): 205–21.

Raina, Muzaffar. 2022. 'After Pandit Exodus, Centre in a Fix over Dalit Killings, Victimhood'. *The Telegraph*, 5 June.

Rajagopal, Krishnadas. 2021. 'Supreme Court Terms Cancellation of 3 Crore Ration Cards as Serious, Seeks Replies from Centre, States'. *Economic Times*, 17 March.

———. 2022. 'Minority Status of Religious, Linguistic Communities Is State-Dependent: SC'. *The Hindu*, 18 July.

Ranade, M. G. 1906. *Essays on Indian Economics: A Collection of Essays and Speeches*, 2nd edition. Madras: G. Natesan & Co.

Rao, C. R. 1999. *Statistics and Truth: Putting Chance to Work*. Singapore: World Scientific.

Ray, Rammohan. 1832. 'Some Remarks in Vindication of the Resolution Passed by the Government of Bengal in 1829 Abolishing the Practice of Female Sacrifices in India'. In *The Essential Writing of Raja Rammohan Ray*, edited by Bruce C. Robertson, 166–72. Delhi: Oxford University Press.

Raza, Moonis, and Aijazuddin Ahmad. 1990. *An Atlas of Tribal India: With Computed Tables of District-Level Data and Its Geographical Interpretation*. New Delhi: Concept Publishing Company.

Reamer, A. 2012. 'Census: Planning Ahead for 2020: Testimony of Andrew Reamer, Subcommittee on Federal Financial Management, Committee on Homeland Security and Governmental Affairs, US Senate'. Washington, DC, 18 July.

Rediff. 2000a. 'J&K Ultras Raid Census Offices, Destroy Records: AFP'. 26 October.

————. 2000b. 'LET Holds Out Death Threat against Census: PTI'. 7 September.

Riaz-ud-din. 2000. 'Kashmiris Not for Mardamshumari but Raishumari'. *Kashmir Times*, 8 October, 5.

Robinson, Cabeiri deBergh. 2013. *Body of Victim, Body of Warrior Refugee: Families and the Making of Kashmiri Jihadists*. Berkeley and Los Angeles: University of California Press.

Root, H. L. 2013. *Dynamics among Nations: The Evolution of Legitimacy and Development in Modern States*. Cambridge, MA: MIT Press.

Rose, Nikolas. 1991. 'Governing by Numbers: Figuring Out Democracy'. *Accounting Organizations and Society* 16(7): 673–92.

Roy, Arundhati. 2008. 'Kashmir: Land and Freedom'. *The Guardian*, 22 August.

Roy Burman, B. K. 1992. *Beyond Mandal and After: Backward Classes in Perspective*. New Delhi: Mittal Publications.

————. 1998. 'Backward Classes and the Census: Putting the Record Straight'. *Economic and Political Weekly* 33(50): 3178–79.

Roy, Shubajit. 2022. 'India, UAE Ink Comprehensive Trade Pact; Vow to Jointly Fight Terrorism'. *Indian Express*, 19 February.

Sahoo, Sarbeswar. 2018. *Pentecostalism and Politics of Conversion in India*. New Delhi: Cambridge University Press.

Sandefur, Justin, and Amanda Glassman. 2015. 'The Political Economy of Bad Data: Evidence from African Survey and Administrative Statistics'. *Journal of Development Studies* 51(2): 116–32.

Sandhu, Kamaljit Kaur. 2020. 'Demographic Change Will Be Good for J&K, Says Dr Jitendra Singh'. *India Today*, 29 October.

*Sangai Express*. 2013. 'Reverification Drive of Head Count at Senapati District State Govt May Recommend RGI Figure'. 13 September.

Sankrityayan, Rahul. 1939. *Meri Ladakh Yatra*. Prayag: Indian Press.

Schofield, Victoria. 2003. *Kashmir in Conflict: India, Pakistan and the Unending War*. London and New York: I.B.Tauris.

Schwartzberg, J. E. 1981. 'Sources and Types of Error'. In *The Census in British India: New Perspectives*, edited by N. G. Barrier, 41–60. New Delhi: Manohar Publications.

Sen, Amartya. 1990. 'More than 100 Million Women Are Missing'. *New York Review of Books*, 20 December.

————. 2001. *Development as Freedom*. New Delhi: Oxford University Press.

Serra, Gerardo. 2014. 'An Uneven Statistical Topography: The Political Economy of Household Budget Surveys in Late Colonial Ghana, 1951–1957'. *Canadian Journal of Development Studies* 35(1): 9–27.

Shah, Syed Amjad. 2021. 'Darbar Move Employees Asked to Vacate Govt Flats within 21 Days'. *Greater Kashmir*, 1 July.

Shaikh, Zeeshan. 2020. 'Muslim–Hindu Demography of Jammu and Kashmir: What Census Data Show'. *Indian Express*, 2 December.

Shani, Ornit. 2018. *How India Became Democratic: Citizenship and the Making of the Universal Franchise.* Cambridge and New York: Cambridge University Press.

Sharma, Arun. 2020. 'End of Article 370 Was Also to Grant Rights to West Pak Refugees, but Not One Claimant Is Stepping Forward'. *Indian Express*, 24 January.

———. 2022. 'UAE Investments, Grassroot Democracy: Twin Messages in PM Narendra Modi's J&K Visit Today'. *Indian Express*, 24 April.

Sharma, Sandipan. 2016. 'UP's Kairana Turning Kashmir? BJP's Communal Spin Based on Distorted Facts'. *Firstpost*, 13 June.

Sharma, Sant K., and Dipankar Sengupta. 2008. *Loaded Dice?: The 2008 Elections in J&K.* Jammu: Taskeen Offset Press.

Shastri, Satyananda. 1947. 'Importance of Kashmir'. *Organiser*, vol. 1, no. 19.

Shelar, Jyoti. 2019. 'Jiyo Parsi's Baby Boom: 207 Born in Five Years, 12 More on the Way'. *The Hindu*, 18 September.

Shetty, S. L. 2012. 'Dealing with a Deteriorating Statistical Base'. *Economic and Political Weekly* 47(18): 41–44.

Shukla, Ajai. 2018. 'India Has 700,000 Troops in Kashmir? False!!!'. Rediff, 17 July.

Sibal, Kapil. 2021. 'What Pushing Delimitation Does to Talks on Peace and Statehood in J&K'. *Indian Express*, 1 July.

Siddiqui, Sabrina, and Tom McCarthy. 2019. 'Trump Abandons Effort to Put Citizenship Question on 2020 Census'. *The Gaurdian*, 11 July.

Sidiq, Nusrat. 2020. 'HC Probes Deeper into Details of Darbar Move'. *Kashmir Reader*, 18 April.

Sikand, Yoginder. 2001. 'Changing Course of Kashmiri Struggle: From National Liberation to Islamist Jihad?' *Economic and Political Weekly* 36(3): 218–27.

———. 2010a. 'Hindu–Muslim Relations in Jammu'. In *Identity Politics in Jammu and Kashmir*, edited by Rekha Chowdhary, 193–228. New Delhi: Vitasta.

———. 2010b. 'Trapped by Competing Narratives'. *Outlook*, 10 August.

———. 2010c. 'The Face of "Azadi"'. *Outlook*, 26 September.

———. 2010d. '"It Is Undoubtedly a Religious Issue"'. *Outlook*, 28 October.

———. 2010e. 'Jihad, Islam and Kashmir: Syed Ali Shah Geelani's Political Project'. *Economic and Political Weekly* 45(40): 125–34.

Singh, Bhopinder. 2018. 'Neglected Jammu'. *Asian Age*, 14 May.

Singh, J. K. 2021. 'Kashmir's Sikh Delegation Meets Union Home Minister, Ask Him to Address Concerns on Priority to Stop Large-Scale Migration of Sikhs from Valley'. *Times of India*, 4 July.

Singh, K. S. 1996. 'Census as Ethnography'. In *Census as Social Document*, edited by S. P. Mohanty and A. R. Momin, 138–46. Jaipur: Rawat Publications.

Singh, Ranjit. 2020. 'Study the Imbrication: A Methodological Maxim to Follow the Multiple Lives of Data'. In *Lives of Data: Essays on Computational Cultures from India*, edited by Sandeep Mertia, 51–59. Amsterdam: Institute of Network Cultures.

Singh, Tavleen. 2000. '"If Delhi Can Discuss Azadi with the Hurriyat, Why Not Autonomy with Farooq?"' *India Today*, 10 July.

Sivaramakrishnan, K. C. 1997. 'Under-Franchise in Urban Areas: Freeze on Delimitation of Constituencies and Resultant Disparities'. *Economic and Political Weekly* 32(51): 3275–81.

Smith, Sara. 2009. 'The Domestication of Geopolitics: Buddhist–Muslim Conflict and the Policing of Marriage and the Body in Ladakh, India'. *Geopolitics* 14(2): 197–218.

———. 2012. 'Intimate Geopolitics: Religion, Marriage, and Reproductive Bodies in Leh, Ladakh'. *Annals of the Association of American Geographers* 102(6): 1511–28.

Snedden, C. 2017. *Understanding Kashmir and Kashmiris*. New Delhi: Speaking Tiger.

SOS International. n.d.1. 'Press Release'. 27 December. Jammu: SOS International.

———. n.d.2. 'Press Release'. 10 September. Jammu: SOS International.

———. 2004. 'POK Refugees Are Warehoused in Camps for Fifty Seven Years or More: Chuni'. 17 November. Jammu: SOS International.

———. 2006. 'Press Conference at Delhi'. 23 January. Jammu: SOS International.

———. 2014a. 'Rajiv Chunni Lambastes Modi Government for Sabotaging Displaced Community. Refugee Mahapanchayat Serves Ultimatum to Centre on Rehabilitation Package'. 16 November. Jammu: SOS International.

———. 2014b. 'Press Note: Time to Heal Wounds Not Figuring Committee: Chuni'. 5 April.

South Asia Terrorism Portal. n.d. 'India: Timeline (Terrorist Activities): 2008'. https://www.satp.org/terrorist-activity/india-Oct-2008. Accessed on 26 December 2022.

Soz, Saifuddin. 2018. *Kashmir: Glimpses of History and the Story of Struggle*. New Delhi: Rupa Publications.

Sri Aurobindo. 1909. 'The Dying Race'. *Karmayogin*, vol. 1, no. 18, 6 November. In *The Complete Works of Sri Aurobindo*, vol. 8, 294. Pondicherry: Sri Aurobindo Ashram Trust.

Srinivasan, Meera. 2019. 'Bifurcation of J&K India's Internal Matter: Ranil Wickremesinghe'. *The Hindu*, 6 August.

Srinivasan, T. N. 1994. 'Data Base for Development Analysis: An Overview'. *Journal of Development Economics* 44: 3–27.

———. 2003. 'India's Statistical System: Critiquing the Report of the National Statistical Commission'. *Economic and Political Weekly* 38(4): 303–06.

Srivastava, Kanchan. 2020. 'Lucknow Parsis: Bombay Punchayet Not Serious about Dwindling Number'. *Free Press Journal*, 18 February.

*State Times*. 2022. 'AIJMS Calls on Delimitation Commission, Seeks De-Reservation of Constituencies with Jat Population'. 5 April, 8.

Stracey, P. D. 1968. *Nagaland Nightmare*. Bombay: Allied Publishers Pvt Ltd.

Subramanian, Nirupama. 2021. 'In Leh and Kargil, Different Reasons to Oppose Ladakh's Current Status'. *Indian Express*, 3 July.

Subramanian, S. 1960. 'A Brief History of the Organisation of Official Statistics in India during the British Period'. *Sankhyā: The Indian Journal of Statistics* 22(1–2): 85–118.

Sundar, N. 2000. 'Caste as Census Category: Implications for Sociology'. *Current Sociology* 48(3): 111–26.

Suru Valley Public School (SVPS). 2019. *Newsletter 2018–19: Suru*. Kargil: Suru Valley Public Higher Secondary School.

Suryanarayana, M. H., and N. S. Iyengar. 1986. 'On the Reliability of NSS Data'. *Economic and Political Weekly* 21(6): 261–64.

Swami, Praveen. 2000a. 'The Game of Numbers'. *Frontline*, 14 October.

———. 2000b. 'The Autonomy Demand'. *Frontline*, 8 July.

———. 2000c. 'Murder in Leh'. *Frontline*, 22 July.

———. 2001. 'Clueless in Kashmir'. *Seminar* 497 (India 2000).

———. 2002. 'The RSS Game Plan.' *Frontline*, 20 July.

———. 2006. 'Snowstorm of Hate'. *Frontline*, 10 March.

———. 2007. *India, Pakistan and the Secret Jihad: The Covert War in Kashmir, 1947–2004*. Oxon: Routledge.

———. 2008a. 'J&K's Holy War'. *Outlook*, 19 August.

———. 2008b. 'Kashmir's Politics of Hate'. *South Asia Intelligence Review* 7(1). https://www.satp.org/south-asia-intelligence-review-Volume-7-No-1. Accessed on 3 January 2023.

———. 2008c. 'Jamaat in Retreat'. *Frontline*, 14 March.

———. 2010. 'Kashmir: Youth Bulge, Peace Deficit'. *The Hindu*, 29 August.

———. 2012. 'In Kashmir, Some Hot Potatoes'. *The Hindu*, 23 April.

———. 2013. 'Ramban Firing Reveals the Fading Communal Politics of Kashmir'. *Firstpost*, 20 July.

———. 2014. 'Demography and Discontent: Crisis of Modernity and Displacement in Undivided Jammu and Kashmir'. In *The Other Kashmir: Society, Culture and Politics in the Karakoram Himalayas*, edited by K. Warikoo, 201–21. New Delhi: Institute for Defence Studies and Analyses and Pentagon Press.

———. 2021. 'Execution-Style Killings of Minorities Far from New to Kashmir'. News18, 8 October.

———. 2022. 'Delimitation Shows India's Democracy Continues to Struggle in the Face of Kashmir Challenge'. *The Print*, 8 May.

Talib, Arjimand Hussain. 2010. 'Understanding Religious Radicalization: Issues, Threats and Early Warnings in Kashmir Valley'. IPCS Issue Brief No. 149, Institute of Peace and Conflict Studies, New Delhi.

Tantry, Ishfaq. 2010. 'Separatists Urge People to Take Part in Census'. *Rising Kashmir*, 27 March.

Taylor, Matthew. 2016. 'The Political Economy of Statistical Capacity: A Theoretical Approach'. Institutional Capacity of the State Division, Discussion Paper No. IDB-DP-471, Inter-American Development Bank, Washington, DC.

Thakur, S. 1985. 'A Disaster Called G. M. Shah'. *Sunday* 12(32): 14–17.

Thapar, Karan. 2019. 'We Cannot Be Selective about the Past in Jammu & Kashmir'. *Hindustan Times*, 15 September.

———. 2020. '"Kashmiris Do Not Feel Indian, Today They'd Rather Have the Chinese Rule Them": Farooq Abdullah'. *The Wire*, 23 September.

*The Economist*. 2011. 'In Search of a Common Denominator'. 17 March.

———. 2019. 'What's in a Number: Why Ethiopia Has Postponed Its Census'. 29 March.

———. 2020. 'Where Economists Focus Their Research'. 10 December.

*The Hindu*. 2012. 'Illegal LPG Connections: Centre Cold to Karnataka Model'. 7 October.

———. 2017. 'Rohingyas Issue Sets Off Alarm Bells'. 4 April.

———. 2021. 'Coronavirus | J&K Defers Darbar Move Due to Spike in Cases'. 15 April.

*The Telegraph*. 2004. 'Count Him Out'. 18 September.

*Times of India*. 2000a. 'Jammu and Kashmir to Go Ahead with Census Despite Threat'. 4 September.

———. 2000b. 'Kashmir Census Operation Fails to Take Off'. 11 September.

———. 2010. 'Delhi Sex Ratio Takes a Nose-Dive'. 15 September.

———. 2014. 'With Just Two Days to Go for the Telangana Survey, Exodus of Telugus from Chennai Peaks'. 17 August.

———. 2019. 'Territories under J&K UT Include PoK: Govt'. 3 December.

———. 2021. 'Assam on Way to Becoming Next Kashmir: Himanta Biswa Sarma'. 19 September.

———. 2022. 'All Legally Bound to Answer Census Questions to the Best of Their Knowledge or Belief, Govt Tells LS'. 22 March.

Times Now. 2021. 'Does Omar Abdullah Want Passport for Stone Pelters? | The Newshour Debate | Exclusive'. https://www.timesnownews.com/videos/

times-now/newshour/does-omar-abdullah-want-passport-for-stone-pelters-the-newshour-debate-exclusive/105163. Accessed on 23 December 2012.

Tinyi, Venusa, and C. Nienu. 2018. 'Making Sense of Corruption in Nagaland: A Culturalist Perspective'. In *Democracy in Nagaland: Tribes, Traditions, Tensions*, edited by Jelle J. P. Wouters and Zhoto Tunyi, 159–80. Kohima: The Highlander Books.

Tooze, Adam J. 2003. *Statistics and the German State, 1900-1945: The Making of Modern Economic Knowledge*. Cambridge, UK: Cambridge University Press.

Tremblay, Reeta Chowdhari. 2009. 'Kashmir's Secessionist Movement Resurfaces: Ethnic Identity, Community Competition, and the State'. *Asian Survey* 49(6): 924–50.

*The Print*. 2021. 'SC Seeks Govt Reply on Transfer Plea against Granting Minority Status to 5 Communities'. 9 February.

———. 2022a. 'Census, NPR Databases Declared as Critical Information Infrastructure'. 9 November.

———. 2022b. 'No Cyber Attack or Hacking of Census Data: Ajay Kumar Mishra Informs Lok Sabha'. 21 December.

*The Tribune*. 2000a. 'Census Begins in J&K'. 18 May.

———. 2000b. 'Hizb Threat to Eliminate Census Officials'. 9 September.

———. 2000c. 'Soz May Form Political Outfit'. 8 September.

———. 2001. 'Top Hizb Militant Killed'. 26 July.

———. 2003. 'Elections to Kargil Council Today'. 10 July.

———. 2006. 'Soz Nomination a Setback to Azad Camp'. 8 January.

———. 2015a. 'Govt to Provide Subsidised Ration as per 2011 Census'. 23 April.

———. 2015b. 'Guv for Better Execution of Welfare Board Programmes'. 17 September.

———. 2017. '5,743 Rohingyas in State, No Instance of Radicalisation: Mehbooba'. 21 January.

———. 2022. 'Proposal to Merge Jammu Areas with Anantnag Draws Criticism'. 5 February.

Urla, Jacqueline. 1993. 'Cultural Politics in an Age of Statistics: Numbers, Nations, and the Making of Basque Identity'. *American Ethnologist* 20(4): 818–43.

Uvin, Peter. 2002. 'On Counting, Categorizing, and Violence in Burundi and Rwanda.'. In *Census and Identity: The Politics of Race, Ethnicity, and Language in National Censuses*, edited by David I. Kertzer and Dominique Arel, 148–75. Cambridge, UK: Cambridge University Press.

———. 2004. 'Dangerous Liaisons: Hindu Nationalism and Buddhist Radicalism in Ladakh'. In *Religious Radicalism and Security in South Asia*, edited by Satu P. Limaye, Mohan Malik and Robert G. Wirsing, 193–218. Honolulu: Asia-Pacific Center for Security Studies.

———. 2008. 'Beyond Identity Fetishism: "Communal" Conflict in Ladakh and the Limits of Autonomy'. *Cultural Anthropology* 15(4): 525–69.

Upadhyaya, Himanshu. 2020. 'Delayed Audit Reports, Lower Output Marked Last Two Years of CAG's Functioning'. *The Wire*, 24 July.

Venkatesan, V. 2001. 'Census 2001: The Head-Count and Some Gaps'. *Frontline*, 3 March.

Verma, A. K. 2013. 'Tribal "Annihilation" and "Upsurge" in Uttar Pradesh'. *Economic and Political Weekly* 48(51): 52–59.

Vidwans, S. M. 2002a. 'Indian Statistical System at the Crossroads I: Ominous Clouds of Centralisation'. *Economic and Political Weekly* 37(37): 3819–29.

———. 2002b. 'Indian Statistical System at the Crossroads II: Expansion of National Sample Survey'. *Economic and Political Weekly* 37(38): 3943–55.

Vinayak, Ramesh. 2020. 'Ultra-Nationalists Call Us Separatists; Treated as Nationalists in J-K: Omar Abdullah'. *Hindustan Times*, 31 August.

Visaria, P. 1977. 'Publications of the 1971 Census of India'. *Population Index* 43(2): 206–12.

Wade, Robert. 2012. 'The Politics behind World Bank Statistics: The Case of China's Income'. *Economic and Political Weekly* 47(25): 17–18.

Wahab, Ghazala. 2021. 'In Modi's Meeting with Kashmiri Leaders: Saying Was the Deal, Listening Was Optional'. *The Wire*, 3 July.

Wahid, Siddiq. 2022a. 'Is Bhoti a Language, Religious Affiliation, Sanskrit Diminutive or Political Tool?'. *Outlook*, 13 May.

———. 2022b. 'Ladakh's Pushback to BJP's "Politics as Usual"'. *The Wire*, 10 October.

Wani, Muhammad Ashraf. 2010. 'Religion, Economy and Political Crisis in Kashmir.' In *Identity Politics in Jammu and Kashmir*, edited by Rekha Chowdhary, 177–91. New Delhi: Vitasta.

Warikoo, K. 2000. 'Tribal Gujjars of Jammu and Kashmir'. *Himalayan and Central Asian Studies* 4(1): 3–27.

Wazir, Muhammad Asif, and Anne Goujon. 2019. 'Assessing the 2017 Census of Pakistan Using Demographic Analysis: A Sub-National Perspective'. Working Paper No. 06/2019, Austrian Academy of Sciences, Vienna Institute of Demography.

Weinraub, Bernard. 1973. 'Planning Controversy in India Intensifies as Economist Quits'. *New York Times*, 10 December.

Weiss, Anita M. 1999. 'Much Ado about Counting: The Conflict over Holding a Census in Pakistan'. *Asian Survey* 39(4): 679–93.

West Pakistan Refugees Action Committee (WPRAC). 2015. 'Memorandum: Charter of Demand of West Pakistan Refugees of 1947'. 18 December. Samba, Jammu and Kashmir: West Pakistan Refugees Action Committee.

Wilkinson, Steven I. 2004. *Votes and Violence: Electoral Competition and Ethnic Riots in India*. New York: Cambridge University Press.

Wines, Michael. 2020. 'Federal Court Rejects Trump's Order to Exclude Undocumented from Census'. *New York Times*, 10 September.

*The Wire*. 2023. 'Meitei "Pride" Group's Threat: "Kukis Mainly Illegal, Modi Must Intervene Or There'll Be Civil War"'. 7 June. https://www.youtube.com/watch?v=VDJnPewUhmU. Accessed on 14 June 2023.

Woolf, Stuart. 1989. 'Statistics and the Modern State'. *Comparative Studies in Society and History* 31(3): 588–604.

Yan, Wang, and Peng Wuqing. 2020. 'Xinjiang Population Growth Best Answers Western Smear Campaign on Uygurs'. *Global Times*, 4 September.

Yasir, Sameer. 2017. 'Kashmir: 377 Ex-Militants Returned from Pakistan via Nepal, Bangladesh since 2010'. *Firstpost*, 12 January.

Yasir, Syed. 2011. 'Second Phase of Census on Expected Lines'. *Kashmir Times*, 17 February, 5.

Yin, Sandra. 2007. 'Objections Surface over Nigerian Census Results'. Connecticut Avenue, Washington, DC: Population Reference Bureau.

Yu, Sun. 2021. 'China Set to Report First Population Decline in Five Decades'. *Financial Times*, 27 April.

Zargar, Safwar. 2021. 'The Resistance Front Have Claimed Responsibility for Civilian Killings in Kashmir. Who Are They?' *Scroll.in*, 11 October.

Zarkovich, S. S. 1989. 'The Overcount in Censuses of Population'. *Jahrbucher fur Nationalokonomie und Statistik* 206(6): 606–07.

Zavier, A. J. Francis, and P. N. Mari Bhat 2007. 'Factors Influencing the Use of Prenatal Diagnostic Techniques and the Sex Ratio at Birth in India'. *Economic and Political Weekly* 42(24): 2292–303.

Zee News. 2022. 'The Kashmir Files: Anupam Kher Reacts to Kerala Congress' Statement on Exodus of Kashmiri Pandits'. 14 March. https://zeenews.india.com/bollywood/the-kashmir-files-anupam-kher-reacts-to-kerala-congress-statement-on-exodus-of-kashmiri-pandits-2445243.html. Accessed on 26 December 2022.

Zhao, Suisheng. 2004. *A Nation-State by Construction: Dynamics of Modern Chinese Nationalism*. Redwood City, CA: Stanford University Press.

Zhao, Zhongwei, and Guangyu Zhang. 2021. 'The Reality of China's Fertility Decline'. East Asia Forum, 8 July. https://www.eastasiaforum.org/2021/07/08/the-reality-of-chinas-fertility-decline. Accessed on 23 December 2022.

Ziipao, Raile Rocky. 2020. *Infrastructure of Injustice: State and Politics in Manipur and Northeast India*. Oxon: Routledge.

# Index